Sixth Edition

Student's Guide
James C. Hill

CHEMISTRY
The Central Science

Theodore L. Brown
University of Illinois

H. Eugene LeMay, Jr.
University of Nevada

Bruce Edward Bursten
Ohio State University

Prentice Hall, Englewood Cliffs, NJ 07632

© 1994 by Prentice-Hall, Inc.
A Simon & Schuster Company
Englewood Cliffs, New Jersey 07632

Printed in the United States of America
10 9 8 7 6 5 4 3

ISBN: 0-13-338740-2

Prentice-Hall International (UK) Limited, *London*
Prentice-Hall of Australia Pty. Limited, *Sydney*
Prentice-Hall Canada Inc., *Toronto*
Prentice-Hall Hispanoamericana, S.A., *Mexico*
Prentice-Hall of India Private Limited, *New Delhi*
Prentice-Hall of Japan, Inc., *Tokyo*
Simon & Schuster Asia Pte. Ltd., *Singapore*
Editora Prentice-Hall do Brasil, Ltda., *Rio de Janeiro*

Contents

To the Student

The sixth edition of the *Student's Guide* to the text *Chemistry: The Central Science* by T. Brown, G. LeMay and B. Bursten maintains the successful structure of the previous editions. Many aids are available to help you be successful: Learning goals for each chapter; suggested material to review; summaries of key ideas and concepts; sample exercises; end-of-chapter problems with detailed solutions; and sectional tests that check your understanding of material in several chapters. Topic summaries are designed with a "bullet" format wherein key or major points and ideas are identified by a · mark.

As you begin your adventure into the chemical world, you will find it necessary to study certain abstract concepts and learn new problem-solving techniques in order to understand chemical phenomena. To use your time more efficiently when studying chemistry and to maximize your learning of ideas and skills, you should develop a strategy for learning chemistry in a classroom situation.

The following is a suggested method for using this book in conjunction with the Brown, LeMay and Bursten text:

1. In the *Student's Guide,* each chapter corresponds to a chapter in the text and is divided into three sections: Overview of the Chapter, Topic Summaries and Exercises, and Self-Study Exercises. By appropriately using these sections, you can facilitate your learning of chemical concepts, theories, facts, and problem-solving techniques.

2. Before attending the first lecture on a particular chapter in the text and before reading the chapter, read the corresponding Overview of the Chapter in the *Student's Guide.* By doing this first, you will gain familiarity with the key topics found in that chapter of the text. This is important because it will help you to keep the major topics in mind while you are reading the more minute details, such as facts, theories, and skills development. Also, the subject matter in each chapter of the text can be learned more easily if you study small segments of interrelated material. In addition to identifying key topics, the overview lists the sections in which a topic is covered; it also lists learning objectives that will guide you in your study and suggests material that should be reviewed to prepare you for the chapter.

3. After reading the Overview of the Chapter in the *Student's Guide,* you should read the appropriate chapter in the text so that you will be prepared for your instructor's lecture; you will also be equipped to ask appropriate and thoughtful questions.

4. During the lecture, take detailed notes. In most chemistry classes, an instructor chooses to emphasize certain key topics and ideas within a chapter,

HOW TO USE THE
STUDENT'S GUIDE

and the instructor will test your knowledge and your problem-solving skills primarily in those areas. Detailed lecture notes will provide you with a complete record of the material covered in class. Often an instructor will also identify the key points. Be sure to note these for later reference. If you have the time, rewrite your lecture notes later that day. This allows you to carefully reorganize the information given in the lecture and to fill in some missing material that you didn't have time to record during the lecture; it also reinforces what you learned that day. To understand a concept thoroughly and retain it in your memory, you will need to reinforce material through repeated learning.

5. Once your instructor has begun to discuss a topic, you should study the text coverage of it in detail. At the same time, use the Topic Summaries and Exercises section in the *Student's Guide.* This section contains a summary of the key concepts, theories, and facts associated with each topic listed in the Overview of the Chapter. Further explanations of key material are also included. After each topic summary, there are exercises with detailed solutions; these are similar in style to the sample exercises found in the text. The solutions to the sample exercises often include further explanations of important or difficult material. Attempt the sample exercises after looking at the sample problems in the text. When doing the sample exercises, cover up the solutions and try to solve the problems. The more you work with pencil and paper, and the less you just read a solution, the better and faster you will learn chemistry and how to solve chemical problems.

6. You can check your understanding by answering questions provided in the Self-Study Exercises. Because instructors may use a variety of question formats on tests, the *Student's Guide* provides questions using three of the most common testing formats: true-false questions, problems and short-answer questions, and multiple-choice questions. The latter two question formats check your understanding and knowledge of material in the entire chapter. @T:The true-false questions focus on your familiarity with the key terms listed at the end of each chapter in the text. Chemistry has a language of its own, a language in which you must become proficient. Without an understanding of the key terms listed in each chapter of the text, you will have difficulty mastering chemistry.

7. *Sectional Tests:* Tests covering several chapters are provided to help you check your readiness to take an examination over the chapters. Take each test as if you were in a classroom. Tear out the test, find a quiet spot to work, and do the questions without referring to the text, unless specifically directed. Grade the test after you are finished. Each question has the appropriate section in the Brown, LeMay and Bursten text referenced; therefore, if you miss a question, study the section noted. Spend your valuable time studying material you don't know, not what you firmly understand. The sectional tests may not cover the specific chapters on an examination you are taking. If this is the case, design your own sectional test by using the section references in the Brown, LeMay and Bursten text on which you are being tested. Cut-up the questions and paste them together to form a sample test appropriate for your situation.

8. How does one take a test? Every student seems to develop his or her own approach; some approaches to test taking are successful, and others are not. Let me suggest a few ideas based on my own teaching experiences. First, try

to study the material being covered on a test early so that you do not have to study late into the night immediately before the test. A tired and foggy mind cannot do its best. Second, organize beforehand the materials you will need to bring with you for the test — pencils, pens, paper, and calculator (don't forget to bring spare batteries or to charge the calculator before the test begins) — and don't forget to bring them. Third, find a place in the testing room that is comfortable and not too hot and that has sufficient light. Finally, when taking the test, read the instructions carefully, read over the questions quickly, and do the questions that are easy for you first. By answering the easy questions first, you will gain confidence and become more relaxed. After you finish the easy questions, begin the harder ones. If you find that you are stumped by a problem, stop working on it and go on to the next one instead of wasting time worrying about it. Later in the test you may remember how to do the problems that you could not complete earlier.

I would like to make a comment about the need for careful use of modern calculators. The numbers you obtain from a calculator often have six or eight digits, usually more digits than there are in any single number used in a calculation. In Chapter 1 of the text, you will learn that you cannot always report all the digits that a calculator shows. It is also easy to make an error entering a number into the calculator. Always check the final answer to see if it is a reasonable one. Is the answer too large or too small? Is the exponent of 10 reasonable? Sometimes a quick and rough hand calculation will tell you if the answer derived using a calculator is correct. Also, remembering that a calculator is only a tool; it cannot provide the logic for solving a problem. Only you can do that.

These suggestions for learning chemistry and doing well in your chemistry course are only that — suggestions. You may already have your own strategy for studying, and if it is successful, by all means use it. Or, you may want to adapt some of the suggested study tactics so that they complement your own learning methods. The key to being a successful student and learner is to have a consistent study plan and use it!

Acknowledgments

The form and content of the sixth edition reflects the significant input of hundreds of students, reviewers of prior editions, and the editorial staff at Prentice Hall through the years, and my ideas as to what makes a successful study guide. The advent of word processors has enabled me to type the manuscript; thus, I feel I should acknowledge the word processing program I use — ChiWriter. I find this software especially useful when typing scientific formulas, subscripts, superscripts and mathematical equations. To those who formally or accidentally inspired me, gave me ideas, or noted errors in prior editions, I thank you. I also acknowledge the Prentice Hall editorial and production staff who assisted in the design, copy editing, and production of this book. The quality of this text results from their highly effective editorial and production skills and expertise.

Any author with a family knows that the support of all family members creates an environment that makes writing less stressful. Without the love, understanding, and patience of my wife, Jan, and my children, Jason and Jeanina, this book would not be possible.

If you have any comments please send them to me:

Professor James C. Hill
Department of Chemistry
California State University, Sacramento
6000 J Street
Sacramento CA 95819

Introduction: Some Basic Concepts 1

OVERVIEW OF THE CHAPTER

Objectives: You should be able to:

1. Distinguish between physical and chemical properties and also between simple physical and chemical changes.
2. Differentiate between the three states of matter.
3. Distinguish between elements, compounds, and mixtures.
4. Give the symbols for the elements discussed in this chapter.

| 1.1, 1.2 MATTER: ELEMENTS, COMPOUNDS, AND MIXTURES |

Review: Concept of fraction; exponential notation (see text: Appendix A).

Objective: You should be able to list the basic SI units and the common metric prefixes and their meanings.

| 1.3 PHYSICAL QUANTITIES AND UNITS |

Review: Exponential notation (see text Appendix A).

Objectives: You should be able to:

1. Determine the number of significant figures in a measured quantity.
2. Express the result of a calculation with the proper number of significant figures.

| 1.4 SIGNIFICANT FIGURES |

Review: Dimensional analysis (1.5).

Objectives: You should be able to:

1. Convert temperatures among the Fahrenheit, Celsius, and Kelvin scales.
2. Perform calculations involving density.

| 1.3 TEMPERATURE AND DENSITY: INTENSIVE PROPERTIES |

1.5 DIMENSIONAL ANALYSIS

Review: Concepts of fraction and ratio.

Objective: You should be able to interconvert metric and English-system measurements by using dimensional analysis.

TOPIC SUMMARIES AND EXERCISES

MATTER: ELEMENTS, COMPOUNDS, AND MIXTURES

Matter is any material that occupies space and has mass. Three phases (states) of matter exist: gas, liquid, and solid. Check your knowledge of their characteristics by doing Exercise 1.

- A sample of matter is either a substance or a mixture.
- **Substances** are either elements or compounds. **Elements** can not be chemically decomposed into simpler new substances. **Compounds** consist of two or more elements chemically combined in a definite ratio. A compound can be chemically decomposed into its elements.
- **Mixtures** are combinations of two or more substances and are either homogeneous or heterogeneous. Note that a *heterogeneous* mixture exhibits more than one phase and possess a nonuniform distribution of substances. A *homogeneous* mixture consists of one phase and a uniform distribution of substances.
- Mixtures can be separated into substances by physical means.

Alterations in matter can involve chemical or physical changes.

- A **chemical change** involves a change in the composition of a substance. A **chemical property** describes the type of chemical change. For example, the property of wood burning is a chemical property.
- A **physical change** does not involve a change in composition but rather a change in a **physical property** such as temperature, volume, mass, pressure or state.

Check your understanding of the new terms you have learned by doing Exercises 2–5.

EXERCISE 1   Match the following characteristics to one or more of the three states of matter: (**a**) has no shape of its own; (**b**) definite shape; (**c**) occupies the total volume of a container; (**d**) partially takes on the shape of a container; (**e**) does not take on the shape of a container; (**f**) readily compressible; (**g**) slightly compressible; (**h**) essentially noncompressible.
   Solution: Gas—(**a**), (**c**), (**f**) Liquid—(**a**), (**d**), (**g**) Solid—(**b**), (**e**), (**h**)

EXERCISE 2   Match the term with the best identifying phrase:

Terms

1. Homogeneous mixture
2. Heterogeneous mixture

3. Mixture
4. Substance
5. Element
6. Compound

Phrases

a. Any kind of matter that is pure and has a fixed composition
b. Cannot be decomposed into simpler substances by chemical changes
c. A solution
d. Can be decomposed into simpler substances by chemical changes
e. Any kind of matter that can be separated into simpler substances by physical means
f. Nonuniform composition

Solution: 1-c; 2-f; 3-e; 4-a; 5-b; 6-d

EXERCISE 3   With the help of the periodic table, write the name or the chemical symbol for each of the following elements: (**a**) F; (**b**) zinc; (**c**) potassium; (**d**) As; (**e**) Al; (**f**) iron; (**g**) helium; (**h**) barium; (**i**) Ne.
   Solution: (**a**) fluorine; (**b**) Zn; (**c**) K; (**d**) arsenic; (**e**) aluminum; (**f**) Fe; (**g**) He; (**h**) Ba; (**i**) neon

EXERCISE 4   Are the following changes physical or chemical: (**a**) the vaporization of solid carbon dioxide; (**b**) the explosion of TNT; (**c**) the aging of an egg with a resultant unpleasant smell; (**d**) the formation of a solid when honey is cooled?
   Solution: (**a**) A physical change. The form of carbon dioxide is changed from solid to gas. There is no change in its chemical composition. (**b**) A chemical change. The explosion results from a change in the chemical composition of TNT. (**c**) A chemical change. A change in the composition of the egg results in the formation of a gas (hydrogen sulfide) that has an unpleasant smell. (**d**) A physical change. The solid results from the crystallization of dissolved sugars; no change occurs in the chemical form of the sugars.

EXERCISE 5   Classify each of the following as an element, compound, or mixture: (**a**) a 100-percent silver bar; (**b**) wine; (**c**) gasoline; (**d**) carbon dioxide (CO_2).
   Solution: (**a**) Silver is an element and cannot be separated by either chemical or physical means into simpler substances. It is listed among the elements in Table 1.2 in the text. (You should know the symbols in Table 1.2.) (**b**) Wine is a mixture of alcohol and water. The fact that wines contain varying percentages of alcohol attests to their having different compositions. (**c**) We know gasoline must be a mixture because it is available with different compositions and properties (no lead, low lead, ethyl, regular, and different brands with different additives). (**d**) CO_2 is a compound because the ratio of carbon and oxygen atoms does not vary. The name also implies that it is a compound because we do not have such systematic names for mixtures.

A property of a sample is measured by comparing it with a standard unit of that property. Measured quantities such as volume, length, mass and temperature require a number and a reference label, called the unit of measurement. Two systems of unit measurements are shown in Table 1.1. The SI system of units is now the preferred one; however, you will find certain metric units still used. *You must become thoroughly familiar with the units in Table 1.1 before starting the next chapter.*

PHYSICAL QUANTITIES AND UNITS

Table 1.1 Metric and SI Units

Physical quantity	Metric unit name	SI unit name[a]
Length	Meter (m)	Meter (m)
Volume	Cubic centimeter (cm^3)[b]	Cubic meter (m^3)
Mass	Gram (g)	Kilogram (kg)
Time	Second (s or sec)	Second (s or sec)
Energy	Calorie (cal)	Joule (J)
Pressure	Atmosphere (atm)	Newton per square meter (N/m^2)

[a]Systeme International d'Unites (SI) or International System of Units.

[b]Chemists commonly use the unit cubic centimeter when dealing with the volume of a solid, but they usually use the unit liter (L) when a substance is a liquid.

Another skill requiring proficiency is to change a number with a unit to one with a different unit. To do this, equivalence relationships must exist between units. Tables 1.2 and 1.3 give some common equivalences that you will need in this chapter.

Table 1.2 Equivalence Relationships Between SI and Metric Units

Physical quantity	Metric unit name	SI unit name	Equivalence
Length	Meter	Meter	Same
Mass	Gram	Kilogram	1000 g = 1 kg
Time	Second	Second	Same
Energy	Calorie	Joule	1 cal = 4.184 J
Volume	Cubic centimeter	Cubic meter	1,000,000 cm^3 = 1 m^3
Volume	Liter	Cubic meter	1000 L = 1 m^3
Pressure	Atmosphere	Newton per square meter	1 atm = 0.1754 N/m^2

Table 1.3 Equivalence Relationships Between Metric and English Units

Physical quantity	English unit symbol	Metric unit symbol	Equivalence
Mass	lb (= 16 oz)	g	1 lb = 453.6 g
Length	ft (= 12 in.)	m	3.272 ft = 1 m
Length	in.	cm	1 in. = 2.54 cm
Length	mi (= 5280 ft)	m	1 mi = 1609 m
Volume	qt	L	1.057 qt = 1 L

Prefixes are used with units to indicate decimal fractions (< 1) or multiples (> 1) of basic units.

- Example of a decimal fraction: The prefix centi- means 1/100 (= 0.01) of a basic unit; thus, 100 cm = 100 × 1/100 m = 1 m.
- Example of a multiple: The prefix kilo- means 10^3 (= 1,000); thus, 1 km = 1 × 1,000 m = 1000 m.

The commonly used prefixes that you must know are shown in Table 1.4. *Memorize them now.*

Table 1.4 Commonly Used Prefixes for Scientific Measurement in Chemistry

Prefix	Fraction or multiple of base unit	Abbreviation
Deci -	$10^{-1}\left(\dfrac{1}{10}\right)$	d
Centi -	$10^{-2}\left(\dfrac{1}{100}\right)$	c
Milli -	$10^{-3}\left(\dfrac{1}{1000}\right)$	m
Micro -	$10^{-6}\left(\dfrac{1}{1,000,000}\right)$	μ
Nano -	$10^{-9}\left(\dfrac{1}{1,000,000,000}\right)$	n
Pico -	$10^{-12}\left(\dfrac{1}{1,000,000,000,000}\right)$	p
Kilo -	$10^3\,(1000)$	k
Mega -	$10^6\,(1,000,000)$	M
Giga -	$10^8\,(1,000,000,000,000)$	G

EXERCISE 6 Which quantity of each pair is larger: (**a**) 1 nm or 1 micrometer; (**b**) 1 picogram or 1 cg; (**c**) 1 megagram or 1 milligram?

    Solution: Change the pairs so that each quantity is represented by either a numerical fraction or a multiple of the same basic metric unit. Then from their relative magnitudes you can determine which of the quantities is larger.

(**a**)
$$1\,\text{nm} = 1\ \text{nanometer} = 10^{-9}\ \text{meter}$$
$$1\ \text{micrometer} = 1\ \mu m = 10^{-6}\ \text{meter}$$

One micrometer is larger in value than one nanometer because the fraction 10^{-6} $\left(\dfrac{1}{1,000,000}\right)$ is larger in magnitude than the fraction 10^{-9} $\left(\dfrac{1}{1,000,000,000}\right)$.

(**b**)
$$1\ \text{picogram} = 1\ \text{pg} = 10^{-12}\ \text{gram}$$
$$1\ \text{cg} = 1\ \text{centigram} = 10^{-2}\ \text{gram}$$

One centigram is larger in value than one picogram because the fraction 10^{-2} $\left(\dfrac{1}{100}\right)$ is larger in magnitude than the fraction 10^{-12} $\left(\dfrac{1}{1,000,000,000,000}\right)$.

(**c**)
$$1\ \text{megagram} = 1\ \text{Mg} = 10^6\ \text{gram}$$
$$1\ \text{mg} = 1\ \text{milligram} = 10^{-3}\ \text{gram}$$

One megagram is larger in value than one milligram because the multiple 10^6 $(1,000,000)$ is larger in magnitude than the fraction 10^{-3} $\left(\dfrac{1}{1,000}\right)$.

EXERCISE 7   With what type of measurement are the following units associated?

$$g, L, m, km, cm, Mg, pg, cm^3$$

Solution: Mass (g, Mg, pg); volume (L, cm^3); length (m, km) Note that the prefixes such as M- and c- do not change the type of unit. However, the type of unit can be changed if it is multiplied, as is the case for cm^3. The unit cm^3 means cm × cm × cm, which is a unit for volume (V = l × w × h).

EXERCISE 8   What is the advantage of the metric system in comparison to the English system?
Solution: In the metric system, all quantities larger or smaller than the basic unit involve multiplication of the basic unit value by some power of 10 (for example, $10^3 = 1000$, $10^{-1} = \dfrac{1}{10}$, and so on). This is not true of the English system. Smaller or larger quantities of the basic unit in the English system are newly defined units. For example, 4000 qt equals 1000 gal, not 4 "kiloquarts." Many more conversion factors are required in the English unit system than in the metric unit system.

EXERCISE 9   Suggest a reason for the fact that 1 μkg (microkilogram) is not accepted as an appropriate SI mass unit expression.
Solution: The expression 1 μkg involves two prefixes, micro- (μ) and kilo- (k), yielding a compound prefix. This can be confusing, particularly if three or four prefixes are used. Thus, we do not use more than one prefix when expressing numbers. Instead of 1 μkg (microkilogram), we write 1 mg (milligram).

UNCERTAINTY IN MEASUREMENT: SIGNIFICANT FIGURES

Quantities in chemistry are of two types:

- **Exact numbers:** These result from counting objects such as coins or occur as defined values (such as 2, which also means 2.0, 2.00, etc.).
- **Inexact numbers:** These are obtained from measurements and require judgment. Uncertainties exist in their values.

Measured quantities (inexact numbers) are reported so that the last digit is the first uncertain digit. All certain digits and the first uncertain digit are referred to as **significant figures.** For example:

- 2.86: 2 and 8 are certain and well known. The number 6 is the first that is subject to judgment and is uncertain. The first uncertain digit is assumed to have an uncertainty of ± 1: 2.86 ± 0.01. The number 2.86 has three significant figures.
- 0.0020: Zeroes to the left of the first nonzero digit are not significant. The first three zeroes are not significant because they are to the left of the 2 and also define the decimal point. The zero to the right of the 2 is significant. This number has only two significant figures.
- 100: Trailing zeroes that define a decimal point may or may not be significant. Unless stated, assume they are not significant. Therefore, 100 has one significant figure unless otherwise stated; if it is determined from counting objects, it has three significant figures.

Scientific notation removes the ambiguity of knowing how many significant figures a number possesses.

- The form of a number in scientific notation is $A.BC \times 10^x$. If $x < 1$, the number is less than 1. If $x > 1$, the number is greater than 1.
- Only significant digits are shown. The number 0.0020 becomes 2.0×10^{-3}.

Calculated numbers must show the correct number of significant figures. The rules for determining the correct number of significant figures in a calculated number are:

1. Addition and subtraction: The final answer should have the same number of decimal places as the number in the calculation with the fewest number of decimal places.

2. Multiplication and division: The rule for rounding off the final calculated answer when multiplying or dividing numbers is to round off the final calculated answer so that it has the same number of significant figures as the least certain number (the one with fewest significant figures) in the calculation. A little caution must be used when applying this rule. For example, to divide 120 by 95, you might be tempted to report the final answer to two significant figures because 95 appears to be the least certain number. Yet 95 has almost three significant figures; there is very little difference in error between 1 in 95 and 1 in 101. Thus, in this case it makes more sense to round off the final answer to three significant figures. Use common sense in problems when a number is close in magnitude to 100, 1000, 10,000, and so on.

3. Exact measured values or defined numbers are not used in determining the number of significant figures in a final answer. If you use the equation $A = 4\pi r^2$ to measure the surface area of a sphere, the number 4 is considered to have an infinite number of significant figures and is therefore not used in determining the relative certainty of the calculated area.

Caution: The final answer of a calculation made using a modern calculator usually has more digits than any of the numbers in the calculation. You must remember to drop nonsignificant digits so that the proper number of significant figures is reported. This requires rounding off the answer.

The rules for rounding off numbers are:

1. When the figure immediately following the last digit to be retained in a number is less than 5, the last digit is retained unchanged. If 6.4362 is to be rounded off to four significant figures, it becomes 6.436.

2. When the figure immediately following the last digit to be retained is greater than 5, increase the last retained figure by 1. If 6.4366 is to be rounded off to four significant figures, it becomes 6.437.

3. When the figure immediately following the last digit to be retained is 5 and there are no other figures beyond the 5, or there are only zeroes beyond the 5; the last digit to be retained is increased by 1 if it is odd and left unchanged if it is even. Thus, 2.135 or 2.1350 becomes 2.14 if three figures are to be retained.

4. When the figure immediately following the last digit to be retained is 5 and there are figures other than zero beyond the 5, the last figure to be retained is increased by 1, whether it is odd or even. Thus, 2.1453 becomes 2.15 when only three figures are to be retained.

Note: Do not round off until you have completed your calculation.

EXERCISE 10 Remembering that measured values are reported to ±1 uncertainty in the last digit, except for those values determined by counting observable objects, or unless otherwise stated, determine the first uncertain digit in each of the following numbers: **(a)** 10.03 kg; **(b)** 5 apples; **(c)** 5.02 ± 0.02 m.

Solution: **(a)** The 3 in 10.03 kg is uncertain to ±1. **(b)** This is an exact measured value determined by counting. There is no uncertain digit. **(c)** The 2 in 5.02 m is uncertain to ±2.

EXERCISE 11 The precision of a measurement is indicated by the number of significant figures associated with the reported value. How many significant figures does each of these numbers have: **(a)** 225; **(b)** 10,004; **(c)** 0.0025; **(d)** 1.0025; **(e)** 0.002500; **(f)** 14,100; **(g)** 14,100.0?

Solution: Try this technique: If a quantity contains a decimal point, draw an arrow starting at the *left* through all zeros up to the first nonzero digit; the digits remaining are significant. If the quantity does not contain a decimal point, draw an arrow starting at the *right* through all zeroes up to the first nonzero digit; the digits remaining are significant.

(a) 225 $\leftarrow$ three significant figures

(b) 10,004 $\leftarrow$ five significant figures

(c) 00.025 two significant figures

(d) $\rightarrow$ 1.0025 five significant figures

(e) 0.002500 four significant figures

(f) 14,100 three significant figures; however, because of our lack of knowledge about the significance of the two trailing zeros, this number could have also four of five significant figures

(g) $\rightarrow$ 14,100.0 six significant figures

EXERCISE 12 Write the numbers in Exercise 11 using exponential notation.

Solution: Move the decimal in the appropriate direction so that it is to the right of the first nonzero digit reported in the number. If the decimal is moved to the left, multiply the resulting quantity by 10 raised to a power that equals the number of digits the decimal was moved past. If the decimal is moved to the right, the power of 10 is again the number of digits the decimal was moved past, but with a negative sign. That is, a number that is greater than 1 will appear as $A.BC \times 10^x$, while one that is less than 1 will appear as $A.BC \times 10^{-x}$. **(a)** $225 = 2.25 \times 10^2$. The decimal is moved two digits to the left, thus the power of 10 is 2. **(b)** $10,004 = 1.0004 \times 10^4 = 1.0004 \times 10^4$. The power of 10 is 4 because the decimal is moved four places to the left. **(c)** $0.0025 = 0002.5 \times 10^{-3} = 2.5 \times 10^{-3}$. The power of 10 is -3 because the decimal is moved three places to the right. Note that the nonsignificant zeros are omitted. **(d)** 1.0025. We do not write 1.0025×10^0. **(e)** $0.002500 = 0002.500 \times 10^{-3} = 2.500 \times 10^{-3}$. The zeroes after the 2.5 are written because they define the number of significant figures. **(f)** $14,100 = 1.4100 \times 10^4 = 1.41 \times 10^4$ if the zeroes in 14,100 are not significant. **(g)** $14,100.0 = 1.41000 \times 10^4 = 1.41000 \times 10^4$.

EXERCISE 13   Round off the answers in the following problems:

$$\text{(a)} \quad 12.25 + 1.32 + 1.2 = 14.770$$

$$\text{(b)} \quad 13.7325 - 14.21 = -0.4775$$

Solution: Use of the rounding-off rules for addition and subtraction yields the following analysis. (a) 1.2 has the fewest number of decimal places (by decimal places we mean the digits after a decimal point). Thus the final answer can have only one digit to the right of the decimal point: 14.770 is rounded off to 14.8. (b) 14.21 has the fewest number of decimal places (two). Thus the final answer is rounded off to −0.48. *Note:* An answer obtained by subtraction *may* have fewer significant figures than either number used in the subtraction. In this case the final answer has two significant figures whereas 14.21 has four.

EXERCISE 14   Round off the final answer in each of the following calculations:

$$\text{(a)} \quad (1.256)(2.42) = 3.03952$$

$$\text{(b)} \quad \frac{16.231}{2.20750} = 7.352661$$

$$\text{(c)} \quad \frac{(1.1)(2.62)(13.5278)}{2.650} = 14.712121$$

Solution: (a) The least certain number in the calculation is 2.42 (three significant figures). The final answer must be rounded off to three significant figures: 3.04. (b) The least certain number in the calculation is 16.231 (five significant figures). The final answer must be rounded off to five significant figures: 7.3527. (c) The least certain number in the calculation is 1.1 (two significant figures). The final answer must be rounded off to two significant figures: 15.

Two important concepts discussed in Chapter 1 are temperature and density.

TEMPERATURE AND DENSITY

- They are both **intensive properties** as their values are independent of the amount of substance present.
- This contrasts with **extensive properties** such as volume and mass, which depend on the amount of substance.

Temperature is a measure of the intensity of heat—the "hotness" or "coldness" of a body.

- Heat is a form of energy. Heat flows from a hot object to a colder one.
- When there is no heat flow between two objects in contact, they have the same temperature.
- Three temperature scales are used: Celsius (°C), Fahrenheit (°F) and Kelvin (K). You need to know how their reference points differ and how to change between them.

	Reference Points		
	Fahrenheit	**Celsius**	**Kelvin**
Freezing point of water	32°F	0°C	273.15 K
Boiling point of water	212°F	100°C	373.15 K

• *Note that a 1-degree interval is the same on both the Celsius and Kelvin scales, but a 1°C interval equals a 1.8°F interval.* The only temperature at which both the Fahrenheit and Celsius scales are equivalent is $-40°$ ($-40°C = -40°F$). This fact enables us to make conversions between the two scales using the following approach, which is different from the one given in the text.

°F → °C	°C → °F
(a) Add 40° to °F = (1)	Add 40° to °C = (1)
(b) $(1) \times \dfrac{1°C}{1.8°F} = (2)**$	$(1) \times \dfrac{1.8°F}{1°C} = (2)**$
(c) Subtract 40° from (2) = °C	Subtract 40° from (2) = °F

• The following relationship is used to convert between Celsius and Kelvin temperatures.

$$K = \frac{1\,K}{1°C}(°C) + 273.15\,K$$

EXERCISE 15   The temperature on a spring day is around 22°C. What is this temperature in degrees Fahrenheit and degrees Kelvin?

Solution: To change 22°C to °F, first add 40°:

$$22°C + 40° = 62°C$$

then multiply by 1.8°F/1°C:

$$(62°C)\left(\frac{1.8°F}{1°C}\right) = 112°F$$

Finally, subtract 40° from 112°F:

$$112°F - 40° = 72°F$$

To calculate degrees Kelvin, write the relationship between the two degrees and substitute for °C:

$$K = \left(\frac{1\,K}{1°C}\right)(°C) + 273.15\,K = \left(\frac{1\,K}{1°C}\right)(22°C) + 273.15\,K$$
$$= 295\,K \text{ (rounded off)}$$

Note: The method presented in the text gives the same answer as follows.

$$°F = \left(\frac{1.8°F}{1°C}\right)(°C) + 32°F$$

Substituting 22°C for °C yields:

$$(°F) = \left(\frac{1.8°F}{1°C}\right)(22°C) + 32°F = 39.6°F + 32°F = 72°F \text{ (rounded off)}$$

**Notice that only step (b) is different. Step (b) converts Fahrenheit to Celsius or Celsius to Fahrenheit using the relationship that a 1-degree Celsius interval equals a 1.8-degree Fahrenheit interval. Also, do not round off until the calculation is finished.

Density (*d*) measures the amount of a substance (*m*) in a given volume (*V*):

■ $d = \dfrac{\text{mass}}{\text{volume}} = \dfrac{m}{V}$

■ Density varies with temperature because volume changes with temperature.

■ Density can be used as a conversion factor to change mass to volume and vice versa for the same substance.

■ Chemists commonly use the following units for density: g/mL for liquids, g/cm^3 for solids, and g/L for gases.

EXERCISE 16   At 20°C, carbon tetrachloride and water have densities of 1.60 g/mL and 1.00 g/mL respectively. When water and carbon tetrachloride are poured into the same container, two layers form, one being water and the other carbon tetrachloride. Based on their densities, which one will occupy the lower layer in a container?

   Solution: Since carbon tetrachloride and water do not mix together permanently, the heavier substance per unit volume will fall to the bottom of the container. Carbon tetrachloride will be that substance because it has a higher mass per unit volume (density), 1.60 g/mL, than does water, 1.00 g/mL.

EXERCISE 17   Which has the greater mass, 2.0 cm^3 of iron ($d = 7.9$ g/cm^3) or 1.0 cm^3 of gold ($d = 19.32$ g/cm^3)?

   Solution: The mass of a substance is related to its density by the equation $d = m/V$. Multiplying both sides of the equation by *V* yields $m = d \times V$. The mass of 2.0 cm^3 of iron is calculated as follows:

$$m = d \times V = \left(7.9 \frac{g}{cm^3}\right)(2.0\ cm^3) = 16\ g$$

The mass of 1.0 cm^3 of gold is calculated similarly:

$$m = d \times V = \left(19.32 \frac{g}{cm^3}\right)(1.0\ cm^3) = 19\ g$$

Thus 1.0 cm^3 of gold has a greater mass than 2.0 cm^3 of iron.

Dimensional Analysis

You need to develop the habit of including units with all measurements in calculations. *Units are handled in calculations as any algebraic symbol:*

■ Numbers added or subtracted must have the same units.

■ Units are multiplied as algebraic symbols: (2 L)(1 atm) = 2 L-atm

■ Units are cancelled in division if they are identical. Otherwise, they are left unchanged: (3.0 m)/(2.0 mL) = 1.5 m/mL.

   Dimensional analysis is the algebraic process of changing from one system of units to another. A fraction, called a unit conversion factor, is used to make the conversion. These fractions are obtained from an equivalence between two units. For example, consider the equality 1 in. = 2.54 cm. This equality yields two conversion factors.

$$\frac{1 \text{ in.}}{1 \text{ in.}} = \frac{2.54 \text{ cm}}{1 \text{ in.}} \qquad \frac{1 \text{ in.}}{2.54 \text{ cm}} = \frac{2.54 \text{ cm}}{2.54 \text{ cm}}$$

$$1 = \frac{2.54 \text{ cm}}{1 \text{ in.}} \qquad \frac{1 \text{ in.}}{2.54 \text{ cm}} = 1$$

Note that conversion factors equal one and that they are the inverse of one another. They enable you to convert between units in the equality. For example, to convert from centimeter to inches or vice versa:

$$5.08 \cancel{\text{cm}} \times \frac{1 \text{ in.}}{2.54 \cancel{\text{cm}}} = 2.00 \text{ in.}$$

$$4.00 \cancel{\text{in.}} \times \frac{2.54 \text{ cm}}{1 \cancel{\text{in.}}} = 10.2 \text{ cm}$$

Note: $\text{given unit} \times \dfrac{\text{new unit}}{\text{given unit}} = \text{new unit}$

EXERCISE 18   Make the conversion of 10.5 L to milliliters.
   Solution: The required operations for converting 10.5 L to milliliters are as follows: (1) State the general relation required to convert units:

? mL = (10.5 L)(conversion factor that changes L to mL)

(2) Find the conversion factor (or factors) that convert L to mL. In this case it is 1000 mL = 1 L. Using this equivalence relation, determine the appropriate ratio of units that converts L to mL. Because we want to change liters to milliliters, we will have to divide by 1 L:

$$\frac{1000 \text{ mL}}{1 \text{ L}}$$

(3) Substitute 1000 mL/1 L for the conversion factor in the equation in step 1 and solve the problem.

$$? \text{ mL} = (10.5 \cancel{\text{L}})\left(\frac{1000 \text{ mL}}{1 \cancel{\text{L}}}\right) = 10.500 \text{ mL} = 1.05 \times 10^4 \text{ mL}$$

Change the final answer to scientific notation, if necessary. It is sometimes more convenient to change all numbers to scientific notation before doing the mathematics of the problem:

$$? \text{ mL} = (1.05 \times 10^1 \cancel{\text{L}})\left(\frac{1 \times 10^3 \text{ mL}}{1 \cancel{\text{L}}}\right) = 1.05 \times 10^4 \text{ mL}$$

(4) Check that the units properly cancel to yield the desired unit. In this problem, if we had used the conversion factor 1 L/1000 mL instead of 1000 mL/1 L, the result would have been:

$$? \text{ mL} \neq (1.05 \times 10^1 \text{ L})\left(\frac{1 \text{ L}}{1 \times 10^3 \text{ mL}}\right) \neq 1.05 \times 10^{-2} \frac{\text{L}^2}{\text{mL}}$$

The unit L^2/mL does not equal mL; therefore, we know we have used the wrong conversion factor.

EXERCISE 19   Convert 6.23 ft^3 to the appropriate SI unit.

Solution: The appropriate SI unit is the meter because ft^3 is an English unit of length, the foot. Use the same method shown in Exercise 18 to convert ft^3 to m^3.

(1) ? m^3 = (6.23 ft^3) (conversion factor that changes ft^3 to m^3)

(2) From Table 1.3 you find that 3.272 ft = 1 m, but there is no unit conversion given for ft^3 to m^3. What you must recognize is that if

$$1 = \frac{1 \text{ m}}{3.272 \text{ ft}}$$

then you can cube both sides of the expression

$$(1)^3 = \left(\frac{1 \text{ m}}{3.272 \text{ ft}}\right)^3 = \frac{1 \text{ m}^3}{(3.272)^3 \text{ ft}^3} = 1$$

(3) Converting units yields:

$$?\,\text{m}^3 = (6.23 \text{ ft}^3)\left(\frac{1 \text{ m}^3}{(3.272)^3 \text{ ft}^3}\right) = 0.178 \text{ m}^3$$

EXERCISE 20   The Empire State Building in New York City was for many years the tallest building in the world. Its height, including a television mast added in 1951, is 490.7 yd. Convert this distance to meters.

Solution: The conversion of 490.7 yd to meters is accomplished by the method used in Exercise 18: (1) State the general relation required to convert units:

? m = (490.7 yd) (conversion factor or series of factors that changes yards to meters)

(2) Look up the required equivalences in appropriate tables. From the tables in this chapter of the *Student's Guide,* we find that 3.272 ft = 1m. Because no conversion between yards and feet is given, you would have to go to another source or remember from our past experience that 1 yd = 3 ft. To convert yards to meters you will have to make a series of unit conversions that will look like

$$(\text{yd})\left(\frac{\text{ft}}{\text{yd}}\right)\left(\frac{\text{m}}{\text{ft}}\right) = \text{m}$$

Thus the required conversion factors are

$$\frac{3 \text{ ft}}{1 \text{ yd}} \quad \text{and} \quad \frac{1 \text{ m}}{3.272 \text{ ft}}$$

(3) You could solve the problem in two steps:

$$(490.7 \text{ yd})\left(\frac{3 \text{ ft}}{1 \text{ yd}}\right) = 1472 \text{ ft}$$

$$(1472 \text{ ft})\left(\frac{1 \text{ m}}{3.272 \text{ ft}}\right) = 449.9 \text{ m}$$

However, it is usually simpler to do it all in one step:

$$?\,\text{m} = (490.7 \text{ yd.})\left(\frac{3 \text{ ft}}{1 \text{ yd}}\right)\left(\frac{1 \text{ m}}{3.272 \text{ ft}}\right) = 449.9 \text{ m}$$

(4) Check that the units properly cancel. Inspection of the previous equation shows that they do.

SELF–TEST QUESTIONS

Key Terms

Having reviewed the key terms listed at the end of Chapter 1 and their applications, identify the following statements as true or false. If a statement is false, indicate why it is incorrect. Key terms are italicized in the statements.

1.1 The *density* of a silver bar increases with increasing weight.

1.2 The number 6.023×10^{23} contains four *significant figures*.

1.3 An example of *matter* is heat emitted from burning wood.

1.4 One liter of water contains 1.00 kg of matter at 20°C. Thus we can say that the *mass* of water is 1 L.

1.5 An example of an *intensive property* is volume.

1.6 Heat radiating from coal burning in a furnace is an example of an *extensive property*.

1.7 The density of a gold bar is an example of a *physical property*.

1.8 The electrolysis of water to form elemental hydrogen and oxygen is an example of a *physical change*.

1.9 The reaction of chlorine gas (Cl_2) with sodium metal (Na) to form sodium chloride (NaCl) is an example of a *chemical change*.

1.10 The reaction of carbon with oxygen to form carbon dioxide is an example of a *chemical property*.

1.11 Carbon monoxide can be separated into the simpler components carbon and oxygen; thus it is an *element*.

1.12 N_2 is a *compound*.

1.13 The *Celsius temperature* scale has more divisions between the freezing point and melting point of water than the Fahrenheit temperature scale does.

1.14 When ice completely melts in a glass of water, there is a change from a heterogeneous *mixture* to a homogeneous one.

1.15 The *SI unit* for mass is the gram.

1.16 A *conversion factor* contains a ratio of units.

1.17 Milk is a *substance*.

1.18 −273.15°C is equivalent to 0 *K*.

1.19 A student measures the mass of an object three times: 3.60 g, 3.90 g, and 3.75 g. The object actually has a mass of 3.75 g. The student's measurements show good *precision*.

1.20 In problem 1.19, the average value is 3.75 g. Therefore, the student's measurements gave good *accuracy*.

1.21 A *solution* shows a homogeneous distribution of two or more substances.

1.22 The basic unit of mass in the *metric system* is the same as in the SI system of units.

1.23 A *chemical reaction* involves a change in the chemical composition of substances.

1.24 A *change of state* of CO_2(s) to CO_2(g) is a chemical change.

1.25 10 g of CO_2 *gas* occupies less space than 10 g of CO_2 solid.

1.26 A *solid* consists of particles closer together than in the gas state.

1.27 A *liquid* is slightly compressible.

1.28 According to the *law of constant composition (definite proportions)*, H_2O and H_2O_2 represent the same substance.

Problems and Short–Answer Questions

1.29 You calculate your income for last year as $12,125.00. However, when you file your income tax form, you report an income of $12,000.00. You do this because you want to report a rounded-off income figure to two significant figures. Why would an IRS agent be unhappy with your use of significant figures?

1.30 How many significant figures do the following numbers possess?
- **(a)** 20.03 kg
- **(b)** 1.90×10^3 L
- **(c)** 120 m
- **(d)** 10 dollar bills
- **(e)** 0.00067 cm³

1.31 Convert the following numbers to exponential notation form with the correct number of significant figures.
- **(a)** 0.00067 cm³
- **(b)** 210.0 m
- **(c)** 0.040 L
- **(d)** 46900.0 g
- **(e)** 200 rattlesnakes

1.32 Complete the following calculations and round off answers to the correct number of significant figures:
- **(a)** $11.020 + 300.0 + 2.0030 =$
- **(b)** $1211 + 2.205 - 1.70 =$
- **(c)** $\dfrac{(1.425 \times 10^2)(2.61 \times 10^3)}{2.89 \times 10^5} =$
- **(d)** $\dfrac{(0.012)(0.100)}{11.0265} =$

1.33 Make the following conversions.
- **(a)** 1 ML to cubic centimeters.
- **(b)** $\dfrac{1}{1000}$ g to the appropriate SI unit
- **(c)** 8.00 qt to milliliters
- **(d)** 16 lb to grams

1.34 Make the following temperature conversions.
- **(a)** −70°F in the Antarctic (it gets cold there) to °C
- **(b)** 480°C at the surface of Venus to °F
- **(c)** 0.95 K, the freezing point of helium gas, to °C

1.35 In the late 1970s, the standard container sizes of wines and spirits were changed to the metric system. For example, the following changes were made:

	Old size	New size
Distilled spirits	1 qt (32 oz)	1.00 L
Wine	Magnum (51.2 oz)	1.50 L

(a) Calculate the volume in fluid ounces of the new containers.

(b) When the new container sizes were introduced, the price of a new container was often kept the same as the price of the equivalent old size. Assuming that a magnum and 1.50 L of wine each sold for $5.00, did the consumer experience price inflation with the introduction of the new containers?

1.36 Calculate the density of bromine, given that a 125.0-mL sample weighs 375.0 g.

1.37 A gold bar has the following dimensions: 2.50 cm × 2.00 cm × 1.50 cm. Assuming that gold can be sold for $500.00/oz, what is the value of the gold bar? The density of gold is 19.32 g/cm³, and 1 lb = 453.6 g.

1.38 A column of mercury is contained in a cylindrical tube. This tube has a diameter of 8.0 mm, and the height of the mercury column is 1.20 m. Given that the density of mercury is 13.6 g/cm³ and that the volume of mercury in the tube can be calculated from the relation $V = \pi r^2 h$ (r = radius of the tube and h = height of mercury column), calculate the mass of mercury present in the cylindrical tube.

Multiple–Choice Questions

1.39 Which of the following is not correct?
(a) There are 1000 mg in a gram.
(b) There are 100 cm in a meter.
(c) There are 1000 mL in a liter.
(d) There are 100 mm in a centimeter.
(e) There are 1000 μJ in a millijoule.

1.40 Which of the following is the standard unit of mass in the SI system?
(a) meter **(d)** liter
(b) pound **(e)** kilogram
(c) gram

1.41 Which prefix means $\dfrac{1}{1,000,000}$ of a unit?
(a) kilo- **(d)** micro-
(b) centi- **(e)** nano-
(c) milli-

1.42 2.5 nm equals:
(a) 2.5×10^{-9} m **(d)** (a) and (b)
(b) 2.5×10^{-4} mm **(e)** (a) and (c)
(c) 2.5×10^{-7} cm

1.43 A number that has precisely three significant figures is:
(a) 0.01 **(d)** 0.001
(b) 100 **(e)** 110
(c) 0.100

1.44 When 105.50 is rounded off to three significant figures, it becomes:
(a) 106 **(d)** 105.5
(b) 105 **(e)** 106.0
(c) 104

1.45 Solid carbon dioxide, dry ice, changes directly from a solid to a vapor at 195 K if left in an open container. What is this temperature in degrees Celsius and Fahrenheit?

(a) −78°C, 468°F **(d)** −108°C, −78°F
(b) −108°C, 468°F **(e)** −78°C, −108°F
(c) 468°C, −108°F

1.46 Density can be thought of as a conversion factor. For example, the density of aluminum is 2.70 g/cm³, which can be interpreted to mean 2.70 g of aluminum = 1 cm³. Using this equivalence, what is the volume occupied by 223.5 g of aluminum?
(a) 603 cm³ **(d)** 0.0270 cm³
(b) 82.8 cm³ **(e)** 223.5 cm³
(c) 0.0121 cm³

1.47 When a solid substance undergoes a physical change to a liquid, which of the following is always true?
(a) A new substance is formed.
(b) Heat is given off.
(c) A gas is given off.
(d) It vaporizes.
(e) It melts.

1.48 An empty container weighs 15.230 g. When filled with water (density = 1.00 g/mL), it weighs 35.920 g. When filled with an unknown liquid to the same mark as it was filled to with the water, it weighs 36.261 g. What is the density of the unknown liquid?
(a) 1.02 g/mL **(d)** 1.20 g/m³
(b) 1.02 g/m³ **(e)** none of the above
(c) 1.20 g/mL

1.49 The speed limit for automobiles on interstate highways is 55 mi/hr. How many kilometers could one travel in 3.5 hr at this speed?
(a) 88 km **(d)** 310 km
(b) 88.49 km **(e)** 30 km
(c) 309.7 km

SELF-TEST SOLUTIONS

1.1 False. Density is an intensive property. **1.2** True. **1.3** False. Matter has mass and volume, which radiation does not have. **1.4** False. The unit of mass is the kilogram in the SI unit system; liter is the unit of volume. **1.5** False. An intensive property does not depend on the amount of substance present; volume depends on the quantity of substance. **1.6** True. Heat evolution depends on the amount of substance being combusted. **1.7** True. **1.8** False. A physical change involves no change in the composition of a substance. **1.9** True. **1.10** True. **1.11** False. Carbon monoxide is a compound because it can be resolved into its elements, carbon and oxygen. **1.12** False. N_2 is an element containing two atoms. A compound contains two or more different elements. **1.13** False. Remember that nine Fahrenheit divisions equals five Celsius divisions. **1.14** True. **1.15** False. It is the kilogram. **1.16** True. **1.17** False. It is a mixture of substances. **1.18** True. **1.19** False. If the student's measurements had shown good precision, the masses would have been closer in value, not showing a range of 0.30 g between low and high values, but a much smaller range. **1.20** True. **1.21** True. **1.22** False. In the metric system it is the gram, and in the SI system it is the kilogram. **1.23** True. **1.24** False. A physical change–its composition does not change. **1.25** False. A gas expands to fill the entire vessel holding it. A liquid only occupies a limited and lesser amount of space. **1.26** True. **1.27** True. **1.28** False. In the case of water the ratio of hydrogen to oxygen is 2:1, whereas in the case of hydrogen peroxide (H_2O_2) the ratio is 2:2 or 1:1. Thus, they are different substances. **1.29** The IRS requires that all digits to the left of the decimal be significant when you report your income. You may round off the cents to the nearest dollar. Your actual income of $12,125.00 has five significant figures to the left of the decimal. Your reported income of $12,000.00 implies that the three zeros to the right of the 2 and to the left of the decimal are significant. The IRS agent is unhappy because he or she found out that only the first two digits of the $12,000.00 are precise, not the first five digits as implied by your reported income. **1.30 (a)** four; **(b)** three; **(c)** two, because the zero defines the decimal; **(d)** infinite, because this is an exact number; **(e)** two, because the zeroes immediately to the right of the decimal are not significant. **1.31 (a)** $0.00067 \text{ cm}^3 = 6.7 \times 10^{-4} \text{ cm}^3$; **(b)** $210.0 = 2.100 \times 10^2$ m; **(c)** $0.040 \text{ L} = 4.0 \times 10^{-2}$ L; **(d)** 46900.0 g $= 4.69000 \times 10^4$ g; **(e)** 200 rattlesnakes $= 2.00 \times 10^2$ rattlesnakes. Because this is an exact quantity, we need only write 2×10^2 rattlesnakes. **1.32 (a)** $11.020 + 300.0 + 2.0030 = 313.0230$. The answer can have only one digit to the right of the decimal because 300.0 has the fewest number of digits to the right of the decimal—that is, one. The answer is rounded off to 313.0. **(b)** $1211 + 2.205 - 1.70 = 1211.505$. The answer can have no digits to the right of the decimal because 1211 has no digits to the right of the decimal. The answer is rounded off to 1212.

$$\text{(c)} \quad \frac{(1.425 \times 10^2)(2.61 \times 10^3)}{2.89 \times 10^5} = 1.2869377$$

The numbers with the fewest significant figures in the calculation are 2.61×10^3 and 2.89×10^5 with three significant figures. The final answer is rounded off to three significant figures: 1.29. **(d)** First convert all numbers to exponential notation and then solve:

$$\frac{(0.012)(0.100)}{11.0265} = \frac{(1.2 \times 10^{-2})(1.00 \times 10^{-1})}{1.10265 \times 10^1}$$

$$= 1.08828 \times 10^{-4}$$

The number with the fewest significant figures in the calculation is 1.2×10^{-2} with two significant figures. The final answer is rounded off to two significant figures: 1.1×10^{-4}.

$$\text{1.33 (a)} \quad (1 \text{ ML})\left(\frac{10^6 \text{ L}}{1 \text{ ML}}\right)\left(\frac{10^3 \text{ mL}}{1 \text{ L}}\right)\left(\frac{1 \text{ cm}^3}{1 \text{ mL}}\right) = 1 \times 10^9 \text{ cm}^3$$

$$\text{(b)} \quad \left(\frac{1}{1000} \text{ g}\right)\left(\frac{1 \text{ kg}}{1000 \text{ g}}\right)$$

$$= (1 \times 10^{-3} \text{g})\left(\frac{1 \times 10^{-3} \text{ kg}}{1 \text{ g}}\right) = 1 \times 10^{-6} \text{ kg}$$

$$\text{(c)} \quad (8.00 \text{ qt})\left(\frac{1 \text{ L}}{1.057 \text{ qt}}\right)\left(\frac{1000 \text{ mL}}{1 \text{ L}}\right) = 7.57 \times 10^3 \text{ mL}$$

$$\text{(d)} \quad (16 \text{ lb})\left(\frac{453.6 \text{ g}}{\text{lb}}\right) = 7.3 \times 10^3 \text{ g}$$

$$\text{1.34 (a)} \quad °C = (-70°F + 40°)\left(\frac{1°C}{1.8°F}\right) - 40° = -57°C$$

$$\text{(b)} \quad °F = (480°C + 40°)\left(\frac{1.8°F}{1°C}\right) - 40° = 896°F$$

$$\text{(c)} \quad K = \left(\frac{1 \text{ K}}{1°C}\right)(°C) + 273.15 \text{ K} \quad \text{or}$$

$$°C = \left(\frac{1°C}{1 \text{ K}}\right)(K - 273.15 \text{ K})$$

$$= \left(\frac{1°C}{1 \text{ K}}\right)(0.95 \text{ K} - 273.15 \text{ K}) = -272.20°C$$

$$\text{1.35 (a)} \quad (1.00 \text{ L})\left(\frac{1.057 \text{ qt}}{1 \text{ L}}\right)\left(\frac{32.0 \text{ oz}}{1 \text{ qt}}\right) = 33.8 \text{ oz}$$

$$(1.50 \text{ L})\left(\frac{1.057 \text{ qt}}{1 \text{ L}}\right)\left(\frac{32.0 \text{ oz}}{1 \text{ qt}}\right) = 50.7 \text{ oz}$$

(b) The price per ounce for each size is calculated as follows:

$$\text{Magnum price per ounce} = \frac{\text{price}}{\text{weight}} = \frac{\$5.00}{51.2 \text{ oz}} = \frac{\$0.0977}{\text{oz}}$$

$$1.50 \text{ liter per ounce} = \frac{\$5.00}{50.7 \text{ oz}} = \frac{\$0.0986}{\text{oz}}$$

As the calculation shows, there was a slight increase in the price per ounce of the magnum when it was changed to 1.50 L. **1.36** Density of bromine = mass/volume = 375.0 g/125.0 mL = 3.000 g/mL. **1.37** The mass of the gold bar is calculated from its density using the relation mass = density × volume. The volume of the gold bar equals the product of its length times width times height: $V = (2.50 \text{ cm})(2.00 \text{ cm})(1.50 \text{ cm}) = 7.50 \text{ cm}^3$. Calculating the mass of the gold bar with this volume: $m = \text{density} \times \text{volume} = (19.32 \text{ g/cm}^3)(7.50 \text{ cm}^3) = 145 \text{ g}$. The value of the gold bar is its selling price times its mass.

$$\text{Price} = (145 \text{ g})\left(\frac{1 \text{ lb}}{453.6 \text{ g}}\right)\left(\frac{16 \text{ oz}}{1 \text{ lb}}\right) \times \left(\frac{500.00 \text{ dollars}}{\text{oz}}\right)$$

$$= 2560 \text{ dollars}$$

1.38 Because the density is expressed in units of cm^3, it is convenient to change all length measurements into that unit. For this problem, the required equivalences are 10 mm = 1 cm and 100 cm = 1 m. Calculating the volume of the tube the mercury occupies in the unit of cm^3:

$$V = \pi r^2 h = \pi \left(\frac{8.00 \text{ mm}}{2}\right)^2 \left(\frac{1 \text{ cm}}{10 \text{ mm}}\right)^2$$

$$\times (1.20 \text{ m})\left(\frac{100 \text{ cm}}{1 \text{ m}}\right)$$

$$= 60.3 \text{ cm}^3$$

Mass of mercury = density × volume

$$= \left(13.6 \frac{\text{g}}{\text{cm}^3}\right)(60.3 \text{ cm}^3)$$

$$= 820 \text{ g}$$

1.39 (d) 100 mm = (100 mm)(1 m/1000 mm) = 1/10 m; 1 cm = 1/100 m. Thus the two quantities are not equal. **1.40 (e) 1.41 (d). 1.42 (e)** 2.5 nm

$$(2.5 \text{ mm}) \left(\frac{10^{-9} \text{ m}}{1 \text{ nm}} \right) = 2.5 \times 10^{-9} \text{ m}$$

$$(2.5 \text{ mm}) \left(\frac{10^{-9} \text{ m}}{1 \text{ nm}} \right) \left(\frac{100 \text{ cm}}{1 \text{ m}} \right) = 250 \times 10^{-9} \text{ cm}$$

$$= 2.5 \times 10^{-7} \text{ cm}$$

1.43 (c) The zeros after the one are significant because they do not define the decimal point. **1.44 (a)** The figure immediately following the third digit is a five, and it has a zero after it; so the third digit is increased by one because it is odd. **1.45 (e) 1.46 (b).** (223.5 g)(1 cm^3/2.70 g) = 82.8 cm^3. **1.47 (e). 1.48 (a)** Mass of water = 35.920 g − 15.230 g = 20.690 g; mass of liquid = 36.261 g − 15.230 g = 21.031 g; volume occupied by liquid = (mass water)(1.00 mL/1 g) = 20.690 mL; density of liquid = m/V = 21.031 g/20.690 mL = 1.02 g/mL. **1.49 (d)** (55 mi/hr)(3.5 hr)(1.609 km/mi) = 309.7 km = 310 km (rounded off).

2 | Atoms, Molecules, and Ions

OVERVIEW OF THE CHAPTER

**2.1, 2.2, 2.3
ATOMS**

Objectives: You should be able to:

1. Describe the composition of an atom in terms of protons, neutrons, and electrons.

2. Give the approximate size, relative mass, and charge of an atom, proton, neutron, and electron.

3. Write the chemical symbol for an element, having been given its mass number and atomic number, and perform the reverse operation.

4. Describe the properties of the electron as seen in cathode rays. Describe the means by which J. J. Thomson determined the ratio *e/m* for the electron.

5. Describe Millikan's oil-drop experiment and indicate what property of the electron he was able to measure.

6. Cite the evidence from studies of radioactivity for the existence of subatomic particles.

7. Describe the experimental evidence for the nuclear nature of the atom.

**2.4, 2.5
MOLECULES AND
IONS:
RELATIONSHIPS IN
THE PERIODIC
TABLE**

Objectives: You should be able to:

1. Write the symbol and charge for an atom or ion, having been given the number of protons, neutrons, and electrons, and perform the reverse operation.

2. Use the periodic table to predict the charges of monatomic ions.

3. Use the periodic table to predict whether an element is a metal or a nonmetal.

4. Distinguish between empirical formulas, molecular formulas and structural formulas.

**2.6
NOMENCLATURE**

Objectives: You should be able to:

1. Write the simplest formula for a compound, having been given the charges of the ions from which it is made.

2. Write the name of a simple inorganic compound, having been given its chemical formula, and perform the reverse operation.

18

TOPIC SUMMARIES AND EXERCISES

This chapter focuses on the structure of atoms, the form atoms exist in substances and naming inorganic compounds. The work of many scientists, such as John Dalton, J. J. Thomson, and R. Millikan, led to the development of our modern understanding of the structure of atoms. You should know the contributions of the key historical figures discussed in this chapter: See Exercise 3.

Atoms are the smallest particles of an element that have all the same chemical properties of that element. You need to know the following information about the structure of atoms.

- The mass of a single atom is extremely small, less than 10^{-21} g
- The center of an atom contains a small **nucleus.** A nucleus contains **protons** and **neutrons,** which make up most of the mass of an atom.
- The volume of an atom mostly consists of **electrons.**
- Protons (positively charged) and neutrons (no charge) have essentially the same mass. Electrons (negatively charged) are significantly less massive.

All atoms of the same element have the same number of protons.

- Atoms that have the same number of protons but differ in their number of neutrons are called **isotopes.**
- The general symbol for an isotope is

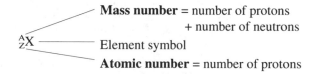

$$^A_Z X$$

Mass number = number of protons + number of neutrons

Element symbol

Atomic number = number of protons

Note: The atomic number Z defines an element; it also tells you how many electrons a neutral atom of that element possesses. Thus, the number of neutrons equals A - Z.

A **nuclide** is a nucleus of a specific isotope; for example a $^2_1$H nuclide(deuterium) or a $^3_1$H nuclide(tritium).

EXERCISE 1 The position of an element in the periodic table is a function of its atomic composition. Elements are arranged in order of increasing atomic number. The atomic number of an atom equals its number of protons (or its number of electrons). By referring to a periodic table, determine which *neutral* atom: **(a)** contains 50 protons; **(b)** contains 17 electrons; **(c)** has an atomic number of 56.
   Solution: **(a)** Tin (Sn). Its atomic number, 50, corresponds to 50 protons. **(b)** Chlorine (Cl). Its atomic number corresponds to 17 protons. This is also the number of electrons in the atom. **(c)** Barium (Ba). Its atomic number is 56.

EXERCISE 2 Write the nuclear-isotope symbols for the four isotopes of sulfur with 16, 17, 18, and 20 neutrons, respectively.
   Solution: The atomic number of sulfur is 16, and all of its isotopes will have 16 protons. The mass number of each isotope is the sum of its number of neutrons plus its number of protons: $16 + 16 = 32$; $16 + 17 = 33$; $16 + 18 = 34$; and $16 + 20 = 36$. The nuclear-isotope symbols are $^{32}_{16}$S, $^{33}_{16}$S, $^{34}_{16}$S, and $^{36}_{16}$S. Note that the subscript before each

elemental symbol (that is, the atomic number) is the same for all isotopes because the number of protons is invariant for sulfur.

EXERCISE 3   The composition of the atom was determined from the experiments of many individuals. For each experiment, state the principal person associated with the experiment, the particle studied, and the property of the particle determined: **(a)** α scattering; **(b)** cathode ray; **(c)** oil drop.

    Solution: **(a)** E. Rutherford determined the relative volume of the nucleus in the atom by observing the scattering of α particles. **(b)** J. J. Thomson determined the *e/m* ratio of the electron by using a cathode-ray tube. **(c)** R. Millikan determined the charge of an electron by observing the motion of charged oil drops.

EXERCISE 4   Find the number of protons, electrons and neutrons in the following isotopes:

$$\text{(a) } {}^{40}_{20}\text{Ca}; \quad \text{(b) } {}^{238}_{92}\text{U}.$$

    Solution: **(a)** The number of protons is the subscript number shown in the isotopic symbol: 20. The number of electrons in a neutral atom is the same as the number of protons: 20. The number of neutrons equals the mass number minus the number of protons. This is the superscript number (mass number) minus the subscript number: $40 - 20 = 20$. **(b)** Preceding as in **(a)**, the number of protons is 92, the number of electrons is 92, and the number of neutrons is $238 - 92 = 146$.

EXERCISE 5   If an electron is added or removed from ^{35}Cl, what changes occur to the isotope?

    Solution: The isotope is still ^{35}Cl because the number of protons and the number of neutrons do not change; however, the number of electrons does. If electrons, which are negatively charged, are added to an isotope, the isotope gains negative charge. Therefore if one electron is added to ^{35}Cl, it becomes $^{35}\text{Cl}^-$. Similarly, if an electron is removed, the isotope loses negative charge and becomes positively charged because there are more protons than electrons: $^{35}\text{Cl}^+$.

MOLECULES AND IONS: RELATIONSHIPS IN THE PERIODIC TABLE

The **periodic table** lists elements in order of increasing atomic number. Learn to use the periodic table: the grouping of elements in vertical columns and horizontal rows helps you to remember that

- *Metals* are on the left side and middle.
- *Nonmetals* are on the right side.
- *Metalloids* have properties of both metals and nonmetals. They are identified by the dark diagonal line toward the right side of the periodic table.
- Elements in vertical columns, called *families* or *groups*, exhibit similar chemical and physical properties, whereas elements in a horizontal row (*period*) exhibit different properties. The names of families in the periodic table and the general location of metals, metalloids, and nonmetals are shown in Figure 2.1.

Molecular compounds consist of small particles called molecules.

- **Molecules** consist of two or more atoms that can have a stable existence as a unit: if the atoms are the same element, a molecule is *homonuclear*; if they are different, it is *heteronuclear*.

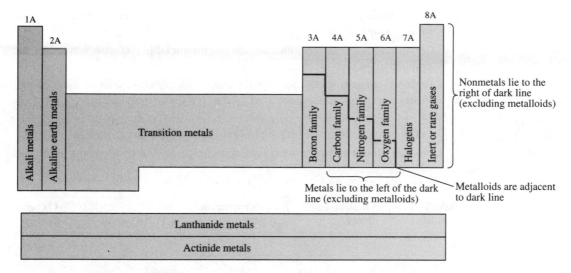

Figure 2.1 Families in the periodic table. Note that hydrogen, the first element in family 1A, is not an alkali metal.

- Typically, molecules consist of nonmetallic elements.
- Elements forming homonuclear diatomic molecules that you should know are: H_2, N_2, O_2, F_2, Cl_2, Br_2, and I_2.
- Subscripts in a **molecular formula** tell you how many atoms are actually present

$$C_2H_4$$

2 atoms of   4 atoms of
carbon   hydrogen

- An **empirical formula** shows only the simplest whole number ratio of atoms. For example, C_2H_4 is a molecular formula; its empirical formula, CH_2, is obtained by dividing the subscripts in the molecular formula by 2:

$$CH_2 = C_{2/2}H_{4/2}$$

- A **structural formula** shows "how atoms are joined together"—that is, the shape, orientation, and position. For example, the structural formula of C_2H_6 shows the following arrangement of atoms:

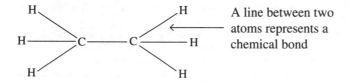

A line between two atoms represents a chemical bond

Ionic compounds consist of positively charged ions (*cations*) and negatively charged ions (*anions*).

- **Cations** are positively charged ions of metallic elements formed by the metal losing electrons; e.g., Na^+.

- **Anions** are negatively charged ions of nonmetallic elements formed by the nonmetal gaining electrons; e.g., S^{2-}.

- Ionic compounds do not contain unique identifiable pairs of ions. Therefore, empirical formulas are used to describe their composition.

- Ionic compounds may contain *polyatomic ions* such as NO_3^- as in $NaNO_3$. Polyatomic ions are like molecules except that they carry a charge.

- You should know the most common charge carried by elements in the following families:

Alkali metals	Alkaline-earth metals	Transition metals	Halogens	Oxygen family
1+	2+	2+,3+	1−	2−

EXERCISE 6 Interpret the following chemical formulas: (a) N_2; (b) NH_3; (c) NH_4^+; (d) H_2SO_4.

Solution: (a) N_2 is a homonuclear (having atoms of the same type) molecule composed of two nitrogen atoms; this is the elemental form of nitrogen. (b) NH_3 is a heteronuclear (having more than one type of atom) molecule composed of one nitrogen atom and three hydrogen atoms. (c) NH_4^+ is an ion that has a 1+ charge and is composed of one nitrogen atom and four hydrogen atoms. (d) H_2SO_4 is a heteronuclear molecule composed of two hydrogen atoms, one sulfur atom, and four oxygen atoms.

EXERCISE 7 (a) Do C_2H_4 and $2CH_2$ represent the same substance? (b) What are the empirical formulas for B_2H_6, H_2O, and $C_2H_2O_4$?

Solution: (a) The formulas do not represent the same substance. C_2H_4 represents the molecular formula for the substance acetylene, a gas used in welding. $2CH_2$ represents two empirical formulas of acetylene. A number in front of a formula tells you how many units of that formula there are; changing subscripts in a formula changes the substance to a different substance or to an empirical formula. (b) To determine an empirical formula divide each subscript of a molecular formula by the largest whole number that goes into each subscript

$$B_{2/2}H_{6/2} = BH_3$$

There is no way to further reduce the subscripts in H_2O—this is the molecular and empirical formula for water.

$$C_{2/2}H_{2/2}O_{4/2} = CHO_2$$

EXERCISE 8 Determine the total number of electrons in each of the following: (a) NH_3; (b) NH_4^+; (c) NH_2^-.

Solution: Determine the number of electrons for each atom in the compound, sum these numbers, and then adjust the number for the charge shown. For each negative charge, one electron is added; for each positive charge, one electron is subtracted. (a) The number of electrons for each atom is the same as the number of protons. Each hydrogen atom has one proton and each nitrogen atom has 7 protons. Thus, for NH_3 the number of electrons is $(1 \times 7) + (3 \times 1) = 10$ electrons. The charge of the compound is zero, thus no adjustment for charge is necessary. (b) The number of electrons for NH_4^+ with no charge is $(1 \times 7) + (4 \times 1) = 11$ electrons. However, the species has a positive charge, thus one electron must be removed so that there are more protons than electrons: $11 - 1 = 10$ electrons for NH_4^+. **(c)** The number of electrons for NH_2^- with no charge is $(1 \times 7) + (2 \times 1) = 9$ electrons. The species has a charge of $1-$, thus one electron must be added so that there are more electrons than protons: $9 + 1 = 10$ electrons.

EXERCISE 9 Write the formula for: **(a)** a neutral polyatomic compound consisting of one barium ion with a 2+ charge (written Ba^{2+}) and chlorine with a 1− charge (written Cl^-). **(b)** A polyatomic ion that has a 1− charge and consists of one boron ion with a 3+ charge (written B^{3+}) and fluorine with a 1− charge (written F^-).

Solution: **(a)** The sum of positive and negative charges for the ions must equal zero in a neutral compound. Two Cl^- ions are required to balance the 2+ charge of Ba^{2+}: $BaCl_2$. **(b)** The sum of positive and negative charges for the ions must equal the 1− charge of the compound. Four F^- ions are required so that, when they are combined with B^{3+}, the sum of the charges is 1−: BF_4^- $[+3 + 4(-1) = -1]$.

Chemists not only write formulas for compounds but also give them reasonably systematic names. Rules for naming inorganic compounds are found in Section 2.6 in the text. *These rules should be committed to memory; and you should then practice them by naming anions, cations, ionic compounds, oxyanions, acids, and simple molecular substances.* The following exercises should be attempted *after* you have reviewed nomenclature rules. The solutions to each exercise provide explanations that will reinforce the nomenclature rules you have learned.

| NOMENCLATURE

EXERCISE 10 The following chemical formulas are written incorrectly. Correct them so that they are in the proper form: **(a)** ClNa; **(b)** CaClCl; **(c)** $NH_4NH_4SO_4$; **(d)** H^2S; **(e)** FXeF.

Solution: **(a)** NaCl. The cation (Na^+) is always placed before the anion (Cl^-). **(b)** $CaCl_2$. A subscript is used to indicate the number of Cl ions present. **(c)** $(NH_4)_2SO_4$. If there is more than one polyatomic ion present, it is enclosed in parentheses, and subscripts are used as necessary. **(d)** H_2S. The 2 must be a subscript when it indicates the number of the same type of atoms or ions present. **(e)** XeF_2.

EXERCISE 11 Name the following cations: **(a)** H^+; **(b)** Na^+; **(c)** Be^{2+}; **(d)** Al^{3+}; **(e)** Co^{2+}; **(f)** Co^{3+}.

Solution: Monatomic cations have the same names as the elements from which they are formed. Some monatomic cations commonly exist in two common ion states; for example, Fe^{2+} and Fe^{3+}. When naming such ions, indicate the charge of the ion in Roman numerals enclosed within parentheses after the name of the element. Sometimes an older method is used, in which the suffix *-ic* indicates the higher charge and *-ous* the lower one. This method is further complicated by the fact that the Latin name of the element is often used. Thus Fe^{2+} is iron(II) or ferrous, and Fe^{3+} is iron(III) or ferric. When in doubt, use the newer method. **(a)** Hydrogen. **(b)** Sodium. **(c)** Beryllium. **(d)** Aluminum. **(e)** Cobalt(II) or cobaltous. **(f)** Cobalt(III) or cobaltic.

EXERCISE 12 Name the following anions: **(a)** F^-; **(b)** S^{2-}; **(c)** OH^-; **(d)** CN^-; **(e)** PO_4^{3-}; **(f)** PO_3^{3-}; **(g)** IO^-; **(h)** IO_2^-; **(i)** IO_3^-; **(j)** IO_4^-; **(k)** HS^-; **(l)** $H_2PO_4^-$.

Solution: **(a)** Fluoride. Monatomic anions have *-ide* added to the stem of the element's name. **(b)** Sulfide. **(c)** Hydroxide. Treated as a monatomic anion. **(d)** Cyanide. Treated as a monatomic anion. **(e)** Phosphate. When two oxyanions exist, as in the cases of PO_4^{3-} and PO_3^{3-}, the one with more oxygen atoms ends in *-ate*. **(f)** Phosphite. The one with fewer oxygen atoms ends in *-ite*. **(g)** Hypoiodite. The prefix *hypo-* indicates one less oxygen atom than in IO_2^-. **(h)** Iodite. **(i)** Iodate. **(j)** Periodate. The prefix *per-* indicates one more oxygen atom than in IO_3^-. **(k)** Hydrogen sulfide. *Hydrogen* is added to the name of the anion to indicate that there is one hydrogen atom present. **(l)** Dihydrogen phosphate. *Dihydrogen* is added to phosphate to indicate that there are two hydrogen atoms present.

EXERCISE 13   Name the following compounds: **(a)** ZnS; **(b)** $(NH_4)_2SO_4$; **(c)** $FeCl_2$; **(d)** $KClO_4$; **(e)** $SnCl_2$; **(f)** PCl_3; **(g)** SF_6; **(h)** CO.
   Solution: Remember: When naming ionic compounds cations are named before anions; in molecular compounds the atoms are named in the order found in the molecular formula. For binary molecular compounds, the first atom is given its elemental name, and the second is named as if it were a negative ion. **(a)** Zinc sulfide. Zinc exists only in the 2+ state. **(b)** Ammonium sulfate. The use of *di-* before *ammonium* to indicate two NH_4^+ ions is not necessary because the 2^- charge of SO_4^{2-} automatically requires two NH_4^+ ions. **(c)** Iron(II) chloride or, using the older method, ferrous chloride. **(d)** Potassium perchlorate. **(e)** Tin(II) chloride or stannous chloride. **(f)** Phosphorus trichloride. A compound formed from two nonmetallic elements is named as if it were an ionic compound, with the appropriate prefixes added to indicate how many atoms are present. **(g)** Sulfur hexafluoride. The prefix *hexa-* indicates six fluorine atoms. **(h)** Carbon monoxide.

EXERCISE 14   Name the following acids in water: **(a)** HCN; **(b)** H_2S; **(c)** H_2CO_3; **(d)** H_3PO_4; **(d)** H_3PO_4; **(e)** H_3PO_3.
   Solution: **(a)** Hydrocyanic acid. For hydrogen acids formed from monatomic anions, *hydro-* is added to the name of the anion—in this case, *cyanide*—and the *-ide* ending of the anion is changed to an *-ic* ending. **(b)** Hydrosulfuric acid. Again, the prefix *hydro-* indicates a hydrogen acid, and *-ic* is added to the end of sulfur. **(c)** Carbonic acid. When an acid is derived from a polyatomic anion whose name ends in *-ate*—*carbonate* (CO_3^{2-}) in this case—the ending *-ic* is added to the name of the central atom of the polyatomic anion, and no prefix is used. **(d)** Phosphoric acid. Its name is derived from the polyatomic anion *phosphate* (PO_4^{3-}). **(e)** Phosphorous acid. When acid is derived from a polyatomic anion ending in *-ite*—in this case, *phosphite* (PO_3^{3-})—the suffix *-ous* is added to the name of the central atom of the polyatomic anion, and no prefix is used.

SELF-TEST QUESTIONS

Key Terms

Having reviewed the key terms listed at the end of Chapter 2 in the text and their applications, identify the following statements as true or false. If a statement is false, indicate why it is incorrect. Key terms are italicized in the statements.

2.1   The addition of an electron to the metal Na forms a *cation*.

2.2   *Anions* are typically formed by nonmetals such as chlorine, sulfur, and oxygen.

2.3   An *electron* is the most massive subatomic particle found in an atom.

2.4   The *proton* and *neutron* are found in the nucleus of an atom, which constitutes only a small part of the total volume of an atom.

2.5   An *alpha particle* contains two protons and one electron.

2.6 $^{85}_{37}Rb$ and $^{87}_{37}Rb$ are *isotopes*. The sum of protons and neutrons for $^{85}_{37}Rb$ is 85 and is called the *mass number*. The number of protons is 37, which is called the *atomic number*.

2.7   NaF is a *diatomic molecule*.

2.8   The *nucleus* of an atom is the same for all atoms of an element.

2.9   An example of an *oxyanion* is NO_3^-.

2.10   The *polyatomic* ion in the compound $Mg_3(PO_4)_2$ is PO_4^{3-}.

2.11   The formula $C_6H_8O_6$ for vitamin C is called its *empirical formula*.

2.12   The *structural formula* for carbon dioxide is CO_2.

2.13   Addition of two protons to an oxygen atom forms the oxide *ion*, O^{2-}.

2.14   The *molecular formula* for the ionic compound calcium phosphate is $Ca_3(PO_4)_2$.

2.15   When we speak of naturally occurring oxygen, we mean it exists as a *diatomic* molecule.

2.16   An example of a *molecule* is $CaCl_2$.

2.17   The *chemical formulas* O_2 and O_3 represent the same chemical species, oxygen.

2.18   *Atoms* of the same element have the same mass number.

2.19   A *cathode ray* is a beam of protons.

2.20 If a particle spontaneously emits alpha particles, we say it is *radioactive*.

2.21 The *atomic number* of a specific element is always constant.

2.22 The phrase "cobalt-60" means an isotope of cobalt that has a *mass number* of 60.

2.23 One *Angstrom* is equivalent to 10^{-8} m.

2.24 Elements in the *periodic table* are sequentially arranged in order of increasing mass number.

2.25 One *electronic charge* is equivalent to one coulomb.

2.26 He, Ne and Ar are all members of a *family* of elements.

2.27 An example of the *law of multiple proportions* is the existence of NO_2 and NO.

2.28 Hg is a *nonmetallic* element.

2.29 Sr is a *metallic* element.

2.30 Antimony is a *metalloid*.

2.31 To specify a *nuclide*, only its atomic number is needed.

Problems and Short-Answer Questions

2.32 The chlorine (Cl) atom occurs in combination with many other atoms.
(a) A $^{37}_{17}Cl^-$ nuclide contains how many protons, electrons, and neutrons?
(b) What is chlorine's elemental form?
(c) In the elemental state, chlorine is a green gas at room temperature. Is this color a physical or chemical property?
(d) Write the formula of a compound containing Cl^- with Cr^{3+}.
(e) 1 g of KCl is placed in 100 mL of water, and it dissolves with stirring. Identify the combination as one or more of the following: element, compound, heterogeneous mixture, homogeneous mixture, or solution.

2.33 Identify the following elements if an atom of each has the following:
(a) A 1+ charge and 11 protons;
(b) a 2+ charge and 36 electrons;
(c) a 2^- charge and an atomic number of 34.

2.34 Complete the following table:

Symbol	Number of protons	Number of neutrons	Number of electrons	Charge
$^{90}_{38}Sr^{2+}$				
	92	143		0
		10	10	1−
	46	60		2+

2.35 Predict the formula of the compound most likely to be formed when
(a) aluminum and sulfur combine, and
(b) aluminum and phosphate ions combine.

2.36 Roentgen discovered X rays. What type of radiation is similar to X rays? What is the penetrating ability of X rays compared to alpha and beta particles?

2.37 $Ba(NO_3)_2$ in the solid state consists of Ba^{2+} and NO_3^- ions in a three-dimensional array. Is the formula $Ba(NO_3)_2$ a molecular or empirical formula?

2.38 Name the following compounds:
(a) $CaSO_4$
(b) PF_5
(c) KBr
(d) $KHSO_4$
(e) Na_2S
(f) H_2SO_4
(g) CO_2
(h) $HClO_4$
(i) $NaClO_3$
(j) $Cu(CN)_2$

2.39 Using the periodic table, determine which ions are not likely:
(a) Cl^+
(b) Cs^+
(c) S^{2-}
(d) Rb^-

2.40 After studying Chapter 2 and doing the sample exercises and problems, you should now be familiar with the more common ions. It is important for you to know the charges carried by common inorganic species because this will help you with nomenclature, learning how atoms bond together, and recognizing the forms of species in solution. State the most common charge carried by the following when they occur as ions: F, O, Na, Ca, S, SO_4, PO_4, Al, H and CO_3.

2.41 Write the formula for each of the following:
(a) tin(IV) chloride;
(b) chromium(III) hydroxide;
(c) cesium cyanide;
(d) dinitrogen trioxide;
(e) cobalt(III) oxide;
(f) calcium phosphate;
(g) osmium tetraoxide;
(h) mercury(II) bromide;
(i) hypobromous acid;
(j) hydroselenic acid.

2.42 Write the chemical formula for or give the name of each of the following oxyanions and oxyacids:
(a) HClO
(b) ClO^-
(c) perchloric acid
(d) perchlorate ion
(e) $HMnO_4$
(f) MnO_4^-
(g) H_2SO_3
(h) SO_3^{2-}

2.43 (a) Explain why $^{27}_{13}Al$ and $^{28}_{13}Al$ have essentially identical chemical properties.
(b) If $^{27}_{13}Al$ is bombarded with neutrons, $^{28}_{13}Al$ is formed. Is this an "ordinary" chemical change?
(c) Explain why a symbol for a neutron could be given as ^1_0n.

2.44 Which of the following are ionic and which are molecular?
(a) P_4
(b) $CaCl_2$
(c) SO_2
(d) HCl(gas)

Multiple-Choice Questions

2.45 Which experimental observation evidence indicates that cathode rays consist of electrons?

    (a) Their path is curved in a magnetic field;
    (b) Gravity significantly affects their path over a short distance;
    (c) They have no effect on an electroscope;
    (d) They are assigned a negative charge;
    (e) All of the above are indications.

2.46 Which particle helped to explain discrepancies in atomic weights observed by early nineteenth century chemists?

    (a) neutron   **(d)** atom
    (b) electron   **(e)** molecule
    (c) proton

2.47 Which charged particle is not found in the 1+ ion of the lightest element known?

    (a) neutron   **(d)** atom
    (b) electron   **(e)** molecule
    (c) proton

2.48 What is the elementary form in which gaseous halogens occur?

    (a) neutron   **(d)** atom
    (b) electron   **(e)** molecule
    (c) proton

2.49 The development of modern atomic theory took a leap forward with the proof that an atom is mostly empty space. Which of the following experiments was responsible for this proof?

    (a) Millikan's oil drop
    (b) cathode-ray deflection in a magnetic field
    (c) the bombardment of gold foil with alpha particles
    (d) separation of gases by gas chromatography
    (e) the Wilson cloud chamber.

2.50 Which of the following cations are *correctly* named: (1) Sn^{2+}—tin(II) ion; (2) NH_4^+—ammonium ion; (3) Ca^{2+}—calcium ion?

    (a) 1 and 2   **(d)** all of them
    (b) 2 and 3   **(e)** none of them
    (c) 1 and 3

2.51 Which of the following anions are *incorrectly* named: (1) IO^-—hyperiodate ion; (2) SO_3^{2-}—sulfite ion; (3) ClO_3^-—chlorate ion; (4) Cl^-—chloride ion; (5) NO_2^-—nitrate ion?

    (a) 1 and 2   **(d)** all of them
    (b) 1 and 3   **(e)** none of them
    (c) 1 and 5

2.52 Which of the following compounds are *incorrectly* named: (1) $FeCl_3$—ferric chloride; (2) $HgCl_2$—mercurous chloride; (3) SnS_2—tin(IV) sulfide; (4) $KClO_2$—potassium chlorate?

    (a) 1 and 2   **(d)** all of them
    (b) 2 and 4   **(e)** none of them
    (c) 3 and 4

2.53 Which of the following acids in water are *correctly* named: (1) H_2S—hydrosulfuric; (2) $HClO_4$—perchloric; (3)

H_3PO_3—phosphorous; (4) HNO_2—nitrous; (5) H_2CO_3—carbonic?

    (a) 1 and 2   **(d)** all of them
    (b) 2, 3, and 4   **(e)** none of them
    (c) 3, 4, and 5

2.54 Which of the following is the formula of phosphoric acid?

    (a) H_2CO_3 **(d)** H_3PO_2
    (b) H_3P **(e)** H_2PO_3
    (c) H_3PO_4

2.55 Which of the following is the most likely formula of the compound formed between Sr and S?

    (a) SrS **(d)** Sr_2S_3
    (b) Sr_2S **(e)** Sr_3S_2
    (c) SrS_2

SELF-TEST SOLUTIONS

2.1 False. The addition of an electron to an atom forms a negatively charged species—an anion. A cation is formed by the removal of electrons from an atom; they are typically formed by metals. **2.2** True. **2.3** False. It is the least massive. **2.4** True. **2.5** False. It is a He^{2+} nuclide; thus it contains two protons and no electrons. **2.6** True. **2.7** False. Although it is diatomic (two elements), it is not a molecule; it is an ionic compound. **2.8** False. Many elements consist of isotopes, which have nuclei that vary in the number of neutrons. **2.9** True. **2.10** True. **2.11** False. An empirical formula shows the smallest whole-number ratio of atoms: $C_3H_4O_3$. **2.12** False. A structural formula shows the arrangement of atoms: O—C—O. **2.13** False. In order to form O^{2-}, two electrons must be added; a proton has a positive charge. **2.14** False. Because $Ca_3(PO_4)_2$ is an ionic compound, there is no molecular unit we can describe with a molecular formula. It is an empirical formula. **2.15** True. **2.16** False. A molecule consists of atoms bonded together to form a discrete particle with its own unique chemical and physical properties. In calcium chloride there are only separate ions, and there are no discrete, unique pairs of ions. **2.17** False. They are allotropes of oxygen, but they are different chemical species with their own physical and chemical properties. **2.18** False. The same atomic number, not mass number. **2.19** False. It is a beam of electrons. **2.20** True. **2.21** True. **2.22** True. **2.23** False. It equals 10^{-10} m. **2.24** False. In order of increasing atomic number. **2.25** False. 1.602×10^{-19}C = 1 electronic charge. **2.26** True. **2.27** True. **2.28** False. A metal. **2.29** True. **2.30.** True. **2.31** False. Both its atomic number and mass number must be specified. **2.32 (a)** $_{17}^{37}Cl^-$ has 18 electrons (17 electrons for a neutral $_{17}^{37}Cl$ and one for the negative charge), 17 protons, and 20 neutrons. **(b)** Cl_2 (similar to F_2, Br_2, and I_2). **(c)** Physical. Its color does not affect its chemical behavior. **(d)** $CrCl_3$. Three Cl^- ions are needed to balance the 3+ charge of Cr^{3+}. **(e)** It is a homogeneous mixture, that is, a solution. **2.33 (a)** The atomic number of the unknown element equals its number of protons: 11. The charge results from a loss of electrons, not the addition of protons. Na has an atomic number of 11. **(b)** The 2+ charge results from the loss of two electrons. Thus in the neutral atom there are 36 + 2 = 38 electrons and 38 protons. It has the atomic number 38, which is the atomic number of Sr. **(c)** Se has an atomic number of 34. **2.34** This table is easily completed if you remember that the superscript number with a nuclide symbol equals the number of protons + the number of neutrons and that the subscript number equals the number of protons. The difference between the former and latter numbers equals the number of neutrons. Also, the number of electrons equals the number of protons adjusted for the charge of the nuclide.

Symbol	Number of protons	Number of neutrons	Number of electrons	Charge
$^{90}_{38}\text{Sr}^{2+}$	38	$90 - 38$ $= 52$	$38 - 2$ $= 36$	2+
$^{235}_{92}\text{U}$	92	143	92	0
$^{19}_{9}\text{F}^{1-}$	$10 - 1$ $= 9$	10	10	1−
$^{106}_{46}\text{Pd}^{2+}$	46	60	$46 - 2$ $= 44$	2+

2.35 (a) Aluminum is a metal and belongs to the Group 3A family, which forms 3+ ions. Sulfur belongs to the Group 6A family, which commonly has ions in the 2− state. The most likely compound is Al_2S_3. **(b)** A phosphate ion has a 3− charge. Therefore the compound is $AlPO_4$. **2.36** X rays are like gamma rays and are more penetrating than alpha or beta particles. **2.37** $Ba(NO_3)_2$ is an empirical formula because there is no discrete $Ba(NO_3)_2$ molecule. Compounds formed from Group 1A and 2A elements are usually composed of ions. **2.38 (a)** calcium sulfate; **(b)** phosphorus pentafluoride; **(c)** potassium bromide; **(d)** potassium hydrogen sulfate; **(e)** sodium sulfide; **(f)** sulfuric acid; **(g)** carbon dioxide; **(h)** perchloric acid; **(i)** sodium chlorate; **(j)** copper(II) cyanide. **2.39 (a)** Chlorine is a nonmetal and thus is expected to form an ion with a negative charge; **(b)** Cs^+ is correct because metals tend to form cations; **(c)** S^{2-} is correct because nonmetals tend to form anions; **(d)** Rb is a metal and thus is expected to form a cation, not an anion. **2.40** F^-, O^{2-} Na^+, Ca^{2+}, S^{2-}, SO_4^{2-}, PO_4^{3-}, Al^{3+}, H^+ and CO_3^{3-}. **2.41 (a)** $SnCl_4$; **(b)** $Cr(OH)_3$; **(c)** $CsCN$; **(d)** N_2O_3; **(e)** Co_2O_3; **(f)** $Ca_3(PO_4)_2$; **(g)** OsO_4; **(h)** $HgBr_2$; **(i)** $HBrO$; **(j)** H_2Se. **2.42 (a)** hypochlorous acid; **(b)** hypochlorite ion; **(c)** $HClO_4$; **(d)** ClO_4^-; **(e)** permanganic acid; **(f)** permanganate ion; **(g)** sulfurous acid; **(h)** sulfite ion. **2.43 (a)** Chemical properties depend primarily upon the number of electrons in an atom and the total energy of those electrons. Since the isotopic nuclides $^{27}_{13}\text{Al}$ and $^{28}_{13}\text{Al}$ have the same number of electrons and therefore essentially the same electronic energy, we would expect similar chemical properties. **(b)** No. An *ordinary* chemical reaction involves decomposition of compounds into elements, formation of compounds from elements, or loss or gain of electrons by atoms to form ions in compounds. The reaction described in this problem involves a nuclear transformation, which is not considered an ordinary chemical reaction. **(c)** Since one neutron has no protons, the subscript 0 in ^1_0n is appropriate, and the superscript 1 indicates a mass number of 1. **2.44 (a)** Molecular, because it consists only of a nonmetal; **(b)** ionic, because it contains a metal, Ca, and a nonmetal, Cl; **(c)** molecular, because both elements are nonmetals; **(d)** molecular. When hydrogen combines with a nonmetal, a molecular-type compound is formed. **2.45 (a). 2.46 (a).** Differing mixtures of isotopic nuclides (which differ in the number of neutrons) in samples of elements led to discrepancies in measurements of atomic weights. **2.47 (b).** H^+ contains one proton, no electrons. **2.48 (e).** For example, Cl_2 and F_2. **2.49 (c). 2.50 (d). 2.51 (c). 2.52 (b). 2.53 (d). 2.54 (c). 2.55 (a).**

3 | Stoichiometry: Calculations with Chemical Formulas and Equations

OVERVIEW OF THE CHAPTER

3.1, 3.2
CHEMICAL
EQUATIONS:
BALANCING AND
PREDICTING
PRODUCTS OF
REACTIONS

Review: Elements and compounds (2.4); formulas (2.5); nomenclature (2.6).

Objectives: You should be able to:

1. Balance chemical equations.
2. Predict the products of a chemical reaction, having seen a suitable analogy.
3. Predict the products of the combustion reactions of hydrocarbons and simple compounds containing C, H, and O atoms.

3.3, 3.4, 3.5
ATOMIC WEIGHT,
MOLECULAR
WEIGHT, AND THE
MOLE

Objectives: You should be able to:

1. Calculate the atomic weight of an element given the abundances and masses of its isotopes.
2. Interconvert number of moles, mass in grams, and number of atoms, ions, or molecules.

3.6
DETERMINATION OF
EMPIRICAL AND
MOLECULAR
FORMULAS

Review: Empirical and molecular formulas (2.5).

Objectives: You should be able to:

1. Calculate the empirical formula of a compound, having been given appropriate analytical data such as elemental percentages or the quantity of CO_2 and H_2O produced by combustion.
2. Calculate the molecular formula, having been given the empirical formula and molecular weight.

3.7 CHEMICAL
EQUATIONS: MASS
AND MOLE
RELATIONSHIPS

Review: Dimensional analysis (1.5); rounding off numbers (1.4).

Objectives: You should be able to:

1. Calculate the mass of a particular substance produced or used in a chemical reaction (mass-mass problem).
2. Determine the limiting reagent in a reaction.

TOPIC SUMMARIES AND EXERCISES

Chemical equation such as $2C + O_2 \rightarrow 2CO$

CHEMICAL
EQUATIONS:
BALANCING AND
PREDICTING
PRODUCTS OF
REACTIONS

- Describe chemical processes involving **reactants** (left side of arrow) to form **products** (right side of arrow).
- Should be balanced.
- Provide a means for calculating mass relationships among products and reactants.

When you balance chemical equations, you must keep the following requirements in mind:

- Formulas of substances must be correctly written.
- The number of atoms of each type of element must be the same on both sides of the arrow.
- Only coefficients in front of substances may be changed to change the number of atoms on the reactant or product side. Subscripts in chemical formulas must not be changed.
- The sum of charges of ions on the left side of the arrow must be the same on the right side.

See Exercise 1 for an approach to balancing chemical reactions.

You also need to develop the ability to predict the outcome of simple reactions knowing only the reactants. At this stage in your chemical education, your best technique for doing this is by referring to analogous reactions. Important types of reactions studied in this chapter are summarized below.

Combustion reactions produce a flame and usually involve oxygen as a reactant. For example:

$$C_2H_2(g) + 3O_2(g) \longrightarrow 2CO_2(g) + 2H_2O(l)$$

| |
Means a | Means a
substance is | substance is
a gas | a liquid

- The text highlights the combustion of **hydrocarbons,** carbon and hydrogen containing compounds. When hydrocarbons are combusted in the presence of oxygen, the carbon is converted to CO_2 and hydrogen is converted to H_2O. If oxygen is also present in a compound containing C and H, for example CH_3OH, the oxygen is used along with O_2 in balancing the equation.

The text also considers combination and decomposition reactions.

Combination reactions involve forming one product from two or more reactants. For example:

$$CaO(s) + CO_2(g) \longrightarrow CaCO_3(s)$$

Decomposition reactions are the reverse of combination reactions: One reactant breaks down (decomposes) into two or more substances. For example:

$$2H_2O(l) \longrightarrow O_2(g) + 2H_2(g)$$

EXERCISE 1   Balance the following reactions:
(a) $P_4(s) + O_2(g) \rightarrow P_4O_{10}(s)$ **(b)** $SF_4(g) + H_2O(l) \rightarrow SO_2(g) + HF(g)$

Solution: We want the smallest whole-number coefficient in front of each formula in the chemical equation so that the number of each kind of atom on the reactant side equals the number of identical atoms on the product side. (a) By inspection we see that the atoms are not balanced in the chemical equation

$$P_4(s) + O_2(g) \longrightarrow P_4O_{10}(s)$$

An inventory of atoms shows:

	No. of P Atoms	No. of O Atoms
Reactants:	4	2
Products:	4	10

Although the number of phosphorus atoms is balanced, the number of oxygen atoms is not. Ten oxygen atoms, that is, $5O_2$, are needed on the reactant side to equal the ten oxygen atoms on the product side. The balanced chemical equation is

$$P_4(s) + 5O_2(g) \longrightarrow P_4O_{10}(s)$$

(b) An inventory of atoms in the chemical equation

$$SF_4(g) + H_2O(l) \longrightarrow SO_2(g) + HF(g)$$

shows:

	No. of S Atoms	No. of F Atoms	No. of H Atoms	No. of O Atoms
Reactants:	1	4	2	1
Products:	1	1	1	2

Starting with the atom that is present in the greatest number on the reactant side, fluorine, we can balance the fluorine atoms on the product side with 4HF. The inventory is now

	No. of S Atoms	No. of F Atoms	No. of H Atoms	No. of O Atoms
Reactants:	1	4	2	1
Products:	1	4	4	2

The hydrogen atoms can be balanced with $2H_2O$ on the reactant side. The balancing of the hydrogen atoms also causes the oxygen atoms to be balanced. The balanced equation is

$$SF_4(g) + 2H_2O(l) \longrightarrow SO_2(g) + 4HF(g)$$

EXERCISE 2   Write the balanced chemical equation for the combustion of butane, C_4H_{10}.

Solution: First write the skeletal equation for the combustion of butane. Oxygen will be a reactant in a combustion reaction involving a hydrocarbon. Carbon dioxide and water will be products because butane contains only hydrogen and carbon atoms.

$$C_4H_{10}(g) + O_2(g) \longrightarrow CO_2(g) + H_2O(l)$$

When balancing combustion reactions of simple hydrocarbons, first balance the carbon and hydrogen atoms without considering oxygen. Then balance the oxygen atoms using $O_2(g)$. There are four carbon atoms, therefore place a four in front of CO_2; there are 10 hydrogen atoms in butane, therefore place a five in front of H_2O:

$$C_4H_{10}(g) + O_2(g) \longrightarrow 4CO_2(g) + 5H_2O(l)$$

The products now show 13 oxygen atoms, eight from four carbon dioxide molecules and 10 from five water molecules. Place a 13/2 in front of $O_2(g)$ to balance the 13 oxygen atoms:

$$C_4H_{10}(g) + \frac{13}{2}\,O_2(g) \longrightarrow 4CO_2(g) + 5H_2O(l)$$

It is more convenient to deal with whole numbers; it is usually customary (but not always required) to clear the fractions by multiplying the entire equation by a number that cancels the denominators. In this case, multiply the entire equation by two to eliminate the denominator in 13/2:

$$2C_4H_{10}(g) + 13O_2(g) \longrightarrow 8CO_2(g) + 10H_2O(l)$$

EXERCISE 3   Complete and balance the following reactions, and indicate the phases of each substance: **(a)** $SbBr_3 + H_2S \rightarrow$ **(b)** $LiH + H_2O \rightarrow$ **(c)** $CO_2 + Na \rightarrow$ **(d)** $FeS + O_2 \rightarrow$ Use the following reactions as examples:
(1) $3CO_2(g) + 4K(s) \rightarrow 2K_2CO_3(s) + C(s)$ (2) $CaH_2(s) + 2H_2O(l) \rightarrow Ca(OH)_2(aq) + 2H_2(g)$
(3) $2SbCl_3(s) + 3H_2S(g) \rightarrow Sb_2S_3(s) + 6HCl(g)$ (4) $2ZnS(s) + 3O_2(g) \rightarrow 2ZnO(s) + 2SO_2(g)$

Solution: The reactions can be completed by finding an analogous reaction among those given. Analogous compounds of elements that are from the same family or that are otherwise similar in nature often exist in the same phase.

Balanced equation	Analogous reaction
(a) $2SbBr_3(s) + 3H_2S(g) \rightarrow Sb_2S_3(s) + 6HBr(g)$	(3)
(b) $LiH(s) + H_2O(l) \rightarrow LiOH(aq) + H_2(g)$	(2)
(c) $3CO_2(g) + 4Na(s) \rightarrow 2Na_2CO_3(s) + C(s)$	(1)
(d) $2FeS(s) + 3O_2(g) \rightarrow 2FeO(s) + 2SO_2(g)$	(4)

We cannot isolate and weigh on a scale a single atom of an element. However, it is possible to measure the relative masses of elements that combine with one another:

ATOMIC WEIGHT, MOLECULAR WEIGHT, AND THE MOLE

- The ^{12}C isotope of carbon is assigned a mass, called **atomic mass,** of exactly 12 atomic mass units (amu).
- The atomic mass of any isotope is then determined relative to ^{12}C. A modern mass spectrometer is used to determine accurate relative masses.
- The relationship between amu and grams is 1.66056×10^{-24} g = 1 amu.

Chemists normally work with large collections of atoms. Thus, an average atomic mass is needed because a large collection of atoms of an element will contain a distribution of naturally occurring isotopes rather than one type of isotope.

- **Average atomic mass** is the weighted average of the masses of all naturally occurring isotopes of an element.
- Average atomic mass is more commonly known as **atomic weight.**

The masses of substances are determined from atomic weights.

- **Formula weight** refers to the sum of atomic weights of the atoms in a substance.
- **Molecular weight** is used in place of formula weight if the substance is molecular in form. Such substances exist as molecules. For example, the molecular weight of water is calculated as follows:

$$2(\text{AW of H}) + 1(\text{AW of O}) = (2 \text{ atoms H}) \left(\frac{1.01 \text{ amu}}{1 \text{ atom H}} \right)$$
$$+ (1 \text{ atom O}) \left(\frac{16.00 \text{ amu}}{1 \text{ atom O}} \right)$$
$$= 18.02 \text{ amu}$$

The percent composition of a substance refers to the percent by mass contributed by each element in the substance.

- The sum of percent by mass of all elements in a substance equals 100%.

$$\% \text{ by mass of an element} = \frac{\text{mass of an element in substance}}{\text{formula weight of substance}} \times 100$$

Chemists do not ordinarily work with single molecules or atoms, but rather with trillions upon trillions of them. To facilitate the counting and weighing of such samples, a quantity called the mole has been defined.

- A **mole** of any type of particle equals the number of ^{12}C atoms in exactly 12 g of ^{12}C. Thus a mole represents a certain number of objects, just like a dozen represents 12.
- In 12 g of ^{12}C, there are 6.022×10^{23} atoms; this number is given the name **Avogadro's number.**
- Thus a mole of water contains 6.022×10^{23} molecules of water and a mole of NaCl contains 6.022×10^{22} sodium ions and 6.022×10^{23} chloride ions.

A mole of a substance has a mass in grams that is numerically equal to its formula weight in amu. The term **molar mass** is used to describe the mass in grams of one mole of a substance.

You can use these mass-quantity relationships as conversion factors. Examples of equivalences that can be used as such are

1 ^{12}C atom = 12 amu	1 mol ^{12}C = 12
1 Cl_2 molecule = 70.90 amu	1 mol Cl_2 = 70.90 g
1 $BaCl_2$ formula = 208 amu	1 mol $BaCl_2$ = 208 g

EXERCISE 4   Calculate the molecular or formula weights for: **(a)** CO_2; **(b)** NO_3^-; **(c)** $C_{21}H_{30}O_2$.

    Solution: Look up the atomic weights of all elements in the substance. Then calculate the molecular or formula weights by adding the total mass of each element:

(a)

$$C: (1 \text{ atom C}) \left(\frac{12.01 \text{ amu}}{1 \text{ atom C}} \right) = 12.01 \text{ amu}$$

$$O: (2 \text{ atoms O}) \left(\frac{16.000 \text{ amu}}{1 \text{ atom O}} \right) = 32.00 \text{ amu}$$

$$\text{Molecular weight of } CO_2 = 44.01 \text{ amu}$$

(b)

$$N: (1 \text{ atom N}) \left(\frac{14.01 \text{ amu}}{1 \text{ atom N}} \right) = 14.01 \text{ amu}$$

$$O: (3 \text{ atoms O}) \left(\frac{16.00 \text{ amu}}{1 \text{ atom O}} \right) = 48.00 \text{ amu}$$

$$\text{Formula weight of } NO_3^- = 62.01 \text{ amu}$$

(c)

$$C: (21 \text{ atoms C}) \left(\frac{12.01 \text{ amu}}{1 \text{ atom C}} \right) = 252.21 \text{ amu}$$

$$H: (30 \text{ atoms H}) \left(\frac{1.01 \text{ amu}}{1 \text{ atom H}} \right) = 30.30 \text{ amu}$$

$$O: (2 \text{ atoms O}) \left(\frac{16.00 \text{ amu}}{1 \text{ atom O}} \right) = 32.00 \text{ amu}$$

$$\text{Molecular weight of } C_{21}H_{30}O_2 = 314.51 \text{ amu}$$

EXERCISE 5   Answer the following with respect to ethanol, C_2H_6O: **(a)** What is its molecular weight? **(b)** What is the weight of 1 mol of ethanol molecules? **(c)** Calculate the number of moles of ethanol in 1.00 g. **(d)** Calculate the number of molecules in 1.00 g of ethanol. **(e)** Calculate the percentage of carbon in one molecule of ethanol.

    Solution: **(a)** The molecular weight of ethanol is calculated as follows:

$$C: (2 \text{ atoms C}) \left(\frac{12.01 \text{ amu}}{1 \text{ atom C}} \right) = 24.02 \text{ amu}$$

$$H: (6 \text{ atoms H}) \left(\frac{1.01 \text{ amu}}{1 \text{ atom H}} \right) = 6.06 \text{ amu}$$

$$O: (1 \text{ atom O}) \left(\frac{16.00 \text{ amu}}{1 \text{ atom O}} \right) = 16.00 \text{ amu}$$

$$\text{Molecular weight of } C_2H_6O = 46.08 \text{ amu}$$

(b) The weight of 1 mol of ethanol is the weight of one molecule expressed in grams 46.08 g. **(c)** In this problem you are asked to determine the number of moles in 1.00 g. Thus you need a conversion factor that will change 1.00 g of ethanol to moles. The molar mass relationship yields the necessary conversion factor:

$$46.08 \text{ g ethanol} = 1 \text{ mol ethanol}$$

Since you want to carry out the conversion *grams* → *moles* (the symbol → indicates a unit conversion), the conversion factor you need to use is 1 mol ethanol/46.08 g ethanol.

$$\text{Moles ethanol} = (1.00 \text{ g ethanol}) \left(\frac{1 \text{ mol ethanol}}{46.08 \text{ g ethanol}} \right) = 0.0217 \text{ mol ethanol}$$

(d) The approach to this problem is similar to that used in **(c)**. The relationships required to carry out the conversion *grams → moles → molecules* are

$$46.08 \text{ g ethanol} = 1 \text{ mol ethanol} \qquad\qquad \text{[Molar Mass]}$$

$$1 \text{ mol ethanol} = 6.022 \times 10^{23} \text{ molecules ethanol} \qquad \text{[Avogadro's number]}$$

$$\text{Molecules ethanol} = (1.00 \text{ g ethanol})\left(\frac{1 \text{ mol ethanol}}{46.08 \text{ g ethanol}}\right)$$

$$\times \left(\frac{6.022 \times 10^{23} \text{ molecules ethanol}}{1 \text{ mol ethanol}}\right)$$

$$= 1.31 \times 10^{22} \text{ molecules ethanol}$$

(e) The percentage of any element in a compound is the mass of that element in one molecule divided by the molecular weight of the molecule and multiplied by 100:

$$\% \text{ C} = \frac{\left(\begin{array}{c}\text{number C atoms} \\ \text{per molecule}\end{array}\right)(\text{AW of C})}{(\text{mass of one C}_2\text{H}_6\text{O molecule})} \times 100$$

$$\% \text{ C} = \frac{\left(\dfrac{2 \text{ atoms C}}{1 \text{ molecule C}_2\text{H}_6\text{O}}\right)\left(\dfrac{12.01 \text{ amu}}{1 \text{ atom C}}\right)}{\dfrac{46.08 \text{ amu}}{1 \text{ molecule C}_2\text{H}_6\text{O}}} \times 100$$

Alternatively, the percentage of carbon in C_2H_6O can be solved for as follows:

$$\left(\frac{2 \text{ mol C}}{1 \text{ mol C}_2\text{H}_6\text{O}}\right)\left(\frac{12.01 \text{ g C}}{1 \text{ mol C}}\right)\left(\frac{1 \text{ mol C}_2\text{H}_6\text{O}}{46.08 \text{ g C}_2\text{H}_6\text{O}}\right) = \frac{0.5213 \text{ g C}}{1 \text{ g C}_2\text{H}_6\text{O}} = \frac{52.13 \text{ g C}}{100 \text{ g C}_2\text{H}_6\text{O}}.$$

The last ratio is equivalent to saying that the percentage of carbon in C_2H_6O is 52.13.

EXERCISE 6   A sample of $Na_2B_4O_7$ contains 0.3478 g of sodium. What is the mass of this sample?

Solution: You can calculate the mass of $Na_2B_4O_7$ by realizing that the mass of the sample is related to the mass of the sodium in the following manner:

$$\text{Mass of sample} = (\text{mass of Na}) \times \left(\begin{array}{c}\text{conversion factors that} \\ \text{change g Na to g Na}_2\text{B}_4\text{O}_7\end{array}\right)$$

To convert from grams Na to grams $Na_2B_4O_7$, you will need to go through the following series of unit conversions:

$$\textit{Grams } \text{Na} \longrightarrow \textit{moles } \text{Na} \longrightarrow \textit{moles } \text{Na}_2\text{B}_4\text{O}_7 \longrightarrow \textit{grams } \text{Na}_2\text{B}_4\text{O}_7$$

The required equivalences are

1 mol $Na_2B_4O_7$ = 2 mol Na	[from formula of $Na_2B_4O_7$]*
1 mol $Na_2B_4O_7$ = 201.24 g $Na_2B_4O_7$	[from molecular weight of $Na_2B_4O_7$]
1 mol Na = 23.00 g Na	[from atomic weight of Na]

---

*The conversion factor 1 mol $Na_2B_4O_7$/2 mol Na is included in the problem to reflect the fact that there are two sodium atoms per $Na_2B_4O_7$ formula unit.

Thus:

$$\text{Mass of sample} = (0.3478 \text{ g Na})\left(\frac{1 \text{ mol Na}}{23.00 \text{ g Na}}\right)$$

$$\times \left(\frac{1 \text{ mol Na}_2\text{B}_2\text{O}_7}{2 \text{ mol Na}}\right)\left(\frac{201.24 \text{ g Na}_2\text{B}_4\text{O}_7}{1 \text{ mol Na}_2\text{B}_4\text{O}_7}\right)$$

$$= 1.522 \text{ g Na}_2\text{B}_4\text{O}_7$$

DETERMINATION OF EMPIRICAL AND MOLECULAR FORMULAS

In Chapter 2, you learned that an empirical formula shows the simplest whole-number ratio of atoms. A molecular formula shows the actual ratio of atoms. For example, CH_2 is the empirical formula for C_2H_4. An empirical formula is typically determined from experimental percent composition data. *The subscripts in an empirical formula are calculated as follows:*

- Convert the mass percent of each element to grams using an arbitrarily chosen sample size, such as 100 g.
- Determine the number of moles of each element.
- Determine the simplest whole-number ratio of atoms in the compound by dividing the moles of each element by the smallest number of moles.
- If the ratios are not whole numbers, multiply the ratios by a numerical factor that clears the denominators of the fractions. For example, the numbers 0.50 and 1.75 are expressed as fractions: $\frac{1}{2}$ and $\frac{7}{4}$ ($1\frac{3}{4}$). If they are multiplied by four, the ratios are converted to whole numbers: $4 \times \frac{1}{2} = 2$ and $4 \times \frac{7}{4} = 7$. If a ratio is very near a whole number or fraction, such as 1.05 or 1.55, assume that they can be expressed as 1.00 and 1.50 because of experimental error.

To determine the molecular formula of a compound, you need its molecular weight.

- Calculate the number of empirical formula units making up the molecular formula by dividing the mass of one mole by the mass of one empirical formula.
- To determine the molecular formula, multiply the subscripts of the empirical formula by the number of empirical formula units making up the molecular formula.

EXERCISE 7   A compound contains only the elements Al and O. Its elemental composition is determined to be 53.0% aluminum and 47.0% oxygen. The mass of one mole of the compound is 86.0 g. What is the empirical formula of the compound? What is the molecular formula?

Solution: Following the outline summarized above, you first convert the weight percentage of each element in an arbitrarily chosen 100-g sample to grams of each element:

$$\text{Gram Al} = (100\text{-g sample})\left(\frac{53.0 \text{ g Al}}{100 \text{ g}}\right) = 53.0 \text{ g Al}$$

$$\text{Grams O} = (100\text{-g sample})\left(\frac{47.0 \text{ g O}}{100 \text{ g}}\right) = 47.0 \text{ g O}$$

Next determine the number of moles of each element in 100 g of the sample:

$$\text{Moles Al} = (53.0 \text{ g Al}) \left(\frac{1 \text{ mol Al}}{27.0 \text{ g Al}} \right) = 1.96 \text{ mol Al}$$

$$\text{Moles O} = (47.0 \text{ g O}) \left(\frac{1 \text{ mol O}}{16.0 \text{ g O}} \right) = 2.94 \text{ mol O}$$

To determine the empirical formula of the compound, calculate the simplest whole-number ratio of atoms in the compound. This is done by dividing the number of moles of each element by the number of moles of the element having the smallest number of moles. In this case Al has the fewer number of moles; thus you divide by 1.96:

$$\text{Al}_{\frac{1.96}{1.96}} \text{O}_{\frac{2.94}{1.96}} = \text{Al}_{1.00} \text{O}_{1.50}$$

The numbers in the mole ratio are not a ratio of integers. You must multiply them by an integer that will convert the 1.50 into an integer. Inspection should convince you that if you multiply by 2 you will convert 1.50 to 3.00 and 1.00 to 2.00. The empirical formula is therefore Al_2O_3.

To determine the molecular formula, divide the mass of one mole by the mass of one empirical formula unit. The resulting number tells you how many empirical formula units make up the molecular formula. The mass of one mole of empirical formula units for Al_2O_3 is 86.0 g, the same as the mass of one mole. Thus, the empirical and molecular formulas are identical, Al_2O_3.

CHEMICAL EQUATIONS: MASS AND MOLE RELATIONSHIPS

A balanced chemical equation provides quantitative mass-mole relationships among the reactants and products. The branch of chemistry that deals with these quantitative relationships is termed **stoichiometry.** When we talk about the "stoichiometric relationships in a chemical reaction," we are referring to the mass-mole relationships given by the balanced chemical equation that describes the reaction of interest.

To understand the above statements, examine the following balanced chemical reaction and see what mass-mole relationships can be derived from it:

$$\text{CH}_4(g) + 2\text{O}_2(g) \longrightarrow \text{CO}_2(g) + 2\text{H}_2\text{O}(l)$$

▪ On the atomic-molecular level, the equation states:

1 molecule CH_4 + 2 molecules $\text{O}_2 \longrightarrow$

1 molecule CO_2 + 2 molecules H_2O

Or, since an Avogadro's number of molecules in all cases is equivalent to a mole:

1 mol CH_4 + 2 mol $\text{O}_2 \longrightarrow$ 1 mol CO_2 + 2 mol H_2O

▪ This last interpretation of the balanced chemical equation is the one that enables you to derive mass-mole relationships among reactants and products. The numerical coefficients in front of the reactants and products give the ratios of moles in which the chemical substances react. For example, since 2 mol of O_2 is required to react with 1 mol of CH_4, 2×2, or 4, mole of O_2 is required to react with 2×1, or 2, mol of CH_4—that is, O_2 always reacts with CH_4 in a 2:1 mole ratio. You can represent this stoichiometrically equivalent ratio by the statement

$$1 \text{ mol } CH_4 \simeq 2 \text{ mol O}$$

where the symbol $\simeq$ means a stoichiometrically equivalent quantity *in terms of the given reaction*. Similarly, you can represent the stoichiometrically equivalent ratio for the formation of products as

$$1 \text{ mol } CO_2 \simeq 2 \text{ mol } H_2O$$

That is, 1 mol of CO_2 forms for every 2 mol of water that forms. Other stoichiometrically equivalent ratios can be derived from the balanced chemical reaction. For example:

$$1 \text{ mol } CH_4 \simeq 1 \text{ mol } CO_2$$
$$2 \text{ mol } O_2 \simeq 1 \text{ mol } CO_2$$
$$1 \text{ mol } CH_4 \simeq 2 \text{ mol } H_2O$$
$$2 \text{ mol } O_2 \simeq 2 \text{ mol } H_2O$$

▪ All of the above stoichiometrically equivalent statements can be converted to mass equivalences by converting a mole of a substance to its formula or molecular weight or molar mass.

Various kinds of chemical problems involving stoichiometry are encountered in chemical practice.

▪ One important type involves a **limiting reactant.** In many reactions, one or more substances are in excess and therefore some will be left over when the reaction is completed. The substance that is completely consumed determines the amount of product formed and is called the limiting reactant. Exercise 11 explores how to solve limiting reactant problems.

▪ Most chemical reactions are not 100 percent efficient; they do not produce as much product as expected from the stoichiometry. The extent of a reaction is given by **percent yield:**

$$\text{Percent yield} = \frac{\text{actual yield}}{\text{theoretical yield}} \times 100$$

See Exercise 11.

EXERCISE 8 How many moles of Al_2O_3 are produced when 0.50 mol of Al reacts with an excess of PbO_2? The pertinent balanced chemical equation is

$$4Al(s) + 3PbO_2(s) \longrightarrow 2Al_2O_3(s) + 3Pb(s)$$

Solution: To solve this problem, you need stoichiometric equivalences that allow you to make the transformation

$$Moles\ Al \longrightarrow moles\ Al_2O_3$$

The stoichiometric equivalence derived from the balanced chemical equation that relates moles of Al to moles of Al_2O_3 is:

$$4\ mol\ Al \simeq 2\ mol\ Al_2O_3$$

The problem states that there is sufficient PbO_2 to react with 0.50 mol of Al. Thus the amount of PbO_2 present does not have to be considered, and so the appropriate calculation is

$$Moles\ Al_2O_3 = (0.50\ mol\ Al)\left(\frac{2\ mol\ Al_2O_3}{4\ mol\ Al}\right) = 0.25\ mol\ Al_2O_3$$

Thus when 0.50 mol of Al reacts with an excess of PbO_2, 0.25 mol of Al_2O_3 is produced.

EXERCISE 9 Determine how many grams of HI are required to form 1.20 mol of H_2 when HI reacts according to the balanced chemical equation

$$2HI(g) \longrightarrow H_2(g) + I_2(g)$$

Solution: You need to determine the stoichiometric equivalences that will be required to make the following transformations:

$$Moles\ H_2 \longrightarrow moles\ HI \longrightarrow grams\ HI$$

From this sequence of proposed conversions, you see that you need both mole and mass stoichiometric equivalences. These are

$$2\ mol\ HI \simeq 1\ mol\ H_2\ \text{(from balanced equation)}$$
$$127.91\ g\ HI = 1\ mol\ HI\ \text{(from molar mass of HI)}$$

The problem can be solved in two steps:

$$Moles\ H_2 \longrightarrow moles\ HI:$$

$$moles\ HI = (1.20\ mol\ H_2)\left(\frac{2\ mol\ HI}{1\ mol\ H_2}\right) = 2.40\ mol\ HI$$

and

$$Moles\ HI \longrightarrow grams\ HI:$$

$$grams\ HI = (2.40\ mol\ HI)\left(\frac{127.91\ g\ HI}{1\ mol\ HI}\right) = 307\ g\ HI$$

Alternatively, the two steps can be combined into one as follows:

$$Grams\ HI = (1.20\ mol\ H_2)\left(\frac{2\ mol\ HI}{1\ mol\ H_2}\right) \times \left(\frac{127.91\ g\ HI}{1\ mol\ HI}\right) = 307\ g\ HI$$

In the *Student's Guide* the latter approach will be the one usually followed when solving stoichiometry problems.

EXERCISE 10   Determine how many grams of PCl_3 are produced when 2.80 g of Cl_2 reacts with a sufficient quantity of P_4 according to the chemical reaction

$$P_4(s) + 6Cl_2(g) \longrightarrow 4PCl_3(l)$$

Solution: The following conversions are required:

$$Grams\ Cl_2 \longrightarrow moles\ Cl_2 \longrightarrow moles\ PCl_3 \longrightarrow grams\ PCl_3$$

The necessary stoichiometric equivalence is

$$6\ mol\ Cl_2 \simeq 4\ mol\ PCl_3$$

Also, you will need the following molar mass equivalences:

$$70.90\ g\ Cl_2 = 1\ mol\ Cl_2$$
$$137.32\ g\ PCl_3 = 1\ mol\ PCl_3$$

The grams of PCl_3 produced are calculated using the sequence of conversions previously given:

$$Grams\ PCl_3 = (2.80\ g\ Cl_2)\left(\frac{1\ mol\ Cl_2}{70.90\ g\ Cl_2}\right)$$
$$\times \left(\frac{4\ mol\ PCl_3}{6\ mol\ Cl_2}\right)\left(\frac{137.32\ g\ PCl_3}{1\ mol\ PCl_3}\right) = 3.62\ g\ PCl_3$$

EXERCISE 11   **(a)** What is the limiting reactant when 10.0 g of C_2H_6 reacts with 50.0 g of O_2 according to the chemical equation

$$2C_2H_6(g) + 7O_2(g) \longrightarrow 4CO_2(g) + 6H_2O(l)$$

(b) Calculate the percent yield of the reaction

$$2Al(OH)_3(s) + 3H_2SO_4(aq) \longrightarrow Al_2(SO_4)_3(s) + 6H_2O(l)$$

given that 205 g of $Al(OH)_3$ reacts with 751 g of H_2SO_4 to yield 252 g of $Al_2(SO_4)_3$.

   Solution: (a) The approach used in this solution is only one of several possibilities. First calculate the exact mass of O_2 required to react with 10.0 g of C_2H_6. If the available mass of O_2 is greater than the calculated mass of O_2, then C_2H_6 is the limiting reagent. Conversely, O_2 is the limiting reagent if its available mass is less than the calculated mass. The sequence of conversions required to make this determination is

$$Grams\ C_2H_6 \longrightarrow moles\ C_2H_6 \longrightarrow moles\ O_2 \longrightarrow grams\ O_2$$

The required stoichiometric and molar mass equivalences needed for these conversions are

$$2\ mol\ C_2H_6 \simeq 7\ mol\ O_2$$
$$30.08\ g\ C_2H_6 = 1\ mol\ C_2H_6$$
$$32.00\ g\ O_2 = 1\ mol\ O_2$$
$$Grams\ O_2 = (10.0\ g\ C_2H_6)\left(\frac{1\ mol\ C_2H_6}{30.08\ g\ C_2H_6}\right)$$
$$\times \left(\frac{7\ mol\ O_2}{2\ mol\ C_2H_6}\right)\left(\frac{32.00\ g\ O_2}{1\ mol\ O_2}\right) = 37.2\ g\ O_2$$

The available mass of O_2, 50.0 g, is greater than the calculated mass (37.2 g) of O_2 required to react completely with 10.0 g of C_2H_6. Therefore, O_2 is in excess, and C_2H_6 is the limiting reagent. **(b)** To calculate the percent yield you need to know what is the theoretical amount of aluminum sulfate produced. Most likely one of the reactants is a limiting reagent, but you do not know which one it is. Determine the theoretical amount of aluminum sulfate produced assuming first that aluminum hydroxide is the limiting reagent and then sulfuric acid. The calculation that yields the smaller number of grams of aluminum sulfate tells us which reactant is the limiting reagent and the theoretical amount of aluminum sulfate produced. If aluminum hydroxide is the limiting reagent:

$$\text{g Al}_2(\text{SO}_4)_3 = (205 \text{ g Al(OH)}_3)\left(\frac{1 \text{ mol Al(OH)}_3}{78.0 \text{ g}}\right)$$

$$\times \left(\frac{1 \text{ mol Al}_2(\text{SO}_4)_3}{2 \text{ mol Al(OH)}_3}\right)\left(\frac{342.3 \text{ g Al}_2(\text{SO}_4)_3}{1 \text{ mol Al}_2(\text{SO}_4)_3}\right) = 450 \text{ g}$$

If sulfuric acid is the limiting reagent:

$$\text{g Al}_2(\text{SO}_4)_3 = (751 \text{ g H}_2\text{SO}_4)\left(\frac{1 \text{ mol H}_2\text{SO}_4}{98.1 \text{ g}}\right)$$

$$\times \left(\frac{1 \text{ mol Al}_2(\text{SO}_4)_3}{3 \text{ mol H}_2\text{SO}_4}\right)\left(\frac{342.3 \text{ g Al}_2(\text{SO}_4)_3}{1 \text{ mol Al}_2(\text{SO}_4)_3}\right) = 873 \text{ g}$$

$Al(OH)_3$ is the limiting reagent because it yields the smaller amount of $Al_2(SO_4)_3$ and thus the theoretical amount of $Al_2(SO_4)_3$ is 450 g. The percent yield is

$$\text{Percent yield} = \frac{252 \text{ g}}{450 \text{ g}} \times 100 = 56.0\%$$

Note: In Exercises 8 through 11 every sequence of conversions involved moles. The unit mole is the central focus point for most stoichiometric conversions. The reason for this is that if you know the number of moles you can transform to either grams or number of particles of a substance.

Number of particles	$\underset{\text{Avogadro's number}}{\rightleftharpoons}$	Moles	$\underset{\text{Molar mass}}{\rightleftharpoons}$	Grams

Avogadro's number is used to convert between number of particles and number of moles. The mass of one mole of a substance (molar mass) is used to convert between number of moles and grams. Keep the above picture in mind when you do stoichiometric problems.

SELF-TEST QUESTIONS

Key Terms

Having reviewed the key terms listed at the end of Chapter 3 and their applications, identify the following statements as true or false. If a statement is false, indicate why it is incorrect. Key terms are italicized in the statements.

3.1 The following reaction is an example of a *balanced chemical equation:*

$$2NH_3 + 4F_2 \longrightarrow N_2F_4 + 6HF$$

3.2 The reaction

$$4CO_2 + 6H_2O \longrightarrow 2C_2H_6 + 7O_2$$

is an example of a *combustion reaction* because O_2 is formed.

3.3 A value of 12 *atomic mass units* is assigned to the isotope of carbon with six protons and six neutrons in the nucleus.

3.4 An *Avogadro's number* of people is 6.022×10^{23} people; it corresponds to one *mole* of people.

3.5 The reported *atomic weight* of carbon in grams is the mass of an Avogadro's number of ^{12}C atoms.

3.6 The *molecular weight* of KCl is 74.55 g.

3.7 The *formula weight* and molecular weight of a molecular substance are always the same.

3.8 A modern analytical instrument that can be used to determine precise masses of atomic ions is a *mass spectrometer*.

3.9 42 g of Cu and 105 g of HNO_3 react according to the equation

$$3Cu + 8HNO_3 \longrightarrow 3Cu(NO_3)_2 + 2NO + 4H_2O$$

HNO_3 is the *limiting reagent*.

3.10 The *law of conservation of mass* is obeyed in the following chemical equation:

$$3NO_2 + 2H_2O \longrightarrow 2HNO_3 + NO$$

3.11 The *theoretical yield* of a product is the experimental mass obtained from a specified reaction.

3.12 The area of study of *stoichometry* involves measuring concentrations and percent yield, for example.

3.13 *Reactants* are gases and liquids.

3.14 *Products* are liquids and solids.

3.15 To determine if a reaction is a *combination reaction*, you would look to see if two or more products are formed.

3.16 To determine if a reaction is a *decomposition reaction*, you would look to see if two or more reactants exist.

3.17 A *mole* is a direct measure of mass.

3.18 The mass of one mole of a substance is known as *molar mass*.

3.19 *Percent yield* is defined as

$$\frac{\text{amount of product formed}}{\text{amount of reactant}} \times 100$$

Problems and Short-Answer Questions

3.20 Balance the following equations:
(a) $AgNO_3(aq) + CaCl_2(aq) \rightarrow AgCl(s) + Ca(NO_3)_2(aq)$
(b) $VO(s) + Fe_2O_3(s) \rightarrow FeO(s) + V_2O_5(s)$
(c) $Na(s) + H_2O(l) \rightarrow NaOH(aq) + H_2(g)$
(d) $NH_4NO_3(s) \rightarrow N_2O(g) + H_2O(l)$
(e) $MnO_2(s) + HCl(ag) \rightarrow Cl_2(g) + MnCl_2(aq) + H_2O(l)$

3.21 Write balanced equations for the following reactions:
(a) aqueous silver nitrate reacts with aqueous copper(II) chloride to form insoluble silver chloride and aqueous copper(II) nitrate;
(b) metallic aluminum reacts with oxygen gas to form solid aluminum oxide;
(c) aqueous barium chloride reacts with aqueous potassium sulfate to form solid barium sulfate and aqueous potassium chloride;

(d) solid magnesium chloride reacts with aqueous sodium hydroxide to yield insoluble magnesium hydroxide and aqueous sodium chloride;
(e) solid potassium chlorate decomposes to solid potassium chloride and oxygen gas.

3.22 Ethylene, C_2H_4, is used to make the plastic polyethylene.
(a) What are its molecular weight and formula weight?
(b) How many moles of C_2H_4 are there in 5.50 g?
(c) How many molecules of C_2H_4 are there in 5.50 g?

3.23 If a certain quantity of $NO(g)$ has a mass of 5.00 g, what is the quantity of H_2O containing the same number of molecules?

3.24 What is the density of $F_2(g)$ if at 0°C one mole of it occupies 22.4 L?

3.25 How many molecules of $CH_4(g)$ are contained in 1.35 g?

3.26 The main constituent of gallstones is cholesterol. Cholesterol may have a role in heart attacks and blood clot formation. Its elemental percentage composition is 83.87% C, 11.99% H, and 4.14% O. It has a molecular weight of 386.64 amu. Calculate its empirical and molecular formulas.

3.27 What is the empirical formula of compound such that 200 g of it contains 87.2 g of P and 112.8 g of O?

3.28 A stimulant of the nervous system found in coffee, tea, and cola is caffeine. Caffeine contains the following weight percentage of elements: 49.5% C; 28.9% N; 16.5% O; 5.2% H. What is the empirical formula of caffeine?

3.29 $H_2C_2O_4$ and $KMnO_4$ react according to the equation

$$5H_2C_2O_4 + 2KMnO_4 \longrightarrow 4H_2O + 10CO_2 + 2MnO_2 + 2KOH$$

How many grams of CO_2 are formed when 10.05 g of $H_2C_2O_4$ and 26.72 g of $KMnO_4$ are mixed together?

3.30 Silicon carbide, SiC, is an important industrial abrasive. It is formed by the reaction of SiO_2 and carbon at high temperatures:

$$SiO_2 + 3C \longrightarrow SiC + 2CO$$

(a) Calculate the moles of silicon carbide formed when 5.00 g of carbon reacts with an excess of SiO_2.
(b) What is the minimum amount of carbon required to react with 25.0 g of SiO_2?

3.31 The reaction for the production of iron from the reduction of the ore hematite, Fe_2O_3, is as follows:

$$Fe_2O_3 + 3CO \longrightarrow 2Fe + 3CO_2$$

(a) If the reaction yields 4.52 g of CO_2, how many grams of Fe are also formed?
(b) How many grams of Fe are formed from 7.25 g of Fe_2O_3 and 6.00 g of CO?

3.32 Bromine has two naturally occurring isotopes: ^{79}Br (78.918 amu) and ^{81}Br (80.916 amu). The atomic weight of Br is 79.904 amu. What are the fractional abundances of ^{79}Br and ^{81}Br?

Multiple-Choice Questions

3.33 Which of the following equations do not obey the law of conservation of mass?

(1) $C_6H_{12}O_6 + 6O_2 \rightarrow 6CO_2 + 6H_2O$
(2) $C_2H_6 + 7O_2 \rightarrow 4CO_2 + 6H_2O$
(3) $BCl_3 + H_2O \rightarrow H_3BO_3 + 3HCl$

(a) (1)   (d) (1) and (2)
(b) (2)   (e) (2) and (3)
(c) (3)

3.34 0.600 mol of a substance weighs 62.5 g. How many moles of it are there in 100.0 g?

(a) 1.04 mol   (d) 0.104 mol
(b) 0.96 mol   (e) 0.600 mol
(c) 0.0096 mol

3.35 What is the weight percentage of Al in Al_2O_3?

(a) 26.46%   (d) 47.08%
(b) 20.93%   (e) 53.21%
(c) 52.92%

3.36 In photosynthesis, $CO_2(g)$ and $H_2O(l)$ are converted into glucose, $C_6H_{12}O_6$ (a sugar), and O_2. If 0.256 mol of $C_6H_{12}O_6$ is formed by the reaction of CO_2 with water, how many grams of CO_2 would be needed?

(a) 67.6 g   (d) 76.6 g
(b) 11.3 g   (e) 0.256 g
(c) 0.0349 g

3.37 Which of the following pairs of substances, with their indicated quantities contain(s) the same number of atoms: (1) 0.50 mol HCl, 0.50 mol He; (2) 0.20 mol H_3PO_4, 0.80 mol N_2; (3) 0.45 mol HNO_3, 0.45 mol HNO_2?

(a) (1)   (d) all of them
(b) (2)   (e) none of them
(c) (3)

3.38 What is the mass in grams of 4.25×10^{20} molecules of H_2O?

(a) 0.127 g   (d) 142 g
(b) 0.0127 g   (e) 4.25×10^{20} g
(c) 1420 g

3.39 Which of the following is the limiting reagent if three moles of H_2O and eight moles of NO_2 are available in the reaction:

$$3NO_2(g) + H_2O(l) \longrightarrow 2HNO_3(l) + NO$$

(a) NO_2 (c) HNO_3
(b) H_2O (d) NO

3.40 What is the percentage yield of C_2H_2 when 50.0 g of $CaC_2(s)$ (molar mass = 64.01 g) reacts with an excess of water to yield 13.5 g of C_2H_2 (molar mass = 26.04 g) according to the following reaction:

$$CaC_2(s) + 2H_2O(l) \longrightarrow Ca(OH)_2(s) + C_2H_2(g)$$

(a) 27.0%   (c) 66.5%
(b) 51.8%   (d) 82.5%

3.41 What is the empirical formula of benzoic acid if it has a mass percentage composition of 69% carbon, 5% hydrogen and 26% oxygen?

(a) $C_7H_6O_2$ (c) C_5H_6O
(b) $C_6H_7O_2$ (d) C_3H_4O

3.42 What is the coeffcient in front of BF_3 when the following equation is balanced?

$$BF_3 + NaBH_4 \longrightarrow NaBF_4 + B_2H_6$$

(a) 2   (c) 4
(b) 3   (d) 5

SELF-TEST SOLUTIONS

3.1 False. The chemical equation shows eight fluorine atoms on the left and ten on the right; it does not satisfy the law of conservation of mass. The correct chemical equation is $2NH_3 + 5F_2 \rightarrow N_2F_4 + 6HF$. **3.2** False. In most combustion reactions, O_2 is consumed. The *reverse* reaction is an example of a combustion type: $2C_2H_6 + 7O_2 \rightarrow 4CO_2 + 6H_2O$. **3.3** True. **3.4** True. **3.5** False. The atomic weight of carbon is the weighted average mass of the atoms of all of carbon's isotopes. **3.6** False. Although 74.55 g is the sum of the weights of a potassium and a chlorine atom, we should call it a *formula weight* because KCl is a salt and does not exist in the molecular form. **3.7** True. **3.8** True. **3.9** True. 2.2 g of the 42 g of Cu is not needed to react completely with the 105 g of HNO_3. **3.10** False. According to the given equation, hydrogen and oxygen atoms decrease in number when the products form and a loss in mass results. This is contradictory to the law of conservation of mass. **3.11** False. It is the maximum mass that can be obtained, assuming perfect conditions. **3.12** True. **3.13** True. Also, solids. **3.14** True. Also gases. **3.15** False. A single product is formed from two reacting substances. **3.16** False. A single reactant decomposes. **3.17** False. It is a counting number—like a dozen. It is not a measure of mass, although if you know the number of moles of a chemical substance and its formula, you can calculate its mass. **3.18** True. **3.19** False. It is

$$\frac{\text{amount of product formed}}{\text{theoretical amount of product}} \times 100$$

3.20 (a) $2AgNO_3(aq) + CaCl_2(aq) \rightarrow$
$\qquad\qquad\qquad\qquad 2AgCl(s) + Ca(NO_3)_2(aq)$
(b) $2VO(s) + 3Fe_2O_3(s) \rightarrow 6FeO(s) + V_2O_5(s)$
(c) $2Na(s) + 2H_2O(l) \rightarrow 2NaOH(aq) + H_2(g)$
(d) $NH_4NO_3(s) \rightarrow N_2O(g) + 2H_2O(l)$
(e) $MnO_2(s) + 4HCl(aq) \rightarrow$
$\qquad\qquad\qquad Cl_2(g) + MnCl_2(aq) + 2H_2O(l)$

3.21 **(a)** $2AgNO_3(aq) + CuCl_2(aq) \rightarrow$
$$2\,AgCl(s) + Cu(NO_3)_2(aq)$$
(b) $4Al(s) + 3O_2(g) \rightarrow 2Al_2O_3(s)$
(c) $BaCl_2(aq) + K_2SO_4(aq) \rightarrow BaSO_4(s) + 2KCl(aq)$
(d) $MgCl_2(s) + 2NaOH(aq) \rightarrow$
$$Mg(OH)_2(s) + 2NaCl(aq)$$
(e) $2KClO_3(s) \rightarrow 2KCl(s) + 3O_2(g)$

3.22 **(a)** Molecular and formula weights are the same; 28.04 amu.

(b) Moles $C_2H_4 = (5.50 \text{ g C}_2\text{H}_4)\left(\dfrac{1 \text{ mol C}_2\text{H}_4}{28.04 \text{ g C}_2\text{H}_4}\right)$

$$= 0.196 \text{ mol C}_2\text{H}_4$$

(c) Molecules C_2H_4

$$= (0.196 \text{ mol C}_2\text{H}_4)\left(\dfrac{6.022 \times 10^{23} \text{ molecules}}{1 \text{ mol}}\right)$$

$$= 1.18 \times 10^{23} \text{ molecules}$$

3.23 Since equal numbers of moles contain equal numbers of molecules, we want a quantity of H_2O such that it is equivalent in number of moles to the number of moles of NO in the given sample weight. The unit conversions required are *grams NO → moles NO → moles H₂O → grams H₂O*. The required equivalences are 1 mol H_2O = 1 mole NO; 30.01 g NO = 1 mol NO; 18.02 g H_2O = 1 mol H_2O. Hence,

Grams H_2O

$$= (5.00 \text{ g NO})\left(\dfrac{1 \text{ mol NO}}{30.01 \text{ g NO}}\right)\left(\dfrac{1 \text{ mol H}_2\text{O}}{1 \text{ mol NO}}\right)$$

$$\times \left(\dfrac{18.02 \text{ g H}_2\text{O}}{1 \text{ mol H}_2\text{O}}\right) = 3.00 \text{ g H}_2\text{O}$$

3.24 The definition of density is: density = mass/volume. The mass of 1 mol of F_2 occupying 22.4 L is the molecular weight of F_2 in grams, which equals 38.0 g. The density of $F_2(g)$ is

$$\text{Density} = \dfrac{38.0 \text{ g}}{(22.4 \text{ L})\left(\dfrac{1000 \text{ mL}}{1 \text{ L}}\right)\left(\dfrac{1 \text{ cm}^3}{1 \text{ mL}}\right)}$$

$$= 1.70 \times 10^{-3} \dfrac{\text{g}}{\text{cm}^3}$$

3.25 We will need to make the conversions *grams CH₄ → moles CH₄ → molecules CH₄*. Hence.

Molecules $CH_4 = (1.35 \text{ g CH}_4)\left(\dfrac{1 \text{ mol CH}_4}{16.05 \text{ g CH}_4}\right)$

$$\times \left(\dfrac{6.022 \times 10^{23} \text{ molecules}}{1 \text{ mol}}\right)$$

$$= 5.07 \times 10^{22} \text{ molecules CH}_4$$

3.26 The number of moles of each element in an arbitrary sample size of 100 g is as follows:

Moles $C = (83.37 \text{ g C})\left(\dfrac{1 \text{ mol C}}{12.01 \text{ g C}}\right) = 6.983 \text{ mol C}$

Moles $H = (11.99 \text{ g H})\left(\dfrac{1 \text{ mol H}}{1.008 \text{ g H}}\right) = 11.89 \text{ mol H}$

Moles $O = (4.14 \text{ g O})\left(\dfrac{1 \text{ mol O}}{16.00 \text{ g O}}\right) = 0.260 \text{ mol O}$

The mole ratio of O:H:C is 0.260/0.260:11:89/0.260: 6.983/0. 260, or 1:45.7:26.9. The empirical formula is $C_{27}H_{46}O$. The empirical formula weight is 386.64 amu. This is the same as the actual molecular weight: thus the molecular formula is $C_{27}H_{46}O$. **3.27** From the masses of P and O in the sample, we can calculate the number of moles of each in the sample:

Moles $P = (87.2 \text{ g P})\left(\dfrac{1 \text{ mol P}}{30.97 \text{ g P}}\right) = 2.82 \text{ mol P}$

Moles $O = (112.8 \text{ g O})\left(\dfrac{1 \text{ mol O}}{16.00 \text{ g O}}\right) = 7.05 \text{ mol O}$

The mole ratio of P to O is 2.82/2.82:7.05/2.82, or 1:2.5, or $1:2\frac{1}{2}$ or 2:5. The empirical formula is P_2O_5. **3.28** The moles of each element in an arbitrary sample size of 100 g are as follows:

Moles $C = (49.5 \text{ g C})\left(\dfrac{1 \text{ mol C}}{12.01 \text{ g C}}\right) = 4.12 \text{ mol C}$

Moles $N = (28.9 \text{ g N})\left(\dfrac{1 \text{ mol N}}{14.01 \text{ g N}}\right) = 2.06 \text{ mol N}$

Moles $O = (16.5 \text{ g O})\left(\dfrac{1 \text{ mol O}}{16.0 \text{ g O}}\right) = 1.03 \text{ mol O}$

Moles $H = (5.2 \text{ g H})\left(\dfrac{1 \text{ mol H}}{1.01 \text{ g H}}\right) = 5.15 \text{ mol H}$

The ratio of moles of O:N:C:H is 1.03/1.03:2.03/1.03: 4.12/1.03:5.15/1.03, or 1:2:4:5. The empirical formula is $C_4H_5N_2O$. **3.29** Reaction equation: $5H_2C_2O_4 + 2KMnO_4 \rightarrow 4H_2O + 10CO_2 + 2MnO_2 + 2KOH$. Let us determine if $KMnO_4$ is in excess. Conversions required: *Grams H₂C₂O₄ → moles H₂C₂O₄ → moles KMnO₄ → grams KMnO₄*. Equivalences needed for this calculation: 5 mol $H_2C_2O_4 \approx 2$ mol $KMnO_4$; 90.4 g $H_2C_2O_4$ = 1 mol $H_2C_2O_4$; 158.0 g $KMnO_4$ = 1 mol $KMnO_4$. Calculating the amount of $KMnO_4$

required to react with 10.05 g $H_2C_2O_4$:

Grams $KMnO_4$

$$= (10.05 \text{ g } H_2C_2O_4) \times \left(\frac{1 \text{ mol } H_2C_2O_4}{90.4 \text{ g } H_2C_2O_4} \right)$$

$$\times \left(\frac{2 \text{ mol } KMnO_4}{5 \text{ mol } H_2C_2O_4} \right) \times \left(\frac{158.0 \text{ g } KMnO_4}{1 \text{ mol } KMnO_4} \right)$$

$$= 7.03 \text{ g } KMnO_4$$

$KMnO_4$ is in excess because 26.72 g is available but only 7.03 g is required. To calculate the mass of CO_2 formed, we must make the following conversions: *grams* $H_2C_2O_4 \rightarrow$ *moles* $H_2C_2O_4 \rightarrow$ *moles* $CO_2 \rightarrow$ *grams* CO_2. Additional equivalences needed are 5 mol $H_2C_2O_4 \stackrel{\frown}{=}$ 10 mol CO_2; 44.01 g CO_2 = 1 mol CO_2. Hence,

$$\text{Grams } CO_2 = (10.05 \text{ g } C_2H_2O_4) \left(\frac{1 \text{ mol } H_2C_2O_4}{90.4 \text{ g } H_2C_2O_4} \right)$$

$$\times \left(\frac{10 \text{ mol } CO_2}{5 \text{ mol } H_2C_2O_4} \right) \left(\frac{44.1 \text{ g } CO_2}{1 \text{ mol } CO_2} \right)$$

$$= 9.785 \text{ g } CO_2$$

3.30 Reaction equation: $SiO_2 + 3C \rightarrow SiC + 2CO$. **(a)** Conversions: *grams* $C \rightarrow$ *moles* $C \rightarrow$ *moles* SiC. Equivalences needed: 12.01 g C = 1 mol C; 3 mol C $\stackrel{\frown}{=}$ 1 mol SiC.

$$\text{Moles } SiC = (5.00 \text{ g } C) \left(\frac{1 \text{ mol } C}{12.01 \text{ g } C} \right) \left(\frac{1 \text{ mol } SiC}{3 \text{ mol } C} \right)$$

$$= 0.139 \text{ mol } SiC$$

(b) Conversions: *Grams* $SiO_2 \rightarrow$ *moles* $SiO_2 \rightarrow$ *moles* $C \rightarrow$ *grams* C. Additional equivalences needed: 1 mol $SiO_2 \stackrel{\frown}{=}$ 3 mol C; 60.09 g SiO_2 = 1 mol SiO_2.

$$\text{Grams } C = (25.00 \text{ g } SiO_2) \left(\frac{1 \text{ mol } SiO_2}{60.09 \text{ g } SiO_2} \right)$$

$$\times \left(\frac{3 \text{ mol } C}{1 \text{ mol } SiO_2} \right) \times \left(\frac{12.01 \text{ g } C}{1 \text{ mol } C} \right)$$

$$= 14.99 \text{ g } C$$

3.31 Reaction equation: $Fe_2O_3 + 3CO \rightarrow 2Fe + 3CO_2$. **(a)** Conversions: *Grams* $CO_2 \rightarrow$ *moles* $CO_2 \rightarrow$ *moles* $Fe \rightarrow$ *grams* Fe. Equivalences needed: 44.01 g CO_2 = 1 mol CO_2; 2 mol Fe $\stackrel{\frown}{=}$ 3 mol CO_2; 55.85 g Fe = 1 mol Fe.

$$\text{Grams } Fe = (4.52 \text{ g } CO_2) \left(\frac{1 \text{ mol } CO_2}{44.01 \text{ g } CO_2} \right)$$

$$\times \left(\frac{2 \text{ mol } Fe}{3 \text{ mol } CO_2} \right) \times \left(\frac{55.85 \text{ g } Fe}{1 \text{ mol } Fe} \right)$$

$$= 3.82 \text{ g } Fe$$

(b) This is another limiting reactant problem. Let us calculate the amount of CO that reacts with 7.25 g of Fe_2O_3. The required conversions are: *Grams* $Fe_2O_3 \rightarrow$ *moles* $Fe_2O_3 \rightarrow$ *moles* $CO \rightarrow$ *grams* CO. Equivalences needed: 159.7 g Fe_2O_3 = 1 mol Fe_2O_3; 1 mol $Fe_2O_3 \stackrel{\frown}{=}$ 3 mol CO; 28.01 g CO = 1 mol CO.

$$\text{Grams } CO = (7.25 \text{ g } Fe_2O_3) \left(\frac{1 \text{ mol } Fe_2O_3}{159.7 \text{ g } Fe_2O_3} \right)$$

$$\times \left(\frac{3 \text{ mol } CO}{1 \text{ mol } Fe_2O_3} \right) \times \left(\frac{28.01 \text{ g } CO}{1 \text{ mol } CO} \right)$$

$$= 3.81 \text{ g } CO$$

Since 6.00 g of CO is available and only 3.82 g is required, CO is in excess. To calculate the amount of Fe formed, we need to make the following conversions. *Grams* $Fe_2O_3 \rightarrow$ *moles* $Fe_2O_3 \rightarrow$ *moles* $Fe \rightarrow$ *grams* Fe. Additional equivalences needed: 1 mol $Fe_2O_3 \stackrel{\frown}{=}$ 2 mol Fe; 55.85 g Fe = 1 mol Fe.

$$\text{Grams } Fe = (7.25 \text{ g } Fe_2O_3) \left(\frac{1 \text{ mol } Fe_2O_3}{159.7 \text{ g } Fe_2O_3} \right)$$

$$\times \left(\frac{2 \text{ mol } Fe}{1 \text{ mol } Fe_2O_3} \right) \times \left(\frac{55.85 \text{ g } Fe}{1 \text{ mol } Fe} \right)$$

$$= 5.07 \text{ g } Fe$$

3.32 Let x and y equal the fractional abundances of ^{79}Br and ^{81}Br, respectively. Then $x + y = 1$, as the sum of the fractional abundances must equal unity. The fractional abundances can be calculated using the following relations: AW of Br = (fractional abundance ^{79}Br) (mass of ^{79}Br) + (fractional abundance ^{81}Br) (mass of ^{81}Br); AW of Br = (x) (mass of ^{79}Br) + (y) (mass of ^{81}Br); 79.904 amu = (x) (78.918 amu) + (y) (80.916 amu). Substituting the relation $y = 1 - x$ (from $x + y = 1$) yields 79.904 amu = (x) (78.918 amu) + ($1 - x$) (80.916 amu). Solving for x gives $x = 0.50651$ and $y = 1 - x = 0.49349$. Thus bromine contains 50.65 percent ^{79}Br and 49.35 percent ^{81}Br. **3.33** (e). $2C_2H_6 + 7O_2 \rightarrow 4CO_2 + 6H_2O$; $BCl_3 + 3H_2O \rightarrow H_3BO_3 + 3HCl$.

3.34 **(b)**

$$\frac{0.600 \text{ mol}}{62.5 \text{ g}} = \frac{\text{moles}}{100 \text{ g}};$$

$$\text{moles} = (100 \text{ g}) \left(\frac{0.600 \text{ mol}}{62.5 \text{ g}} \right)$$

$$= 0.96 \text{ mol}$$

3.35 **(c)**

$$\% \text{ Al} = \left(\frac{(2 \text{ mol Al}) \left(\frac{26.98 \text{ g Al}}{1 \text{ mol Al}} \right)}{101.96 \text{ g } Al_2O_3} \right) \times 100$$

$$= 52.92 \%$$

3.36 **(a)** Chemical equation: $6CO_2 + 6H_2O \rightarrow C_6H_{12}O_6 + 6O_2$.

$$\text{Grams } CO_2 = (0.256 \text{ mol } C_6H_{12}O_2)$$
$$\times \frac{(6 \text{ mol } CO_2)}{(1 \text{ mol } C_6H_{12}O_2)} \times \frac{(44.01 \text{ g } CO_2)}{(1 \text{ mol } CO_2)}$$
$$= 67.6 \text{ g } CO_2.$$

3.37 **(b)** $\left(\dfrac{8 \text{ mol of atoms of } H_3PO_4}{1 \text{ mol } H_3PO_4}\right) \times (0.20 \text{ mol } H_3PO_4)$

$= 1.6 \text{ mol of atoms}$

$\left(\dfrac{2 \text{ mol of atoms in } N_2}{1 \text{ mol } N_2}\right)(0.80 \text{ mol } N_2)$

$= 1.6 \text{ mol of atoms}$

3.38 **(b)** Grams $H_2O = (4.25 \times 10^{20} \text{ molecules})$

$$\times \left(\frac{1 \text{ mol}}{6.022 \times 10^{23} \text{ molecules}}\right)\left(\frac{18.02 \text{ g } H_2O}{1 \text{ mol } H_2O}\right)$$
$$= 0.0127 \text{ g } H_2O$$

3.39 **(a)**. NO_2 and H_2O react in a 3:1 ratio of moles. The given moles are in the ratio 8:3, or $2\frac{2}{3}$:1. The last ratio shows that there is an insufficient quantity of NO_2 to react completely with water. **3.40 (c)** 50.0 g × (1 mol/64.1 g) × (26.04 g C_2H_2/1 mol) = 20.3 g. (13.5 g/20.3 g) × 100 = 66.5%. **3.41 (a)**. **3.42 (c)**.

4 Aqueous Reactions and Solution Stoichiometry

OVERVIEW OF THE CHAPTER

4.1 SOLUTIONS: MOLARITY AND DILUTION

Review: Solutions (1.1)

Objectives: You should be able to:

1. Calculate molarity, solution volume, or number of moles of solute given any two of these quantities.
2. Calculate the volume of a more concentrated solution that must be diluted to obtain a given quantity of a more dilute solution.

4.2, 4.3 ELECTROLYTE SOLUTIONS AND ACIDS, BASES AND SALTS

Review: Nomenclature of acids, bases and salts (2.6)

Objectives: You should be able to:

1. Identify substances as acids, bases, or salts.
2. Predict whether a substance is a nonelectrolyte, strong electrolyte, or weak electrolyte from its chemical formula.

4.4, 4.5 IONIC REACTIONS: NET IONIC EQUATIONS

Objectives: You should be able to:

1. Predict the ions formed by electrolytes when they dissociate or ionize.
2. Identify the spectator ions and write the net-ionic equations for solution reactions starting with their molecular equations.
3. Predict the products of metathesis reactions (including both neutralization and precipitation reactions), and write balanced chemical equations for them.
4. Identify the driving force for any metathesis reaction.
5. Use solubility rules to predict whether a precipitate will form when electrolyte solutions are mixed.

4.6 OXIDATION OF METALS AND THE ACTIVITY SERIES

Objectives: You should be able to: Use the activity series to predict whether a reaction will occur when a metal is added to an aqueous solution of either a metal salt or an acid; write the balanced molecular and net-ionic equations for the reaction.

Objectives: You should be able to: Calculate the concentration or mass of solute in a sample from titration data.

4.7 SOLUTION STOICHIOMETRY: TITRATIONS

TOPIC SUMMARIES AND EXERCISES

SOLUTIONS: MOLARITY AND DILUTION

In Chapter 2 of the text, you learned that a solution is a homogeneous mixture of two or more substances.

- A solution contains a solvent and solute(s)
- The **solvent** is the substance in a mixture that maintains its physical state and usually is in the greatest amount.
- Whatever else is dissolved in the solution is called the **solute.** A solute may or may not maintain its physical state.
- For example, water is the most common solvent you will encounter. An aqueous solution of NaOH contains water as the solvent and NaOH as the solute. At room temperature, NaOH is a solid, but in solution it takes on the phase of water, a liquid.

Quantitative measurements of concentration require accurate determination of the amounts of solvent and solute present in a solution.

- **Molarity** (*M*) is used to measure the amount of solute in a solution.
- $M = \dfrac{\text{number of moles of solute}}{\text{volume of the solution in } liters}$
- Molarity, like density, can be used as a conversion factor to change between volume and number of moles.

Solutions of known concentration may be diluted with the solvent to produce a more diluted (less concentrated) solution.

- Note that the *number of moles of solute does not change* when only solvent is added to a solution. Therefore,

  Moles of solute before dilution = moles of solute after dilution
- Because Moles solute = Molarity × Volume, the above relationship becomes

$$M_iV_i = M_fV_f$$

The subscript i refers to the undiluted solution and the subscript f refers to the final diluted solution. This relationship is often used in solution-type problems. *Be sure you understand its use.* See Exercises 3 and 4.

EXERCISE 1 What is the molarity of an ethanol (C_2H_6O) solution containing 10.0 g of ethanol in water with a total volume of 100 mL?

   Solution: To calculate the molarity of the solution, you need to calculate the number of moles of ethanol and the volume in liters. You will need the equivalence relations between moles and grams of ethanol and between milliliters and liters:

$$46.07 \text{ g } C_2H_6O = 1 \text{ mol } C_2H_6O \qquad \text{[Molar Mass]}$$
$$1000 \text{ mL} = 1 \text{ L}$$

Using the relation for molarity,

$$M = \frac{\text{moles of solute}}{\text{volume of solution in liters}}$$

you can solve for the molarity of the solution:

$$M \, C_2H_6O = \frac{(10.0 \text{ g } C_2H_6O)\left(\dfrac{1 \text{ mol } C_2H_6O}{46.07 \text{ g } C_2H_6O}\right)}{(100 \text{ mL})\left(\dfrac{1 \text{ L}}{1000 \text{ mL}}\right)}$$

$$= \frac{0.217 \text{ mol } C_2H_6O}{0.100 \text{ L}}$$

$$= 2.17 \, M$$

EXERCISE 2 How many grams of HCl are contained in 500 mL of a 0.250 M HCl solution?

   Solution: To do this problem, you need to make the following conversions:

$$\textit{Milliliters } HCl \longrightarrow \textit{moles } HCl \longrightarrow \textit{grams } HCl$$

using the relationships

$$0.250 \text{ mol HCl} = 1 \text{ L HCl solution} \qquad \text{[from molarity definition]}$$
$$1000 \text{ mL} = 1 \text{ L}$$
$$36.45 \text{ g HCl} = 1 \text{ mol HCl} \qquad \text{[Molar Mass]}$$

Calculating the grams of HCl in 500 mL of solution:

$$\text{Grams HCl} = (500 \text{ mL})\left(\frac{1 \text{ L}}{1000 \text{ mL}}\right)\left(\frac{0.250 \text{ mol HCl}}{1 \text{ L}}\right)\left(\frac{36.45 \text{ g HCl}}{1 \text{ mol HCl}}\right)$$

$$= 4.56 \text{ g HCl}$$

EXERCISE 3 You have 1.00 L of 6.00 M HCl, and you want to use it to prepare 100 mL of 1.00 M HCl. How many milliliters of the 6.00 M HCl must be diluted with water in order to prepare the 1.00 M HCl solution:

   Solution: This is a typical dilution problem that is often encountered in the laboratory. The key to this problem is to understand that the number of moles of HCl before dilution must equal the same number after dilution because only water is added. The relation between the molarity of a solute before dilution and after dilution is

$$\text{Moles solute}_i = \text{moles solute}_f$$

or

$$M_i V_i = M_f V_f$$

The subscript i means the initial conditions before dilution, and f means the final conditions after dilution. The volumes may be expressed in units of milliliters or liters providing both volumes have the same unit. The initial volume of 6.00 M HCl required for dilution is

$$V_i = \frac{M_f V_f}{M_i} = \frac{(1.00\ M)(100\ \text{mL})}{6.00\ M} = 16.7\ \text{mL}$$

Thus, 16.7 mL of 6.00 M HCl diluted to 100 mL with H_2O produces a 1.00 M HCl solution.

EXERCISE 4　What is the molarity of a solution of NaOH formed by diluting 125 mL of a 3.0 M NaOH solution to 500 mL?

　　Solution:　The number of moles of NaOH present in the 125 mL of 3.0 M NaOH solution does not change upon dilution. Therefore, you can use the relation

$$M_i V_i = M_f V_f$$

to solve for the molarity of the diluted solution:

$$M_f = M_i \left(\frac{V_i}{V_f} \right)$$

$$M_f = (3.0\ M) \left(\frac{125\ \text{mL}}{500\ \text{mL}} \right) = 0.75\ M$$

Note:　*That the molarity of a diluted solution must always be less than that of the initial solution because the ratio V_f / V_f is always smaller than one.*

You will encounter acids, bases and salts throughout your general chemistry course. Therefore, you must become thoroughly familiar with common acids and bases and know their common properties.

- **Acids** contain one or more hydrogen atoms that dissociate to form $H^+(aq)$ in water. You can often recognize an acid from its formula: The first element listed is hydrogen. For example, the following are acids: HCl (hydrochloric), HNO_3 (nitric), $HC_2H_3O_2$ (acetic) and H_2SO_4 (sulfuric). The first three are monoprotic because they have only one hydrogen that ionizes whereas sulfuric acid is diprotic because both hydrogens ionize.
- **Bases** are substances that accept H^+ ions in chemical reactions. The common bases you will encounter are typically metallic hydroxides such as NaOH and ammonia, NH_3.
- **Neutralization** is a reaction in which an acid reacts with a base to form water (H_2O) and a *salt* (cation of base + anion of acid). For example:

Acid　　　Base　　　　　　　　　Salt
$$HBr(aq) + KOH(aq) \longrightarrow H_2O(l) + KBr(aq)$$

Means a substance
is dissolved in water

Means water is liquid
and in this example
it is also the solvent

ELECTROLYTE SOLUTIONS AND ACIDS, BASES AND SALTS

An **electrolyte** is a substance that causes a solution to be a better electrical conductor than the pure solvent by forming ions. Many acids, bases and salts are electrolytes. Acids and bases are called strong or weak depending on their extent of ionization. Thus, a strong acid is also a strong electrolyte because it ionizes 100% in solution. The different types of electrolytes and acids and bases are summarized below.

1. Strong Electrolytes: Effectively ionize 100 percent in a solvent. Examples are:
   - Most ionic compounds (salts).
   - Strong acids and bases: HCl, HBr, HNO_3, $HClO_4$, $NaOH$, KOH, for example.
2. Weak Electrolytes: Incompletely ionize (< 100%) in a solvent. Examples are:
   - Weak acids: HF, H_2S, $HC_2H_3O_2$, for example.
   - Weak bases: NH_3 is the most common example.
3. Nonelectrolytes: These do not ionize in a solvent.
   - If a substance is not one of the above types, you can assume it is a nonelectrolyte in water.
   - Sugar and alcohol are examples.

Note: Do not confuse solubility with whether a substance is a weak or strong electrolyte.

   - Solubility refers to the quantity of a solute dissolved in a solvent.
   - The form of a solute in solution determines whether it is an electrolyte or not. $BaSO_4$ is only very slightly soluble in water. However, essentially all of the $BaSO_4$ that dissolves is present in solution as $Ba^{2+}(aq)$ and $SO_4^{2-}(aq)$; therefore it is a strong electrolyte.

When writing the chemical reactions of strong and weak electrolytes, you need to use single and double arrows:

   - Single arrow: Strong electrolyte or strong acid.

$$HCl(g) \xrightarrow{H_2O} H^+(aq) + Cl^-(aq)$$

   - Double arrows: Weak electrolytes or weak acids.

$$HC_2H_3O_2(aq) \rightleftarrows H^+(aq) + C_2H_3O_2^-(aq)$$

EXERCISE 5   (a) Identify each of the following as an acid or base and write its reaction with water: $HF(g)$; $H_2SO_4(l)$; $NaOH(s)$; $Ba(OH)_2(s)$. (b) Write the neutralization reaction between $Ba(OH)_2(s)$ and $HNO_3(aq)$.
   Solution: (a) The acids have a hydrogen atom listed as the first element in the chemical formulas for the compounds given, and the bases contain one or more hydroxide units.

Compound	Type	Reaction with Water
HF	Acid	$HF(g) \xrightarrow{H_2O} H^+(aq) + F^-(aq)$
H_2SO_4	Acid	$H_2SO_4(l) \xrightarrow{H_2O} H^+(aq) + HSO_4^-(aq)$
		$HSO_4^-(aq) \longrightarrow H^+(aq) + SO_4^{2-}(aq)$
NaOH	Base	$NaOH(s) \xrightarrow{H_2O} Na^+(aq) + OH^-(aq)$
$Ba(OH)_2$	Base	$Ba(OH)_2(s) \xrightarrow{H_2O} Ba^{2+}(aq) + 2OH^-(aq)$

Note that H_2SO_4 has two acidic protons and thus two reactions are written.
(b) A neutralization reaction between an acid and a base yields a salt and water:

$$2HNO_3(aq) + Ba(OH)_2(s) \longrightarrow 2H_2O(l) + Ba(NO_3)_2(aq)$$
$$\text{Acid} \qquad\qquad \text{Base} \qquad\qquad\qquad \text{Water} \qquad \text{Salt}$$

Note that the salt formed consists of the cation of the base (Ba^{2+}) and the anion of the acid (NO_3^-). Also notice that because Ba has a 2+ charge we need two NO_3^- ions to balance the 2+ charge when writing the formula of $Ba(NO_3)_2$.

EXERCISE 6 Classify the following as a strong electrolyte, weak electrolyte or nonelectrolyte: HBr; H_2S; NH_3; $Ba(OH)_2$; KCl; C_6H_6; I_2.
Solution: If a substance is a salt (containing a metallic cation and a nonmetallic anion) it should be a strong electrolyte. $Ba(OH)_2$ and KCl both fit this requirement; thus they should be strong electrolytes. HBr is a strong acid, and thus it is a strong electrolyte. How would you know HBr is a strong acid? You must memorize the strong and weak acids given in the text and in the Student Guide. H_2S is a weak acid and NH_3 is a weak base: thus both are weak electrolytes. C_6H_6 is a hydrocarbon and I_2 is a homonuclear diatomic element; they are not salts and are not listed as acids or bases—thus we may conclude that they are nonelectrolytes.

EXERCISE 7 What is the molarity of Na^+ ions in a 0.02 *M* Na_3PO_4 solution?
Solution: Na_3PO_4 is a salt and a strong electrolyte. It ionizes completely in H_2O according to the reaction:

$$Na_3PO_4 \longrightarrow 3Na^+ + PO_4^{3-}$$

For every 1 mol of Na_3PO_4 that ionizes, 3 mol of Na^+ ions form. Thus, a 0.020 *M* Na_3PO_4 solution is 3(0.02 *M*) = 0.06 *M* in Na^+ ions.

An important class of reactions are ionic reactions in aqueous solution. In Sections 4.4 and 4.5 you should focus on learning the general solubility rules for salts given in Table 4.3 in the text and how to write net ionic equations.

Net ionic equations show only ions and weak or nonelectrolytes involved in chemical reactions.

IONIC REACTIONS: NET IONIC EQUATIONS

- An equation showing *complete* chemical formulas of reactants and products, such as $HNO_3(aq) + NH_3(aq) \rightarrow NH_4NO_3(aq)$, is called a **molecular equation.**
- To form a net ionic equation from a molecular equation, rewrite the molecular equation so that all *strong* electrolytes are written in their ion form in solution. Nonelectrolytes, weak electrolytes, gases, and insoluble and slightly soluble salts are not changed. Eliminate all ions common to both reactants and products; these are termed **spectator ions.** The prior example of a mole-

cular equation becomes

$$\underbrace{H^+(aq) + \cancel{NO_3^-}(aq)}_{\substack{HNO_3 \text{ is a} \\ \text{strong electrolyte}}} + \underbrace{NH_3(aq)}_{\substack{NH_3 \text{ is a} \\ \text{weak electrolyte} \\ \text{and is not altered}}} \rightarrow \underbrace{NH_4^+(aq) + \cancel{NO_3^-}(aq)}_{\substack{NH_4NO_3 \text{ is a} \\ \text{strong electrolyte}}}$$

▪ The ionic reaction that remains after removing spectator ions is the net ionic equation:

$$H^+(aq) + NH_3(aq) \longrightarrow NH_4^+(aq)$$

You are asked to predict the products of ionic reactions, given the reactants. Here are some suggested steps in doing this.

1. If two salts are the reactants, carry out a **metathesis** (double displacement) reaction by switching ions between the reactants.

$$AX + BY \longrightarrow AY + BX$$

2. If one reactant is a salt containing the bicarbonate ion HCO_3^-) or the carbonate ion (CO_3^{2-}) and the other is a strong acid, the products will be a salt containing the anion of the acid, water, and carbon dioxide. A similar situation occurs if a reactant contains the sulfite ion (SO_3^{2-}) or bisulfite ion (HSO_3^-): The products will be a salt containing the anion of the acid, water and sulfur dioxide gas.

3. If the reactants are an acid and a base, carry out a neutralization reaction. If both reactants are weak acids/bases, then no reaction occurs.

4. Insoluble metal oxides react with strong acids to form a salt containing the anion of the acid and water.

5. Determine if a reaction is likely by analyzing the characteristics of the products. If any product has one of the following characteristics, then a reaction is likely:

    a. Insoluble (precipitate)—Use Table 4.3 in the text

    b. Weak or nonelectrolyte (remember water is a nonelectrolyte)

6. If all products are strong electrolytes, then no reaction occurs.

EXERCISE 8 Without referring to Table 4.3 in the text, identify the following salts as soluble or insoluble in water: **(a)** $Fe(NO_3)_3$; **(b)** $PbCl_2$; **(c)** $CaBr_2$; **(d)** $BaSO_4$; **(e)** Na_2SO_4; **(f)** K_2CO_3.
    Solution: **(a)** Soluble. All nitrate salts are soluble; **(b)** Insoluble. This is one of the three exceptions to the rule that chloride salts are soluble. $AgCl$, Hg_2Cl and $PbCl_2$ are insoluble. **(c)** Soluble. All bromide salts are soluble except $AgBr$, Hg_2Br_2, $PbBr_2$ and $HgBr_2$; **(d)** Insoluble. $BaSO_4$ is one of the exceptions to the rule that sulfate salts are soluble; **(e)** Soluble. All alkali metal salts are soluble; **(f)** Soluble. All alkali metal salts are soluble.

EXERCISE 9 Write net ionic equations for the following reactions:

(a) $NaCl(aq) + AgNO_3(aq) \longrightarrow AgCl(s) + NaNO_3(aq)$;

(b) $BaCO_3(s) + 2HCl(aq) \longrightarrow BaCl_2(aq) + CO_2(g) + H_2O(l)$

Solution: Rewrite each equation so that strong electrolytes are in an ion form. Then cross out spectator ions.

(a) $\cancel{Na^+}(aq) + Cl^-(aq) + Ag^+(aq) + \cancel{NO_3^-}(aq) \longrightarrow AgCl(s) + \cancel{Na^+}(aq) + \cancel{NO_3^-}(aq)$

AgCl is an insoluble salt and remains in a combined form. All other species are strong electrolytes. Na^+ and NO_3^- ions are spectator ions. The net ionic equation is

$$Ag^+(aq) + Cl^-(aq) \longrightarrow AgCl(s)$$

(b) $BaCO_3(s) + 2H^+(aq) + \cancel{2Cl^-}(aq) \longrightarrow Ba^{2+}(aq) + \cancel{2Cl^-}(aq) + CO_2(g) + H_2O(l)$

$BaCO_3$ is an insoluble salt, and $CO_2(g)$ and $H_2O(l)$ are a gas and a nonelectrolyte and remain unchanged. HCl and $BaCl_2$ are strong electrolytes; $BaCl_2$ is a soluble ionic salt. Cl^- ions are the only spectator ions.

$$BaCO_3(s) + 2H^+(aq) \longrightarrow Ba^{2+}(aq) + CO_2(g) + H_2O(l)$$

EXERCISE 10 Complete the following ionic reactions, and write net ionic equations for them.

(a) $Pb(NO_3)_2(aq) + KBr(aq) \longrightarrow$

(b) $NiCl_2(aq) + Na_3PO_4(aq) \longrightarrow$

Solution: For each chemical equation, predict the products of the reaction by switching ions between the reactants. Then identify the phase of each product. Use Table 4.3 in the text to help you identify insoluble salts.

(a) $Pb(NO_3)_2(aq) + 2KBr(aq) \longrightarrow PbBr_2(s) + 2KNO_3(aq)$

$PbBr_2$ is an insoluble salt, (s); KNO_3 is a strong electrolyte and a soluble salt, (aq).

$Pb^{2+}(aq) + \cancel{2NO_3^-}(aq) + \cancel{2K^+}(aq) + 2Br^-(aq) \longrightarrow PbBr_2(s) + \cancel{2K^+}(aq) + \cancel{2NO_3^-}(aq)$

Net ionic equation:

$$Pb^{2+}(aq) + 2Br^-(aq) \longrightarrow PbBr_2(s)$$

(b) $3NiCl_2(aq) + 2Na_3PO_4(aq) \longrightarrow Ni_3(PO_4)_2(s) + 6NaCl(aq)$

$Ni_3(PO_4)_2$ is an insoluble salt; NaCl is a strong electrolyte and a soluble salt.

$3Ni^{2+}(aq) + \cancel{6Cl^-}(aq) + \cancel{6Na^+}(aq) + 2PO_4^{3-}(aq) \longrightarrow$
$$Ni_3(PO_4)_2(s) + \cancel{6Na^+}(aq) + \cancel{6Cl^-}(aq)$$

Net ionic equation:

$$3Ni^{2+}(aq) + 2PO_4^{3-}(aq) \longrightarrow Ni_3(PO_4)_2(s)$$

EXERCISE 11 Predict the products of the following reactions, state whether a reaction occurs and why, and write a net ionic equation for each reaction that occurs:

(a) $Pb(NO_3)_2(aq) + Na_2SO_4(aq) \longrightarrow$

(b) $KCl(aq) + AgNO_3(aq) \longrightarrow$

(c) $CaCO_3(s) + HBr(aq) \longrightarrow$

(d) $NaCl(aq) + Ba(NO_3)_2(aq) \longrightarrow$

(e) $HCl(aq) + KOH(aq) \longrightarrow$

    Solution: **(a)** The molecular equation is obtained by writing the complete reaction equation and identifying the phases of the products using solubility rules:

$$Pb(NO_3)_2(aq) + Na_2SO_4(aq) \longrightarrow PbSO_4(s) + 2NaNO_3(aq)$$

The formation of the insoluble $PbSO_4$ causes the reaction to proceed. The net ionic equation is obtained by rewriting the molecular equation so that all strong electrolytes are written in the form of their separated ions, and all other substances are left in a molecular form:

$$Pb^{2+}(aq) + 2NO_3^-(aq) + 2Na^+(aq) + SO_4^{2-}(aq) \longrightarrow PbSO_4(aq) + 2Na^+(aq) + 2NO_3^-(aq)$$

All ions common to reactants and products are eliminated, and what remains is the net ionic equation:

$$Pb^{2+}(aq) + SO_4^{2-}(aq) \longrightarrow PbSO_4(s)$$

(b) Following the procedures in (a) results in the molecular equation

$$KCl(aq) + AgNO_3(aq) \longrightarrow AgCl(s) + KNO_3(aq)$$

The reaction proceeds because of the formation of a precipitate. Rewriting the equation so that all strong electrolytes are written as separated ions yields

$$K^+(aq) + Cl^-(aq) + Ag^+(aq) + NO_3^-(aq) \longrightarrow AgCl(s) + K^+(aq) + NO_3^-(aq)$$

Eliminating ions common to both sides of the arrow yields the net ionic equation:

$$Ag^+(aq) + Cl^-(aq) \longrightarrow AgCl(s)$$

(c) HBr is a strong acid. The acidic H^+ ion of HBr will react with the basic CO_3^{2-} ion of $CaCO_3$ to form $CO_2(g)$ and $H_2O(l)$. The molecular equation is

$$CaCO_3(s) + 2HBr(aq) \longrightarrow CO_2(g) + H_2O(l) + CaBr_2(aq)$$

The reaction proceeds because a volatile gas, CO_2, forms. Rewriting strong electrolytes into their separated ions yields

$$CaCO_3(s) + 2H^+(aq) + 2Br^-(aq) \longrightarrow CO_2(g) + H_2O(l) + Ca^{2+}(aq) + 2Br^-(aq)$$

The net ionic equation is

$$CaCO_3(s) + 2H^+(aq) \longrightarrow CO_2(g) + H_2O(l) + Ca^{2+}(aq)$$

(d) The predicted products $BaCl_2(aq)$ and $NaNO_3(aq)$ are both strong electrolytes and soluble in water. There are no insoluble salts, volatile gases, or water formed from the reaction; therefore, there is no reaction. **(e)** This is the reaction between a strong acid and base, referred to as a neutralization reaction:

$$HCl(aq) + KOH(aq) \longrightarrow H_2O(l) + KCl(aq)$$

The reaction proceeds because H_2O, a nonelectrolyte, is formed in the net ionic equation. Thus

$$H^+(aq) + Cl^-(aq) + K^+(aq) + OH^-(aq) \longrightarrow H_2O(l) + K^+(aq) + Cl^-(aq)$$

Net ionic equation: $H^+(aq) + OH^-(aq) \longrightarrow H_2O(l)$

Section 4.6 focuses on the reactions of metals with acids and salts. Metals in their reactions tend to lose electrons to form positive ions, called cations.

- **Oxidation** occurs when a substance loses electrons. A substance that has lost electrons is said to be *oxidized*. The oxidation of a metal can be shown by the generalized reaction:

$$M \longrightarrow M^{n+} + ne^-$$

- Oxidation does not occur by itself. Another substance must gain electrons. **Reduction** is the gain of electrons; when a substance gains electrons, it is said to be *reduced*.

Metals participate in displacement reactions.

-
$$A + BX \longrightarrow AX + B$$

$\uparrow$ metal $\qquad$ $\uparrow$ metallic compound

A displaces B from BX to form AX.

- Many metals react with acids such as HCl, HNO_3 and H_2SO_4 to form a salt and $H_2(g)$. Examples of such displacement reactions are

$$Zn(s) + H_2SO_4(aq) \longrightarrow ZnSO_4(aq) + H_2(g)$$
$$Fe(s) + 2HCl(aq) \longrightarrow FeCl_2(aq) + H_2(g)$$

Metals may also displace another metal from a salt.

- For example, $Zn(s) + CuSO_4(aq) \rightarrow Cu(s) + ZnSO_4(aq)$
- Displacement reactions of this type occur if

$$A + BX \longrightarrow AX + B$$

$\uparrow$ must be more easily oxidized than B

- An **activity series** arranges metals in order of decreasing ease of oxidation. Refer to Table 4.5 in the text. A metallic element is able to displace elements below it from their compounds.
- Note in Table 4.5 in the text that Li is the most active element, whereas Au is the least active. Li can displace all elements below it from their compounds. Au is essentially nonreactive.
- Also note H_2 is low in the activity series. Metals above it can displace H^+ from acids to form $H_2(g)$. Refer to Table 4.5 in the text for the activity of metals in displacing hydrogen from water. Only Li, K, Ba, Ca, and Na will react with cold water: others require steam.

EXERCISE 12 Write a balanced single displacement equation illustrating each of the following: **(a)** a metal displacing H^+ from H_2O; **(b)** a metal displacing H^+ from an acid; and **(c)** a metal replacing copper in a copper (II) nitrate solution.

Solution: To answer this type of question, you can either refer to prior analogous reactions or use an activity series. The most general approach is the latter. Refer to Table 4.5 in the text for needed information. **(a)** Choose a metal in the activity series that is above hydrogen. Li, K, Ba, Sr, Ca and Na will liberate H_2 in cold water (Mg, Al, Mn, Zn and Fe require steam). For example,

$$Mg(s) + 2H_2O(g) \longrightarrow Mg(OH)_2(s) + H_2(g)$$

(b) Also choose a metal that is higher in the activity series than hydrogen. For example,

$$Mg(s) + H_2SO_4(aq) \longrightarrow MgSO_4(aq) + H_2(g)$$

(c) Choose an element higher in the activity series than copper. For example,

$$Zn(s) + CuSO_4(aq) \longrightarrow ZnSO_4(aq) + Cu(s)$$

EXERCISE 13   Use the activity series to predict which of the following reactions will occur:

(a) $Hg(l) + MnSO_4(aq) \longrightarrow HgSO_4(s) + Mn(s)$

(b) $2Ag(s) + H_2SO_4(aq) \longrightarrow Ag_2SO_4(aq) + H_2(g)$

(c) $Ca(s) + 2H_2O(l) \longrightarrow Ca(OH)_2(aq) + H_2(g)$

Solution: All of the reactions are examples of displacement reactions:

$$M + M'X \longrightarrow MX + M'.$$

For such a reaction to occur, M must be higher in the activity series than M'. Refer to Table 4.5 in the text for required information. **(a)** Hg lies below Mn in the activity series; thus, the reaction does not occur. **(b)** Ag lies below hydrogen in the activity series; thus, the reaction does not occur. **(c)** Ca lies above hydrogen in the activity; thus, the reaction occurs.

SOLUTION STOICHIOMETRY: TITRATIONS

Molarity can be used as a conversion factor to change between the volume of a solution and number of moles of a solute:

$$Molarity \times Volume\ (L) = number\ of\ moles.$$

This relationship is useful in working titration problems.

Titration is a procedure for determining the concentration of a solution.

- The volume of a solution of known concentration (called a *standard solution*) required to react completely with a given volume of an unknown solution is experimentally determined. Molarity and volume relationships are used to calculate the number of moles of reactants.

- The **equivalence point** of a titration is the point in the titration when chemically equivalent, or stoichiometric, amounts of reactants have reacted.

- In acid-base titrations, the mole ratios of acid and base that react determine the relative amounts of acid and base required to reach the equivalence point. For example, a one to one mole ratio of the acid to base is required in the titration of HCl with NaOH. However, in the titration of H_2SO_4 with NaOH, each mole of sulfuric acid requires *two* moles of NaOH:

$$H_2SO_4(aq) + 2NaOH(aq) \longrightarrow Na_2SO_4(aq) + 2H_2O(l)$$

EXERCISE 14   What volume of 0.250 M HCl is required to react completely with 25.00 mL of 0.500 M NaOH?

   Solution: First write the balanced chemical equation for the reaction between HCl and NaOH. Since HCl is an acid and NaOH is a base, the reaction type is neutralization, in which water and a salt are formed:

$$HCl(aq) + NaOH(aq) \longrightarrow NaCl(aq) + H_2O(l)$$

According to the chemical reaction equation

$$1 \text{ mol HCl} \approx 1 \text{ mol NaOH}$$

We also have the following equivalences from the molarity statements:

$$0.250 \text{ mol HCl} = 1 \text{ L HCl solution}$$
$$0.500 \text{ mol NaOH} = 1 \text{ L NaOH solution}$$

To solve this problem we will need to carry out the following unit conversions:

Milliliters NaOH → *moles* NaOH → *moles* HCl → *milliliters* HCl

$$\text{Milliliters HCl} = (25.00 \text{ mL})\left(\frac{0.500 \text{ mol NaOH}}{1 \text{ L}}\right)$$
$$\times \left(\frac{1 \text{ mol HCl}}{1 \text{ mol NaOH}}\right)\left(\frac{1 \text{ L}}{0.250 \text{ mol HCl}}\right)$$
$$= 50.00 \text{ mL HCl}$$

Therefore, 50.00 mL of 0.250 M HCl reacts with 25.00 mL of 0.500 M NaOH.

EXERCISE 15   What is the molarity of a solution of H_2SO_4 if 20.00 mL of a 0.1000 M NaOH solution is required to react completely with 25.00 mL of the H_2SO_4 solution?

   Solution: The chemical equation describing the reaction between H_2SO_4 and NaOH is

$$H_2SO_4(aq) + 2NaOH(aq) \longrightarrow Na_2SO_4(aq) + 2H_2O(l)$$

Note that H_2SO_4 has two ionizable H^+ ions that react with OH^- ions from NaOH. To solve for the molarity of H_2SO_4 that is used in this titration, calculate the number of moles of H_2SO_4 that react with the number of moles of NaOH available in the 20.00 mL of 0.1000 M NaOH solution. The following conversion sequence will be required:

Volume NaOH $\longrightarrow$ *moles NaOH* $\longrightarrow$ *moles* H_2SO_4

You have available the following equivalences:

$$1 \text{ mol } H_2SO_4 \approx 2 \text{ mol NaOH} \qquad \text{[from balanced equation]}$$
$$0.1000 \text{ mol NaOH} = 1 \text{ L NaOH solution} \qquad \text{[from molarity of NaOH solution]}$$

$$\text{Moles } H_2SO_4 = (20.00 \text{ mL})\left(\frac{1 \text{ L}}{1000 \text{ mL}}\right)$$
$$\times \left(\frac{0.1000 \text{ mol NaOH}}{1 \text{ L}}\right)\left(\frac{1 \text{ mol } H_2SO_4}{2 \text{ mol NaOH}}\right)$$
$$= 1.000 \times 10^{-3} \text{ mol } H_2SO_4$$

Thus the molarity of a solution with 1.000×10^{-3} mol of H_2SO_4 and a volume of 25.00 mL is

$$M\ H_2SO_4 = \frac{\text{moles } H_2SO_4}{\text{volume in liters of } H_2SO_4}$$

$$= \frac{1.000 \times 10^{-3}\ \text{mol } H_2SO_4}{(25.00\ \text{mL})\left(\dfrac{1\ L}{1000\ \text{mL}}\right)}$$

$$= \frac{0.04000\ \text{mol } H_2SO_4}{1.000}$$

$$= 0.04000\ M$$

Note: In Exercises 2 and 14 through 15 the unit mole is again central in the sequence of conversions.

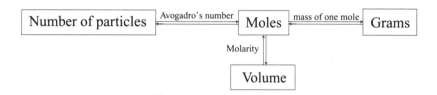

SELF-TEST QUESTIONS

Key Terms

Having reviewed the key terms listed at the end of Chapter 4 and their applications, identify the following statements as true or false. If a statement is false, indicate why it is incorrect. Key terms are italicized in the statements.

4.1 A metal high in the *activity series* is easily oxidized.

4.2 Metals typically undergo *reduction* by gaining electrons.

4.3 *Oxidation* occurs when an element becomes more positively charged.

4.4 HNO_3 is a strong *base*.

4.5 An *acid* decreases the concentration of $H^+(aq)$ in solution.

4.6 Molarity is an example of a measure of *concentration*.

4.7 NaCl is a *weak electrolyte* in water.

4.8 Sulfuric acid is a *strong electrolyte* in water.

4.9 Electrolytes are substances that contain ions because electrolytes form ions in solution.

4.10 The *equivalence point* in the titration of H_2SO_4 with NaOH occurs when equal molar amounts of the two have reacted.

4.11 A *standardized solution* is one whose volume is accurately determined before the start of a titration.

4.12 In order to determine the concentration of a solution of HNO_3, we could use a *titration* procedure to measure accu-rately the number of milliliters of 0.15 M NaOH that reacts with 50.00 mL of the nitric acid solution.

4.13 An *indicator* is a dye that changes color near the equivalence point in a titration.

4.14 *Spectator ions* appear in a net-ionic equation.

4.15 Weak electrolytes appear in a molecular form in a *net-ionic equation*.

4.16 Strong electrolytes and weak electrolytes are left in an unionized form in a *molecular equation*.

4.17 A *neutralization reaction* between an acid and a base produces water and an insoluble salt.

4.18 The *molarity* of a solution containing 7.46 g of KCl in 0.500 liters of a solution is 0.200 M.

4.19 An example of a *metathesis reaction* is

$$HNO_3(aq) + KOH(aq) \longrightarrow H_2O(l) + KNO_3(aq)$$

4.20 In the following *precipitation reaction*

$$CaI_2 + 2AgNO_3 \longrightarrow Ca(NO_3)_2 + 2AgI$$

the insoluble material (the precipitate) is $Ca(NO_3)_2$.

4.21 The term *aqueous solution* means water is the solvent.

4.22 3 g of sodium metal and a large amount of liquid ammonia form a solution. Thus, sodium is the *solvent*.

4.23 A solution can have only one *solute*.

4.24 When a solution is *diluted* with a solvent, the number of moles of solute decrease.

4.25 Sugar is a *nonelectrolyte* in water.

4.26 In a state of *chemical equilibrium,* the concentrations of reactants and products remain constant.

4.27 A *base* in water always forms hydroxide ion.

4.28 Nitric acid is a *strong acid.*

4.29 Hydroiodic acid is a *weak acid.*

4.30 Barium hydroxide is a *weak base.*

4.31 When a strong acid reacts with a strong base, a *salt* is produced.

4.32 The reaction of lead nitrate with sodium chloride in water produces a white *precipitate.*

4.33 The *solubility* of iron(III) phosphate in water is high.

Problems and Short-Answer Questions

4.34 Determine the following for a sucrose $(C_{12}H_{22}O_{11}$, molar mass = 342 g/mol) solution:
  (a) The number of moles of sucrose in 200 mL of a 0.100 *M* solution.
  (b) The volume of a 1.25 *M* solution containing 0.100 moles of sucrose.
  (c) The molarity of a solution containing 10 g of sucrose in 500 mL.

4.35 Answer the following questions about a 1.50 *M* HCl solution:
  (a) What is the new molarity if 200 mL of the solution is diluted to 1000 mL?
  (b) How many milliliters of the solution must be diluted to 500 mL to form a 0.100 *M* HCl solution?
  (c) How many moles of HCl are present in a diluted solution if 100 mL of the original solution are diluted to 2000 mL?

4.36 Identify the acid and base in each of the following reactions:
  (a) $NH_4^+(aq) + H_2O(l) \rightleftharpoons NH_3(aq) + H^+(aq)$
  (b) $HC_2H_3O_2(aq) + H_2O(l) \rightleftharpoons$
$$H^+(aq) + C_2H_3O_2^-(aq)$$
  (c) $HF(aq) + NaOH(aq) \rightarrow NaF(aq) + H_2O(l)$

4.37 A basic solution can be formed by the addition of which of the following to water: (1) HCl; (2) KOH; (3) NH_3, (4) H_2SO_4?

4.38 Complete and balance the following neutralization reactions. Underline the acid reactant with one line and the base reactant with two lines.
  (a) $Ba(OH)_2(aq) + HBr(aq) \rightarrow$
  (b) $NH_3(aq) + H_2SO_4(aq) \rightarrow$

4.39 Classify each of the following substances as an acid or a base and identify each substance as a nonelectrolyte, weak electrolyte, or strong electrolyte:
  (a) CsOH   **(d)** H_2O
  (b) N_2 **(e)** CCl_4
  (c) $HClO_4$ **(f)** HF

4.40 Write balanced net ionic equations for each of the following reactions:
  (a) $NaC_2H_3O_2(aq) + HCl(aq) \rightarrow$
  (b) $ZnS(s) + HBr(aq) \rightarrow$
  (c) $KCl(aq) + Hg_2(NO_3)_2(aq) \rightarrow$
  (d) $Ni(NO_3)_2(aq) + CuSO_4(aq) \rightarrow$
  (e) $Li(s) + H_2O(l) \rightarrow$
  (f) $H_2SO_4(aq) + KOH(aq) \rightarrow$
  (g) $HC_2H_3O_2(aq) + NaOH(aq) \rightarrow$

4.41 For each reaction in question 4.40 that occurs, indicate the nature of the driving force.

4.42 Using solubility rules or reasonable extensions of them, predict whether each of the following compounds is soluble in water:
  (a) $PbBr_2$ **(f)** ZnS
  (b) CsCl   **(g)** Hg_2Cl_2
  (c) $Cu(C_2H_3O_2)_2$ **(h)** Ag_2SO_4
  (d) $Mn(OH)_2$ **(i)** Sr_2SO_3
  (e) $Ca_3(PO_4)_2$

4.43 Identify the following substances as strong or weak acids or bases in water:
  (a) HCl   **(e)** $HClO_4$
  (b) H_2S **(f)** HF
  (c) NH_3 **(g)** H_3PO_4
  (d) KOH

4.44 What is the molarity of a nitric acid solution if 36.00 mL of it reacts completely with 2.00 g of NaOH?

4.45 What is the molarity of a solution of H_3PO_4 if 50.00 mL of it is titrated with 25.86 mL of 0.1201 *M* NaOH? Assume that all three hydrogens in H_3PO_4 react with NaOH.

4.46 Using the activity series (Table 4.5 in the text), write balanced chemical equations for the following reactions: (a) Mg is added to a solution of silver nitrate; (b) iron metal is added to a solution of copper (II) sulfate; (c) silver is added to a solution of zinc nitrate; (d) hydrogen gas is passed over a sample of zinc oxide; (e) hydrogen gas is passed over mercuric oxide.

Multiple-Choice Questions

4.47 Which of the following represents the net ionic equation for $NH_3(aq) + HBr(aq) \rightarrow$?
  (a) $NH_3(aq) + HBr(aq) \rightarrow NH_4^+ (aq) + Br^-(aq)$
  (b) $NH_3(aq) + HBr(aq) \rightarrow NH_4Br(aq)$
  (c) $NH_3(aq) + H^+(aq) + Br^-(aq) \rightarrow NH_4Br(aq)$
  (d) $NH_3(aq) + H^+(aq) \rightarrow NH_4^+ (aq)$

4.48 Which are the spectator ions in the following aqueous reaction (phases are not indicated): $MgCO_3 + HCl \rightarrow$?
  (a) Mg^{2+}
  (b) Cl^-
  (c) CO_3^{2-}
  (d) there are no spectator ions.

4.49 Which of the following aqueous solutions has the greatest total concentration of ions present?
  (a) 0.01 *M* HCl   **(c)** 0.02 *M* $Ca(NO_3)_2$
  (b) 0.05 *M* $CaBr_2$ **(d)** 0.02 *M* $Al(NO_3)_3$

4.50 Which of the following metal ions form chloride compounds that are insoluble in water: (1) K^+, (2) Ba^{2+}, (3) Ag^+, (4) Zn^{2+}, (5) Pb^{2+}?

(a) (1) and (2) **(d)** (3) and (5)
(b) (3) and (4) **(e)** (3), (4) and (5)
(c) (1) and (3)

4.51 Which of the following metal ions form sulfate salts that are insoluble in water: (1) Na^+, (2) Ba^{2+}, (3) Fe^{3+}, (4) Pb^{2+}, (5) Cd^{2+}?

(a) (1) and (2) **(d)** (3) and (5)
(b) (3) and (4) **(e)** all of them
(c) (2) and (4)

4.52 Which of the following substances is a weak electrolyte in water?

(a) HF **(d)** KNO_3
(b) H_2SO_4 **(e)** KCH_3CO_2
(c) $CaCl_2$

4.53 Which of the following describes an atom that is reduced?

(a) loses electrons and becomes more positively charged

(b) loses electrons and becomes more negatively charged

(c) gains electrons and becomes more positively charged

(d) gains electrons and becomes more negatively charged

4.54 Which of the following pairs of substances will *not* react, based on the activity series?

(a) Ca, $Cu(NO_3)_2$ **(c)** Li, $Fe(NO_3)_3$
(b) Pt, $Mn(NO_3)_2$ **(d)** Mn, $Ni(NO_3)_2$

4.55 Which of the following metals *cannot* displace hydrogen from steam?

(a) Au **(c)** Mg
(b) Zn **(d)** Sr

4.56 What are the products of the neutralization reaction between hydrogen iodide and calcium hydroxide in water?

(a) calcium and water **(c)** CaI_2 and water
(b) CaI_2 **(d)** water

4.57 What is the molarity of an aqueous HBr solution if 35.00 mL is neutralized with 70.00 mL of a 0.500 M NaOH solution?

(a) 0.250 *M* **(c)** 0.750 *M*
(b) 0.500 *M* **(d)** 1.00 *M*

4.58 How many grams of KBr (molar mass = 74.6 g/mol) are contained in 500 mL of a 0.250 *M* KCl aqueous solution?

(a) 0.125 **(c)** 125
(b) 9.33 **(d)** 9330

4.59 What is the molarity of a solution consisting of 1.25 g of NaOH in enough water to form 250 mL of solution?

(a) 1.25 *M* **(d)** 1.25×10^{-4} *M*
(b) 0.800 *M* **(e)** 0.125 *M*
(c) 8.00 *M*

4.60 A 300-mL solution of 3.0 *M* HCl is diluted to 2.0 L. What is the final molarity?

(a) 0.45 *M* **(d)** 0.05 *M*
(b) 450 *M* **(e)** 0.50 *M*
(c) 20 *M*

4.61 A 3.000-g sample of a soluble chloride is titrated with 52.60 mL of 0.2000 *M* $AgNO_3$. What is the percentage of chloride in the sample?

(a) 37.29% **(d)** 18.65%
(b) 24.86% **(e)** 37.29%
(c) 12.43%

SELF-TEST SOLUTIONS

4.1 True. **4.2** False. Metals typically undergo oxidation. **4.3** True. **4.4** False. It is a strong acid. **4.5** False. It increases the $H^+(aq)$ concentration. **4.6** True. **4.7** False. It is a strong electrolyte. **4.8** True. **4.9** False. Not all electrolytes are ionic. For example, $HCl(g)$ dissolves in water to form ions; $HCl(g)$ is covalent. **4.10** False. Two moles of NaOH react with one mole of H_2SO_4. **4.11** False. Its concentration is accurately known at the start. **4.12** True. **4.13** True. **4.14** False. They do not appear. **4.15** True. **4.16** True. **4.17** False. A soluble salt is usually formed. **4.18** True. **4.19** True. **4.20** False. All nitrate salts are soluble in water; silver iodide is not soluble. **4.21** True. **4.22** False. Ammonia is the solvent because it is in greatest quantity and has the phase of the solution. **4.23** False. One or more solutes. **4.24** False. The number of moles of solvent increase upon dilution, but the number of moles of solute does not change. **4.25** True. **4.26** True. **4.27** True. **4.28** True. **4.29** False. It is a strong acid. **4.30** False. It is a strong base. Metallic hydroxides generally form strong bases. **4.31** True. **4.32** True. **4.33** False. According to solubility rules, most metallic phosphates are insoluble, and iron is not one of the exceptions. **4.34** (a) mol sucrose = MxV(liters) = (0.100 M)(0.200 L) = 2.00×10^{-2} mol. (b) V = mol/M = 0.100 mol/1.25 M = 0.0800 L. (c) M = (g/molar mass)/V = (10 g/342 g/mol)/0.500 L = 0.058 molar. **4.35** (a) $M_f = M_i(V_i/V_f) = 1.50$ *M* (200 mL/1000 mL) = 0.300 *M*. (b) $V_i = V_f(M_f/M_i) = 500$ mL (0.100 *M*/1.50 *M*) = 33.3 mL. (c) The number of moles does not change upon dilution, only the concentration. Moles = $M \times V$ = (1.50 *M*)(0.100 *L*) = 0.150 mol HCl. **4.36** (a) acid: NH_4^+; base: H_2O. (b) acid: $HC_2H_3O_2$; base: H_2O. (c) acid: HF; base: NaOH. **4.37** (2) and (3); KOH is a strong base and NH_3 is a weak base. **4.38** (a) $\underline{Ba(OH)_2}(aq) + \underline{2HBr}(aq) \rightarrow BaBr_2(aq) + 2H_2O(l)$. (b) $\underline{2NH_3}(aq) + \underline{H_2SO_4}(aq) \rightarrow (NH_4)_2SO_4(aq)$. **4.39** (a) base, strong electrolyte; (b) neither acid nor base, nonelectrolyte; (c) acid, strong electrolyte; (d) acid and base, weak electrolyte; (e) neither acid nor base nonelectrolyte; (f) acid, weak electrolyte. **4.40** (a) $H^+(aq) + C_2H_3O_2^-(aq) \rightarrow HC_2H_3O_2(aq)$; (b) $ZnS(s) + 2H^+(aq) \rightarrow H_2S(g) + Zn^{2+}(aq)$; (c) $Hg_2^{2+}(aq) + 2Cl^-(aq) \rightarrow Hg_2Cl_2(s)$; (d) No reaction—predicted products $NiSO_4$ and $Cu(NO_3)_2$ are both soluble and strong electrolytes in water; (e) $2Li(s) + 2H_2O(l) \rightarrow 2Li^+(aq) + 2OH^-(aq) + H_2(g)$; (f) $H^+(aq) + HSO_4^-(aq) + 2OH^-(aq) \rightarrow 2H_2O(l) + SO_4^{2-}(aq)$; (g) $HC_2H_3O_2(aq) + OH^-(aq) \rightarrow H_2O(l) + C_2H_3O_2^-(aq)$. **4.41** (a) Formation of weak electrolyte, $HC_2H_3O_2$; (b) Formation of a gas, H_2S; (c) Formation of a precipitate, $Hg_2Cl_2(s)$. This chloride salt is an exception to the rule that chloride salts are soluble; (d) No reaction; (e) Formation of a gas, H_2; (f) Formation of a nonelectrolyte, H_2O; (g) Formation of a nonelectrolyte, H_2O. **4.42** Table 4.3 is used to reach the following conclusions: (a) insoluble, an exception to rule that all bromide salts are soluble; (b) soluble; (c) soluble; (d) insoluble; (e) insoluble; (f) insoluble; (g) insoluble; (h) insoluble, an exception to the rule that sulfate salts are soluble; **(i)** soluble. Group 1A salts are soluble. **4.43 (a)** strong acid; **(b)** weak acid; **(c)** weak base; **(d)** strong base; **(e)** strong acid; **(f)** weak acid; **(g)** somewhat weak acid. **4.44** Reaction equation: $HNO_3 + NaOH \rightarrow NaNO_3 + H_2O$. Conversions required: *grams NaOH → moles NaOH → moles HNO₃ → molarity HNO₃.* Equivalences: 40.00 g NaOH = 1 mol NaOH; 1 mol NaOH ≏ 1 mol HNO₃ (from neutralization reaction).

$$M\ HNO_3 = \frac{\text{moles } HNO_3}{\text{volume in liters}}$$

$$= \frac{(2.00\ \text{g NaOH})\left(\dfrac{1\ \text{mol NaOH}}{40.00\ \text{g NaOH}}\right)\left(\dfrac{1\ \text{mol } HNO_3}{1\ \text{mol NaOH}}\right)}{(36.00\ \text{mL})\left(\dfrac{1\ \text{L}}{1000\ \text{mL}}\right)}$$

$$= \frac{1.39\ \text{mol } HNO_3}{1\ \text{L}} = 1.39\ \text{molar}$$

4.45 Reaction equation:
$H_3PO_4 + 3NaOH \rightarrow Na_3PO_4 + 3H_2O$. Conversions required: *milliliters NaOH → moles NaOH → moles H₃PO₄.* Equivalences: 0.2101 mol NaOH = 1 L of NaOH solution; 1 mol H₃PO₄ ≏ 3 mol NaOH.

$$\text{Moles } H_3PO_4 = (25.86\ \text{mL NaOH})\left(\frac{1\ \text{L}}{1000\ \text{mL}}\right)$$

$$\times \left(\frac{0.1201\ \text{mol NaOH}}{1\ \text{L}}\right) \times \left(\frac{1\ \text{mol } H_3PO_4}{3\ \text{mol NaOH}}\right)$$

$$= 1.035 \times 10^{-3}\ \text{mol}$$

$$M\ H_3PO_4 = \frac{\text{moles } H_3PO_4}{\text{volume in liters}}$$

$$= \frac{1.035 \times 10^{-3}\ \text{mol}}{0.0500\ \text{L}}$$

$$= 0.02070\ M$$

4.46 (a) $Mg(s) + 2AgNO_3(aq) \rightarrow Mg(NO_3)_2(aq) + 2Ag(s)$; **(b)** $Fe(s) + CuSO_4(aq) \rightarrow FeSO_4(aq) + Cu(s)$; **(c)** no reaction. Ag lies below zinc in the activity series. **(d)** no reaction. H₂ gas lies below Zn in the activity series. **(e)** $H_2(g) + HgO(s) \rightarrow H_2O(l) + Hg(s)$. **4.47 (d)** HBr and NH₄Br are both strong electrolytes. **4.48 (b)** $MgCO_3(s) + 2H^+(aq) \rightarrow Mg^{2+}(aq) + H_2O(l)$

$+ CO_2(g)$ is the net ionic equation. **4.49 (b)** $3 \times 0.05\ M = 0.15$ M in total ions. Note that although **(d)** has four ions per formula, the total ion concentration is 0.08 M. **4.50 (d). 4.51 (c). 4.52 (a). 4.53 (d). 4.54 (b). 4.55 (a). 4.56 (c). 4.57 (d). 4.58 (b). 4.59**

(e) $$M\ NaOH = \frac{\text{mol NaOH}}{\text{volume in liters}}$$

$$= \frac{(1.25\ \text{g NaOH})\left(\dfrac{1\ \text{mol NaOH}}{40.01\ \text{g NaOH}}\right)}{(250\ \text{mL})\left(\dfrac{1\ \text{L}}{1000\ \text{mL}}\right)}$$

$$= 0.125\ M$$

4.60 (a) $M_i V_i = M_f V_f$

$$M_f = M_i \frac{M_i V_i}{V_f}$$

$$= (3.0\ M)\left[\frac{300\ \text{mL}}{(2.0\ \text{L})\left(\dfrac{1000\ \text{mL}}{\text{L}}\right)}\right]$$

$$= 0.45\ M$$

4.61 (c) The reaction of interest is $Cl^-(aq) + AgNO_3(aq) \rightarrow$ $AgCl(s) + NO_2^-(aq)$. The metal ion associated with the Cl⁻ does not react; therefore, we do not need to know what it is. We will need to make the following conversion: *Milliliters AgNO₃ → moles AgNO₃ → moles Cl⁻ → grams Cl⁻.* Hence,

$$\text{Grams } Cl^- = (52.60\ \text{mL } AgNO_3)\left(\frac{1\ \text{L}}{1000\ \text{mL}}\right)$$

$$\times \left(\frac{0.2000\ \text{mol } AgNO_3}{1\ \text{L } AgNO_3}\right)\left(\frac{1\ \text{mol } Cl^-}{1\ \text{mol } AgNO_3}\right)$$

$$\times \left(\frac{35.45\ \text{g } Cl^-}{1\ \text{mol } Cl^-}\right) = 0.3729\ \text{g } Cl^-$$

$$\%\ Cl^- \text{ in sample} = \left(\frac{\text{weight of } Cl^- \text{ in sample}}{\text{sample weight}}\right) \times 100$$

$$= \left(\frac{0.3729\ \text{g } Cl^-}{3.000\ \text{g sample}}\right) \times 100$$

$$= 12.43\%$$

5 Thermochemistry

OVERVIEW OF THE CHAPTER

5.1, 5.2 THERMODYNAMICS: THE FIRST LAW AND INTERNAL ENERGY CHANGES

Review: Dimensional Analysis (1.5)

Objectives: You should be able to:

1. Give examples of different forms of energy.

2. List the important units in which energy is expressed and convert from one to another.

3. Define the first law of thermodynamics both verbally and by means of an equation.

4. Describe how the change in internal energy of a system is related to the exchanges of heat and work between the system and its surroundings.

5. Define the term *state function* and describe its importance in thermochemistry.

5.2, 5.3 ENTHALPY: HEATS OF REACTIONS

Review: Meaning of chemical equations (3.1); stoichiometric calculations (3.7).

Objectives: You should be able to:

1. Define enthalpy, and relate the enthalpy change in a process occurring at constant pressure to the heat added to or lost by the system during the process.

2. Sketch an energy diagram such as that shown in Figure 5.13 of the text, given the enthalpy changes in the processes involved, and associate the sign of ΔH with whether the process is exothermic or endothermic.

3. Calculate the quantity of heat involved in a reaction at constant pressure given the quantity of reactants and the enthalpy change for the reaction on a mole basis.

5.5 HESS'S LAW

Objective: You should be able to:

1. State Hess's law, and apply it to calculate the enthalpy change in a process, given the enthalpy changes in other processes that could be combined to yield the reaction of interest.

Objectives: You should be able to:

1. Define and illustrate what is meant by the term *standard state,* and identify the standard states for the elements carbon, hydrogen, and oxygen.

2. Define the term *standard heat of formation,* and identify the type of chemical reaction with which it is associated.

3. Calculate the enthalpy change in a reaction occurring at constant pressure, given the standard enthalpies of formation of each reactant and product.

5.6 HEATS OF FORMATION: CALCULATING HEATS OF REACTIONS

Objectives: You should be able to:

1. Define the terms *heat capacity* and *specific heat.*

2. Calculate any one of the following quantities given the other three: heat, quantity of material, temperature change, and specific heat.

3. Calculate the heat capacity of a calorimeter, given the temperature change and quantity of heat involved; also calculate the heat evolved or absorbed in a process from a knowledge of the heat capacity of the system and its temperature change.

4. Define the term *fuel value;* calculate the fuel value of a substance given its heat of combustion or estimate the fuel value of a material given its composition.

5. List the major sources of energy on which humankind must depend, and discuss the likely availability of these for the foreseeable future.

5.7, 5.8 CALORIMETRY: FUEL VALUES

TOPIC SUMMARIES AND EXERCISES

In Chapter 3 of the text, you learned about the law of conservation of mass and how applying it to chemical reactions enables you to solve stoichiometric problems. In this chapter, you will learn about energy, how it is conserved, and its measurement.

What is energy? Energy can be conceived as the ability to do work or to transfer heat.

THERMODYNAMICS: THE FIRST LAW AND INTERNAL ENERGY CHANGES

▪ **Mechanical work**(w) = force (f) applied to an object times the distance (d) the object is moved by the force: $w = f \times d$. Energy enables you to supply the required force.

▪ Chemical compounds possess energy in two principal forms: potential and kinetic energy. **Potential energy** $(= mgh)$ is energy stored by an object as a result of its position and mass. **Kinetic energy** $(= 1/2mv^2)$ is associated with motion. Except in the case of a solid at 0 K, the particles that make up a substance (gas, liquid, or solid) undergo motion.

▪ Kinetic and potential energy are interconvertible; however, their total energy is constant during all changes as required by the first law of thermodynamics.

▪ Energy is conserved in physical and chemical processes—that is, in any change that occurs in nature the total energy of the universe remains constant. This is a statement of the law of conservation of energy, which is also called the first law of thermodynamics.

- The term **thermodynamics** refers to the study of the forms of energy and the changes energy undergoes.

The universe is divided for purposes of studying energy changes into a system and its surroundings.

- The **system** is arbitrarily defined, but it is usually that part of the universe that is being studied. Everything else is termed the **surroundings.**
- A system can interact with its surroundings in two ways: (1) by doing work (w) on the surroundings; and (2) by exchanging heat (q) with the surroundings.

The total change of all forms of energy in a system is $\Delta \mathbf{E}$ (Δ means a change in), the **internal energy** change.

- $\Delta E = E_{final} - E_{initial}$

  internal energy   internal energy
  at end of change   at start of change
- ΔE = positive number—system has gained energy.
- ΔE = negative number—system has lost energy.

Energy is exchanged between the system and surroundings as heat (q) and work (w):

$$\Delta E = q + w \quad \text{(a statement of the first law of thermodynamics)}$$

	+	−
q (heat)	added to system	given off by system
w (work)	done on system	done by system

The internal energy of a substance is a state function. You need to know the properties of state functions:

- A **state function** depends only on its present condition or state (specified by temperature, pressure, moles, etc.) and not on the manner in which the state of the system was reached.
- Δ (state function) = final state value − initial state value.

You need to remember the following units of energy:

- **Calorie:** The amount of energy required to raise the temperature of 1 g of water by 1°C from 14.5° to 15.5°C.

▪ **Joule:** 1 cal = 4.184 J. The kilojoule is used ordinarily as the unit of energy in the text.

EXERCISE 1   Classify the following as possessing kinetic energy, potential energy, or both: **(a)** two charged particles that are adjacent to one another; **(b)** an arrow moving through the air.

    Solution: **(a)** The particles have only potential energy because they are stationary. **(b)** The arrow has kinetic energy because it is not stationary and potential energy because of its position with respect to the surface of the earth.

EXERCISE 2   Calculate the velocity (v) of an electron whose mass is 9.107×10^{-28} g and whose kinetic energy (E_k) is 1.585×10^{-17} J (1 J = 1 kg-m^2/sec^2).

    Solution: The kinetic energy of a moving particle is $E_k = \frac{1}{2}mv^2$, where the unit for E_k is joule, for m, kilogram, and for v, meter/second. Substituting the mass (converted to kilograms) and the energy of the electron into this equation yields

$$E_k = \tfrac{1}{2}mv^2$$

$$1.585 \times 10^{-17}\ \text{J} = \tfrac{1}{2}(9.107 \times 10^{-25}\ \text{kg})v^2$$

Solving for v yields

$$v = \sqrt{\frac{2E_k}{m}} = \sqrt{\frac{(2)(1.585 \times 10^{-17}\ \text{kg-m}^2/\text{sec}^2)}{9.107 \times 10^{-25}\ \text{kg}}} = 5.899 \times 10^3\ \text{m/sec}$$

EXERCISE 3   Calculate the internal energy change in a system in which the following changes occur: **(a)** 100 g of $MgO(s)$ is heated from 50°C to 100°C, a process that requires 870 J; no work is done; and **(b)** a gas expands slowly while heating, and does 200 J of work and gains 350 J of heat.

    Solution: Both problems require substituting the appropriate data in $\Delta E = q + w$. To do this correctly, you must determine the signs of q and w are from the given data. **(a)** Raising the temperature of MgO requires heat; thus the sign of q is positive:

$$\Delta E = q + w = 870\ \text{J} + 0 = 870\ \text{J}$$

(b) Since the system gains heat, q is positive. Because the gas does work on the suroundings, w is negative.

$$\Delta E = q + w = 350\ \text{J} + (-200\ \text{J}) = 150\ \text{J}$$

Chemical reactions in the laboratory or in the environment commonly occur at constant external pressure, most often under the constant pressure of the atmosphere. **Enthalpy change** (ΔH) is the amount of energy exchanged between a system and its surroundings at constant pressure.

ENTHALPY: HEATS OF REACTIONS

▪ $\Delta H = \Delta E + P\,\Delta V = q_p$

▪ $\Delta H = H_{\text{final}} - H_{\text{initial}}$ (state function)

▪ Remember the meaning of the sign of ΔH: ΔH = positive number—a process gains heat: **endothermic;** ΔH = negative number—a process gives off heat: **exothermic.**

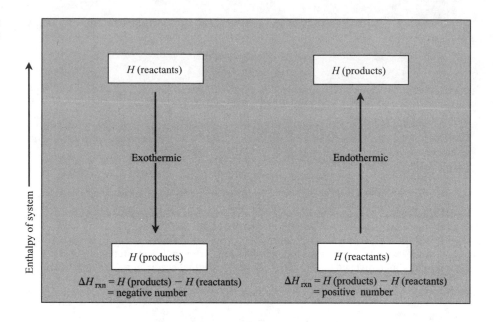

Figure 5.1 Energy diagram for endothermic and exothermic reactions.

- For a chemical reaction

$$\Delta H_{rxn} = H \text{ (products)} - H \text{ (reactants)}$$

See Figure 5.1.

The physical states of reactants and products must be specified when you give an enthalpy change for a particular reaction.

- The enthalpy content of a substance varies as its physical state changes.

- A thermochemical equation is used to describe the physical states of reactants and products and the associated enthalpy change. The enthalpy change usually is written to the right. For example,

$$CaO(s) + CO_2(g) \rightarrow CaCO_3(s) \quad \Delta H_{298} = -178 \text{ kJ}$$

The subscript with ΔH specifies the temperature in degrees Kelvin at which the reaction is carried out. Notice that the negative sign of ΔH tells you that the reaction is exothermic.

What are the relationships between (1) the sign of ΔH and the direction of a reaction and between (2) the value of ΔH and the stoichiometry of a reaction?

- The enthalpy change for a reaction is equal in magnitude but opposite in sign to ΔH for the reverse reaction. For example,

$$CaCO_3(s) \rightarrow CaO(s) + CO_2(g) \quad \Delta H_{298} = 178 \text{ kJ}$$
$$CaO(s) + CO_2(g) \rightarrow CaCO_3(s) \quad \Delta H_{298} = -178 \text{ kJ}$$

- Enthalpy is an extensive property of a system, just as is mass. Therefore, if you change the stoichiometry of a reaction, you also change the enthalpy. For example, if the number of moles of $CaCO_3(s)$ decomposing is changed to two, then the enthalpy must also be multiplied by two:

$$2 \times (CaCO_3(s) \rightarrow CaO(s) + CO_2(g) \quad \Delta H_{298} = 2 \times 178 \text{ kJ}$$
$$2CaCO_3(s) \rightarrow 2CaO(s) + 2CO_2(g) \; \Delta H_{289} = 356 \text{ kJ}$$

EXERCISE 4   You observe an ice cube melting in a glass of water. What are the system and the surroundings for this observed change? Is the melting of ice an exothermic or endothermic process?

Solution: We can label what we observe, the glass containing the water and ice, as the system and everything else as the surroundings. We know that the application of heat to ice causes it to melt. Because the melting of ice requires heat, the change is said to be endothermic. By contrast, an exothermic change releases heat.

EXERCISE 5   For which one of the following processes does the heat evolved or absorbed equal the enthalpy change: the combustion of gasoline in a closed container or the boiling of water at 100°C and at atmospheric pressure?

Solution: Only for a process at constant pressure and temperature does the enthalpy change equal q. The boiling of water occurs at a constant atmospheric pressure and temperature; thus the heat change equals q_p, enthalpy. As the gasoline is combusted in the closed container, the pressure changes because gases are evolved during combustion.

EXERCISE 6   The reaction

$$2H_2(g) + O_2(g) \rightarrow 2H_2O(l)$$

occurs at 25°C with a ΔH value of −571.70 kJ. **(a)** Is this an exothermic or endothermic reaction? **(b)** What is ΔH for the reaction

$$2H_2O(l) \rightarrow 2H_2(g) + O_2(g)$$

(c) What is ΔH for the reaction when 0.50 mol of $H_2O(l)$ is formed?

Solution: **(a)** This is an exothermic reaction because the sign of ΔH is negative. **(b)** 571.70 kJ. This reaction is the reverse of the first one, and so the enthalpy change is equal in magnitude but opposite in sign to that of the original reaction. **(c)** The heat evolved in the original reaction is for the formation of 2 mol of $H_2O(l)$; the heat evolved in (c) is for the formation of 0.50 mol of $H_2O(l)$:

$$\Delta H = \left(\frac{-571.70 \text{ kJ}}{2 \text{ mol } H_2O} \right) (0.50 \text{ mol } H_2O) = -140 \text{ kJ}$$

| HESS'S LAW

Hess's law is a statement of an important principle of thermochemistry: *the enthalpy change for a chemical reaction is the same whether the reaction takes place in one step or in several steps.* ΔH_{rxn} will equal the sum of enthalpy changes for each step. Use the following procedure to calculate an enthalpy change for a reaction, having been given known thermochemical equations:

1. Write the thermochemical equation for the desired chemical process.

2. Carefully examine the given thermochemical equations to find the reactants and products of the desired thermochemical equation.

3. If a desired reactant or product is on the wrong side of a given thermochemical equation, reverse the thermochemical equation. You also must change the sign of ΔH.

4. Multiply the known thermochemical equations by the appropriate coefficients so that the reactants and products have the same coefficients as they have in the desired thermochemical equation.

5. Algebraically add the thermochemical equations after step 4. If an unwanted species does not disappear, you must now include another thermochemical equation such that when it is added to the others the unwanted species cancels. Be sure that the reactants and products remain balanced.

6. Add ΔH's for each thermochemical equation to obtain ΔH for the desired reaction.

EXERCISE 7   Calculate the enthalpy change for the reaction

$$C(s) + \frac{1}{2}O_2(g) \rightarrow CO(g)$$

given the following information:

(a) $C(s) + O_2(g) \rightarrow CO_2(g)$ $\Delta H = -393.5$ kJ

(b) $CO(g) + \frac{1}{2}O_2(g) \rightarrow CO_2(g)$ $\Delta H = -283.0$ kJ

   Solution: You can apply Hess's law to this problem if you add equations **(a)** and **(b)** to yield the desired reaction. Since you want $CO(g)$ as a product, you can see by inspection of reaction **(b)** that it will have to be reversed so that $CO(g)$ is changed from a reactant to a product. If you add equation **(a)** and the reverse of equation **(b)**, the desired equation is obtained:

$$C(s) + O_2(g \rightarrow CO_2(g) \qquad\qquad\qquad \Delta H = -393.5 \text{ kJ}$$

$$CO_2(g) \rightarrow CO(g) + \frac{1}{2}O_2(g) \qquad\qquad\qquad \Delta H = 283.0 \text{ kJ}$$

$$\overline{\qquad\qquad\qquad\qquad\qquad\qquad\qquad\qquad\qquad\qquad\qquad\qquad\qquad\qquad}$$

$$C(s) + \frac{1}{2}O_2(g) \rightarrow CO(g) \qquad\qquad\qquad \Delta H = -110.5 \text{ kJ}$$

Note that the sign of ΔH for reaction **(b)** must be changed because the reaction has been reversed.

EXERCISE 8   Calculate ΔH for the reaction

$$C_2H_4(g) + H_2(g) \rightarrow C_2H_6(l)$$

given the following data:
(a) $C_2H_4(g) + 3O_2(g) \rightarrow 2CO_2(g) + 2H_2O(l)$ $\Delta H = -845.2$ kJ
(b) $2H_2(g) + O_2(g)$ $\rightarrow 2H_2O(l)$ $\Delta H = -573.2$ kJ
(c) $2C_2H_6(l) + 7O_2(g) \rightarrow 4CO_2(g) + 6H_2O(l)$ $\Delta H = -1987.4$ kJ

   Solution: The three given equations must be written so that when they are added the required chemical equation results. Equation **(c)** must be reversed and divided by 2 so that 1 mol of $C_2H_6(l)$ is produced, since it is the product of the required reaction. Chemical equation **(b)** must be divided by 2 so that only 1 mol of $H_2(g)$ appears as a reactant as in the required reaction. This gives us:

$$C_2H_4(g) + 3O_2(g) \rightarrow 2CO_2(g) + 2H_2O(l) \qquad \Delta H = -845.2 \text{ kJ}$$

$$H_2(g) + \frac{1}{2}O_2(g) \rightarrow H_2O(l) \qquad \Delta H = -286.6 \text{ kJ}$$

$$2CO_2(g) + 3H_2O(l) \rightarrow \frac{7}{2}O_2(g) + C_2H_6(l) \qquad \Delta H = 993.7 \text{ kJ}$$

$$C_2H_4(g) + H_2(g) \rightarrow C_2H_6(l) \qquad \Delta H = -138.1 \text{ kJ}$$

Note: The ΔH value associated with equation **(b)** has been divided by 2 because equation **(b)** has been divided by 2. Similarly, the ΔH value associated with equation **(c)** has been divided by 2; its sign has also changed because equation **(c)** has been reversed.

EXERCISE 9 Explain how Hess's law of heat summation depends on the fact that enthalpy is a state function.

   Solution: The enthalpy change for a reaction depends only on the amount of matter that undergoes a change and the initial states of the reactants and the final states of the products. Thus $\Delta H = H_{\text{final}} - H_{\text{initial}}$, and the value does not depend on any intermediate steps that might be involved. Therefore, if a particular reaction is carried out in a series of steps, the sum of ΔH's for the series of reactions must be the same as the enthalpy change for the given one-step reaction.

You can use standard heats of formation to calculate enthalpy changes of reactions. Before doing such calculations, you need to know the following about heats of formation:

- The enthalpy change associated with the formation of one mole of a compound from its constituent elements is termed **heat of formation** (ΔH_f). For example, at 100°C and standard atmospheric pressure, the thermochemical equation for the formation of $H_2O(g)$ is

$$H_2(g) + \tfrac{1}{2}O_2(g) \rightarrow H_2O(g) \quad \Delta H_f = -242.3 \text{ kJ/mol}$$

- The value of ΔH_f for a compound depends on the physical state of the compound, pressure, and temperature.

To enable chemists to make appropriate comparisons of ΔH_{rxn} values, a standard set of conditions is defined.

- The **standard state** for a compound is its most stable form at the given temperature and pressure. The pressure usually is standard atmospheric pressure and the temperature is 25°C; however, other pressures and temperatures can be specified.
- The **standard heat of formation** of a compound, ΔH_f°, is the heat of formation of *one mole* of the compound in its *standard state* formed from its constituent elements in their standard states.
- The standard heat of formation of any element in its most stable form is by convention taken to be zero: $\Delta H_{f\,298}^\circ (O_2) = 0$ kJ/mol, but $\Delta H_{f\,298}^\circ (O_3) = 142.7$ kJ/mol because O_3 is not the most stable form of elemental oxygen at 25°C and 1 atm.

A simple method for determining heats of reactions from standard heats of formations is

- ΔH_{rxn}° = (sum of heats of formation of each product) –

                              (sum of the heats of formation of each reactant)

which is sometimes written as

$$\Delta H_{rxn}^{\circ} = \Sigma n \Delta H_f^{\circ} \text{(products)} - \Sigma m \Delta H_f^{\circ} \text{(reactants)}$$

where Σ (sigma) means "the sum of" and n and m are the stoichiometric coefficients of the chemical equation.

- Keep in mind that $\Delta H_f^{\circ} = 0$ for elements in their standard states.

EXERCISE 10   The standard state of an element or compound is its most stable form at 25°C and standard atmospheric pressure. Write the reaction associated with the standard heat of formation for each of the following: **(a)** $AgCl(s)$; **(b)** $NH_3(g)$; **(c)** $Na_2CO_3(s)$.

    Solution: The standard heat of formation (ΔH_f°) is the enthalpy change for the formation of 1 mol of a compound from its elements in their standard states of 25°C and standard atmospheric pressure. Develop balanced chemical reactions that have as reactants the elements comprising each compound. Each element should be in its most stable form at 25°C and at 1 atmosphere pressure.

(a) $Ag(s) + \frac{1}{2}Cl_2(g) \rightarrow AgCl(s)$
(b) $\frac{1}{2}N_2(g) + \frac{3}{2}H_2(g) \rightarrow NH_3(g)$
(c) $2Na(s) + C(graphite) + \frac{3}{2}O_2(g) \rightarrow Na_2CO_3(s)$

EXERCISE 11   Calculate the standard enthalpy change for the reaction

$$CH_2N_2(s) + O_2(g) \rightarrow CO(g) + H_2O(l) + N_2(g)$$

at 25°C and standard atmospheric pressure, given the following standard heats of formation: $CO(g)$, −110.0 kJ/mol; $CH_2N_2(s)$, +62.4 kJ/mol; H_2O (l), −285.9 kJ/mol. (Remember that the standard heat of formation for any element in its standard state is zero.)

    Solution: The standard enthalpy for a reaction at 25°C and 1 atm can be calculated using the relation

$$\Delta H_{rxn}^{\circ} = \Sigma n \Delta H_f^{\circ} \text{(products)} - \Sigma m \Delta H_f^{\circ} \text{(reactants)}$$

Using this relation, the standard enthalpy change is calculated as follows:

$$\Delta H_{rxn}^{\circ} = [\Delta H_f^{\circ}(CO) + \Delta H_f^{\circ}(H_2O) + \Delta H_f^{\circ}(N_2)] - [\Delta H_f^{\circ}(CH_2N_2) + \Delta H_f^{\circ}(O_2)]$$

Substituting for the values of ΔH_f° yields

$$\Delta H_{rxn}^{\circ} = \left[(1\text{ mol})\left(-110.0\ \frac{kJ}{mol}\right) + (1\text{ mol})\left(-285.9\ \frac{kJ}{mol}\right) \right.$$

$$\left. + (1\text{ mol})\left(0\ \frac{kJ}{mol}\right) \right] - \left[(1\text{ mol})\left(+62.4\ \frac{kJ}{mol}\right) + (1\text{ mol})\left(0\ \frac{kJ}{mol}\right) \right]$$

$$\Delta H_{rxn}^{\circ} = -458.3\text{ kJ}$$

CALORIMETRY: FUEL VALUES

Calorimetry is an experimental technique for measuring heat flow in a thermochemical change. Temperature changes are accurately measured and converted to heat energy by using heat capacities.

- **Heat capacity** (C) represents the amount of heat required to raise the temperature of a given amount of a substance by 1 K.

- $C = \dfrac{q}{\Delta T}$ $\leftarrow$ heat flow
  $\phantom{C = \dfrac{q}{\Delta T}}$ $\leftarrow$ temperature change $= T_{\text{final}} - T_{\text{initial}}$

- Heat flow $= q = C \times \Delta T$ $\leftarrow$ *Key concept in calorimetry*

The text discusses two types of heat capacities: Molar heat capacity and specific heat.

- **Molar heat capacity** is the heat capacity of one mole of a substance. It has the units of joules per mole per degree (J/mol-K).

$$\text{Molar heat capacity} = \dfrac{q}{\Delta T} \leftarrow \text{J/mol}$$

- **Specific heat** is the heat capacity for one gram of a substance. It has the units of joules per gram per degree (J/g-K).

$$\text{Specific heat capacity} = \dfrac{q}{\Delta T} \leftarrow \text{J/g}$$

A key equation in calorimetry is:

$$\text{Heat flow for a given mass } (q) = m \times \text{specific heat} \times \Delta T$$

One type of calorimeter for measuring heat flow is a **bomb calorimeter.**

- Heat flow (q) is measured under *constant volume* conditions. Therefore,

$$q_v = \Delta E.$$

- Heat flow from the reaction vessel, called the bomb, raises the temperature of the bomb and the surrounding water bath. Therefore, the heat evolved is related to the temperature increase as follows:

$$q_{\text{evolved}} = -C_{\text{calorimeter}} \times \Delta T$$

where $C_{\text{calorimeter}}$ is the heat capacity of the calorimeter.

- $C_{\text{calorimeter}}$ is determined by measuring ΔT for combustion of a given mass of a standard (i.e., a reaction whose q_v is known.)

Reactions at constant pressure can be studied in a General Chemistry laboratory using an insulated styrofoam cup as an inexpensive calorimeter.

- At *constant pressure* conditions enthalpy is measured.
- Because the calorimeter is insulated; heat is not transferred from the reaction to the surroundings. Thus, all of the heat change is shown as a temperature change in the calorimeter contents.

$$\text{Heat change} = (\text{mass})(\text{specific heat})(\text{temperature change})$$
$$q_p = m \times (\text{specific heat}) \times \Delta T$$

where m is the mass of the contents of the calorimeter and ΔT is the temperature change of the contents.

The foods we eat and the fuels we use to run our machines are potential energy sources. When they are combusted, the heat released can be used by our bodies or by machines to do work. The heat evolved during the combustion of a gram of food or fuel is known as the **fuel value** of the substance.

▪ Fuel values are always exothermic quantities. They are reported as positive quantities in tables and thus must have a negative sign included when numbers are used in calculations.

EXERCISE 12   The specific heat of $NH_3(l)$ is 4.381 J/g-K. Calculate the heat required to raise the temperature of 1.50 g of $NH_3(l)$ from 213 K to 218.0 K.

Solution: The heat required to raise the temperature of any substance can be calculated using the following relation:

Heat = (mass of substance)(specific heat)(temperature change)

Heat = (mass)(specific heat)(final temperature − initial temperature)

Substituting the appropriate quantities into this equation:

$$\text{Heat} = (1.50 \text{ g})\left(4.381 \frac{J}{g\text{ - }^\circ K}\right)[218.0 \text{ K} - (213.0 \text{ K})]$$

$$= (1.50 \text{ g})\left(4.381 \frac{J}{g\text{ - }^\circ K}\right)(5.0 \text{ K}) = 33 \text{ J}$$

EXERCISE 13   0.80 g of sulfur is combusted to form SO_2. All of the heat evolved is used to raise the temperature of 100.0 g of water by 17.8 K. **(a)** Write the combustion reaction for sulfur. **(b)** Calculate the heat of combustion of 1 mol of sulfur.

Solution: **(a)** The combustion of sulfur involves its reaction with O_2 to form an oxide of sulfur: $S(s) + O_2(g) \rightarrow SO_2(g)$. **(b)** The heat of combustion of sulfur is exactly equal to the heat required to raise the temperature of 100.0 g of H_2O by 17.8 K. Therefore, we can derive the following relation:

Heat of combustion of sulfur = (mass of H_2O)

$$\times \text{ (specific heat of } H_2O)\text{(temperature change)}$$

$$\text{Heat of combustion} = (100.0 \text{ g})\left(4.184 \frac{J}{g\text{-}K}\right)(17.8 \text{ K})$$

$$= 7450 \text{ J } = 7.45 \text{ kJ}$$

This is the heat of combustion per 0.80 g of sulfur. To determine the molar heat of combustion, we must make the following conversion:

$$\text{Molar heat of combustion} = \left(\frac{7.45 \text{ kJ}}{0.80 \text{ g}}\right)\left(\frac{32.0}{1 \text{ mol}}\right) = 298 \frac{kJ}{mol}$$

EXERCISE 14   A 1.05 g sample of benzoic acid is combusted completely in a bomb calorimeter. Calculate the temperature change given that: the heat capacity of the calorimeter is 1.80 kJ/K; and the heat of combustion of benzoic acid is 26.4 kJ/g.

Solution: The heat evolved in the combustion of 1.05 g of benzoic acid is given by

$$q_{\text{benzoic acid}} = \Delta H_{\text{combustion}} \times m_{\text{benzoic acid}} = (-26.4 \text{ kJ/g})(1.05 \text{ g}) = -27.7 \text{ kJ}$$

The heat evolved raises the temperature of the calorimeter.

$$q_{\text{benzoic acid}} = -C_{\text{calorimeter}} \times \Delta T$$
$$-27.7 \text{ kJ} = -(1.80 \text{ kJ/K}) \times \Delta T$$

Solving for ΔT,

$$\Delta T = \frac{-27.7 \text{ kJ}}{-1.80 \text{ kJ/K}} = 15.4 \text{ K}$$

EXERCISE 15 Fudge is about 2 percent protein, 11 percent fat, and 81 percent carbohydrate. The average fuel values of these substances are as follows: protein, 17 kJ/g; fat, 38 kJ/g; carbohydrate, 17 kJ/g. What is the fuel value of 10 g of fudge?

Solution: The fuel value of fudge is a weighted average of each component's fuel value.

Fuel value = (mass of fudge)[(protein fraction × its average fuel value)
+ (fat fraction × its average fuel value)
+ (carbohydrate fraction × its average fuel value)]

$$\frac{\text{Fuel value}}{\text{of fudge}} = (10 \text{ g})\left[(0.02)\left(17\,\frac{\text{kJ}}{\text{g}}\right) + (0.11)\left(38\,\frac{\text{kJ}}{\text{g}}\right) + (0.81)\left(17\,\frac{\text{kJ}}{\text{g}}\right)\right]$$

$$\frac{\text{Fuel value}}{\text{of fudge}} = (10 \text{ g})\left(18\,\frac{\text{kJ}}{\text{g}}\right) = 180 \text{ kJ}$$

The food value of 10 g of fudge in calories is therefore as follows:

$$(180 \text{ kJ})\left(\frac{10^3 \text{ J}}{1 \text{ kJ}}\right)\left(\frac{1 \text{ cal}}{4.184 \text{ J}}\right) = 4.3 \times 10^4 \text{ cal}$$

Or since 1 kcal = 1 Calorie, the food value is 43 Calories. (The Calorie, capitalized, is the unit used in nutrition.)

SELF-TEST QUESTIONS

Key Terms

Having reviewed the key terms listed at the end of Chapter 5 and their applications, identify the following statements as true or false. If a statement is false, indicate why it is incorrect. Key terms are italicized in the statements.

5.1 Before a pitcher in baseball throws a baseball to the batter, the ball has *kinetic energy.*

5.2 A person standing on the fifth floor of a chemistry building has more *potential energy* than a person standing on the first floor.

5.3 The explosion of a hydrogen bomb is accompanied by the release of heat; thus it is an *endothermic process.*

5.4 The universe can be considered to be composed of two parts when we do an experiment: the *system,* that which we focus our attention upon, and the *surroundings,* the rest of the universe.

5.5 The *energy* within a battery that enables it to power a radio is a form of potential energy.

5.6 In the SI system of units the unit of heat is the *joule.*

5.7 The heat measured in a bomb calorimeter is termed *enthalpy.*

5.8 When the *standard heat of formation* of a substance is measured, both the elements and substances must be at the same temperature and at the standard atmospheric pressure of 1 atm.

5.9 The *first law of thermodynamics* requires that energy be conserved in all processes.

5.10 The heat measured using a *bomb calorimeter* is ΔE.

5.11 ΔH for the physical change $H_2O(l) \rightarrow H_2O(g)$ at 1 atm pressure and 100°C is an example of a state function.

5.12 The *specific heat* of Al(s) is 0.89 J/g-°C. A 10°C rise in the temperature of a 90-g block of Al requires 801J.

5.13 *Thermodynamics* is the study of heat and processes that involve energy changes.

5.14 If q is + and w is +, ΔE (*internal energy* change) is *always* positive.

5.15 Thermodynamics is concerned with *macroscopic* properties.

5.16 Heats determined by *calorimetry* experiments involve calculations that assume the specific heat of substance does not vary with temperature.

5.17 The quantity of *heat* flow between substances of differing temperatures depends only on their individual temperatures.

5.18 At 25°C and 1 atm the *standard state* of carbon dioxide is a solid.

5.19 *Syngas* is primarily a mixture of H_2 and CO.

5.20 The validity of *Hess's law* arises from the fact that enthalpy is a state function.

5.21 The *specific heat* capacity of water is 4.18 J/g-°C. To convert the specific heat capacity of water to molar heat capacity, you would divide the specific heat capacity by the molar mass of water.

5.22 Proteins and carbohydrates have similar average *fuel values*, about 17 kJ/g.

5.23 *Thermochemistry* is the study of chemical reactions and their energy changes.

5.24 The greater the *force* applied to an object when it is moved, the greater the work.

5.25 In one minute, a 100-watt light bulb requires a *power* input of 100 J to operate.

5.26 One *calorie* is the same as the Calorie.

5.27 In an insulated container, an *exothermic* reaction would increase the temperature of the system.

5.28 The value of a *state function* depends on the particular history of the sample.

5.29 *Heats of formations* for compounds are generally exothermic quantities.

5.30 An example of a constant pressure *calorimeter* is the "coffee cup" calorimeter described in the text.

5.31 *Heat capacity* is a negative number.

5.32 The *molar heat capacity* of water is calculated by taking its specific heat capacity and dividing by 18 g, the molar mass of water.

5.33 *Specific heat* of a substance is based on the heat capacity of 1 g of the material.

5.34 *Fossil fuels* include gasoline.

5.35 A major component of *natural gas* is methane.

5.36 *Coal* contains hydrocarbons of low molecular weight.

Problems and Short-Answer Questions

5.37 Determine the enthalpy of reaction for

$$CH_3COCH_3(l) + 2O_2 \rightarrow CH_3COOH(l) + CO_2(g) + H_2O(g)$$

using Hess's law of constant heat summation and given:

$$CH_3COCH_3(l) + 4O_2(g) \rightarrow 3CO_2(g) + 3H_2O(l)$$
$$\Delta H = -1787.0 \text{ kJ}$$

$$H_2O(l) \rightarrow H_2O(g) \qquad \Delta H = -44.1 \text{ kJ}$$

$$CH_3COOH(l) + 2O_2(g) \rightarrow 2CO_2(g) + 2H_2O(l)$$
$$\Delta H = -835 \text{ kJ}$$

5.38 Calculate the standard heat of formation of $PCl_5(g)$, given that $\Delta H° = -136.0$ kJ for the reaction

$$PCl_5(g) + H_2O(g) \rightarrow POCl_3(g) + 2HCl(g)$$

and the following standard heats of formation: $H_2O(g)$, −241.8 kJ/mol; $POCl_3(g)$, −592 kJ/mol; $HCl(g)$, −92.5 kJ/mol.

5.39 The combustion of 4.086 g of In increased the temperature of a calorimeter by 1.626°C. The energy equivalent (the amount of energy required to raise the temperature by 1°C) of the calorimeter is 10.095 kJ. What is the heat of combustion of 1 mol of In?

5.40 Calculate the heat evolved when 0.500 g of $Cl_2(g)$ reacts with an excess of HBr(g) to form HCl(g) and $Br_2(l)$, given the following standard heats of formation (kJ/mol): HCl(g), −92.30; HBr(g), −36.20.

5.41 The fuel value for bituminous coal is 32 kJ/g. The fission fuel value of ^{235}U is 3.17×10^{-11} kJ/atom. How many grams of coal must be combusted to equal the energy released by the fission of 1 g of ^{235}U in a nuclear power plant?

5.42 Benzene, C_6H_6, is a chemical that has been used extensively in chemical laboratories. However, recent toxicological evidence suggests that it is capable of causing cancer, and thus chemists are limiting its use. Calculate the heat of combustion of $C_6H_6(l)$, given the following heats of formation: $C_6H_6(l)$, 49.0 kJ/mol; $CO_2(g)$, −395.5 kJ/mol; $H_2O(l)$, −285.9 kJ/mol.

5.43 In some processes, the system is totally isolated from the surroundings. If such a process is exothermic, what do you think happens to the heat energy? What device commonly found in a house enables one to approximate these conditions?

5.44 In some processes, the temperature of the system does not change. In an exothermic process that occurs under such conditions, what happens to the energy released?

5.45 A small "coffee cup" calorimeter contains 110 g of H_2O at 22.0°C. A 100-g sample of lead is heated to 90.0°C and then placed in the water. The contents of the calorimeter come to a temperature of 23.9°C. What is the specific heat of lead?

Multiple-Choice Questions

5.46 Given the following information:

$$AgBr(s) + \tfrac{1}{2}Cl_2(g) \rightarrow AgCl(sl) + \tfrac{1}{2}Br_2(l)$$
$$\Delta H = -27.6 \text{ kJ/mol}$$

which of the following statements concerning the reaction are true: (1) heat is released; (2) heat is absorbed; (3) reaction is exothermic; (4) reaction is endothermic; (5) products have higher enthalpy content than reactants; (6) reactants have higher enthalpy content than products?

(a) 1, 3, and 5   (d) 1 and 3
(b) 2, 4, and 6   (e) 2, 4, and 5
(c) 1, 3, and 6

5.47 Given the chemical equation

$$S(s) + O_2(g) \rightarrow SO_2(g) \quad \Delta H = -297.1 \text{ kJ/mol}$$

how much heat is associated with the burning of 4.00 g of sulfur?

(a) −586 kJ   (d) −36.4 kJ
(b) 36.4 kJ   (e) −293 kJ
(c) 293 kJ

5.48 How much heat is required to convert 100 g of water at 40°C to water vapor at 100°C? The heat capacity of water is 4.184 J/g-°C, and the heat required to vaporize water is 2.26 kJ/g.

(a) 227 kJ   (d) 25.1 kJ
(b) 418 kJ   (e) 251 kJ
(c) 226 kJ

5.49 In a hydrogen bomb, the requisite high temperature for initiating the nuclear fusion comes from the following exothermic reaction, where ${}^{1}_{0}n$ represents a neutron:

$${}^{2}_{1}H + {}^{3}_{1}H \rightarrow {}^{4}_{2}He + {}^{1}_{0}n$$

During the above reaction, mass is converted to energy. For each 1 g of reactants (${}^{2}_{1}H$ and ${}^{3}_{1}H$), 0.0188 g of mass is converted to energy. Using Einstein's equation $E = mc^2$, calculate how much energy is released when 0.0188 g of mass is converted to energy. (The symbol c represents the speed of light; it has a value of 3.00×10^8 m/sec).

(a) 5.67×10^{11} J   (d) 1.70×10^{15} kJ
(b) 5.67×10^{12} J   (e) 1.70×10^{9} kJ
(c) 1.70×10^{12} J

5.50 If it takes 46 cal to vaporize 1 g of a substance at 106°C, then which of the following is true: (1) 46 cal will be required to vaporize 1 g of the substance at 53°C; (2) 46 cal will be required to change 1 g of the vapor to liquid at 106°C; (3) the temperature of the vapor must be higher than that of the liquid; (4) 46 cal will vaporize 1 g of the substance at 100°C; (5) 46 cal of heat will be given off when 1 g of the vapor condenses at 106°C?

(a) 1   (d) 4
(b) 2   (e) 5
(c) 3

5.51 If 1 g of kerosene liberates 46.0 kJ of heat when it is burned, to what temperature can 0.250 g of kerosene heat 75.00 cm³ of water at 25°C?

(a) 62.0°C   (d) 61.6°C
(b) 6.10°C   (e) 63.0°C
(c) 62.6°C

5.52 Given the following data:

$$2SO_2(g) + O_2(g) \rightarrow 2SO_3(g) \quad \Delta H = -196.7 \text{ kJ/mol}$$
$$SO_3(g) + H_2O(l) \rightarrow H_2SO_4(l) \quad \Delta H = -130.1 \text{ kJ/mol}$$

what is the heat of reaction for

$$2SO_2(g) + O_2(g) + 2H_2O(l) \rightarrow 2H_2SO_4(l)$$

(a) 66.6 kJ   (d) 456.9 kJ
(b) 326.7 kJ   (e) −456.9 kJ
(c) −326.7 kJ

5.53 Given the following:

$$C_2H_5OH(l) + 3O_2(g) \rightarrow 2CO_2(g) + 3H_2O(l)$$
$$\Delta H = -1366.9 \text{ kJ}$$
$$CH_3CO_2H(l) + 2O_2(g) \rightarrow 2CO_2(g) + 2H_2O(l)$$
$$\Delta H = -869.9 \text{ kJ}$$

what is the heat of reaction for

$$C_2H_5OH(l) + O_2(g) \rightarrow CH_3CO_2H(l) + H_2O(l)$$

(a) 497.0 kJ   (c) 2237 kJ
(b) −497.0 kJ   (d) −2237 kJ
(e) insufficient data provided to answer question

5.54 Which of the following is *not* in a standard state at 25°C and standard atmospheric pressure?

(a) $H_2O(l)$ (d) $F_2(g)$
(b) $O_2(g)$ (e) $H_2(g)$
(c) $Fe(l)$

5.55 Which of the following has a standard heat of formation of 0 kJ/mol at 25°C and standard atmospheric pressure?

(a) $H_2O(l)$ (d) $HCl(l)$
(b) $O_2(g)$ (e) $SiO_2(s)$
(c) $Fe(l)$

SELF-TEST SOLUTIONS

5.1 False. It has potential energy because the ball is stationary before it is thrown. **5.2** True. **5.3** False. It is an exothermic process because heat is released. An endothermic process involves the absorption of heat. **5.4** True. **5.5** True. **5.6** True. **5.7** False. Enthalpy is the heat change measured under constant pressure conditions, not under constant volume conditions as in a bomb calorimeter. **5.8** True. **5.9** True. **5.10** True. **5.11** True. **5.12** True. **5.13** True. **5.14** True. **5.15** True. **5.16** True. Over a small temperature range the assumption is valid. **5.17** False. The quantity of energy transferred also depends on the masses of the substances and their specific heats. **5.18** False. At 25°C, CO_2 is a gas. **5.19** True. **5.20** True. **5.21** False. The specific heat capacity is multiplied by 18 g/1 mol. Note that grams cancel and mole is now a unit in the denominator. **5.22** True. **5.23** True. **5.24** True. **5.25** False. A watt is 1 J/s. Thus, in one minute, 100-watt light bulb requires 100 J/s × 60 s or 6,000 J. **5.26** False. One Calorie (a big calorie) is 1000 calories. **5.27** True. **5.28** False. It does not depend on how the state of the system was achieved. **5.29** True. **5.30** True. **5.31** False. It is a positive number because heat is required to raise the temperature of a substance. **5.32** False. The unit of molar heat capacity is J/mol-g. Thus, to convert J/g-K to J/mol-K, you have to multiply by 18 g/1 mol. **5.33** True. **5.34** False. Fossil fuels include coal and natural gas. **5.35** True. **5.36**

False. Coal contains hydrocarbons of high molecular weight, otherwise, coal would be a liquid or gas if the molecular weights were low. **5.37** The appropriate arrangement of the three reactions to yield the desired reaction is

$CH_3COCH_3(l) + 4O_2(g) \rightarrow 3CO_2(g) + 3H_2O(l)$
$$\Delta H = -1787 \text{ kJ}$$

$H_2O(l) \rightarrow H_2O(g) \qquad\qquad\qquad \Delta H = -44.1 \text{ kJ}$

$2CO_2(g) + 2H_2O(l) \rightarrow CH_3COOH(l) + 2O_2(g)$
$$\Delta H = +835 \text{ kJ}$$

$CH_3COCH_3(l) + 2O_2(g) \rightarrow CO_2(g) + H_2O(g)$
$+ CH_3COOH(l) \qquad \Delta H = -966 \text{ kJ}$

The enthalpy of reaction is −996 kJ. **5.38** The heat of reaction for $PCl_5(g) + H_2O(g) \rightarrow POCl_3(g) + 2HCl(g)$ is related to the standard heats of formation of the products and reactants by the relation: $\Delta H^\circ_{rxn} = [\Delta H^\circ_f(POCl_3) + 2\Delta H^\circ_f(HCl)] - [\Delta H^\circ_f(PCl_5) - \Delta H^\circ_f(H_2O)]]$. Substituting the given data into this equation yields: $-136.0 \text{ kJ} = -592 \text{ kJ} + 2(-92.5 \text{ kJ}) - \Delta H^\circ_f(PCl_5) - (-241.8 \text{ kJ})$. Solving for $\Delta H^\circ_f(PCl_5)$ gives $\Delta H^\circ_f = -399 \text{ kJ}$. **5.39** The combustion of In in a calorimeter is described by the reaction $2In(s) + \frac{3}{2}O_2(s) \rightarrow In_2O_3(s)$. The heat produced by the combustion of 4.086 g of In is: heat = (energy equivalent) (temperature rise) = $(10.095 \text{ kJ}/^\circ C)(1.626^\circ C) = 16.41 \text{ kJ}$. The heat of combustion of 1 mol of In is the heat produced per gram of indium times the mass of a mol of indium:

$$\text{Heat per mole} = \left(\frac{16.41 \text{ kJ}}{4.086 \text{g}}\right)\left(\frac{114.82 \text{ g}}{1 \text{ mol}}\right)$$

$$= 461.1 \text{ kJ/mol}$$

5.40 The chemical reaction is described by $Cl_2(g) + 2HBr(g) \rightarrow 2HCl(g) + Br_2(l)$. The heat evolved per mole of Cl_2 is: $\Delta H^\circ_{rxn} = \Delta H^\circ_f(Br_2) + 2\Delta H^\circ_f(HCl) - \Delta H^\circ_f(Cl_2) - 2\Delta H^\circ_f(HBr) = 0 \text{ kJ} + 2(-92.30 \text{ kJ}) - 0 \text{ kJ} - 2(-36.20 \text{ kJ}) = -112.2 \text{ kJ}$. The heat evolved when 0.500 g of Cl_2 reacts is

$$\Delta H^\circ = (\Delta H^\circ_{rxn})(\text{moles } Cl_2)$$

$$\Delta H^\circ = \left(\frac{-112.2 \text{ kJ}}{1 \text{ mol}}\right)\left(\frac{0.500 \text{ g}}{70.90 \text{ g/mol}}\right) = 0.791 \text{ kJ}$$

5.41 We first need to calculate the heat evolved when 1 g of ^{235}U undergoes fission. The fission fuel value is given as 3.17×10^{-11} kJ/atom. The necessary equivalences that convert this number to kJ/g are: 6.022×10^{23} atoms = 1 mol ^{235}U; 1 mol ^{235}U = 235 g. This fission fuel value for ^{235}U, expressed in kJ/g, is

$$\left(3.17 \times 10^{-11} \frac{\text{kJ}}{\text{atom}}\right)\left(6.022 \times 10^{23} \frac{\text{atoms}}{\text{mol}}\right)$$

$$\times \left(\frac{1 \text{ mol}}{235 \text{ g}}\right) = 8.12 \times 10^{10} \text{ kJ/g}$$

The fission fuel value for 1 g is 8.12×10^{10} kJ. The mass of coal required to produce the same energy is calculated as follows: 8.12×10^{10} kJ = (fuel value of coal) (mass of coal) = (32 kJ/g) × (mass of coal). Solving for the mass of coal gives a mass of 2.5×10^9 g (which equals 5.5×10^6 lb). The high fuel value per gram for ^{235}U versus the low fuel value for coal is one reason for the great interest in using ^{235}U rather than coal to generate electricity. **5.42** The thermochemical equation for the combustion of benzene is $2C_6H_6(l) + 15O_2(g) \rightarrow 12CO_2(g) + 6H_2O(l)$. We can calculate the heat of combustion of C_6H_2 using the relation

$$\Delta H^\circ_{rxn} = [12\Delta H^\circ_f(CO_2) + 6\Delta H^\circ_f(H_2O)]$$
$$- [2\Delta H^\circ_f(C_6H_6) + 15\Delta H^\circ_f(O_2)]$$

$$\Delta H^\circ_{rxn} = \left[(12 \text{ mol})\left(-395.5 \frac{\text{kJ}}{\text{mol}}\right) + (6 \text{ mol})\left(-285.9 \frac{\text{kJ}}{\text{mol}}\right)\right]$$
$$- \left[(2 \text{ mol})\left(49.0 \frac{\text{kJ}}{\text{mol}}\right) + (15 \text{ mol})\left(0 \frac{\text{kJ}}{\text{mol}}\right)\right]$$

$$\Delta H^\circ_{rxn} = (-4746 \text{ kJ} - 1715 \text{ kJ}) - (98.0 \text{ kJ} + 0)$$

$$= -6559 \text{ kJ}$$

This is the heat of combustion for 2 mol of C_6H_6. For 1 mol of C_6H_6, the enthalpy is $-6559 \text{ kJ}/2 = -3280 \text{ kJ}$. **5.43** Since the heat energy released is contained in the system, the heat must be absorbed by the substances present in the system in such a way that their kinetic energy is increased. This increased kinetic motion is reflected by a temperature rise within the system. A closed thermos jug is a device that approximates this type of system. **5.44** For a process of this sort to occur, all the heat released during the exothermic change must be given off to the surroundings. Thus the temperature of the system remains constant, but the temperature of the surroundings increases.

5.45 $\qquad$ Heat lost = heat gained $\quad or$

$$q_{\text{lead}} = -q_{H_2O} \text{ where } q = mC\Delta T$$

$(100 \text{ g})(C)(90.0^\circ C - 23.9^\circ C) =$
$\qquad\qquad (110 \text{ g}) \times (4.184 \text{ J/g-}^\circ C)(23.9^\circ C - 22.0^\circ C)$

Solving for C gives $C = 0.134 \text{ J/g-}^\circ C$.

5.46 (c).

5.47 (d) $(4.00 \text{ g S})\left(\frac{1 \text{ mol S}}{32.06 \text{ g S}}\right)\left(\frac{-291.7 \text{ kJ}}{1 \text{ mol S}}\right) = 36.4 \text{ kJ}$

5.48 (e) Total heat = heat required to raise temperature of water + heat required to vaporize water

$$= (100 \text{ g})\left(4.184 \frac{\text{J}}{\text{g -}^\circ C}\right)\left(\frac{1 \text{ kJ}}{1000 \text{ J}}\right)(100^\circ C - 40^\circ C)$$

$$+ (100 \text{ g})\left(2.26 \frac{\text{kJ}}{\text{g}}\right) = 251 \text{ kJ}$$

5.49 (c) $E = mc^2$

$$= (0.0188 \text{ g}) \left(\frac{1 \text{ kg}}{1000 \text{ g}} \right) \left(3.00 \times 10^8 \frac{\text{m}}{\text{sec}} \right)^2$$

$$= (1.88 \times 10^{-5} \text{ kg}) \left(9.00 \times 10^{16} \frac{\text{m}^2}{\text{sec}^2} \right)$$

$$= 1.70 \times 10^{12} \frac{\text{kg} - \text{m}^2}{\text{sec}^2} = 1.70 \times 10^{12} \text{ J}$$

5.50 (e)

5.51 (d) Mass of 75.00 cm^3 of water = 75.00 g
Heat liberated by kerosene

$$= (0.250 \text{ g}) \left(46.0 \frac{\text{kJ}}{\text{g}} \right) \left(\frac{1000 \text{ J}}{1 \text{ kJ}} \right) = 11,500 \text{ J}$$

Heat flow = (mass of water) (C of water)(ΔT)

$$11,500 \text{ J} = (75.0 \text{ g}) \left(4.184 \frac{\text{J}}{\text{g-}^\circ\text{C}} \right) (T - 25^\circ\text{C})$$

$$T - 25^\circ\text{C} = 36.6^\circ\text{C} \qquad T = 61.6^\circ\text{C}$$

5.52 (e)

$$2SO_2(g) + O_2(g) \rightarrow 2SO_3(g) \qquad \Delta H = -196.7 \text{ kJ}$$
$$2SO_3(g) + 2H_2O(l) \rightarrow 2H_2SO_4(l) \qquad \Delta H = -260.2 \text{ kJ}$$

$$\overline{2SO_2(g) + O_2(g) + 2H_2O(l) \rightarrow 2H_2SO_4(l)}$$
$$\Delta H = -456.9 \text{ kJ}$$

5.53 (b)

$$C_2H_5OH(l) + 3O_2(g) \rightarrow 2CO_2(g) + 3H_2O(l)$$
$$\Delta H = -1366.9 \text{ kJ}$$
$$2CO_2(g) + 2H_2O(l) \rightarrow CH_3CO_2H(l) + 2O_2(g)$$
$$\Delta H = 866.9 \text{ kJ}$$

$$\overline{C_2H_5OH(l) + O_2(g) \rightarrow CH_3CO_2H(l) + H_2O(l)}$$
$$\Delta H = -497.0 \text{ kJ}$$

5.54 (c) Fe(s), not Fe(l), is the normal, most stable form of iron at this temperature and pressure. **5.55 (b)** For an element in its standard state, $\Delta H_f = 0$ kJ/mol.

Sectional Test 1

CHAPTERS 1–5

You should make this test as realistic as possible. **Tear out this test so that you do not have access to the information in the** *Student's Guide*. Use only data provided in the questions and a periodic table. Do not check your answers until you are finished. If you answer questions incorrectly, review the section in the text indicated after each question.

Choose the best response for each question.

1. Which prefix matches the fraction or multiple of a basic unit?
   (1) centi-: 10^{-2} (2) milli-: 10^{-3} (3) micro-: 10^{-6} (4) kilo-: 10^{3}
   (a) (1), (2) (b) (1), (2), (3) (c) (1), (3), (4) (d) all
   [1–3]

2. Which pair consists of equalities?
   (a) $-117°C$, $-179°F$ (b) $-117°C$, 390 K (c) $-32°C$, $0°F$
   (d) $0°F$, 243 K
   [1–3]

3. Which pair consists of equalities?
   (a) 6×10^{-2} mm, 6×10^{-5} m (b) 3×10^{-4} µg, 3×10^{-7} g
   (c) 3×10^{-3} mL, 3 L (d) 7×10^{-2} g/mL, 7×10^{-9} kg/L
   [1–3]

4. Which number contains exactly four significant figures?
   (a) 1240 (b) 124.0 (c) 0.0124 (d) 0.124
   [1–4]

5. How many liters will 1760 g of benzene occupy at 20°C if $d_{20°C}$ = 0.88 g/mL?
   (a) 0.25 L (b) 0.50 L (c) 1.0 L (d) 2.0 L
   [1–3, 1–5]

6. How many grams of sugar are contained in 1.00 L of a sugar solution that has a density of 1.15 g/mL and contains 52.0 percent sugar?
   (a) 1150 g (b) 598 g (c) 552 g (d) 115 g
   [1–3, 1–5]

7. What is the mass in grams of a 5.0 grain aspirin tablet, given the following Apothecaries' Weight System: 60 grains = 1 dram and 96 drams = 1 lb? (454 g = 1 lb.)
   (a) 42.6 g (b) 37.8 g (c) 1.39 g (d) 0.39 g
   [1–5]

8. What is the correct sum for the following measurements: 3.65×10^1 cm, 4.26×10^2 cm, 1.32×10^2 cm?

    (a) 9.23×10^1 cm **(b)** 9.23×10^2 cm **(c)** 5.95×10^2 cm
    (d) 5.945×10^2 cm
    [1–4]

9. How many significant figures should the result of the following calculation possess?

$$(2.65)(3.002)(38.26 - 1.2) =$$

    (a) 1 **(b)** 2 **(c)** 3 **(d)** 4
    [1–4]

10. Which of the following statements is true, given the following information?

    Density of A = 2.0 g/mL; Density of B = 3.5 g/mL; Density of C = 0.10 g/mL; All densities are at 20°C
    (a) 1 mL of A weighs more than 1 mL of B.
    (b) 10 g of C has a smaller volume than 10 g of B.
    (c) A, B, and C all have densities greater than water at 20°C.
    (d) One liter of B has a greater mass than 1 liter of C.
    [1–3]

11. Which of the following names of elements is *incorrectly* matched with its chemical symbol?

    (a) sodium–Na **(b)** phosphorus–P **(c)** magnesium–Mn **(d)** tin–Sn
    [2–6]

12. Which of the following is *incorrectly* named?

    (a) H_3PO_4–phosphoric acid **(b)** PO_3^{3-}–phosphate ion
    (c) HCl–hydrochloric acid **(d)** Cl^-–chloride ion
    [2–6]

13. Which of the following is *incorrectly* named?

    (a) CO–carbon monoxide **(b)** SO_2–sulfur dioxide
    (c) PCl_3–phosphorus trichloride **(d)** N_2O–nitrogen oxide
    [2–6]

14. Which of the following is *incorrectly* named?

    (a) $CaCl_2$–calcium chloride **(b)** Na_3PO_4–sodium phosphate
    (c) $KClO_4$–potassium chlorate **(d)** $Mg(NO_2)_2$–magnesium nitrite
    [2–6]

15. Which of the following is true about Dalton's Atomic Theory?

    (a) Atoms were viewed as indivisible.
    (b) It was the first statement on the atomic characteristics of matter.
    (c) It was immediately and widely accepted.
    (d) All of his postulates are still true based on today's information.
    [2–1]

16. Which of the following is the atomic mass of silver given the following distribution and masses of naturally occurring isotopes of silver: $^{107}_{47}Ag$: 51.88%, 106.906 amu; and $^{109}_{47}Ag$: 48.12%, 108.905 amu.

    (a) 107.868 amu **(b)** 106.926 amu **(c)** 108.235 amu
    (d) 108.642 amu
    [3–3]

17. Which of the following is *not* an empirical formula?

    (a) N_2O **(b)** N_2O_4 **(c)** NO_2 **(d)** HNO_3
    [2–5]

18. Which of the following is an ionic compound?

(a) CO_2 (b) PCl_3 (c) NO_2 (d) $BaCl_2$

[2–5]

19. Which of the following is true about $^{63}_{29}Cu$? It has

(a) 63 protons. (b) 29 electrons. (c) 29 neutrons. (d) 63 neutrons.

[2–3]

20. How many electrons does $^{24}_{12}Mg^{2+}$ possess?

(a) 10 (b) 12 (c) 14 (d) 24

[2–3]

21. Which of the following reactions represents a decomposition reaction?

(a) $P_4O_{10} + 6H_2O \rightarrow 4H_3PO_4$ (b) $2NaN_3 \rightarrow 2Na + 3N_2$
(c) $2H_2O + O_2 \rightarrow 2H_2O_2$ (d) $HCl + NaOH \rightarrow NaCl + H_2O$

[3–2]

22. Which of the following chemical equations is not balanced?

(a) $CaF_2 + H_2SO_4 \rightarrow 2HF + CaSO_4$
(b) $Ca_3(PO_4)_2 + 4H_3PO_4 \rightarrow 3Ca(H_2PO_4)_2$
(c) $NaNO_3 + H_2SO_4 \rightarrow Na_2SO_4 + 2HNO_3$
(d) $P_4O_6 + 6H_2O \rightarrow 4H_3PO_3$

[3–1]

23. What is the coefficient in front of MnO_2 when the following equation is balanced?

$$MnO_2 + Al \longrightarrow Mn + Al_2O_3$$

(a) 1 (b) 2 (c) 3 (d) 4

[3–1]

24. Which of the following contains 6.00×10^{16} atoms?

(a) 6.00×10^{16} H_2O molecules
(b) 3.00×10^{16} Cl_2 molecules
(c) 2.00×10^{16} P_4 molecules
(d) 1.50×10^6 $CaSO_4$ empirical units

[3–5]

25. How many atoms are there in 36.20 g of P_4 (mass of one mole P = 30.97 g)

(a) 1.746×10^{22} atoms (b) 1.746×10^{23} atoms
(c) 7.040×10^{22} atoms (d) 7.040×10^{23} atoms

[3–5]

26. What is the percentage of Ca in one mole of $Ca_3(PO_4)_2$? (Mass of one mole: Ca, 40.1 g; P, 31.0 g; O, 16.0 g)

(a) 9.9% (b) 20.0% (c) 12.9% (d) 38.8%

[3–5]

27. What is the empirical formula of caffeine if it contains 5.19% H, 28.85% N, 16.48% O, and 49.48% C by weight?

(a) $C_4H_5NO_2$ (b) $C_2H_6NO_2$ (c) C_3H_5NO (d) $C_4H_5N_2O$

[3–6]

28. How many milliliters of 1.00 *M* NaOH are required to *completely* neutralize 100.0 mL of 0.500 *M* H_3PO_4?

(a) 50.0 mL (b) 100.0 mL (c) 150.0 mL (d) 200.0 mL

[3–7, 4–7]

29. What is the molarity of a $NaNO_3$ solution if 25.0 mL of a 0.200 *M* $NaNO_3$ solution is diluted to 100.0 mL?

(a) 0.0500 M **(b)** 0.100 M **(c)** 0.150 M **(d)** 0.200 M
[4–1]

30. What is the mass of Cu produced if 10.0 g of Cu_2S reacts with 16.0 g of O_2 as follows: (Mass of one mole: Cu, 63.5 g; O, 16.0 g; S, 32.1 g.)

$$Cu_2S + O_2 \longrightarrow 2Cu + SO_2$$

(a) 31.8 g **(b)** 63.5 g **(c)** 3.99 g **(d)** 7.98 g
[3–7, 3–8]

31. What is the percent yield if 25.0 g of I_2 is formed when 130.0 g of HNO_3 reacts with 285.0 g of HI as follows: (Mass of one mole: H, 1.01 g; N, 14.0 g; O, 16.0 g; I, 126.9 g.)

$$2HNO_3 + 6HI \longrightarrow 2NO + 3I_2 + 4H_2O$$

(a) 92 g **(b)** 142 g **(c)** 283 g **(d)** 566 g
[3–7, 3–8]

32. How many moles of $KClO_3$ are required to produce 2.51 g of O_2? (Mass of one mole: K, 39.1 g; Cl, 35.5 g; O, 16.0 g.)

$$2KClO_3 \longrightarrow 2KCl + 3O_2$$

(a) 1.18×10^{-1} mol **(b)** 5.23×10^{-2} mol **(c)** 2.35×10^{-1} mol
(d) 1.57×10^{-2} mol
[3–7, 3–8]

33. What is the molarity of a solution that contains 32.0 g of HCl (molar mass = 36.46 g) in 2.50 L of solution?
(a) 0.00351 M **(b)** 0.0128 M **(c)** 0.351 M **(d)** 12.8 M
[4–1]

34. How many moles of H_2SO_4 are present in 250 mL of a 3.00 M H_2SO_4 solution?
(a) 0.150 moles **(b)** 0.750 moles **(c)** 1.50 moles **(d)** 750 moles
[4–1]

For questions 35 through 42, identify each substance as a **(a)** strong electrolyte, **(b)** weak electrolyte, or **(c)** nonelectrolyte in water.

35. HCl   **36.** HF   **37.** HCN   **38.** NH_3
39. Sugar   **40.** I_2 **41.** $HClO_4$ **42** $MgCl_2$

43. Which of the following is *not* a strong base?
(a) NaOH **(b)** $Ca(OH)_2$ **(c)** LiOH **(d)** NH_3
[4–3]

44. When HBr is neutralized with KOH, what salt is produced?
(a) H_2O **(b)** KH **(c)** KBr **(d)** K_2O
[4–3]

45. When the following is completed and balanced,

$$Al(OH)_3(s) + H_2SO_4(aq)$$

the products of the reaction are
(a) $AlSO_4(aq) + 6H_2O(l)$ **(b)** $Al_2(SO_4)_3(aq) + 6H_2O(l)$
(c) $AlSO_4(s) + H_2O(l)$ **(d)** $Al_2(SO_4)_3(aq) + H_2O(l)$
[4–3]

46. What are the spectator ions in the reaction

$$BaCl_2(aq) + Na_2SO_4(aq) \longrightarrow 2NaCl(aq) + BaSO_4(s)$$

(a) Ba^{2+}, Cl^- (b) Na^+, Cl^- (c) Ba^{2+}, SO_4^{2-} (d) Na^+, SO_4^{2-}
[4–4]

47. Which of the following is the net ionic equation for the reaction

$$Fe(NO_3)_3(aq) + K_3PO_4(aq) \longrightarrow FePO_4(s) + 3KNO_3(aq)$$

(a) $K^+(aq) + NO_3^-(aq) \to KNO_3(aq)$
(b) $Fe^{3+}(aq) + PO_4^{3-}(aq) \to FePO_4(aq)$
(c) $3K^+(aq) + PO_4^{3-}(aq) \to K_3PO_4(aq)$
(d) $Fe^{3+}(aq) + PO_4^{3-}(aq) \to FePO_4(s)$
[4–4, 4–5]

For questions 48 through 53, identify each species as (a) soluble or (b) insoluble in water. [4–4]

48. Na_3PO_4 **49.** $AgBr$ **50.** NH_4NO_3 **51.** $Ni(OH)_2$
52. $CaCO_3$ **53.** $BaCl_2$
54. What is the net ionic equation for the reaction between $Cu(NO_3)_2(aq)$ and $NaOH(aq)$?
(a) no reaction (b) $Na^+(aq) + NO_3^-(aq) \to NaNO_3(s)$
(c) $Cu^{2+}(aq) + 2OH^-(aq) \to Cu(OH)_2(s)$
(d) $Cu^{2+}(aq) + 2NO_3^-(aq) \to Cu(NO_3)_2(s)$
[13–1]
55. What is the net ionic equation for the reaction between $Pb(NO_3)_2(aq)$ and $CaCl_2(aq)$?
(a) no reaction (b) $Pb^{2+}(aq) + 2Cl^-(aq) \to PbCl_2(s)$
(c) $Pb^{2+}(aq) + 2NO_3^-(aq) \to Pb(NO_3)_2(s)$
(d) $Ca^{2+}(aq) + 2NO_3^-(aq) \to Ca(NO_3)_2(s)$
[4–5]
56. Which of the following is most easily oxidized?
(a) Na (b) Cl_2 (c) Au (d) S
[4–6]
57. When a strong acid reacts with a metal, which of the following is formed?
(a) $H_2(g)$ (b) $H_2O(l)$ (c) an element of the acid
(d) another acid
[4–6]
58. Magnesium is more active than cobalt and hydrogen lies below both of them in the activity series. This means that
(a) cobalt is the most easily oxidized.
(b) magnesium is the most easily oxidized.
(c) neither metal reacts with acids.
80 water is produced when the metals react with hydrogen ion.
[4–6]
59. Copper does not react with hydrochloric acid whereas manganese does. This means that
(a) copper is more active than hydrogen

(b) manganese is less active than hydrogen

(c) chloride ion will react with copper.

(d) manganese is higher on the activity series than copper

[4–6]

60. Which of the following metals *cannot* displace hydrogen from water (steam or liquid)?

(a) Mg **(b)** Ba **(c)** Li **(d)** Ag

[4–6]

61. Use the activity series in your text to predict which of the following reactions will occur:

(a) $Cu(s) + 2AgNO_3(aq) \rightarrow Cu(NO_3)_2(aq) + 2Ag(s)$

(b) $3Fe(s) + Al_2(SO_4)_3(aq) \rightarrow 2Al + 3FeSO_4(aq)$

(c) $H_2(g) + LiOH(aq) \rightarrow 2Li(s) + 2H_2O(l)$

(d) $Cu(s) + ZnSO_4(aq) \rightarrow Zn(s) + CuSO_4(aq)$

[4–6]

62. Which of the following will *always* decrease the internal energy of a gaseous system?

(a) an endothermic process and a corresponding decrease in volume

(b) an endothermic process and a corresponding increase in volume

(c) an exothermic process and a corresponding increase in volume

(d) an exothermic process and a corresponding decrease in volume

[5–2]

63. What is ΔH_f° for one mole of $C_2H_5OH(l)$ given the following data at 25°C:

$$C_2H_5OH(l) + 3O_2(g) \rightarrow 2CO_2(g) + 3H_2O(l)$$

$$\Delta H_{rxn}^\circ = -1366.0 \text{ kJ}; \quad \Delta H_f^\circ(CO_2(g)) = -393.5 \text{ kJ/mol};$$

$$\Delta H_f^\circ(H_2O(l)) = -285.8 \text{ kJ/mol}$$

(a) +362.6 kJ **(b)** +277.8 kJ **(c)** −362.6 kJ **(d)** −278.4 kJ

[5–6]

64. What is the heat of hydrogenation of acetylene, at 25°C and 1 atm,

$$C_2H_2(g) + 2H_2(g) \longrightarrow C_2H_6(g)$$

given the following thermochemical equations:

$$2C_2H_6(g) + 7O_2(g) \longrightarrow 4CO_2(g) + 6H_2O(l) \quad \Delta H = -3123 \text{ kJ}$$

$$2C_2H_2(g) + 5O_2(g) \longrightarrow 4CO_2(g) + 2H_2O(l) \quad \Delta H = -2602 \text{ kJ}$$

$$H_2(g) + 1/2O_2(g) \longrightarrow H_2O(l) \quad \Delta H = -286 \text{ kJ}$$

(a) −312 kJ **(b)** +312 kJ **(c)** +1613 kJ **(d)** −1613 kJ

[5–5]

65. What is the heat of combustion for one mole of benzene at 25°C and 1 atm,

$$2C_6H_6(l) + 15O_2(g) \longrightarrow 12CO_2(g) + 6H_2O(l)$$

given the following data: $\Delta H_f^\circ(CO_2(g)) = -394$ kJ/mol; $\Delta H_f^\circ(C_6H_6(l)) = 49$ kJ/mol; $\Delta H_f^\circ(H_2O(l)) = -286$ kJ/mol.

(a) +3271 kJ **(b)** −3271 kJ **(c)** +1636 kJ **(d)** −1636 kJ

[5–6]

66. What is ΔE for a system when it does 230 kJ of work on its surroundings and 130 kJ of heat is removed from the system?

   (a) +100 kJ (b) −100 kJ (c) +360 kJ (d) −360 kJ

   [5–3]

67. When is ΔH approximately equal to ΔE?

   (a) at constant pressure (b) $P\Delta V$ is large (c) $P\Delta V$ is small

   (d) pressure is large

   [5–3]

68. What is the final temperature of 30 g of Al if 2000 J of heat is added to a sample at 25.0°C? (specific heat of Al = 0.902 J/K-g)

   (a) 32.6°C (b) 98.9°C (c) 120.2°C (d) 2192°C

   [5–7]

69. What is the specific heat of copper if 90.0 g of Cu at 39.1°C is placed in water at 30.0°C, the calorimeter gains 263 J, and the final temperature of the mixture is 31.5°C?

   (a) 0.126 J/K-g (b) 0.385 J/K-g (c) 10.2 J/K-g (d) 34.65 J/K-g

   [5–7]

70. What is the final temperature in degrees centigrade of 100.0 g of water at 30.0°C if it is mixed with 50.0 g of water at 0.0°C?

   (a) 40.0 (b) 20.0 (c) 15.0 (d) 10.0

   [5–7]

ANSWERS

1. (d). **2.** (a). **3.** (a). **4.** (b). **5.** (d). **6.** (b). **7.** (d). **8.** (c). **9.** (c).
10. (d). **11.** (c). **12.** (b). **13.** (d). **14.** (c). **15.** (a). **16.** (a). **17.** (b). **18.** (d).
19. (b). **20.** (a). **21.** (b). **22.** (c). **23.** (c). **24.** (b). **25.** (b). **26.** (d). **27.** (d).
28. (c). **29.** (a). **30.** (d). **31.** (c). **32.** (b). **33.** (c). **34.** (b). **35.** (a). **36.** (b).
37. (b). **38.** (b). **39.** (c). **40.** (c). **41.** (a). **42.** (a). **43.** (d). **44.** (c). **45.** (b).
46. (b). **47.** (d). **48.** (a). **49.** (b). **50.** (a). **51.** (b). **52.** (b). **53.** (a). **54.** (c).
55. (b). **56.** (a). **57.** (a). **58.** (b). **59.** (d). **60.** (d). **61.** (a). **62.** (c). **63.** (d).
64. (a). **65.** (b). **66.** (d). **67.** (c). **68.** (b). **69.** (b). **70.** (b).

Electronic Structures of Atoms | 6

OVERVIEW OF THE CHAPTER

Objectives: You should be able to:

1. Describe the wave properties and characteristic speed of propagation of radiant energy (electromagnetic radiation).

2. Use the relationship $\lambda v = c$, which relates the wavelength (λ) and the frequency (v) of radiant energy to its speed (c).

Objectives: You should be able to:

1. Explain the essential feature of Planck's quantum theory, namely, that the smallest increment, or quantum, of radiant energy of frequency, v, that can be emitted or absorbed is hv, where h is Planck's constant.

2. Explain how Einstein accounted for the photoelectric effect by considering the radiant energy to be a stream of particle-like photons striking a metal surface. In other words, you should be able to explain all the observations about the photoelectric effect using Einstein's model.

Objectives: You should be able to:

1. Explain the origin of the expression *line spectra*.

2. List the assumptions made by Bohr in his model of the hydrogen atom.

3. Explain the concept of an allowed energy state and how this concept is related to the quantum theory.

4. Calculate the energy differences between any two allowed energy states of the electron in hydrogen.

5. Explain the concept of ionization energy.

6.4, 6.5, 6.6
PRINCIPLES OF
MODERN QUANTUM
THEORY

Objectives: You should be able to:

1. Calculate the characteristic wavelength of a particle from a knowledge of its mass and velocity.

2. Describe the uncertainty principle and explain the limitation it places on our ability to define simultaneously the location and momentum of a sub-atomic particle, particularly an electron.

3. Explain the concepts of orbital, electron density, and probability as used in the quantum-mechanical model of the atom: explain the physical significance of Ψ^2.

4. Describe the quantum numbers, n, l, m_l used to define an orbital in an atom and list the limitations placed on the values each may have.

5. Describe the shapes of the s, p, and d orbitals.

6.7 ENERGIES OF ORBITALS IN MANY-ELECTRON ATOMS

Objectives: You should be able to:

1. Explain why electrons with the same value of principal quantum number (n) but different values of the azimuthal quantum number (l) possess different energies.

6.8 ELECTRONIC STRUCTURE OF MANY-ELECTRON ATOMS

Objectives: You should be able to:

1. Explain the concepts of electron spin and the electron spin quantum number.

2. State the Pauli exclusion principle and Hund's rule, and illustrate how they are used in writing the electronic structures of the elements.

3. Write the electron configuration for any element.

4. Write the orbital diagram representation for electron configurations of atoms.

6.9 THE PERIODIC TABLE: PERIODIC ARRANGEMENT OF ELECTRON CONFIGURATIONS AND VALENCE ELECTRONS

Objectives: You should be able to:

1. Describe what we mean by the s, p, d, and f blocks of elements.

2. Write the electron configuration and valence electron configuration for any element once you know its place in the periodic table.

TOPIC SUMMARIES AND EXERCISES

ELECTROMAGNETIC
RADIATION

When you sense the warmth of a fire in a fireplace, you are feeling what scientists call **radiation,** or **electromagnetic radiation.** The fire gives off light (visible radiation) and heat (thermal radiation). Both of these types of electromagnetic radiation exist in the form of **electromagnetic waves.**

▪ Note in Figure 6.2 in the text the following characteristics of electromagnetic waves: wavelength (λ) and amplitude.

▪ Electromagnetic waves move through a vacuum at 3×10^8 m/s, the "speed of light."

▪ Electromagnetic waves differ from one another by their **frequency** (v): The number of cycles per second (1 herz = cycle/s).

Chemists often characterize electromagnetic radiation by its **wavelength, λ.**

▪ λ measures the distance between two adjacent maxima (peaks) in a periodic wave.

▪ Remember this key relationship between wavelength, λ, and frequency:

$$v = \frac{c}{\lambda}$$

where c is the speed of light, 3.00×10^8 m/s.

▪ Remember these wavelength units:

Unit	Symbol	Length
Nanometer	nm	1×10^{-9} m
Micrometer (micron)	μ	1×10^{-6} m
Angstrom	Å	1×10^{-10} m
Picometer	pm	1×10^{-12} m

EXERCISE 1   Figure 6.1 shows two periodic waves, 1 and 2. **(a)** What do the distances a and b correspond to? **(b)** Which electromagnetic radiation, 1 or 2, has the greater frequency?

Solution: **(a)** The distance a corresponds to the wavelength for wave 1, and similarly, the distance b corresponds to the wavelength for 2. **(b)** Frequency is defined as the number of times per second that maxima of a periodic wave pass a given point. We can see by inspection of Figure 6.1 that the adjacent maxima of wave 2 are closer together than those of wave 1. Because of this closer proximity of adjacent maxima, wave 2 will have more maxima passing a given point in one second than will wave 1. Thus wave 2 has the higher frequency.

EXERCISE 2   The wavelength of a periodic wave is often reported in a variety of units based on fractions of a meter. Complete the following table by converting a 2×10^{-10} m wavelength into the other indicated units and by writing their unit symbols.

	Meter	Nanometer	Micron	Millimicron
Symbol	m			
Length	2×10^{-10}			

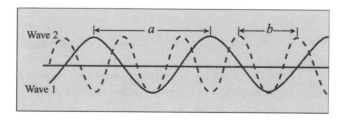

Figure 6.1   Two periodic waves.

	Meter	**Nanometer**	**Micron**	**Millimicron**
Symbol	m	nm	μ	mμ
Length	2×10^{-10}	2×10^{-1}	2×10^{-4}	2×10^{-1}

The unit conversions are made as follows: A nanometer is equal to 1×10^{-9} meter; therefore, the conversion from meter to nanometer is

$$\text{Nanometers} = (2 \times 10^{-10} \text{ m})\left(\frac{1 \text{ nm}}{1 \times 10^{-9} \text{ m}}\right) = 2 \times 10^{-1} \text{ nm}$$

A micron is equal to 1×10^{-6} meter; therefore, the conversion from meter to micron is

$$\text{Micron} = (2 \times 10^{-10} \text{ m})\left(\frac{1 \text{ μ}}{1 \times 10^{-6} \text{ m}}\right) = 2 \times 10^{-4} \text{ μ}$$

A millimicron ($10^{-3} \times 10^{-6}$ m) is equivalent to a nanometer.

EXERCISE 3   Calculate the wavelength of electromagnetic radiation that has a frequency of 9.22×10^{17}/s.
   Solution: The relation between the frequency and wavelength of electromagnetic radiation is

$$\nu = \frac{c}{\lambda}$$

where the value for ν is in units of s^{-1}. We want to calculate the wavelength, λ; thus it is convenient to rearrange the equation so the term λ is on the left-hand side of the equation. Substituting the known values of c and ν:

$$\lambda = \frac{c}{\nu} = \frac{3.00 \times 10^{8} \text{ m/s}}{9.22 \times 10^{17} \text{/s}} = 3.25 \times 10^{-10} \text{ m}$$

QUANTIZATION OF ENERGY

During the nineteenth century, theories of classical physics were unable to explain certain physical phenomena. For example, the color of an extremely hot iron bar was predicted to be blue; yet it is red. In 1900, Max Planck revised classical ideas by proposing that light, which before then had been thought as containing a continuous collection of electromagnetic waves, consisted, in fact, of bundles of energy.

- Each bundle of energy has a discrete (individually distinct or separate) value.
- The smallest allowed increment of energy gained or lost is called a **quantum.**
- *Planck's relationship* between the smallest amount of energy gained or lost and the frequency of the associated radiation is

$\Delta E = h\nu$
$h = 6.63 \times 10^{-34}$ J–s and is called Planck's constant.

Acceptance of Planck's quantum theory was slow until Albert Einstein used it successfully in 1905 to explain the **photoelectric effect.** This effect occurs when a light shines on the surface of a clean metal and electrons are ejected from the surface.

- Einstein proposed that a beam of light consists of a collection of small particles of energy, called **photons.**

- A photon can interact with an electron on a metal surface and transfer its energy to the electron.

- An electron is ejected if it gains sufficient energy to overcome the forces binding the electron to the surface. Energy beyond the binding energy appears as the kinetic energy of the emitted electron

$$E_k = v - E_b$$

Kinetic energy   photon   binding
of electron   energy   energy

EXERCISE 4   (a) Does the energy of a photon increase or decrease with increasing wavelength? (b) In Exercise 1, which electromagnetic wave, 1 or 2, has the greater energy?

    Solution: (a) The relation between energy and wavelength is derived from the following two equations:

$$E = hv \qquad v = \frac{c}{\lambda}$$

Substituting c/λ for v in Planck's equation yields

$$E = hv = h\frac{c}{\lambda}$$

Since the energy of a photon is inversely proportional to its wavelength, its energy decreases with increasing wavelength. (b) In Exercise 1, we concluded that electromagnetic wave 2 has the greater frequency. According to Planck's equation, the energy of a wave is directly proportional to frequency; therefore, electromagnetic wave 2 has the greater energy.

EXERCISE 5   Calculate the energy of an X-ray photon with a wavelength of 3.00×10^{-10} m.

    Solution: The energy of the X-ray photon is calculated by using Planck's equation and substituting the known values of h, c, and λ:

$$E = \frac{hc}{\lambda} = \frac{(6.63 \times 10^{-34}\,\text{J-sec})(3.00 \times 10^{8}\,\text{m/sec})}{3 \times 10^{-10}\,\text{m}} = 6.63 \times 10^{-16}\,\text{J}$$

In the 1800s, Josef Fräunhofer and others had shown that light emitted from the sun, or from hydrogen atoms heated in a partial vacuum, was not a continuum.

LINE SPECTRA AND THE BOHR MODEL

- A continuum is a collection of electromagnetic waves of all possible wavelengths in a given energy region.

- A prism separates light emitted from the sun, or from heated atoms, into discrete and noncontinuous wavelength components.

- A collection of discrete lines or wavelengths on a photographic plate is called a **line spectrum.** See Figure 6.12 in your text.

- Each gaseous monatomic element has its own characteristic line spectra.

In 1914, Niels Bohr proposed a model for the hydrogen atom that enabled scientists to explain hydrogen line spectra. Key ideas in **Bohr's model** of the hydrogen atom are:

- An electron moves in a circular path, called an **orbit.**
- Only orbits restricted to certain energies and radii are permitted.
- Each allowed orbit is assigned an integer n, known as the **principal quantum number:**

$$n = \underbrace{1,\ 2,\ 3,\ 4,\ \ldots,\ \infty}_{\text{range of values of } n}$$

- The energy of an orbit is quantized:

$$E_n = -R_H \left(\frac{1}{n^2} \right) \qquad n = \text{principal quantum number of an orbit}$$

$$R_H = \text{Rydberg constant} = 2.18 \times 10^{-18}\ \text{J}$$

- Note that as the principal quantum number increases, the energy of an orbit increases (becomes a smaller negative number).
- The orbit of lowest energy, $n = 1$, is called the **ground state.** All others are termed excited states.

Emission or absorption of radiant energy from a hydrogen atom occurs when an electron moves from one orbit to another.

$$\Delta E = h\nu \qquad\qquad \text{Emission: electron radius decreases: } n_i > n_f$$

$$\Delta E = R_H \left(\frac{1}{n_i^2} - \frac{1}{n_f^2} \right) \qquad \text{Absorption: electron radius increases: } n_f > n_i$$

energy difference between two orbits   starting (initial) orbit of electron   final orbit of electron

EXERCISE 6   What are the postulates of Niels Bohr's model for the hydrogen atom?

Solution: The postulates are: **(a)** An electron moves in a circular path about the nucleus and does not collapse into the nucleus. This movement about the nucleus is characterized by the electron's orbit. **(b)** The energy of an electron can have only certain allowed values; its energy is quantized. **(c)** An electron can move from one circular path to another circular path only when it absorbs or emits a photon corresponding to the exact energy difference between the two quantized energy states.

EXERCISE 7   What is the meaning of the negative sign in the equation $E_n = -R_H(1/n^2)$? What is an excited state? When an electron is ionized, what is its final principal quantum number?

Solution: The negative sign means that energy is required to remove the electron from the nucleus—that is, when the electron and nucleus are combined, energy is released. An alternative explanation is that the electron and nucleus in the atom are more stable than the separated electron and nucleus. Any orbit of the hydrogen atom with $n \geq$

2 is considered a higher energy orbit and is said to be an "excited state." Ionization corresponds to a transition to a final state of $n = \infty$.

EXERCISE 8 An electron moves from an $n = 1$ Bohr hydrogen orbit to an $n = 2$ Bohr hydrogen orbit. **(a)** What is the energy associated with this transition? **(b)** Is this electronic transition accompanied by the emission or absorption of energy?

Solution: **(a)** The energy transition associated with an electron moving from one Bohr orbit to another is given by the relation

$$\Delta E = R_H \left(\frac{1}{n_i^2} - \frac{1}{n_f^2} \right)$$

where n_i is the principal quantum number of the initial orbit, n_f is the principal quantum number of the final orbit, and R_H is the Rydberg constant, 2.18×10^{-18} J. In order to solve for the energy of electronic transition, substitute into the previous equation the values $n_i = 1$ (the initial orbit) and $n_f = 2$ (the final orbit):

$$\Delta E = R_H \left(\frac{1}{n_i^2} - \frac{1}{n_f^2} \right) = (2.18 \times 10^{-18} \text{ J}) \left(\frac{1}{1^2} - \frac{1}{2^2} \right)$$

$$= (2.18 \times 10^{-18} \text{ J}) \left(1 - \frac{1}{4} \right) = (2.18 \times 10^{-18} \text{ J}) \left(\frac{3}{4} \right)$$

$$= 1.63 \times 10^{-18} \text{ J}$$

(b) The sign of the value for ΔE is positive; thus, the transition requires energy. For this transition to occur, a photon with an energy of at least 1.63×10^{-18} J would have to interact with the electron.

As discussed in the previous topic summary sections, electromagnetic radiation has both wave and particle aspects. In 1924, Louis de Broglie postulated that all material particles have wave properties.

PRINCIPLES OF MODERN QUANTUM THEORY

- The result of his conjecture was the **de Broglie relation:**

$$\lambda = \frac{h}{p} = \frac{h}{mv}$$

where λ is the wavelength of the matter wave, h is Planck's constant, p is the momentum of the moving particle (which equals $m \times v$) m is the mass of the particle, and v is the velocity of the particle.
- Note that for large particles, such as a ball, m is very large; therefore λ is extremely small and not measurable. For small particles, such as an electron, λ is larger and measurable.

Application of wave-equation principles, the de Broglie postulate, and the uncertainty principle to the properties of an electron moving about a hydrogen nucleus eventually led to the derivation of mathematical functions (given the symbol ψ) that describe the wave behavior of electrons.

Table 6.1   Relationship among quantum numbers n, l and m$_l$

Quantum Number	Name	Dependence on Other Quantum Numbers	Range of Values
n	Principal	None	$1, 2, 3, \ldots, \infty$
l	Azimuthal	Integral values that depend on the value of n. For each n quantum number, there are n number of l values.	$0, 1, 2, \ldots, n - 1$ for each n value
m_l	Magnetic	Integral values that depend on the value of l. For each l quantum number there are $2l + 1$ possible m_l values.	$l, l - 1, l - 2, \ldots 0, \ldots,$ $-(l - 1), -l$ for each l value

- See Figure 6.19 in the text. Large values of ψ^2 represent a high probability of finding an electron at a particular point in space. This density of dots is referred to as an electron-density distribution.

- An **orbital** is a one-electron wave function and has an associated allowed energy state. An orbital also is represented as an electron-density distribution in space.

Each hydrogen-like orbital is described by three characteristic **quantum numbers:** n, l, m_l.

- All three are required to completely characterize a hydrogen orbital. Their names and their relationships to one another are summarized in Table 6.1.
- A collection of orbitals having the same value of n is called an **electron shell.**
- A **subshell** consists of all orbitals with the same n and l values.

A shorthand notation has been devised to describe each hydrogen orbital in terms of its n and l quantum numbers.

- In this notation system, the principle quantum number, n, appears in front of the l value for the orbital, where the value of l is designated by a letter:

Quantum number l	0	1	2	3
Designation of orbital	s	p	d	f

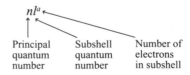

Thus, the shorthand notation for a hydrogen orbital with quantum numbers $n = 2$ and $l = 1$ is 2p, where 2 is the n quantum number, which defines the energy level, and p is the designation for the value of the l quantum number. Note that the m_l value of an orbital is not indicated.

- Note that all orbitals with the same n and l values, such as the $2p$ orbitals, form a subshell.

- Carefully examine Table 6.2 in the text. Learn the subshells for each n level.

- Remember that each n level contains n^2 orbitals; each l subshell level contains $2l + 1$ orbitals. For example, the $n = 2$ principle quantum level contains $2^2 = 4$ orbitals: one $2s$ orbital ($2l + 1 = 2 (0) + 1 = 1$ orbital) and three $2p$ orbitals ($2l + 1 = 2 (1) + 1 = 3$ orbitals).

Graphical representations of the $1s$, $2s$, $2p$ and $3d$ orbitals provide you with an understanding of the properties of orbitals.

- You need to be familiar with the graphical representations of the following orbitals: $1s$, $2s$, $2p_x$, $2p_y$, $2p_z$, $3d_z^2$, $3d_{x^2-y^2}$, $3d_{xy}$, $3d_{xz}$, $3d_{yz}$.

- Be able to reproduce Figures 6.20, 6.22, and 6.23 in your text. Note that the letter subscripts do not necessarily equal a given m_l value. They are only an aid in helping you know an orbital's orientation in space.

- Graphical representations of $1s$, $2s$, $2p$, and other orbitals provide you with an understanding of the properties of orbitals. In graphing wave functions, particularly ψ^2, it is important to remember that you cannot accurately and simultaneously measure or describe the position of an electron and its direction of movement (a component of velocity) as stated by Heisenberg's **uncertainty principle.** Since you cannot measure simultaneously all properties of an electron, we can only talk about its probability of being at a particular point in space. A common graphical representation is a 90-percent boundary plot, such as the one shown in Figure 6.2.

EXERCISE 9   Each wave function (ψ) describing an orbital has three quantum numbers, n, l, and m_l, associated with it. What do these quantum numbers determine in terms of orbital properties?

Solution: The principal quantum number, n, defines the size of an orbital and its energy. As the value of n increases, so does the energy of the orbital with respect to the ground-state energy. That is, the energy of an orbital becomes a smaller negative quantity as the value of n increases. The shape of an orbital is related to the azimuthal quantum number, l. For a given l value for an orbital, the magnetic quantum number, m_l, determines the orientation of the orbital in space with respect to an axis in the presence of a magnetic field.

EXERCISE 10   How many hydrogen orbitals exist with the same $n = 3$ value but with differing l and m_l values?

Solution: For $n = 3$, there exist hydrogen orbitals with azimuthal quantum numbers $l = 0, 1, \ldots, n - 1$. In this case the values are 0, 1, and 2. For each l value, there exist hydrogen orbitals with the following range of m_l values: $l, l - 1, \ldots 0, \ldots, -(l - 1), -l$. These hydrogen orbitals are tabulated as follows:

l	m_l	Number of orbitals
0	0	1
1	1, 0, −1	3
2	2, 1, 0, −1, −2	5

The total number of hydrogen orbitals that can have an $n = 3$ value is therefore 9.

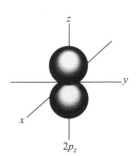

Figure 6.2   A 90-percent boundary plot for a $2p_z$ orbital. The spherical surfaces enclose 90 percent of the total ψ^2 for the orbital. Note that there is no electron density in the xy plane; such a surface is called a *node.*

EXERCISE 11 Write the shorthand notation (for example, $2s$) for the orbitals described by the following quantum numbers: (a) $n = 2$, $l = 1$, $m_l = 0$; (b) $n = 3$, $l = 2$, $m_l = 1$; (c) $n = 4$, $l = 3$, $m_l = 2$.
   Solution: For each case, determine the letter designation for the given l value. Remember that in the shorthand notation system the value of m_l is not indicated. (a) $l = 1$ corresponds to a p orbital; thus the orbital designation is $2p$. (b) $l = 2$ corresponds to a d orbital; thus the orbital designation is $3d$. (c) $l = 3$ corresponds to an f orbital; thus the orbital designation is $4f$.

EXERCISE 12 Why is it not possible for a hydrogen orbital to have the quantum numbers $n = 3$, $l = 2$, $m_l = 3$ associated with it?
   Solution: For a given l value, only certain m_l values are allowed. The maximum m_l value for $l = 2$ is 2. Thus, $m_l = 3$ is not possible for a hydrogen orbital with an $l = 2$ value.

EXERCISE 13 How many hydrogen orbitals are possible for the quantum state designated by the following shorthand notations: (a) $2s$; (b) $3d$?
   Solution: (a) The letter s corresponds to an orbital with $l = 0$. For $l = 0$, there is only one possible orbital with $m_l = 0$. (b) The letter d corresponds to an orbital with $l = 2$. For $l = 2$, there are five orbitals with five different orientations in space determined by the m_l values, which equal 2, 1, 0, −1, and −2.

EXERCISE 14 (a) What is a nodal surface? (b) What quantum number determines the number of nodal surfaces?
   Solution: (a) A nodal surface is a region in space where ψ^2 goes to zero. (b) The number of nodal surfaces, or nodes, depends on the principal quantum number, n. For an orbital with a given n value, there are $n - 1$ nodal surfaces. Thus a $2p$ orbital has one node; a $3d$ orbital has two nodes; and a $1s$ orbital has no nodes. For example, look at Figure 6.20 in your text and Figure 6.2 in the *Student's Guide*. The xy plane is a nodal surface for a $2p_z$ orbital.

EXERCISE 15 Label the following orbitals as to type:

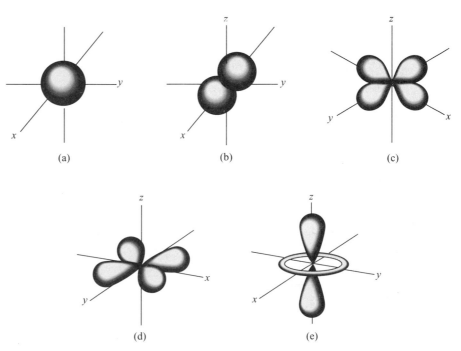

(a)   (b)   (c)

(d)   (e)

   Solution: (a) s; (b) p_x; (c) $d_{x^2-y^2}$ (d) d_{xy}; (e) d_z^2. (See Figures 6.20, 6.22, and 6.23 in the text.)

The electronic structure of the hydrogen atom forms the basis of electronic structure for atoms with two or more electrons.

- Orbitals are described like those for hydrogen, and have similar shapes.
- You can continue to use the orbital designations such as $1s$, $2s$, $2p_x$, $2p_y$, and $3d_z^2$.

However, energies of orbitals in many-electron atoms differ from those in a hydrogen atom.

- Atoms with many electrons have orbitals that possess energies that depend not only on n and the nuclear charge, Z, but also on the azimuthal quantum number l, not just the value of n and Z as in the case for hydrogen orbitals.
- In many-electron atoms, inner electrons shield outer electrons from the full nuclear charge. This effect reduces the nuclear charge experienced by outer electrons and is called the screening effect.
- Outer electrons partially penetrate the electron clouds of inner electrons and this also alters the nuclear charge experienced by outer electrons.
- The net positive nuclear charge, resulting from screening and penetration effects, experienced by an electron is greater than 1+, but less than the full nuclear charge. This net effective nuclear charge experienced by an electron is termed the effective nuclear charge, Z_{eff}.

In the 1920s, S. A. Goudsmit and B. E. Uhlenbeck proposed that electrons possess intrinsic spin as well as orbital motion.

- Electrons in an atom show two intrinsic spin orientations in the presence of a magnetic field. The two spin orientations are in opposite directions and are quantized.
- Electron spin is defined by the **electron-spin quantum number,** m_s, which has two possible values, $m_s = +\frac{1}{2}$ and $m_s = -\frac{1}{2}$.

There is a further limitation placed on quantum numbers for electrons in an atom. In 1924, Wolfgang Pauli proposed what has become known as the **Pauli exclusion principle.**

- According to this principle, *no two electrons in an atom can have the same set of four quantum numbers, n, l, m$_l$, and m$_s$.*
- The effect of the Pauli exclusion principle is that an orbit can hold a maximum of two electrons and these electrons must have opposite spins.

The order of energies of orbitals in nonhydrogen atoms is shown in Figure 6.24 in the text.

- Note that subshells (for example, $2s$ and $2p$) do *not* have the same energies as they do in a hydrogen atom.
- Note that orbitals *within* a subshell (for example, $2p_x$, $2p_y$, $2p_z$) possess the same energy.
- Use the mnemonic method illustrated in Figure 6.3 of the *Student's Guide* to help you remember the relative order of energies. To determine the correct order of energies, simply follow each arrow from tail to head, and then go to the next arrow.

ENERGIES OF ORBITALS IN MANY-ELECTRON ATOMS

Figure 6.3 A technique to aid memorization of the order of orbital energy levels in many-electron atoms. Each horizontal row in the diagram contains all the orbitals for a given n quantum number. The rows follow a sequential listing of n quantum numbers starting from $n = 1$. Parallel arrows are then drawn as shown.

- Remember that this ordering of energies is only used to "construct" the electronic structure of atoms.

EXERCISE 16   Will a 2p electron in a hydrogen atom experience the same nuclear charge as a 2p electron in an atom that has five protons and also two 2s and two 1s electrons? Explain.

Solution: No. The 2p electron in a hydrogen atom experiences the full 1+ nuclear charge because there are no inner electrons to shield the 2p electron from the nucleus. The 2p electron in the many-electron atom is shielded from the nuclear charge by the four inner 2s and 1s electrons. This shielding of the 2p electron by the inner 2s and 1s electrons is called the screening effect. However, the 2p electron is not fully shielded from the nucleus by the inner electrons. The 2p electron is able to penetrate into the 2s and 1s electron clouds, and as a result it experiences an average nuclear charge, called the effective nuclear charge, which is greater than 1+ but less than 5+. Thus, a 2p electron in a hydrogen atom experiences a smaller nuclear charge than one in a many-electron atom.

EXERCISE 17   What other factor besides the penetration effect described in Exercise 16 causes an electron in a many-electron atom to experience an effective nuclear charge different from the actual charge of the nucleus?

Solution: An electron in a many-electron atom interacts with other electrons and, therefore, experiences electron-electron repulsions. Those electrons that are at any instant between a particular electron and the nucleus exert a repulsive force that cancels part of the attraction of the electron for the nucleus. The effective nuclear charge experienced by an electron is the net result of the full attraction of the electron for the nucleus and the averaged electron-electron repulsions.

EXERCISE 18   The 4s and 4p orbitals in the hydrogen atom have the same energy. In contrast, experiments have shown that the 4s and 4p orbitals in many-electron atoms do not have the same energy. Why do the energies of these orbitals differ in a many-electron atom but not in a hydrogen atom?

Solution: An electron in either the 4s or 4p orbital in a hydrogen atom experiences the same 1+ nuclear charge because there are no electron-electron repulsions, there being only one electron. In a many-electron atom, a 4s electron penetrates closer to the nucleus than a 4p electron does. Thus a 4s electron experiences a greater effective nuclear charge than a 4p electron in a many-electron atom. This results in the 4s orbital having a lower energy than the 4p orbitals.

EXERCISE 19   In a many-electron atom, which of the following orbitals, described by their three quantum numbers, have the same energy in the absence of magnetic and electric fields: **(a)** $n = 1$, $l = 0$, $m_l = 0$; **(b)** $n = 2$, $l = 1$, $m_l = 1$; **(c)** $n = 2$, $l = 0$, $m_l = 0$; **(d)** $n = 3$, $l = 2$, $m_l = 1$; **(e)** $n = 3$, $l = 2$, $m_l = 0$?

Solution: Orbitals with both the same n quantum number and the same l quantum number have the same energy in many-electron atoms when electric or magnetic fields are not present. Thus orbitals **(d)** and **(e)**—which are both 3d orbitals—are the same.

EXERCISE 20   What is the maximum number of electrons that can occupy each of the following subshells: 1s, 2p, 3d, and 4f?

Solution: To determine the maximum electron occupancy for each subshell, you need to know how many degenerate orbitals exist for each subshell. The number of degenerate orbitals equals the total number of m_l values for the l value of the subshell. Each orbital can hold a maximum of two electrons. The maximum occupancy for each subshell is shown on the next page:

Subshell	l Value	m_l Values	Number of m_l Values	Maximum Number of Electrons in Subshell
$1s$	0	0	1	$2 \times 1 = 2$
$2p$	1	1, 0, −1	3	$2 \times 3 = 6$
$3d$	2	2, 1, 0, −1, −2	5	$2 \times 5 = 10$
$4f$	3	3, 2, 1, 0, −1, −2, −3	7	$2 \times 7 = 14$

At this point, you should review the rules governing the possible values of the four quantum numbers for electrons:

1. The principal quantum number, n, has integral values of 1, 2, 3, . . .
2. The integral values of the l quantum number for a given n are $l = 0, 1, 2, 3, . . . , n − 1.$
3. The integral values of the m_l quantum number for a given l are $m_l = l, l − 1, l − 2, . . . , 0, . . . , −1 + l, −l.$
4. The values of the m_s quantum number for a given m_l are $m_s = +\frac{1}{2}, −\frac{1}{2}.$

With this information you can now specify the four quantum numbers for all electrons in an atom. This description of how electrons are arranged in orbitals is called an **electron configuration.** The procedure for writing the electron configuration of an element is summarized as follows:

1. Electrons occupy orbitals in order of increasing energy. All orbitals that are equal in energy (degenerate) are filled with electrons first, before the next level of energy begins to fill up. (A few exceptions exist; for example, Cr and Cu.)
2. An orbital can have a maximum of two electrons with *opposite* spins. This is an application of the Pauli exclusion principle.
3. When several orbitals of equal energy exist (for example, $2p_x$, $2p_y$, $2p_z$), the filling of orbitals with electrons follows **Hund's rule:** *Electrons enter orbitals of equal energy singly and with the same spins, until all orbitals have one electron each.* Then electrons with opposite spins pair with the electrons in the half-filled set of equal-energy orbitals.

Three methods are used to describe the electron configuration for an element:

1. Assignment of the four quantum numbers (n, l, m_l, m_s) for each electron;
2. a shorthand system using the orbital system notation described in Section 6.5 of the text with the number of electrons indicated (as a superscript) for each set of degenerate orbitals;
3. an orbital diagram (see Exercise 22).

These methods are described in Exercises 21 through 23.

EXERCISE 21 Using the four quantum numbers, describe the electron configuration for each of the following atoms: **(a)** B; **(b)** N. Assume that an electron first enters an orbital in the spin state described by the spin quantum number $m_s = +\frac{1}{2}$

Solution: **(a)** Boron has five electrons. The first two electrons enter the lowest energy orbital, 1s, with opposite spins as per the Pauli exclusion principle. The next energy level, 2s, also has two electrons with opposite spins. The last electron enters the 2p energy level. The electron configuration for boron is as follows:

n	l	m_l	m_s	Orbital notation
1	0	0	$+\frac{1}{2}$	1s
1	0	0	$-\frac{1}{2}$	1s
2	0	0	$+\frac{1}{2}$	2s
2	0	0	$-\frac{1}{2}$	2s
2	1	1	$+\frac{1}{2}$	2p
2	1	1	$+\frac{1}{2}$)^a	
2	1	-1	$+\frac{1}{2}$)	

^{a}These two orbitals are equivalent to the one immediately above, and the fifth electron can be placed in any of the three.

(b) Nitrogen has seven electrons. The first five electrons enter orbitals as for boron. The last two electrons must also enter 2p orbitals because they have a maximum occupancy of six electrons. According to Hund's rule, they enter singly until all degenerate orbitals are half filled:

n	l	m_l	m_s	Orbital notation
1	0	0	$+\frac{1}{2}$	1s
1	0	0	$-\frac{1}{2}$	1s
2	0	0	$+\frac{1}{2}$	2s
2	0	0	$-\frac{1}{2}$	2s
2^a	1	1	$+\frac{1}{2}$	2p
2^a	1	0	$+\frac{1}{2}$	2p
2^a	1	-1	$+\frac{1}{2}$	2p

^{a}These three 2p orbitals are referred to as a half-filled set of orbitals.

EXERCISE 22 Another approach to describing the electron configuration of an atom is an orbital diagram, in which a box represents an orbital. Electrons are represented by arrows, with an arrow that points up (↑) corresponding to the spin state $m_s = +\frac{1}{2}$ and an arrow that points down (↓) corresponding to the spin state $m_s = -\frac{1}{2}$. For example, the electron configuration for boron can be represented as

↑↓		↑		↑		

$\quad$ 1s $\qquad$ 2s $\qquad\quad$ 2p

What are the orbital diagrams for: **(a)** O; **(b)** Sr; **(c)** V?
$\quad$ *Solution:* **(a)** Oxygen has eight electrons:

↑↓		↑↓		↑↓	↑	↑

$\quad$ 1s $\qquad$ 2s $\qquad\quad$ 2p

The electron configuration shown for the 2p orbitals conforms to Hund's rule, which requires that electrons remain unpaired until all degenerate (equal energy) orbitals are filled with one electron. The 2p orbitals are degenerate. Thus, the first three 2p electrons are placed singly in orbitals, and the fourth one is paired. **(b)** Strontium has 38 electrons:

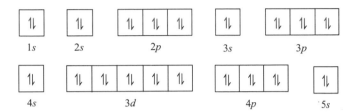

When *forming* an electron configuration, fill the $4s$ orbitals before the $3d$ orbitals. As a check, you should count the number of electrons. **(c)** Vanadium has 23 electrons:

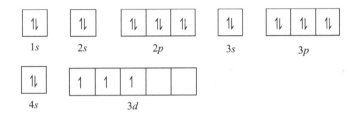

EXERCISE 23   An orbital diagram is useful and informative, but cumbersome. More commonly, a shorthand notation system is also used to describe electron configurations. This notation uses the nl symbol for each subshell (such as $1s$, $2s$, $2p$, $3s$, $3d$), with the number of electrons occupying the orbital set indicated with a superscript. For example, boron has five electrons; its electron configuration is $1s^2 2s^2 2p^1$. Refer to Exercise 20 for the maximum occupancy of s, p, d, and f subshells. Write the shorthand configuration for the atoms in Exercise 20.

Solution: **(a)** O: $1s^2 2s^2 2p^4$; or $[He]2s^2 2p^4$, where [He] is used to describe the He core electron configuration $1s^2$. **(b)** Sr: $1s^2 2s^2 2p^6 3s^2 3p^6 4s^2 3d^{10} 4p^6 5s^2$; or $[Kr]5s^2$, where [Kr] is used to describe the Kr core electron configuration $1s^2 2s^2 2p^6 3s^2 3p^6 4s^2 3d^{10} 4p^6$. **(c)** V: $1s^2 2s^2 2p^6 3s^2 3p^6 4s^2 3d^3$ or $[Ar]4s^2 3d^3$, where [Ar] is used to describe the Ar core electron configuration $1s^2 2s^2 2p^6 3s^2 3p^6$.

The electronic configuration of an element can be constructed from its location in the periodic table.

THE PERIODIC TABLE: PERIODIC ARRANGEMENT OF ELECTRON CONFIGURATIONS AND VALENCE ELECTRONS

- Elements in **families** (*vertical columns*) possess the same number and type of outer electrons beyond the inner-core electron configuration. These outer-shell electrons are referred to as **valence electrons.**

- Valence electrons are involved in ion formation and chemical reactions.

- If you know the valence electron configuration for the first member of a family, you then know the valence electron configuration for the other members. For example, the electron configuration for oxygen is

O: $1s^2$ $2s^2 2p^4$

inner-core noble-gas electron configuration: $\boxed{\text{He}}$ ⟵⟶   valence electrons

Any other member of the oxygen family will have a similar valence electron configuration:

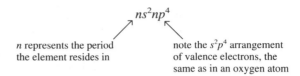

ns^2np^4

n represents the period
the element resides in

note the s^2p^4 arrangement
of valence electrons, the
same as in an oxygen atom

A knowledge of the period in which an element resides will help you write complete electron configurations. Note the following key points shown in Figure 6.30 in the text:

- The first three periods consist only of **representative elements:** only s and p orbitals are involved. The period number corresponds to the principal quantum number of the valence electrons.

- Periods 4–7 also contain **transition elements.** The $(n - 1)d$ outer-orbitals are occupied in transition elements, where n is the period number. Thus, the valence orbitals of transition elements are the ns $(n - 1)d$ orbitals. Note that beyond the transition elements in a period, the d electrons form part of the inner-core of electrons and are no longer valence electrons.

- The **inner-transition elements** have f orbitals occupied.

EXERCISE 24   (a) Write the valence-shell electron configuration for the elements of the second period. (b) How, in general, do the valence-shell electron configurations of the other members of each family relate to the first member's valence-shell electron configuration? (c) How does the group number of a representative family relate to its valence-shell electron configuration?
   Solution: (a) The second period of elements begins after $H(1s^1)$ and $He(1s^2)$. The next quantum level to be filled is $n = 2$. Thus the valence shell of each element contains electrons in the energy level $n = 2$. Using Hund's rule and the Pauli exclusion principle, you should obtain the following valence-electron configurations:

Family:	1A	2A	3A	4A
Element:	Li	Be	B	C
Electron configuration:	$2s^1$	$2s^2$	$2s^22p^1$	$2s^22p^2$

Family	5A	6A	7A	8A
Element:	N	O	F	Ne
Electron configuration:	$2s^22p^3$	$2s^22p^4$	$2s^22p^5$	$2s^22p^6$

(b) Any other element within a given family of representative elements will have the same number and type of valence-shell electrons, except for the n quantum number, which will reflect the period the element is in. For example, in the halogen family, the next member after fluorine is chlorine, and its valence-shell electron configuration is $3s^23p^5$. Chlorine is in the third period, so $n = 3$. (c) The group number of a representative family equals the number of valence-shell electrons for the family.

EXERCISE 25   Use the periodic table to write the complete electron configuration of P and Fe.
   Solution: Phosphorus is in the 5A family and in the third period. Since it is in the 5A family it has five valence electrons in orbitals having a principal quantum number of 3, the period number. Construct the complete electron configuration by writing the electron occupancies of orbitals row by row:

Period 1: $1s^2$

Period 2: $2s^2 2p^6$

Period 3: $3s^2 3p^3$ (5 valence electrons)

TOTAL: $1s^2 2s^2 2p^6 3s^2 3p^3$

Iron is a transition element, so you should recognize that a d orbital will be occupied in the valence shell. Repeating the above procedure:

Period 1: $1s^2$

Period 2: $2s^2 2p^6$

Period 3: $3s^2 3p^6$

Period 4: $4s^2 3d^6$

TOTAL: $1s^2 2s^2 2p^6 3s^2 3p^6 4s^2 3d^6$

EXERCISE 26 Copper and chromium are two elements in the periodic table whose electron configurations are not correctly predicted by application of Hund's rule and Pauli's exclusion principle. Their valence-shell electron configurations are: Cu, $3d^{10}4s^1$; and Cr, $3d^5 4s^1$. You should memorize these. What do you expect for the valence-shell electron configurations for Mo and Ag?

   Solution: Other members of the copper and chromium family should possess the same valence-shell electron configuration, but with different n quantum numbers for the d and s orbitals. Ag is a member of the Cu family; thus its valence-shell electron configuration is $4d^{10}5s^1$. For Mo, a member of the chromium family, it is $4d^5 5s^1$. Note that the quantum numbers for the d and s orbitals are increased by 1 because Ag and Mo are in the period immediately following that containing Cu and Cr.

SELF-TEST QUESTIONS

Key Terms

Having reviewed the key terms listed at the end of Chapter 6 and their applications, identify the following statements as true or false. If a statement is false, indicate why it is incorrect. Key terms are italicized in the statements.

6.1   In the expression $\Delta E = h\nu$, the term $h\nu$ is called a *quantum* of energy.

6.2   When hydrogen gas is placed under reduced pressure (partial vacuum) and a high voltage is passed through it, a continuous band of colors, referred to as a continuous *spectrum*, is produced.

6.3   Green *electromagnetic radiation* with a wavelength of 500 nm has an energy that is directly proportional to its wavelength.

6.4   An example of a *line spectrum* is the white light emitted from a tungsten light bulb commonly used in household fixtures.

6.5   According to *Heisenberg's uncertainty principle,* it is not possible to measure simultaneously and accurately the position and velocity of a particle the size of a baseball.

6.6   The *subshells* of an electron shell with $n = 2$ are $2s$ and $2p$, respectively.

6.7   The $2p_x$ and $2p_y$ orbitals in a hydrogen atom are *degenerate*.

6.8   An example of a *transition element* is the element with atomic number 27.

6.9   An example of a *lanthanide element* is the element with atomic number 62.

6.10   Boron is an example of a *representative element*.

6.11   In iodine, the *effective nuclear charge* experienced by electrons increases with the increasing value of their n quantum number.

6.12   The values for the *electron-spin quantum number* are $+\frac{1}{2}$ and $-\frac{1}{2}$.

6.13   In explaining the photoelectric effect, we assume that *photons* can transfer their kinetic energy to electrons in a metal's surface.

6.14   The periodic properties of an electron moving about a hydrogen nucleus can be described by the *wave function* ψ.

6.15   A hydrogen $1s$ orbital has a small *electron density* at the nucleus of the hydrogen atom.

6.16   In modern quantum mechanics, the shorthand notation for a hydrogen *orbital* with the associated quantum number $n = 3$ and $l = 1$ is $3p$.

6.17   A $2p$ hydrogen orbital has a *node* at the coordinates $x = y = z = 0$, whereas a $1s$ hydrogen orbital possesses electron density at these coordinates.

6.18 According to Louis de Broglie's postulate, the wavelength of a *matter wave* associated with an electron increases with the velocity of an electron as it moves about a nucleus.

6.19 The 3s and 3p orbitals comprise the third *electron shell*.

6.20 Actinide is an *f-block element*.

6.21 The *electron configuration* for carbon is $1s^2 2s^2 2p^3$.

6.22 The *valence-shell electrons* for fluorine are $2p^5$.

6.23 According to the *Pauli exclusion principle*, two electrons in an atom can have the following set of quantum numbers: $n = 2, l = 1, m_l = 0, m_s = +\frac{1}{2}$ and $n = 2, l = 1, m_l = 0, m_s = +\frac{1}{2}$.

6.24 In titanium, the two 3d electrons are predicted to be unpaired by application of *Hund's rule*.

6.25 Because of the *screening effect*, a 2s electron in Mg experiences a smaller effective nuclear charge than a 1s electron.

6.26 The *electronic structure* of an atom refers to the number of electrons possessed by it.

6.27 *Electromagnetic radiation* moves at the speed of light in a vacuum.

6.28 The longer the *wavelength* of radiant energy, the higher the energy.

6.29 *Frequency* of a wave is directly proportional to wavelength.

6.30 When sunlight is passed through a prism, a *continuous spectrum* of all wavelengths is produced.

6.31 The *ground state* for a helium atom is $1s^1 2s^1$.

6.32 When a 2s electron of a carbon atom gains energy and occupies a 3s orbital, an *excited state* of carbon is produced.

6.33 According to de Broglie, the greater the *momentum* of a particle, the longer its wavelength.

6.34 All materials possess a *matter wave*.

6.35 The *probability density* for an electron in an atom is related directly to its wave function.

6.36 There are four orientations for *electron spin*.

6.37 *Core electrons* are those electrons in the shell closest to the nucleus.

6.38 The fourth row of the periodic table contains *transition metals*.

6.39 Calcium is an *active metal*.

Problems and Short-Answer Questions

6.40 Calculate the energy of a photon with a wavelength of 420 nm.

6.41 Using the Bohr model, calculate the wavelength of radiant energy associated with the transition of an electron in the hydrogen atom from the $n = 4$ orbit to the $n = 2$ orbit. Is energy emitted or absorbed? Indicate the type of radiation emitted.

6.42 One of Bohr's postulates is that an electron moves in a well-defined path about the nucleus. What principle later negated this postulate? Why?

6.43 Differentiate between the functions ψ and ψ^2.

6.44 For each of the following pairs of hydrogen orbitals, indicate which orbital has the higher energy:
   (a)   2s or 3s   (b)   3p or 4d   (c)   2s or 2p

6.45 A hydrogen orbital is designated by the quantum numbers $n = 3, l = 2,$ and $m_l = 1$.
   (a)   Write the shorthand notation for this orbital.
   (b)   Is there any other hydrogen orbital that has the same shorthand notation? If so, which one(s)?

6.46 Write the shorthand notation for all hydrogen orbitals that belong to the $n = 4$ principal quantum number energy level. What property do they all have in common?

6.47 Give the values of n, l, and m_l for each hydrogen orbital in the $n = 2$ principal quantum number energy level.

6.48 What differences in orbital size and number of nodes do you expect between a 3s and a 4s hydrogen orbital?

6.49 Draw the contour representations for the $3d_{z^2}$, $3d_{x^2-y^2}$, and $3d_{xy}$ hydrogen orbitals. In your sketches, not all of the orbitals will look identical. Does this indicate that these 3d orbitals are all different in energy?

6.50 Louis de Broglie showed that an electron's wavelength and momentum are related by the equation $\lambda = h/mv$. Why is the relationship $E = hc/\lambda$ not valid for a matter wave?

6.51 Microwave ovens are now commonly used to heat foods. Most microwave ovens have warnings posted to the effect that microwave radiation is harmful to humans. A typical microwave frequency is 20,000 megacycles per second. Calculate the wavelength and energy of this radiation. Show that this energy is significantly less in magnitude than that of visible radiation with a frequency of 100,000,000 megacycles per second, and, thus, the danger of microwave radiation cannot be attributed to its energy content.

6.52 (a)   In what orbitals and in what order are electrons entering the atoms of the second and fifth periods?
   (b)   Electrons are entering the 4s, 3d, and 4p orbitals, in that order, in the atoms of fourth period elements. Why isn't the fourth period composed of elements with only outer 4s and 4p orbitals?

6.53 Write the shorthand notation and the box-diagram electron-configuration representation for each of the following elements and indicate the outer orbital(s):
   (a)   N   (c)   Cu
   (b)   K   (d)   Al

6.54 Each of the following electron configurations represents the valence-shell configuration for an atom of an element in the periodic table. For each, what is the element, and in what period does it belong?
   (a) $4s^2 4p^2$ (d) $2s^2 2p^5$
   (b) $3d^6 4s^2$ (e) $3d^{10} 4s^2$
   (c) $6s^1$

6.55 An atom contains layers or shells of electrons. A completely filled layer or shell of electrons in an atom is a very stable electron configuration.

(a) Both Ne and Xe contain completely filled shells of electrons. What electrons comprise the shell that gets filled between Ne and Ar?

(b) When a representative element forms an ion, it usually forms one whose electron configuration corresponds to the electron configuration for the nearest inert gas. What electrons must be added to or removed from each of the following atoms to form an ion with the nearest inert gas electron configuration: Li; Sr; O?

6.56 Which of the following electron configurations for neutral atoms correspond to ground states, and which correspond to excited states?

(a) $1s^1 2s^1$ **(c)** $[Ar]3d^6$

(b) $[Ar]4s^1$ **(d)** $[Xe]6s^5 d^{10} 6p^3$

Multiple-Choice Questions

6.57 A quantum of electromagnetic radiation has a wavelength equal to 7.52×10^6 Å. What is the frequency of this radiation in cycles/sec?

(a) 1.13×10^{-12} **(d)** 9.45×10^{13}
(b) 8.80×10^{-26} **(e)** 2.69×10^{-15}
(c) 3.99×10^{11}

6.58 What is the energy of radiation that has a frequency of 9.00×10^{11} cycles/sec? (Remember that Planck's constant, h, has a value of 6.63×10^{-34} J-sec.)

(a) 1.66×10^{-45} J   **(d)** 5.00×10^{-22} J
(b) 5.97×10^{-22} J   **(e)** 3.32×10^{-45} J
(c) 4.99×10^{-27} J

6.59 Calculate the wavelength of an electron traveling with a velocity of 4.0×10^9 cm/sec in an electron microscope. The mass of an electron is 9.1×10^{-28} g; $h = 6.63 \times 10^{-34}$ J-sec; and 1 J = 1 kg-m^2/sec^2.

(a) 0.18 Å   **(d)** 1.5×10^8 cm
(b) 0.67×10^{-8} cm   **(e)** 1.1×10^{-38} Å
(c) 1.5 Å

6.60 In the photoelectric effect, in order for an electron to be released from the surface of a clean metal, which one of the following conditions must exist?

(a) The metal must have a low temperature;
(b) the metal must have a high temperature;
(c) the kinetic energy of photons striking the metal's surface must equal that of the emitted electron;
(d) the kinetic energy of photons striking the metal's surface must be less than that of the emitted electrons;
(e) the kinetic energy of photons striking the metal's surface must be greater than or equal to that of the emitted electrons plus the binding energy holding the electron in the metal.

6.61 Passing an electrical charge through argon gas contained in a partially evacuated vessel yields which of the following?

(a) a continuous spectrum;
(b) a line spectrum;
(c) white light;
(d) no visible change;
(e) **(a)** and **(c)**.

6.62 According to Bohr's model of the atom, which of the

following characteristics of metallic elements explains the fact that many of these elements can easily form positively charged ions?

(a) few electrons;
(b) many electrons;
(c) electrons in lowest-energy n level having high energies;
(d) electrons having momentum;
(e) relatively small ionization energies in their ground states.

6.63 According to the Bohr model of the atom, emission of electromagnetic radiation by heated atoms in a vacuum is directly due to which of the following?

(a) photons absorbed by atoms;
(b) particle emission from the nucleus;
(c) momentum possessed by electrons;
(d) electrons being excited from an inner to outer orbit;
(e) electrons falling from an outer to inner orbit.

6.64 Which of the following is *not* true about the Bohr model of the hydrogen atom?

(a) Electrons do not decay into the nucleus;
(b) the model cannot account for the ionization of an electron;
(c) an electron in a stable Bohr orbit does not emit radiation continuously;
(d) an electron may remain in an orbit indefinitely;
(e) the hydrogen atom absorbs radiant energy in multiples of $h\nu$.

6.65 Which of the following statements is *not* true about the principal quantum number, n?

(a) It is related to the spin of an electron;
(b) it is related to the energy of an electron;
(c) the larger the n value of an electron, the higher its energy;
(d) the larger the n value of an electron, the larger the value of its Bohr radius;
(e) the lowest energy state for an electron in a Bohr atom corresponds to $n = 1$.

6.66 For an orbital with $l = 4$, what are the possible numerical values of m_l?

(a) 1, 0, −1
(b) 3, 2, 0, −1, −2, −3
(c) 3, 2, 1, 0
(d) 4, 3, 2, 1, 0, −1, −2, −3, −4
(e) 4, 3, 2, 1, 0

6.67 Which of the following combinations of quantum numbers for an electron is *not* permissible?

(a) $n = 5, l = 2, m_l = 0$
(b) $n = 3, l = 2, m_l = 3$
(c) $n = 4, l = 3, m_l = -2$
(d) $n = 1, l = 0, m_l = 0$
(e) $n = 2, l = 1, m_l = -1$

6.68 Which of the following statements is the most correct and complete with reference to an electron transition from a Bohr $n = 2$ to $m = 4$ hydrogen orbit?

(a) Energy is released during the electronic transition;
(b) energy is absorbed during the electronic transition;

(c) the energy of the transition is proportional to $\frac{3}{16}$;

(d) both **(b)** and **(c)**.

6.69 A set of hydrogen $3d$ orbitals can contain which of the following maximum number of orbitals?

(a) 1   **(d)** 4

(b) 2   **(e)** 5

(c) 3

6.70 For the principal quantum number 2, what are the allowed values for the associated azimuthal quantum number?

(a) 0, 1, and 2   **(d)** 2, 1, 0, −1, −2

(b) 0, −1, and −2   **(e)** none of the above

(c) −1, 0, +1

6.71 What designation is used to describe the following orbital?

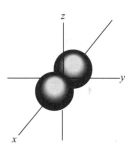

(a) s **(d)** d_z^2

(b) p_z **(e)** d_{xz}

(c) p_x

6.72 In iron, which of the following electrons, characterized by the four quantum numbers, has the lowest energy?

(a) $n = 4, l = 0, m_l = 0, m_s = +\frac{1}{2}$

(b) $n = 3, l = 2, m_l = 1, m_s = -\frac{1}{2}$

(c) $n = 3, l = 2, m_l = 0, m_s = +\frac{1}{2}$

(d) $n = 3, l = 1, m_l = 0, m_s = +\frac{1}{2}$

(e) Electrons **(b)**, **(c)**, and **(d)** are all degenerate and lowest in energy.

6.73 What is the electron configuration of the element Xe?

(a) $1s^2 2s^2 2p^6 3s^2 3p^6$

(b) $1s^2 2s^2 2p^6 3s^2 3p^6 3d^{10} 4s^1$

(c) $1s^2 2s^2 2p^6 3s^2 3p^6 3d^{10} 4s^2 p^6$

(d) $1s^2 2s^2 2p^6 3s^2 3p^6 3d^{10} 4s^2 4p^6 5s^2$

(e) none of the above

6.74 How many electrons are there in the outermost shell of arsenic?

(a) 3   **(d)** 6

(b) 4   **(e)** 7

(c) 5

6.75 Which element possesses the electron configuration $1s^2 2s^2 2p^6 3s^2 3p^6 3d^5 4s^1$?

(a) V   **(d)** Cu

(b) Cr   **(e)** Zn

(c) Mn

6.76 How many electrons in an atom can possess an $n = 3$ principal quantum number?

(a) 3   **(d)** 10

(b) 6   **(e)** 18

(c) 8

6.77 Which of the following is the correct order of increasing energy of atomic orbitals when forming a many-electron atom?

(a) $1s2s2p3s3d$ **(d)** $1s2s2\ p3d4s$

(b) $1s2s3s3p4s$ **(e)** none of the above

(c) $1s2s2p3s3p$

SELF-TEST SOLUTIONS

6.1 True. **6.2** False. Under these conditions, hydrogen gas produces not a continuous spectrum, but a line spectrum, which consists of radiation of only certain wavelengths. **6.3** False. The energy of radiation is directly proportional to the inverse of its wavelength ($E \propto 1/\lambda$). **6.4** False. White light consists of a continuous spectrum in the visible range of radiation. **6.5** False. Heisenberg's uncertainty principle applies to very small particles, those on the subatomic level, not large particles. **6.6** True. **6.7** True. **6.8** True. The element is cobalt. **6.9** True. The element is samarium. **6.10** True. **6.11** False. As the value of the n quantum number increases, there are more inner electrons to shield outer electrons, and the effective nuclear charge is thereby reduced. **6.12** True. **6.13** True. **6.14** True. **6.15** False. The $1s$ electron has a high probability (large electron density) of existing at the hydrogen nucleus. **6.16** True. **6.17** True. **6.18** False. $\lambda = h/mv$. The value of λ decreases with increasing velocity. **6.19** False. An electron shell consists of a complete collection of orbitals that have the same value of n. The $3s$, $3p$ and $3d$ comprise the third electron shell. **6.20** True. **6.21** False. The electron configuration for carbon is $1s^2 2s^2 2p^2$. **6.22** False. The valence-shell electrons for fluorine are $2s^2 2p^5$. **6.23** False. They must have different values for m_s because their values for n, l, and m_l are the same. **6.24** True. **6.25** True. **6.26** False. Not only does it include the number of electrons, but also their energies. **6.27** True. **6.28** False. Energy and wavelength are inversely related. The longer the wavelength, the smaller the energy. **6.29** False. Frequency and wavelength are inversely related. **6.30** True. **6.31** False. It is $1s^2$. **6.32** True. **6.33** False. The wavelength associated with matter is inversely proportional to momentum. **6.34** True. **6.35** False. Probability density is proportional to the square of a wave function. **6.36** False. There are two spin orientations. **6.37** False. Core electrons are those that are not valence electrons. **6.38** True. **6.39** True. Group IA and IIA elements are active metals. **6.40** The relation between the energy of a photon and its wavelength is $E = hc/\lambda$. Substituting the appropriate values for h, c, and the wavelength (converted to the unit meter) yields

$$E = \frac{(6.63 \times 10^{-34} \text{ J-sec})(3.00 \times 10^8 \text{ m/sec})}{(4.20 \times 10^2 \text{ nm})(1 \times 10^{-9} \text{ m/1 nm})}$$

$$= 4.74 \times 10^{-19} \text{ J}$$

6.41 The wavelength of radiation is related to the energy of the electron transition by the equation $\Delta E = hc/\lambda$, where ΔE is the energy difference between the $n = 4$ and $n = 2$ orbits. This energy difference is calculated as follows:

$$\Delta E = R_H\left(\frac{1}{n_i^2} - \frac{1}{n_f^2}\right) = (2.18 \times 10^{-10}\ \text{J})\left(\frac{1}{4^2} - \frac{1}{2^2}\right)$$

$$= -4.09 \times 10^{-19}\ \text{J}$$

Since the value of ΔE is negative, energy is released. The absolute value of ΔE is used in the calculation of the value of λ:

$$\lambda = \frac{hc}{\Delta E}$$

$$= \frac{(6.63 \times 10^{-34}\ \text{J-sec})(3.00 \times 10^8\ \text{m/sec})}{4.09 \times 10^{-19}\ \text{J}}$$

$$= 4.86 \times 10^{-7}\ \text{m} = 4860\ \text{Å}$$

A wavelength of 4.86×10^{-7} m or 4860 Å belongs in the visible range of light. **6.42** Heisenberg's uncertainty principle negated this postulate. It states that it is impossible to measure the position and velocity of a small particle simultaneously and accurately. In order to describe an electron as having motion in a well-defined circular path, we would have to specify its velocity and position accurately at all times, which, according to the uncertainty principle, is impossible. **6.43** ψ is a mathematical expression that describes the amplitude and the motion of the matter wave of an electron as a function of time. ψ^2 is the probability of finding the electron at some point in space. **6.44** The energy of a hydrogen orbital increases as the value of the principal quantum number increases. Thus, among the pairs, the following have the higher energy: **(a)** $3s$; **(b)** $4d$; **(c)** both orbitals, because both have the same principal quantum number. *Note:* The answer to **(c)** is true only for the hydrogen atom. **6.45 (a)** The shorthand notation for this orbital is $3d$. **(b)** Yes. Any other orbital with the quantum numbers $n = 3$, $l = 2$ will have the same notation. The $l = 2$ azimuthal quantum number has associated with it orbitals with m_l values of 2, 1, 0, −1, −2, not just $m_l = 1$. Therefore, five orbitals can have the designation $3d$. **6.46** For the $n = 4$ principal quantum energy level, hydrogen orbitals with $l = 0, 1, 2, 3$ values can exist. Thus, the $n = 4$ principal quantum energy level possesses the following subshells: $4s$, $4p$, $4d$, and $4f$. They all possess the same energy because they have the same principal quantum number, 4. **6.47** For the $n = 2$ principal quantum energy level, hydrogen orbitals with $l = 0, 1$ values can exist. For each l value, there exists a further subset of orbital types:

l	m_l	Shorthand notation
0	0	$2s$
1	+1	$2p$
1	0	$2p$
1	−1	$2p$

6.48 The size of a hydrogen orbital increases as the value of the principal quantum number increases. A $4s$ orbital is thus larger than a $3s$ orbital. For each orbital with a given n principal quantum number, the number of nodes is the same—one

less than the value of n. Therefore, the $4s$ orbital has three nodes, and the $3s$ orbital has two nodes. **6.49** See Figure 6.23 in the text for these contour representations. These orbitals all have the same energy in a hydrogen atom because they all have the same principal quantum number value, 3. **6.50** The expression $E = hc/\lambda$ is valid for electromagnetic radiation; matter is not such. Various matter and their associated matter waves cannot move at the speed of light and thus are not electromagnetic radiation. **6.51** First, convert megacycles per second into cycles per second using the equivalence 1 megacycle $= 1 \times 10^6$ cycles:

$$\nu_{\text{microwave}} = \left(20{,}000\ \frac{\text{megacycles}}{\text{sec}}\right)$$

$$\times \left(\frac{1 \times 10^6\ \text{cycles}}{1\ \text{megacycle}}\right)$$

$$= 2 \times 10^{10}\ \text{cycles/sec}$$

$$\nu_{\text{visible}} = \left(100{,}000{,}000\ \frac{\text{megacycles}}{\text{sec}}\right)$$

$$\times \left(\frac{1 \times 10^6\ \text{cycles}}{1\ \text{megacycle}}\right)$$

$$= 1 \times 10^{14}\ \text{cycles/sec}$$

Substituting these values into the relation $E = h\nu$ yields

$$E_{\text{microwave}} = h\nu$$
$$= (6.63 \times 10^{-34}\ \text{J-sec})(2 \times 10^{10}/\text{sec})$$
$$= 1.33 \times 10^{-23}\ \text{J}$$

$$E_{\text{visible}} = h\nu$$
$$= (6.63 \times 10^{34}\ \text{J-sec})(1 \times 10^{14}/\text{sec})$$
$$= 6.63 \times 10^{-20}\ \text{J}$$

The value of the ratio $E_{\text{microwave}}/E_{\text{visible}}$ shows the relative magnitude of their energies:

$$\frac{E_{\text{microwave}}}{E_{\text{visible}}} = \frac{1.33 \times 10^{-23}\ \text{J}}{6.63 \times 10^{-20}\ \text{J}} = 2.01 \times 10^{-3}$$

From this we see that $E_{\text{microwave}} = 2.01 \times 10^{-3}\ E_{\text{visible}}$. Thus the energy content of microwave radiation is significantly less than that of visible radiation. **6.52 (a)** Electrons enter, in the order stated, the $2s$ and $2p$ orbitals in the atoms of the second-row elements and the $5s$, $4d$, and $5p$ orbitals in the atoms of the fifth period. **(b)** The energy of a $3d$ orbital is lower than the energy of a $4p$ orbital and must be filled with electrons first when atoms of the fourth period are formed. In order for the fourth row to be composed of elements sequentially increasing by one atomic number, elements with $3d$ electrons, the transition elements, must be included.

6.53 (a) $1s^2\ \boxed{2s^2 2p^3}$—outer orbital set

$1s$ $2s$ $2p$

(b) $1s^2 2s^2 2p^6 3s^2 3p^6 \boxed{4s^1}$—outer orbital set

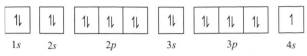

(c) $1s^2 2s^2 2p^6 3s^2 3p^6 \boxed{4s^1 3d^{10}}$—outer orbital set

The 4s and 3d orbitals are very similar in energy in the atom once the orbitals are filled with electrons. Thus, both orbitals comprise the outer set of orbitals. The outer orbital electron configuration is $4s^1 3d^{10}$ instead of the expected $4s^2 3d^9$ because a half-filled 4s orbital and a filled 3d orbital appear to be more stable than a partially filled 3d orbital.

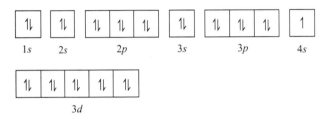

(d) $1s^2 2s^2 2p^6 \boxed{3s^2 3p^1}$—outer orbital set

6.54 (a) The element with the highest-energy $4s^2 4p^2$ electron configuration belongs to the fourth period and the carbon family because $n = 4$ (fourth period) and the valence-shell configuration of the carbon family is $s^2 p^2$. The element is germanium. **(b)** The element with the highest-energy $3d^6 4s^2$ electron configuration belongs to the fourth period and to the transition element family because $n = 4$ for the s orbital and transition elements have incompletely filled d orbitals. The element is iron ($1s^2 2s^2 2p^6 3s^2 3p^6 3d^6 4s^2$). **(c)** The only family that has an ns^1 electron configuration is the alkali metal family, group 1A. The element belongs to the sixth period because $n = 6$. It is cesium. **(d)** This element belongs to the second period, as $n = 2$ for the valence electrons: F. **(e)** See (b) for rationale: Zn. **6.55 (a)** Neon has the electron configuration $1s^2 2s^2 2p^6$; argon's is $1s^2 2s^2 2p^6 3s^2 3p^6$. The electrons that comprise the shell from Ne to Ar are the $3s^2 3p^6$ electrons. **(b)** Li has the electron configuration $1s^2 2s^1$. The nearest inert gas in the periodic table is He, with a $1s^2$ electron configuration. Li loses the $2s^1$ electron to form Li^+, with a $1s^2$ electron configuration. Sr has the electron configuration $[Kr]5s^2$. Sr loses the two 5s electrons to form Sr^{2+} with a [Kr] electron configuration.

Oxygen has the electron configuration [He] $2s^2 2p^4$. The nearest inert gas in the periodic table is Ne. Oxygen adds two 2p electrons to form O^{2-} with the neon electron configuration $1s^2 2s^2 2p^6$. **6.56** The electron configurations shown in **(a)** and **(c)** correspond to excited states. The ground-state electron configuration for **(a)** is $1s^2$, and for **(c)**, [Ar] $3d^4 4s^2$. The electron configurations in **(b)** and **(d)** correspond to ground states.

6.57 (c) $v = \dfrac{c}{\lambda}$

$$= \frac{3.00 \times 10^8 \text{ m/sec}}{(7.52 \times 10^6 \text{ Å})(1 \times 10^{-10} \text{ m/Å})}$$

$$= 3.99 \times 10^{11}/\text{sec}$$

6.58 (b) $E = hv$

$$= (6.63 \times 10^{-34} \text{ J-sec})(9.00 \times 10^{11} \text{ sec})$$

$$= 5.97 \times 10^{-22} \text{ J}$$

6.59 (a) $\lambda = \dfrac{h}{mv}$

$$= \left(\frac{6.63 \times 10^{-34} \text{ J-sec}}{(9.1 \times 10^{-28} \text{ g})(4.00 \times 10^9 \text{ cm/sec})} \right)$$

$$\times \left(\frac{1 \text{ kg-m}^2/\text{sec}^2}{1 \text{ J}} \right) \left(\frac{100 \text{ cm}}{\text{m}} \right)^2 \left(\frac{10^3 \text{ g}}{\text{kg}} \right)$$

$$= 1.8 \times 10^{-9} \text{ cm}$$

The answer can be converted to angstroms as follows:

$$\lambda = (1.8 \times 10^{-9} \text{ cm}) \left(\frac{1 \text{ Å}}{1 \times 10^{-8} \text{ cm}} \right)$$

$$= 0.18 \text{ Å}$$

6.60 (e). 6.61 (b). 6.62 (e). 6.63 (e). 6.64 (b). 6.65 (a). 6.66 (d). 6.67 (b). The value of m_l cannot be greater than the value of l. **6.68 (e).** $\Delta E \propto (\frac{1}{2^2} - \frac{1}{4^2}) \frac{3}{16}$. Energy is absorbed because ΔE is positive. **6.69 (e).** The range of m_l values for a d orbital ($l = 2$) is 2, 1, 0, −1, −2; thus there is a maximum of five orbitals. **6.70 (e). 6.71 (c). 6.72 (d).** In many-electron atoms, the energy of an electron depends on both n and l. The electron with the lowest n value along with the lowest l value has the lowest energy. **6.73 (e).** $1s^2 2s^2 2p^6 3s^2 3p^6 3d^{10} 4s^2 4p^6 4d^{10} 5s^2 5p^6$. **6.74 (c). 6.75 (b). 6.76 (e).** For $n = 3$, the maximum possible outer electron configuration is $3s^2 3p^6 3d^{10}$. **6.77 (c).**

Periodic Properties of the Elements | 7

Review: Bohr radius (6.3); ionization energy (6.3); ions (2.5).

Objectives: You should be able to:

1. Explain the effect of increasing nuclear charge on the radial density function in many-electron atoms.

2. Explain the variations in atomic radii among the elements, as shown in Figures 7.5 and 7.6, and predict the relative sizes of atoms based on their positions in the periodic table.

3. Explain the observed changes in values of the successive ionization energies for a given atom.

4. Explain the general variations in first ionization energies among the elements, as shown in Figure 7.7, and relate these variations to variations in atomic radii.

5. Explain the variations in electron affinities among the elements.

7.1, 7.2, 7.3, 7.4, 7.5 ATOMIC PROPERTIES: ATOMIC RADIUS, IONIZATION ENERGY AND ELECTRON AFFINITY

Review: Anions (2.5); cations (2.5); family (2.4); period (2.4); metals (2.4, 4.6); nonmetals (2.4); semimetals (2.4).

Objectives: You should be able to:

1. Describe the periodic trends in metallic and nonmetallic behavior.

2. Describe the general differences in chemical reactivity between metals and nonmetals.

7.6 OVERVIEW: METALS AND NONMETALS

Review: Alkali and alkaline-earth metals (2.4); concept of family (2.4).

Objectives: You should be able to:

1. Describe the general physical and chemical behavior of the alkali metals and alkaline earth metals, and explain how their chemistry relates to their position in the periodic table.

7.7 GROUP TRENDS EXEMPLIFIED: THE ACTIVE METALS

2. Write balanced equations for the reaction of hydrogen with metals to form metal hydrides.

3. Write balanced equations for simple reaction between the active metals (groups 1A and 2A) and the nonmetals in groups 6A and 7A.

7.8 GROUP TRENDS EXEMPLIFIED: SELECTED NONMETALS

Review: Nonmetals (2.4).

Objectives: You should be able to:

1. Write balanced equations for the reaction of hydrogen with nonmetals such as oxygen and chlorine.

2. Describe the allotropy of oxygen.

3. Explain the dominant chemical reactions of oxygen and relate this behavior to its position in the periodic table.

4. Describe the physical states and colors of the halogens, and explain the trends in reactivity with increasing atomic number in the family.

5. Explain the very low chemical reactivity of the noble gas elements.

TOPIC SUMMARIES AND EXERCISES

ATOMIC PROPERTIES: ATOMIC RADIUS, IONIZATION ENERGY AND ELECTRON AFFINITY

Ionization energy, electron affinity, and atomic radius are three physical properties of atoms that are important in the interpretation of the chemical behaviors of elements. Their magnitudes and periodic trends can usually be rationalized and explained by analyzing the electron configurations of the family or element.

Accurate calculations of electron-charge distributions in many-electron atoms have shown that the distribution of electronic charge in an atom is not continuous but rather occurs in layers or shells. Thus, atomic size is only an approximation.

- Each shell corresponds to a collection of electrons with the same principal quantum number; for example $1s^2$ forms a shell different from a $2s^2 2p^6$ shell.
- The distribution of electron density as a function of the radial distance of the electron from the nucleus is termed radial electron density. The atomic sizes of atoms are approximated by **atomic radii.**
- There is a decrease in atomic radii from left to right across a period and an increase in atomic radii with increasing atomic number within a family. Be sure you can use the concepts of effective nuclear charge and energy levels of electrons to explain periodic trends in atomic radii.

The energy required to remove an electron from a gaseous atom or ion so that the electron and resulting species are an infinite distance apart is termed **ionization energy.**

- An example of this process is the loss of one electron by a sodium atom:

$$Na(g) \longrightarrow Na^+(g) + e^-(g) \quad I_1 = 490 \text{ kJ/mol}$$

Energy is always added because an electron is being removed from a positively charged nucleus.

- The symbol I is used for ionization energy, and the subscript 1 in I_1 means that the first electron from a gaseous atom is removed.

- Within each period there is a gradual increase in first ionization energy from left to right. Within a family there is a decrease in ionization energy with increasing atomic number. Be sure that you can use the concepts of effective nuclear charge and atomic radii trends to explain periodic trends in ionization energies. Also, you should note the irregularities in the general trend of first ionization energy across a period.

When an electron is added to a gaseous atom or ion, energy is either absorbed or released. The energy change during the process of adding an electron to an atom or ion is termed **electron affinity** (E).

- An example of such a process is the addition of one electron to iodine:

$$I(g) + e^-(g) \longrightarrow I^-(g) \quad E = -295 \text{ kJ/mol}$$

Energy is released during this process; the sign is negative.

- There is a rough trend of electron affinities becoming more negative from left to right across a period; in going down a family only small changes are usually observed. Carefully note rationales given in the text for trends and irregularities in trends.

EXERCISE 1   The energy added in the reaction

$$Ca(g) \longrightarrow Ca^+(g) + e^-(g)$$

is the first ionization energy (I_1) for calcium. What ionization reaction is associated with the second ionization energy (I_2) for calcium? Will it be endothermic or exothermic?

   Solution: The second ionization energy for calcium is the energy required to remove the highest-energy electron from $Ca^+(g)$. The ionization reaction is

$$Ca^+(g) \longrightarrow Ca^{2+}(g) + e^-(g)$$

Energy is always required to remove an electron: endothermic.

EXERCISE 2   The first ionization energy for $Si(g)$ is 780 kJ/mol; the second ionization energy is 1575 kJ/mol. Why is the second ionization energy greater than the first?

   Solution: The first ionization energy is the energy required for the ionization reaction

$$Si(g) \longrightarrow Si^+(g) + e^-(g)$$

The second ionization energy is the energy required for the ionization reaction

$$Si^+(g) \longrightarrow Si^{2+}(g) + e^-(g)$$

Note that in the first reaction the electron is removed from a noncharged $Si(g)$ atom, whereas in the second reaction the electron is removed from a positive $Si^+(g)$ ion. According to Coulomb's law, more energy is required to remove an electron from a positive ion than from a noncharged ion; thus the second ionization energy is greater than the first.

EXERCISE 3 What is the ionization energy for $F^-(g)$ if the electron affinity for $F(g)$ is -330 kJ/mol?
Solution: When $F(g)$ accepts an electron according to the reaction

$$F(g) + e^-(g) \longrightarrow F^-(g)$$

330 kJ/mol of energy is released. The ionization energy for $F^-(g)$ is the energy required for the reaction

$$F^-(g) \longrightarrow F(g) + e^-(g)$$

Inspection of these two reactions shows that the reaction associated with the ionization energy of $F^-(g)$ is the reverse of the reaction associated with the electron affinity of $F(g)$. Reversing the direction of either reaction causes the heat for the reversed reaction to be the negative of the heat of the original reaction on the basis of the first law of thermodynamics. Thus the ionization energy of $F^-(g)$ is $+330$ kJ/mol.

EXERCISE 4 Which element has the higher electron affinity, sulfur or chlorine? Why?
Solution: Chlorine. More energy is released when chlorine accepts an electron than when sulfur accepts one. This occurs because the effective nuclear charge experienced by the electron being added to a chlorine atom is greater than in the case of an electron that is added to a sulfur atom.

EXERCISE 5 Describe the limitations that must be placed on the concept of atomic radii.
Solution: The electronic-charge distribution in an atom is not sharply defined because the electron cloud extends through all space. We can define an approximate atomic radius in terms of bond distances by apportioning the bond distance between two atoms of a compound to two atomic radii. The atomic radius of an atom in a homonuclear diatomic molecule, a molecule that consists of two atoms of the same element, such as Br_2, is approximated as one-half the distance between the two bonded atoms.

EXERCISE 6 Arrange the following atoms in order of increasing atomic size: **(a)** Mg, Ca, Sr; **(b)** B, F, Ge, Pb.
Solution: **(a)** Within a family, atomic size increases with increasing atomic number: Mg < Ca < Sr. **(b)** Boron and fluorine are smaller than germanium and lead because they are in the second period, whereas the other two are in the fourth and sixth periods, respectively. Boron is larger than fluorine because atomic size tends to decrease across a period. Lead is larger than germanium because it is in a period with a higher number: F < B < Ge < Pb.

OVERVIEW: METALS AND NONMETALS

Metals comprise roughly 70 percent of the known elements, and are situated to the left of the dark line in Figure 7.16 of the text.

- All except mercury are solids at room temperature and pressure.
- They exhibit good electrical and thermal conductivity.
- Many metals melt only at high temperatures.

▪ Metals tend to lose electrons in chemical reactions and become positively charged ions, called **cations.**

$$2Ca(s) + O_2(g) \longrightarrow 2CaO(s)$$

no charge 2+ ion 2− ion

　　▪ Compounds consisting of a metal and an element on the right side of the periodic table tend to be **ionic**: the metal has a positive charge and the other element a negative charge.

Oxides of metals form an important class of metallic compounds.

▪ The **oxide ion** is O^{2-}.
▪ *Metal oxides are basic:* they dissolve in water to form metal hydroxides (containing a OH^- unit) or react with acids to form salts and water.

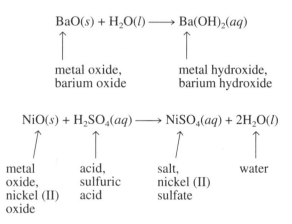

$$BaO(s) + H_2O(l) \longrightarrow Ba(OH)_2(aq)$$

metal oxide, metal hydroxide,
barium oxide barium hydroxide

$$NiO(s) + H_2SO_4(aq) \longrightarrow NiSO_4(aq) + 2H_2O(l)$$

metal   acid,   salt,   water
oxide,   sulfuric   nickel (II)
nickel (II) acid   sulfate
oxide

Important trends in metallic character are

▪ Metallic character is strongest on the left side of a row (period) and tends to decrease to the right.
▪ Metallic character in a family tends to increase from top to bottom.
▪ Group 1A and 2A elements tend to lose electrons very readily.

Nonmetals show a variety of physical and chemical properties.

▪ They are found to the right of the dark line in Figure 7.16 of the text.
▪ H_2, N_2, O_2, F_2, Cl_2, Br_2, and I_2 exist as diatomic molecules.
▪ They are generally poor conductors of heat and electricity.
▪ When they react with metals, they tend to gain electrons and become negatively charged (**anions**).
▪ Compounds consisting only of nonmetals are **molecular substances.**

Most oxides of nonmetals are acidic.

- Some dissolve in water, for example CO_2, P_4O_{10}, and SO_3, to form acidic solutions:

$$CO_2(g) + H_2O(l) \longrightarrow H_2CO_3(aq)$$

- They may also dissolve in base to form salts and water.

$$CO_2(g) + 2KOH(s) \longrightarrow K_2CO_3(aq) + H_2O(l)$$

- CO and N_2O do not react with water or bases.

EXERCISE 7   Which of the following are characteristic properties of metallic compounds: low luster; malleable; poor conductor of heat; form basic oxides; form positive ions?
    Solution: Refer to Table 7.4 in the text. The following are characteristic of metals: malleable, form basic oxides and form positive ions.

EXERCISE 8   Which member of each pair is expected to have the more metallic character: **(a)** Na or Rb; **(b)** K or Ca; **(c)** Pb or F; **(d)** C or Sn?
    Solution: **(a)** Rb. Metallic character increases down a family. **(b)** K. Metallic character tends to decrease from left to right across a periodic table. **(c)** Pb. Metallic character increases to the left and bottom of the periodic table. **(d)** Sn. Metallic character increases down a family.

EXERCISE 9   Which of the following are basic oxides: **(a)** P_2O_5; **(b)** MgO; **(c)** CO_2; **(d)** Na_2O?
    Solution: Basic oxides consist of metals combined with oxygen: **(b)** MgO and **(d)** Na_2O are basic oxides. The others do not contain metals.

EXERCISE 10   Which oxide would you expect to be more acidic, Al_2O_3 or MgO? Explain.
    Solution: Metallic oxides are basic oxides. MgO is a basic oxide because it contains the metal magnesium. Al_2O_3 possesses properties of both basic and acidic oxides. Therefore, Al_2O_3 is more acidic.

EXERCISE 11   Which oxides react with water to form the acids H_3PO_3, H_2CO_3, and H_2SO_4? Write reactions showing how they are formed.
    Solution: The acids contain nonmetals. This suggests that they are formed by the reaction of a nonmetal oxide with water.

$$P_4O_6(s) + 6H_2O(l) \longrightarrow 4H_3PO_3(aq)$$

(**Note:** P_4O_{10} forms H_3PO_4.)

$$CO_2(g) + H_2O(l) \longrightarrow H_2CO_3(aq)$$

$$SO_3(g) + H_2O(l) \longrightarrow H_2SO_4(aq)$$

The text uses the concepts of atomic properties to examine the chemistry of **alkali metals** (Group 1A) and the **alkaline-earth metals** (Group 2A). Also, the chemistry of the groups 1B and 2B is compared to those of groups 1A and 2B. Both group 1A and 1B elements form ions that have a noble-gas electron configuration. Thus, their compounds are colorless or white unless the anion has its own characteristic color.

Be sure you can explain the following general observations about the physical and chemical properties of group 1A and 2A elements. Pertinent concepts that can help you are noted:

- Alkali metals readily form 1+ cations: Explained by their ns^1 valence electron configuration and effect of low first ionization energies of the active metals.
- Alkali metals combine directly with most nonmetals and water: Alkali metals are high in the activity series and have low first ionization energies.
- Alkali metals react with oxygen to form three different types of compounds: *Oxides* (O^{2-} ion); *Peroxides* (O_2^{2-} ion): and *Superoxides* (O_2^- ion). Note the differences among the alkali metals as shown in equations [7.21–7.23] in the text.
- Lithium frequently shows significantly different properties than Na, K, Rb, or Cs: It has an extremely small size compared to the other alkali metals.
- Alkaline earth metals readily lose two electrons to form 2+ cations: Explained by their ns^2 valence electron configuration and relatively low first and second ionization energies.
- Alkaline earth metals are less reactive than alkali metals: A ns^2 outer electron configuration is more stable than ns^1 and alkaline earth metals have relatively higher ionization energies than alkali metals.
- Beryllium, like lithium, shows properties frequently different from those shown by other alkaline earth metals: Note the extremely small size of Be compared to other group 2A elements.

Group 1A elements, the **alkali metals,** consist of Li (lithium), Na (sodium), K (potassium), Rb (rubidium), Cs (cesium), and Fr (francium).

- The elements are soft, have relatively low densities and melting points, and readily lose one electron in chemical reactions.
- Metallic hydrides contain the H^- ion:

$$2Na(s) + H_2(g) \longrightarrow 2NaH(s)$$

1+ ion   1− ion

- Halogens (F_2, Cl_2, Br_2, I_2) readily react with alkali metals to form halide-ion containing salts.

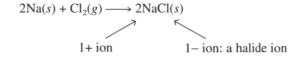

$$2Na(s) + Cl_2(g) \longrightarrow 2NaCl(s)$$

1+ ion   1− ion: a halide ion

- Alkali metals are usually kept under a hydrocarbon solvent such as kerosene to prevent their reaction with oxygen or water.

$$2\text{Cs}(s) + 2\text{H}_2\text{O}(l) \longrightarrow 2\text{CsOH}(aq) + \text{H}_2(g)$$

note hydrogen gas is produced

Group 2A metals, the **alkaline-earth metals,** consist of Be (beryllium), Mg (magnesium), Ca (calcium), Sr (strontium), Ba (barium), and Ra (radium).

- They are harder, more dense and melt at higher temperatures than Group 1A elements.
- They tend to undergo the same reactions as Group 1A, but are less reactive. For example, Mg will not react with cold H_2O, but Na, which is adjacent to it in the periodic table, will.
- They react to form 2+ ions.

How do the characteristics of group 1B and 2B elements compare to those of the group 1A and 2A family members?

- The coinage metals (Cu, Ag, Au), group 1B elements, are far less reactive than the alkali metals, group 1A: The valence electron configuration of group 1B is $d^{10}s^1$ compared to s^1 for group 1A. The effect of the presence of ten d electrons in coinage metals (and also in group 2B) is to increase the effective nuclear charge experienced by outer-shell electrons, resulting in higher ionization energies and thus less reactive metals.
- Group 2B elements (Zn, Cd, Hg) are also less reactive than group 2A elements for similar reasons.

EXERCISE 12   Arrange the following elements in an activity series without referring to a table of activities for metals: Li, Mg, H_2, Ag.

Solution: Li, a group 1A element, is more active than Mg, a group 2A element; both are more active than H_2 because active metals react with hydrogen. Ag is a coinage metal and is thus considerably less reactive than Li, Mg, and H_2. The observed activity series is: Ag < H_2 < Mg < Li.

EXERCISE 13   Write balanced chemical reactions for the preparation of the following alkali metal compounds: **(a)** KOH; **(b)** Cs_2O; **(c)** NaH; **(d)** KBr.

Solution: By referring to the discussion of chemical reactions of alkali metals in Section 7.7, you can write the following by analogy. **(a)** To form a hydroxide, react an alkali metal with water:

$$2\text{K}(s) + 2\text{H}_2\text{O}(l) \longrightarrow 2\text{KOH}(aq) + \text{H}_2(g)$$

(b) To form an oxide, react an alkali metal with oxygen:

$$4\text{Cs}(s) + \text{O}_2(g) \longrightarrow 2\text{Cs}_2\text{O}(s)$$

(c) To form a hydride, react an alkali metal with hydrogen:

$$2\text{Na}(s) + \text{H}_2(g) \longrightarrow 2\text{NaH}(s)$$

(d) To form a bromide salt, react an alkali metal with liquid bromine:

$$2K(s) + Br_2(l) \longrightarrow 2KBr(s)$$

EXERCISE 14   How does magnesium compare with sodium in terms of the following properties: **(a)** atomic size; **(b)** number of outer-shell electrons; **(c)** ionization energy; **(d)** formula of bromide salt?

Solution: **(a)** Mg is smaller; size decreases across a period. **(b)** Mg has two $3s$ electrons, whereas Na has one $3s$ electron. **(c)** The first ionization energy of Mg is greater than that of Na. **(d)** The formula of magnesium bromide is $MgBr_2$, and that of sodium bromide is NaBr. Note that magnesium exists as a 2+ ion, whereas sodium exists as a 1+ ion.

Hydrogen is a unique element and doesn't belong to any family. It is discussed in this section for convenience.

GROUP TRENDS EXEMPLIFIED: SELECTED NONMETALS

- The majority of its compounds contain hydrogen in the 1+ state.
- When it combines with very active metals, such as Na and Ca, a salt containing the hydride (H^-) ion is formed.
- Small quantities can be prepared in the lab by the reaction between zinc and HCl to form $ZnCl_2(aq)$ and $H_2(g)$.

The **oxygen family** (Group 6A) consists of O (oxygen), S (sulfur), Se (selenium), Te (tellurium), and Po (polonium).

- O_2 is a colorless gas at room temperature; the others are solids.
- Oxygen exists in two allotropic forms, O_2 and O_3 (ozone). **Allotropes** are different chemical forms of the same element.
- Review combustion reactions involving oxygen (Section 3.2 of the text).
- Small amounts of oxygen are produced by heating $KClO_3(s)$ in the presence of a small amount of $MnO_2(s)$.

$$2KClO_3(s) \xrightarrow[\Delta]{MnO_2} 2KCl(s) + 3O_2(g)$$

The **halogen family** (Group 7A) consists of F (fluorine), Cl (chlorine), Br (bromine), I (iodine), and At (astatine).

- All form diatomic molecules.
- Remember their physical states: $F_2(g)$, $Cl_2(g)$, $Br_2(l)$, and $I_2(s)$.
- All react with metals to form ionic halide (X^-) salts.
- They react with hydrogen gas to form gaseous HX compounds.

The **noble-gas family** (Group 8A) consists of He (helium), Ne (neon), Ar (argon), Kr (krypton), Xe (xenon), and Rn (radon).

- All are monatomic gases at room temperature.

- They have completely filled *s* and *p* subshells.
- They have low reactivities. Kr, Xe, and Rn react directly with fluorine to form fluorides such as XeF_2, XeF_4, and XeF_6.
- Radon is radioactive.

EXERCISE 15   A flask contains a colorless gas at room temperature. The gas extinguishes a flame. When the gas dissolves in water it forms an acidic solution. Which of the following substances could the unknown gas be? O_2, NaCl, CO_2, I_2. Explain.

  Solution: NaCl and I_2 are solids at room temperature and can be eliminated. $O_2(g)$ would not extinguish a flame as it is a reactant in a combustion reaction. Also, it does not react with water to form an acidic solution. $CO_2(g)$ dissolves in water to form $H_2CO_3(aq)$ and extinguishes a flame.

EXERCISE 16   Identify each of the following unknown elements from the information given: **(a)** This gas is produced when an active metal is added to a strong acid. The gas is colorless and reacts with active metals to form a salt in which it is an anion; **(b)** This gas is highly reactive and has a light blue color and pungent smell. It is an allotrope of a Group 6A element; **(c)** It is a dark violet solid that readily sublimes to form vapors. When it reacts with hydrogen gas, it forms an acid. This element is added to alcohol to form an antiseptic. **(d)** It is a monatomic gas that is radioactive and chemically unreactive: **(e)** It is a pale yellow diatomic gas that reacts with almost all substances. It has a lower electron affinity than chlorine. **(f)** It is a solid that exhibits several allotropic forms; the most common and stable one is a yellow solid with molecular formula X_8. **(f)** This gas is produced in a laboratory by the heating a mixture of potassium chlorate and manganese dioxide. **(g)** This element is a brown liquid that contains the element in a diatomic state and is highly corrosive. Upon contact with the skin it forms severe blisters and burns. **(h)** This gas was once thought to be unreactive; however it is now known to form several compounds with fluorine and oxygen.

  Solution: **(a)** $H_2(g)$. Active metals react with strong acids to form this gas. It also reacts with active metals to form a hydride ion, H^-. **(b)** $O_3(g)$. Ozone is an allotrope of oxygen. **(c)** $I_2(s)$. When iodine reacts with hydrogen gas it forms $HI(g)$. **(d)** Rn. Radon is a noble gas that is radioactive. **(e)** $F_2(g)$. Fluorine is one of the most reactive elements; it is diatomic in its naturally occurring state. It has a smaller electron affinity than chlorine because of its extremely small size. **(f)** $S(s)$. Sulfur occurs in several structural forms; its most common one is yellow S_8. **(f)** $O_2(g)$. **(g)** $Br_2(l)$. **(h)** $Xe(g)$. At one time it was believed that all noble gases were inert. However, Xe was the first noble gas shown to be chemically reactive.

EXERCISE 17   Write the expected formula when the following elements combine: **(a)** Al and O; **(b)** B and Cl; **(c)** Ca and F; **(d)** Na and I; **(e)** S and O; and **(f)** H and Ca.

  Solution: **(a)** Al forms the 3+ state in salts. Oxygen forms the oxide ion, O^{2-}. Al_2O_3. **(b)** Boron is a member of the family containing aluminum; thus, you should expect it to behave in a similar manner to aluminum. BCl_3. **(c)** Calcium forms Ca^{2+} and fluorine forms F^-. CaF_2. **(d)** Na forms Na^+ and iodine in the presence of metals forms I^-. NaI. **(e)** Sulfur and oxygen, both nonmetals, form two nonmetal oxides: $SO_2(g)$ and $SO_3(g)$. **(f)** Ca is an active metal. In the presence of an active metal hydrogen forms the hydride ion, H^-: CaH_2.

SELF-TEST QUESTIONS

Key Terms

Having reviewed the key terms listed at the end of Chapter 7 and their applications, identify the following statements as true or false. If a statement is false, indicate why it is incorrect. Key terms are italicized in the statements.

7.1   *Halogens* readily react with active metals to form salts.

7.2   The *atomic radius* of fluorine is larger than that of nitrogen.

7.3   Ozone is an *allotrope* because it has a different chemical and physical form of the element oxygen than dioxygen.

7.4   The *alkaline-earth metals* are more reactive than group 1A elements.

7.5 The *electron affinity* of fluorine is more negative than that of chlorine.

7.6 *Noble gas elements* tend to be unreactive because their outermost shell of electrons is filled.

7.7 The *first ionization energy* of nitrogen is smaller than that of carbon.

7.8 The formula of *ozone* is O_4.

7.9 Potassium is an element of group 1A and thus it is an *alkali metal*.

7.10 A *valence orbital* of sulfur is the 2*p*.

7.11 In metallic hydrides, the *hydride* ion carries a negative one charge.

7.12 *Metallic* character increases down a family of elements.

Problems and Short-Answer Questions

7.13 The atomic radii for the halogen family are as follows:

Element	Radius (Å)	Atomic number
F	0.64	9
Cl	0.99	17
Br	1.14	35
I	1.33	53
At	1.4	85

Explain the observed trend in atomic radii.

7.14 Does the addition of an electron to $O^-(g)$ to form $O^{2-}(g)$ require or evolve energy (that is, is it an endothermic or exothermic process)?

7.15 Based on their positions in the periodic table, select the atom of each pair that has the larger value of the indicated atomic property:

    (a) ionization energy, Na or Mg
    (b) ionization energy, Mg or Cl
    (c) electron affinity, Cl or Br
    (d) atomic radius, K or Cs
    (e) atomic radius, Se or Br

7.16 Identify each of the following compounds as a halide, oxide, peroxide, superoxide, carbonate, or hydroxide:

    (a) $Ca(OH)_2$ **(d)** Rb_2O_2
    (b) $CsCl$ **(e)** CsO_2
    (c) K_2O **(f)** K_2CO_3

7.17 List three properties that distinguish nonmetals from metals.

7.18 In what way does Na(*s*) metal differ from transition metals, such as iron.

7.19 Classify each as a metal, metalloid or nonmetal:
    (a) Si   **(d)** Xe
    (b) Ba   **(e)** I
    (c) Au

7.20 Identify the most metallic element in each of the following sets
    (a) Li, Al, P, Ar
    (b) Mg, Ca, Sr, Ba
    (c) C, Si, Ge, Sn, Pb

7.21 Write chemical reactions to describe the reaction between:
    (a) Na(*s*) and $H_2(g)$
    (b) Ca(*s*) and HCl(*g*)
    (c) $Al_2O_3(s)$ and HCl(*aq*)
    (d) Al(*s*) and NaOH(*aq*)
    (e) $H_2(g)$ and $Cl_2(g)$
    (f) Ba(*s*) and $Br_2(l)$
    (g) Li(*s*) and $H_2O(l)$

7.22 What is the charge of each of the following ions:
    (a) hydride   **(d)** oxide
    (b) calcium   **(e)** potassium
    (c) chloride

7.23 A student had three test tubes containing the gases O_2, CO and Cl_2. The gas in test tube *A* extinguishes a flame, does not react with aluminum and when dissolved in water forms neither an acidic nor basic solution. A burning splint is inserted into test tube *B*. The splint bursts into flame. The gas is colorless. The gas in test tube *C* has a yellow-green color and reacts with hydrogen gas to form an acid. Identify the gases.

7.24 Which of the following are solids at room temperature, and which are gases?
    (a) CO_2 **(d)** F_2
    (b) BaO   **(e)** NO
    (c) CuO

7.25 Which of the following compounds are ionic and which are molecular?
    (a) N_2O **(d)** PF_3
    (b) K_2O **(e)** $CaSO_4$
    (c) I_2

Multiple-Choice Questions

7.26 Which elements have the lowest ionization energies?
    (a) those on the right side of the periodic table;
    (b) nonmetals;
    (c) those on the left side of the periodic table;
    (d) halogens;
    (e) **(a)** and **(b)**.

7.27 Which element has the largest atomic radius?
    (a) B   **(d)** In
    (b) Al   **(e)** Tl
    (c) Ga

7.28 Why doesn't sodium ordinarily occur in the 2+ ion state?
    (a) low electron affinity;
    (b) high second ionization energy;
    (d) large atomic radius;
    (e) small density.

7.29 What is the cause of the lower activity of copper compared to potassium?

(a) The presence and effect of 3*d* electrons in copper;
(b) the higher ionization energy of copper compared to that of potassium;
(c) the smaller size of the potassium atom compared to that of copper;
(d) (a) and (b):
(e) (b) and (c).

7.30 Based on its position in the periodic table, which atom has the largest ionization energy?

    (a) Li   **(d)** N
    (b) B   **(e)** Ne
    (c) C

7.31 Based on its position in the periodic table, which atom has the largest atomic radius?

    (a) Li   **(d)** Rb
    (b) Na   **(e)** Cs
    (c) K

7.32 Which of the following represents the chemical reaction for the heating of Bi_2O_3 in a hydrogen environment?

    (a) $Bi_2O_3 \longrightarrow 2Bi + 3/2O_2$
    (b) $Bi_2O_3 + 3H_2 \longrightarrow 2Bi + 3H_2O$
    (c) $2Bi_2O_3 + 6H_2 \longrightarrow 4Bi + 3OH^-$
    (d) $H_2O + Bi_2O_3 + H_2 \longrightarrow 2Bi(OH)_2$

7.33 What are the expected ions when $K_2O(s)$ is dissolved in water?

    (a) K^+, H^+ **(c)** K^+, OH^-
    (b) K^+, O^{2-} **(d)** No reaction occurs

7.34 Which of the following elements exhibit allotropic forms?

    (a) hydrogen   **(c)** sodium
    (b) oxygen   **(d)** xenon

7.35 CaO is
    (a) a molecular substance
    (b) a molecular oxide
    (c) an ionic oxide
    (d) an acidic oxide

7.36 Which of the following is true for Group 4A?
    (a) consists only of nonmetals
    (b) contains no metalloids
    (c) contains an active metal
    (d) metallic character increases from top to bottom

7.37 Which of the following descriptions of elements is *not* true?

    (a) Rb, an active metal that forms a +1 cation in aqueous solution
    (b) I_2, a grayish-black solid that readily forms a purple vapor
    (c) Pb, a bluish-gray, soft and dense metal
    (d) H, shows only nonmetallic properties

7.38 Which element has its most common ion state as 2+ and its physical properties are partially determined by *d* electrons.

    (a) Ca
    (b) K
    (c) Cu
    (d) Tl

7.39 Which substance is expected to be a gas at room temperature?

    (a) BaO
    (b) $CaCl_2$
    (c) $SnCl_2$
    (d) NO

7.40 Which substance should form an acidic solution in water?

    (a) Na_2O
    (b) SO_2
    (c) CO
    (d) BaO

SELF-TEST SOLUTIONS

7.1 True. **7.2** False. It is smaller; atomic radii decrease across a given period. **7.3** True. **7.4** False. They tend to be less reactive because of the effect of the second ionization energy. **7.5** False. The small size of the fluorine atom results in a smaller than anticipated electron affinity. **7.6** True. **7.7** False. The s^2p^3 valence electron configuration of nitrogen exhibits a p subshell that is half-filled; this causes nitrogen to have a higher first ionization energy than either of its adjacent elements in the same period. **7.8** False. It is O_3. **7.9** True. **7.10** False. The valence orbitals of sulfur lie in the third shell; therefore a $3p$ orbital is a valence orbital. **7.11** True. **7.12** True. **7.13** The larger the value of the *n* quantum number for the highest-energy orbital(s) of an atom, the larger the size of the atom. Since the value of the *n* quantum number of the outermost orbital increases with increasing atomic number in a family of elements, the atomic radii of the halogen family increase in the order F < Cl < Br < I < At. **7.14** According to Coulomb's law, the addition of a negatively charged electron to a negatively charged $O^-(g)$ ion requires overcoming the repulsion forces between the electron and O^- ion. This requires energy. The process

$$O^-(g) + e^- \longrightarrow O^{2-}(g)$$

is thus endothermic, and its enthalpy change is the electron affinity for the O^- ion. **7.15 (a)** Mg. The outer electron configuration of Mg is $3s^2$, whereas the outer electron configuration of Na is $3s^1$. Because it is more difficult to remove an electron from $3s^2$ than $3s^1$, Mg has the higher ionization energy. **(b)** Cl. Ionization energy generally increases across a period. **(c)** Cl. Within a family, electron affinity generally decreases with increasing atomic number. **(d)** Cs. Atomic radii within a family increase with increasing atomic number. **(e)** Se. Atomic radii decrease across a period of representative elements. **7.16 (a)** hydroxide; **(b)** chloride; **(c)** oxide; **(d)** peroxide; **(e)** superoxide; **(f)** carbonate. **7.17** Nonmetals tend to be less lustrous, are poorer conductors of heat and electricity, and form covalent instead of ionic substances. **7.18** Na is a soft, highly reactive metal that exists only in the 1+ ion state. Transition metals tend to be more dense, less reactive and exhibit more than one ion state. **7.19 (a)** metalloid: **(b)** metal; **(c)** transition metal; **(d)** nonmetal; **(e)** nonmetal. **7.20 (a)** Metallic character decreases left to right across a period: Li; **(b)** Metallic character tends to increase down a family: Ba: **(c)** same reason as **(b)**: Pb. **7.21 (a)** $2Na(s) + H_2(g) \longrightarrow 2NaH(s)$;

(b) $Ca(s) + 2HCl(g) \longrightarrow CaCl_2(s) + H_2(g)$; (c) $Al_2O_3(s) + 6HCl(aq) \longrightarrow 2AlCl_3(s) + 3H_2O(l)$; (d) $2Al(s) + 6NaOH(aq) \longrightarrow 2Na_3AlO_3(aq) + 3H_2(g)$; (e) $H_2(g) + Cl_2(g) \longrightarrow 2HCl(g)$; (f) $Ba(s) + Br_2(l) \longrightarrow BaBr_2(s)$; (g) $2Li(s) + 2H_2O(l) \longrightarrow 2LiOH(aq) + H_2(g)$. **7.22** (a) H^-; (b) Ca^{2+}; (c) Cl^-; (d) O^{2-}; (e) K^+. **7.23** CO is in test tube A. It is a relatively nonreactive nonmetal oxide. When placed in water it does not act as an acidic oxide. O_2 is in test tube B. It is fuel for combustion and is colorless. Cl_2 is in test tube C. It has a yellow-green color and forms HCl in the presence of hydrogen gas. **7.24** In general, metallic compounds are solids whereas nonmetallic compounds occur commonly in the gaseous or liquid form. (a) gas; (b) solid; (c) solid; (d) gas; (e) gas. **7.25** Compounds formed from a metal and nonmetal tend to be ionic whereas if both elements are nonmetallic the compound is molecular: (a) molecular; (b) ionic; (c) molecular; (d) molecular; (e) ionic. **7.26 (c).** Ionization energies increase across periods from left to right. **7.27 (e).** Radii increase down a family. **7.28 (c).** Formation of Na^{2+} from Na^+ requires removal of an electron from a 1+ ion, which according to Coulomb's law, requires a large expenditure of energy. **7.29 (d).** In Cu, the $3d$ electrons do not effectively shield the outer $4s$ electrons, and the effective nuclear charge experienced by them is thereby increased. This effect leads to a higher ionization energy for copper, which reduces the reactivity of copper compared to potassium. **7.30 (e). 7.31 (e). 7.32 (b). 7.33 (c). 7.34 (b). 7.35 (c). 7.36 (d). 7.37 (d). 7.38 (c). 7.39 (d). 7.40 (b).** CO is a nonmetallic oxide, but is not acidic.

8 | Basic Concepts of Chemical Bonding

OVERVIEW OF THE CHAPTER

8.1 LEWIS SYMBOLS—VALENCE ELECTRONS

Review: Concept of outermost electron shell (6.8, 6.8, 6.9); electron configurations (7.2).

Objectives: You should be able to:

1. Determine the number of valence electrons for any atom, and write its Lewis symbol.
2. Recognize when the octet rule applies to the arrangement of electrons in the valence shell for an atom.

8.2, 8.3 IONIC COMPOUNDS: BONDS AND RADII

Review: Atomic radii (7.2, 7.3); electron configurations (6.8, 6.9); ions (2.5); periodic table (2.4, 6.9)

Objectives: You should be able to:

1. Describe the origin of the energy terms that lead to stabilization of ionic lattices.
2. Predict on the basis of the periodic table the probable formulas of ionic substances formed between common metals and nonmetals.
3. Describe how the radii of ions relate to those of atoms.
4. Explain the concept of an isoelectronic series and the origin of changes in ionic radius within such a series.

8.4, 8.6, 8.7 THE LEWIS MODEL FOR COVALENT BONDING

Review: Electron configurations (6.5, 7.3); periodic table (2.4).

Objectives: You should be able to:

1. Describe the basis of the Lewis theory, and predict the valence of common nonmetallic elements from their positions in the periodic table.
2. Write the Lewis structures for molecules and ions containing covalent bonds, using the periodic table.

120

3. Write resonance forms for molecules or polyatomic ions that are not adequately described by a single Lewis structure.

4. Describe the general differences in physical properties between substances with ionic bonds and those with covalent bonds.

Objectives: You should be able to write the Lewis structure for molecules and ions containing covalent molecules that have an odd number of electrons, a deficiency of electrons or an expanded octet.

8.8 EXCEPTIONS
TO THE OCTET RULE

Review: Energy changes in chemical reactions (4.2, 4.4); enthalpy (4.3, 4.4); Hess's law (4.5); ionization energy (7.7); electron affinity (7.8).

8.5, 8.9 BOND
PARAMETERS:
BOND ENERGY,
BOND LENGTH,
ELECTRONEGATIVITY,
AND BOND
POLARITY

Objectives: You should be able to:

1. Explain the significance of electronegativity and in a general way relate the electronegativity of an element to its position in the periodic table.

2. Predict the relative polarities of bonds using either the periodic table or electronegativity values.

3. Relate bond energies to bond strengths and use bond energies to estimate ΔH for reactions.

Review: Chemical equation (3.1, 3.2); nomenclature rules (2.6).

8.10 OXIDATION
NUMBERS:
NOMENCLATURE

Objectives: You should be able to:

1. Assign oxidation numbers to atoms in molecules and ions.

2. Assign acceptable names to simple inorganic compounds and ions.

TOPIC SUMMARIES AND EXERCISES

In Chapter 6, the term *outermost electrons* is used in describing electrons in the highest-energy electron shell of atoms. Outermost electrons that are involved in chemical bonding are referred to as **valence electrons.**

The number of valence electrons for an atom is commonly shown by its **Lewis symbol,** or electron-dot formula.

LEWIS SYMBOLS—
VALENCE
ELECTRONS

• The electron-dot formula for an element consists of the chemical symbol for the element plus a dot for each valence electron. For example, the Lewis symbol for hydrogen ($1s^1$) is H· and for F ($2s^2 2p^5$) it is :F̈·

• Note that pairs of electrons are kept together and that each quadrant about the element's symbol can hold up to two electrons. Also, the placement of electrons can be in any of the quadrants. For example, equally valid Lewis structures are H· or Ḣ or ·H.

During chemical reactions, atoms often achieve a noble-gas electron configuration equivalent to the noble-gas closest to them in the periodic table.

- Since all noble gases except He have eight valence electrons, many atoms undergoing chemical reactions end up with eight valence electrons. This is known as the **octet rule.**
- You should note that some elements do not obey this rule. Some achieve a helium electron configuration, with two s electrons. Others can expand their valence shell to 10 or 12 electrons. The latter commonly occurs with the heavier, nonmetallic, representative elements.

EXERCISE 1   Write the Lewis symbols for the third-row representative elements (Na, Mg, Al, Si, P, S, Cl, and Ar).

Solution: In order to write the Lewis symbol for any element, you must first determine how many valence electrons are associated with it. To do this, write the shorthand notation description of the element's valence-electron configuration (see Section 6.9 of the text). For the third-row elements, the valence electrons are those beyond the neon electron configuration ($1s^2 2s^2 2p^6$):

Group number	Valence-electron configuration[a]	Lewis symbol
1A	$3s^1$	Na ·
2A	$3s^2$	Mg :
3A	$3s^2 3p^1$	A̤l ·
4A	$3s^2 3p^2$	S̤i ·
5A	$3s^2 3p^3$	· P̤ ·
6A	$3s^2 3p^1$	· S̤ ·
7A	$3s^2 3p^5$	: C̤l ·
8A	$3s^2 3p^6$	: A̤r :

[a]Note that the number of valence electrons for a representative element equals its group number.

EXERCISE 2   In the compounds SF_6 and PF_5, the elements sulfur and phosphorus expand their valence shells to 12 and 10 electrons, respectively. Write the Lewis symbols for sulfur and phosphorus with these valence shells.

Solution: To write the Lewis symbols for S and P with 12 and 10 electrons respectively, you will need to use more than the four valence orbitals you have been using. For S you will need to use six atomic orbitals, and for P you will need to use five, since each valence orbital can only hold a maximum of two electrons. Using lines to represent valence orbitals, you can write the Lewis symbols as:

Many inorganic salts, such as table salt (NaCl), are composed of a three-dimensional array of ions. The forces holding the ions together are largely electrostatic (coulombic attractions).

- A measure of the stability of a solid ionic substance is its **lattice energy.** This quantity is the energy required for one mole of a solid ionic substance to be separated completely into ions far removed from one another:

$$CaCl_2(s) \longrightarrow Ca^{2+}(g) + 2Cl^-(g)$$

- Lattice energy is approximated by the product of ion charges $(Q_1 Q_2)$ and to the reciprocal of the distance between two adjacent ions of opposite charge $(1/d)$:

$$E_{lattice} \propto \frac{Q_1 Q_2}{d}$$

- Note that *large* ion charges and *small* distances between oppositely charged ions yield the most stable ionic solids.

When atoms form ions in a crystal, they tend to achieve an octet by forming a valence-electron configuration equal to that of the *nearest* noble gas in the periodic table.

- Atoms that lose electrons to form positive ions (**cations**) are found on the left side of the periodic table: these are the metallic elements.
- Atoms that gain electrons to form negative ions (**anions**) are found on the right side: these are the nonmetallic elements.
- Transition elements usually do not obey the octet rule when they form ions. The outer electron configurations of these elements are either $(n - 1)d^x ns^2$ or $(n - 1)d^x ns^1$ ($n = 4, 5, 6$ and $x = 1 - 10$). When they form ions, the ns electrons are lost *first,* and then the necessary number of $(n - 1)d$ electrons are lost to form an ion of a particular charge. For example,

$$Fe\ (3d^6 4s^2) \longrightarrow Fe^{3+}\ (3d^5) + 3e^-$$
$$Cu\ (3d^{10} 4s^1) \longrightarrow Cu^+\ (3d^{10}) + e^-$$

- Note that an **isoelectronic series** of ions consists of ions having the same number of electrons.

You need to be able to predict the formula of simple ionic compounds.

- The cation is a metal and the anion is a nonmetal in simple ionic substances.
- The sum of positive charges must equal the sum of negative charges. Thus, the total charge of cations is exactly balanced by the total charge of anions. For example, if an ionic substance is composed of Ba^{2+} and I^- ions, the formula of the ionic substance is BaI_2. Two iodide ions form an overall 2– charge that is balanced exactly by one Ba^{2+} ion with a 2+ charge.

It is possible to measure the distance between ions in a crystal. By appropriate calculation methods, an average ion size can be assigned to cations and anions.

- When electrons are removed from an atom to form a cation, the cation is *smaller* than the corresponding neutral atom from which it is derived.

- Addition of electrons to an atom yields an anion that is *larger* than the corresponding neutral atom. You should note the trends in ion size across a row and within a family of the periodic table (see Section 8.3 in the text). Exercise 6 checks your understanding of such trends in ionic sizes.

- Ions in an isoelectronic series have the same number of electrons. The size of an ion in an isoelectronic series decreases with increasing atomic number. You should be able to explain this trend based on the attraction of an increasing number of protons in a nucleus for the same number of valence electrons.

EXERCISE 3   For the 1A, 2A, 3A, 6A, 7A, and 8A families, predict the most common ion that would occur in ionic compounds. Give an example of an ion for each family and also write the complete electron configuration for each example.

Solution: First, write the general description of the valence-electron configuration for each family. Then identify the family as typically metallic or nonmetallic. *Metallic elements lose electrons to form positive ions* (**cation**). *Nonmetallic elements gain electrons to form negative ions* (**anion**). The elements will lose or gain electrons to form the electron configuration of the *nearest* noble gas. All of this is summarized in the following table.

	1A	2A	3A	6A	7A	8A
Electron configuration	ns^1	ns^2	ns^2np^1	ns^2np^4	ns^2np^5	ns^2np^6
Type	Metal	Metal	Metal	Nonmet	Nonmet	Nonmet
Change in number of electrons to form ion	$-1e^-$	$-2e^-$	$-3e^-$	$+2e^-$	$+1e^-$	$+0e^-$
Ion charge	1+	2+	3+	2–	1–	0
Example	Li^+	Mg^{2+}	Al^{3+}	O^{2-}	F^-	He
Electron configuration of ion	$1s^2$	$1s^22s^22p^6$	$1s^22s^22p^6$	$1s^22s^22p^6$	$1s^22s^22p^6$	$1s^2$

EXERCISE 4   Write the electron configuration for each of the following ions and indicate which ions possess a noble-gas electron configuration: (**a**) S^{2-}; (**b**) Be^{2+}; (**c**) Cl^-; (**d**) Fe^{3+}.

Solution: (**a**) Sulfur has the electron configuration $[Ne]3s^23p^4$. Two electrons are added to the $3p$ orbitals to form the S^{2-} ion, which has the noble-gas electron configuration $[Ne]3s^23p^6$ or $[Ar]$. (**b**) Be has the electron configuration $[He]2s^2$. The two $2s$ electrons are lost to form the Be^{2+} ion, which has the noble-gas electron configuration $1s^2$ or $[He]$. (**c**) Cl has the electron configuration $[Ne]3s^23p^5$. One electron is added to a $3p$ orbital to form the Cl^- ion, which has the noble-gas electron configuration $[Ne]3s^23p^6$ or $[Ar]$. (**d**) Fe has the electron configuration $[Ar]3d^64s^2$. We find experimentally that when a transition element loses electrons to form an ion, the ns electrons are lost before the $(n-1)d$ electrons. The two $4s$ electrons and one $3d$ electron are lost to form the Fe^{3+} ion, which has the electron configuration $[Ar]3d^5$; this is not a noble-gas electron configuration.

EXERCISE 5   The cation of an ionic compound is usually from the left side or middle section of the periodic table, and the anion is usually from the right side of the periodic table. Predict the formula of the ionic compounds formed from the following pairs of elements: (**a**) Zn and Cl; (**b**) Al and I; (**c**) Rb and S.

Solution: (a) Zn occurs as a 2+ ion, and Cl occurs as a 1– ion with metals. The formula of neutral ionic compound must have a net zero charge, which requires that the sum of positive ion charges equals the sum of negative ion charges. The predicted formula contains one Zn^{2+} ion and two Cl^- ions: $ZnCl_2$. (b) Al occurs as a 3+ ion with the neon electron configuration, and iodine, like chlorine, occurs as a 1– ion. The predicted formula is AlI_3. (c) Rb occurs as a 1+ ion, and sulfur usually occurs as a 2– ion. The predicted formula is Rb_2S.

EXERCISE 6 Some general trends are observed for the radii of ions. In each of the following pairs of series, which series is arranged in order of *decreasing* radius? In each case state the general rule that the series illustrates.

(a) Mg^{2+}, Na^+, F^-, O^{2-} or O^{2-}, F^-, Na^+, Mg^{2+}

(b) F^-, Cl^-, Br^-, I^- or I^-, Br^-, Cl^-, F^-

(c) Mn^{2+}, Mn^{3+} or Mn^{3+}, Mn^{2+}

(d) Ca^{2+}, Cu^{2+} or Cu^{2+}, Ca^{2+}

Solution: (a) O^{2-}, F^-, Na^+, Mg^{2+}. These five ions form an **isoelectronic series** (a series of ions with the same electron configuration $1s^22s^22p^6$). The radii of ions in an isoelectronic series decrease with increasing atomic number. (b) I^-, Br^-, Cl^-, F^-. These five ions all belong to the halogen family. The radii of ions of the same charge in a family increase with increasing atomic number. (c) Mn^{2+}, Mn^{3+}. The radii of cations of the same element decrease with increasing charge. (d) Ca^{2+}, Cu^{2+}. The radii of Group 1B or 2B ions are smaller than the radii of 1A or 2A ions if the ions have the same charge and are in the same period.

EXERCISE 7 Arrange the oxides of magnesium, calcium and barium in order of increasing lattice energies, without referring to a table of data.

Solution: The trend in magnitude of the lattice energies can be estimated by using the relationship $Q(+)Q(-)/d$, where $Q(+)$ is the charge of the cation, $Q(-)$ is the charge of the anion and d is the closest distance between a cation and anion. For the oxides of magnesium, calcium and barium, the product of charges is a constant because the cations are all 2+, and the oxide ion is −2. Thus, the factor that causes the lattice energies to be different is the distance term, d. The trend in sizes of the cations is Mg < Ca < Ba which means that the trend in Metal—O distances is Mg—O < Ca—O < Ba—O. The trend in lattice energies should be MgO > CaO > BaO because lattice energy increases in magnitude with decreasing distance between the ions.

Most compounds do not show ionic properties; instead the combined atoms act as a unit and are not easily separated. These nonionic compounds have bonds between atoms, referred to as **covalent bonds,** which result from the *sharing* of one or more pairs of electrons.

The Lewis model for covalent bonds emphasizes the idea that atoms share valence electrons with other atoms in order to achieve a noble-gas electron configuration.

THE LEWIS MODEL FOR COVALENT BONDING

- **Lewis structures** show the arrangement of covalent bonds and unshared valence electron pairs about each atom in a compound.

- An example of the formation of a Lewis structure is shown for water:

$$2H\cdot + \cdot\overset{\cdot\cdot}{\underset{\cdot\cdot}{O}}\cdot \longrightarrow H - \overset{\cdot\cdot}{\underset{\cdot\cdot}{O}} - H$$

- Each dash represents a pair of bonding electrons in a covalent bond, and each dot represents an unshared valence electron.

In many molecules, the only way atoms can attain an octet is by sharing more than one pair of electrons between them: two and three electron pairs may be shared.

- In the example of H_2O, the H—O bond is referred to as a **single bond** because one pair of bonding electrons holds the two atoms together.
- **Multiple bonds** occur when there are insufficient valence electrons for all atoms to achieve an octet using single bonds only. For example, if you wrote the Lewis structure for ethylene (C_2H_4), using single bonds only, you would form structure (2):

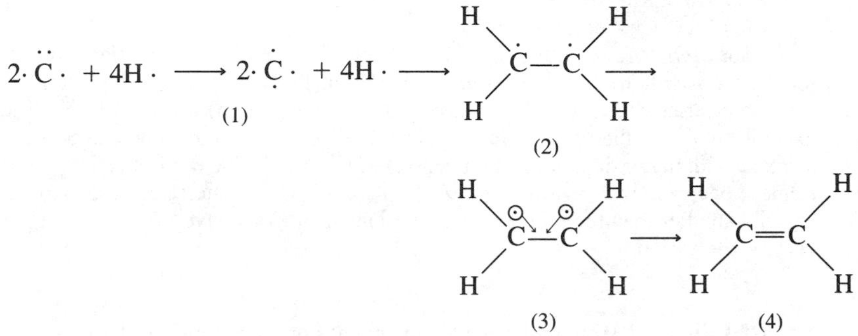

$$2 \cdot \ddot{C} \cdot + 4H \cdot \longrightarrow 2 \cdot \dot{C} \cdot + 4H \cdot \longrightarrow$$

(1)

(2)

(3)   (4)

- When atoms form bonds, valence electrons can unpair as shown in structure (1). Note in structure (2) that both carbons have only seven valence electrons. To achieve an octet about carbon, the electrons on the carbon atoms pair [see (3)] to form an *additional* C—C bond as shown in structure (4). This multiple bond between the carbon atoms in C_2H_4 is termed a **double bond.**
- Acetylene, is an example of a compound in which a **triple bond** binds two atoms together. The formation of multiple bonds occurs when atoms need to share extra electrons to form an octet.

$$H—C\equiv C—H$$

- The relative strengths of bonds in a series of related compounds is:
single < double < triple.

There are molecules or ions for which drawing one Lewis structure provides an incomplete picture of bonding. An example of this situation is the NO_3^- ion:

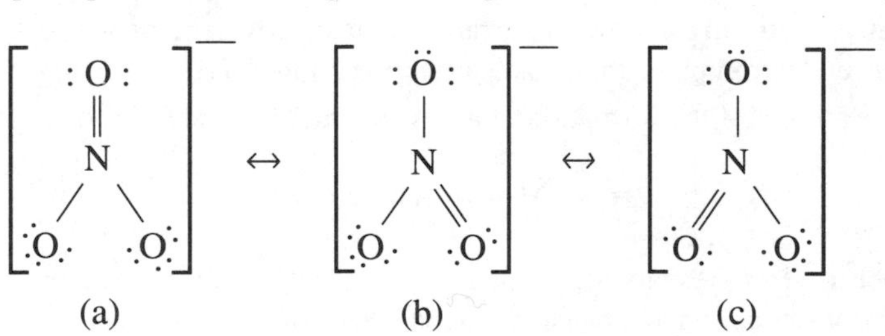

(a)   (b)   (c)

▪ According to experimental evidence, all bond lengths between N and O in the NO_3^- ion are equivalent. But any single Lewis structure for NO_3^- for example, Lewis structure (a) above—shows two kinds of bonds, single and double. As you learned earlier, double bonds are shorter than single bonds. To describe the structure of NO_3^- properly, you write all three Lewis structures and indicate that the real molecule is described by a blend of the three.

▪ All of the Lewis structures shown for NO_3^- are equivalent except for the placement of electrons. (Note that the nuclei are in the same position in each Lewis structure and that the number of unshared pairs is the same.)

▪ Equivalent Lewis structures of this sort are called **resonance structures** or **resonance forms.** Double-headed arrows are used to indicate resonance forms.

The following rules summarize the procedures for writing Lewis structures:

1. Determine the arrangement of the atoms in the compound with respect to each other. In simple molecules or ions, such as CCl_4, NO_3^-, HCN, or SO_3^{2-}, there is usually one central atom to which all other atoms are bonded. If the compound is binary (a compound containing only two kinds of elements), the first element given, such as C in CCl_4, is the central one. When a ternary compound (one with three kinds of elements), such as HCN, is written, the middle atom in the formula is usually the central one. If the compound is ionic, for example, NH_4Cl, the cation is separated from the anion and a Lewis structure is written for each ion.

2. Determine the total number of valence-shell electrons available for bonding by adding the number of valence-shell electrons for each atom. Also, if the compound has a negative charge, add one electron for each negative charge to the previous number of valence-shell electrons. If the compound has a positive charge, subtract one electron for each positive charge.

3. Place the valence-shell electrons about the atoms so that each atom achieves an octet (8) of electrons. Common exceptions to atoms achieving an octet are H (achieves 2 electrons), Be (achieves 4 electrons), B (achieves 6 electrons), and elements in the second, third, and fourth rows, such as P, S, and Si, which *may* achieve 10 to 12 electrons.

4. If there is an insufficient number of valence-shell electrons for atoms in the compound to achieve their required number (see 3), form double or triple bonds between appropriate atoms. Multiple bonds often occur among the atoms C, N, and O.

5. If there are more than the required number of valence-shell electrons for atoms to achieve their required number, expand the octets of the appropriate atoms (see 3).

6. If there is more than one way of arranging bonds and unshared electrons pairs to form equivalent Lewis structures, write all resonance forms.

EXERCISE 8 Write the Lewis structures for: **(a)** HOCl; **(b)** NH_4Cl; **(c)** SiH_4; **(d)** BH_4^-; **(e)** H_2CO

Solution: **(a)** When hydrogen and a halide such as chlorine occur with oxygen, oxygen is the central element. The total number of valence-shell electrons for HOCl is calculated as follows:

	No. of atoms	×	No. of valence electrons	=	Total no. of valence electrons
O:	1	×	6	=	6
H:	1	×	1	=	1
Cl:	1	×	7	=	7
			Total	=	14

Place the 14 electrons (7 pairs) about H, O, and Cl so that O and Cl each have an octet and H has two electrons:

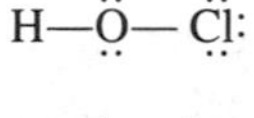

$$\text{H—}\ddot{\text{O}}\text{—}\ddot{\text{C}}\text{l:}$$

Finally, check your work to make sure that the total number of electrons used in your Lewis structure equals the total number available. **(b)** NH_4Cl is an ionic compound containing the cation NH_4^+ and the anion Cl^-. The single atom in compounds of type AB_x is usually the central atom. The total number of valence electrons for NH_4^+ is calculated as follows:

	No. of atoms	×	No. of valence electrons	=	Total no. of valence electrons
N:	1	×	5	=	5
H:	4	×	1	=	4
Subtract one electron for positive charge of ion				=	−1
			Total	=	8

For Cl^-, the number of valence electrons is 1 chlorine atom × 7 valence electrons $(3s^2 3p^5)$ + 1 electron for negative charge = 8 electrons (or 4 pairs). Place an octet of electrons about N and two electrons about each H. Separate the cation from the anion:

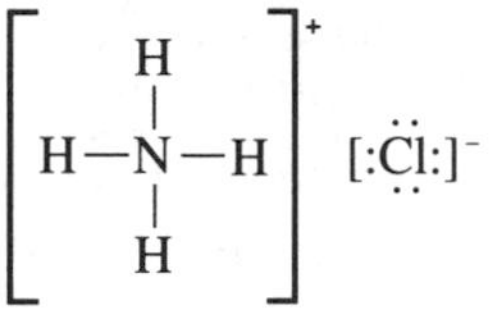

$$\left[\begin{array}{c} \text{H} \\ | \\ \text{H—N—H} \\ | \\ \text{H} \end{array}\right]^+ \quad [:\ddot{\text{C}}\text{l:}]^-$$

(c) Silicon is the central atom because it is the single atom in the type AB_x compound. The total number of valence electrons is

	No. of atoms	×	No. of valence electrons	=	Total no. of valence electrons
Si:	1	×	4	=	4
H:	4	×	1	=	4
			Total	=	8

Eight valence electrons represents an octet. Place four single bonds (four pairs of electrons) about silicon.

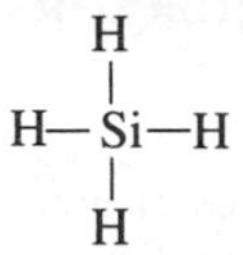

$$\begin{array}{c} H \\ | \\ H-Si-H \\ | \\ H \end{array}$$

Note: SiH_4 has the same Lewis structure as CH_4 which is shown in the text. This is another example of periodic relationships. **(d)** Boron is the central atom because it is the single atom in the type AB_x compound. The total number of valence electrons is

	No. of atoms	×	No. of valence electrons	=	Total no. of valence electrons
B:	1	×	3	=	3
H:	4	×	1	=	4
Add one electron for the negative charge of BH_4^-				=	1
			Total	=	8

As in **(c)** this represents an octet of electrons.

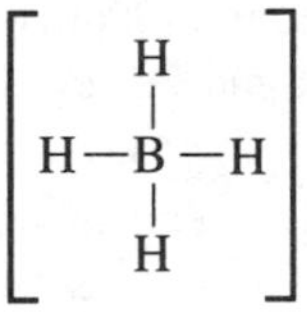

$$\left[\begin{array}{c} H \\ | \\ H-B-H \\ | \\ H \end{array} \right]$$

(e) In simple carbon-containing compounds, such as H_2CO, carbon is the central element. The total number of valence electrons for H_2CO is calculated as follows:

	No. of atoms	×	No. of valence electrons	=	Total no. of valence electrons
H:	2	×	1	=	2
C:	1	×	4	=	4
O:	1	×	6	=	6
			Total	=	12

The following Lewis structures use 12 electrons each, but in both cases the octet rule is not followed for the atoms circled.

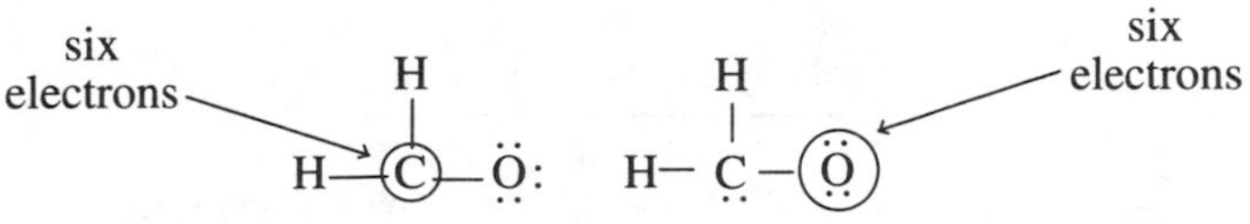

six
electrons

$$H-\overset{H}{\underset{}{C}}-\ddot{O}:$$

$$H-\overset{H}{\underset{}{C}}-\overset{..}{\underset{..}{O}}$$

six
electrons

In a case like this, you need to form a multiple bond between carbon and oxygen so both have a sufficient number of electrons to achieve an octet:

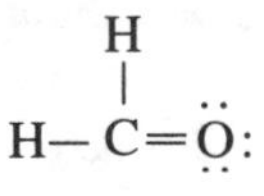

$$H-\overset{H}{\underset{}{C}}=\ddot{O}:$$

EXERCISE 9   From experiments it is known that the S—O bonds in SO_2 are equivalent. A Lewis structure for SO_2 is

$$:\ddot{O}-S=\ddot{O}$$

It shows that one S—O bond is a single bond and that the other S—O bond is a double bond. How does the Lewis model for covalent bonding explain this discrepancy?

Solution: There are, in fact, two Lewis structures for SO_2 that can be written:

$$:\ddot{O}-\ddot{S}=\ddot{O} \longleftrightarrow \ddot{O}=\ddot{S}-\ddot{O}:$$

not just one. The true structure of SO_2 is not represented by either structure, but has the character of both structures. Each S—O bond has some single-bond and some double-bond character. The two Lewis structures are called resonance structures. The overall effect of more than one valid Lewis structure is called resonance.

EXCEPTIONS TO THE OCTET RULE

Many molecules and ions contain atoms in which there are *not* eight electrons in their valence shell. The text discusses three exceptions to the octet rule.

- A few molecules and ions contain an *odd* number of electrons: ClO_2, NO and NO_2 are three examples you should know. Place the odd electron in a multiple bond between two atoms.

- A few molecules and ions contain atoms with *fewer* than eight electrons in their valence shell: H, Be, B and Al are the primary examples of elements which exhibit this behavior.

- The largest class of exceptions consists of molecules and ions which contain atoms having *more* than eight electrons in their valence shell. These atoms are from the third period and beyond; they can use their empty *d* orbitals to accept more than eight valence electrons. In general, expansion of octet occurs when the central atom is large and the surrounding atoms are small and strongly electron-attracting, such as F, Cl, and O.

EXERCISE 10   Write the Lewis structure for the odd-electron molecule NO.

Solution: The procedure for writing Lewis structures for molecules that do not obey the octet rule is given in the prior section. First, determine the number of valence electrons:

	No. of atoms	×	No. of valence electrons	=	Total no. of valence electrons
N:	1	×	5	=	5
O:	1	×	6	=	6
			Total	=	11

NO possesses an odd number of valence electrons. The first ten electrons can be used to form five pairs of electrons. Five pairs of electrons are insufficient to form an octet about each atom in NO. The extra (eleventh) electron is added in such a manner as to best agree with experimental data. When such data is not available, as in this exercise, place the odd electron in a multiple bond between the atoms.

$$:N\dot{=}O:$$

EXERCISE 11   Write the Lewis structures for **(a)** BI_3 and **(b)** PF_5.

   Solution: **(c)** Boron is the single atom in the type-AB_x compound and should be the central atom. The total number of valence-shell electrons for BI_3 is calculated as follows:

	No. of atoms	×	No. of valence electrons	=	Total no. of valence electrons
B:	1	×	3	=	3
I:	3	×	7	=	21
			Total	=	24

Place an octet of electrons about each I and six electrons about B (an exception to the octet rule):

$$:\ddot{I}—B—\ddot{I}:$$
$$|$$
$$:\ddot{I}:$$

(d) Phosphorus is the single atom and therefore should be the central one. The total number of valence-shell electrons for PF_5 is calculated as follows:

	No. of atoms	×	No. of valence electrons	=	Total no. of valence electrons
P:	1	×	5	=	5
F:	5	×	7	=	35
			Total	=	40

Phosphorus must expand its octet to accommodate five fluorine atoms. Place an octet about each F. This results in the presence of ten electrons about P:

$$
\begin{array}{c}
:\ddot{F}: \\
:\ddot{F}\diagdown \,|\,\diagup\ddot{F}: \\
P \\
:\ddot{F}\diagup \quad \diagdown\ddot{F}:
\end{array}
$$

Sections 8.5 and 8.9 of the text discuss three physical parameters and their utility in helping you to understand chemical bonds: bond energies, electronegativity and bond polarity.

   The first of these physical parameters is **bond energy.**

BOND PARAMETERS: BOND ENERGY, BOND LENGTH, ELECTRONEGATIVITY, AND BOND POLARITY

▪ Bond energy is defined as the enthalpy change required to break one mole of bonds in the gas state. When gaseous H_2 molecules dissociate to form gaseous hydrogen atoms, energy is required:

$$H_2(g) \longrightarrow 2H(g) \qquad \Delta H = 436 \text{ kJ/mol}$$

▪ In the above example, 436 kJ of energy is required when one mole of gaseous H—H bonds are broken.

▪ Another way of representing this energy change is $\Delta H = D(\text{H}\!-\!\text{H}) = 436$ kJ/mol where $D(\text{H}\!-\!\text{H})$ means the energy required to break or dissociate one mole of H—H bonds.

▪ In most molecules the bond energy of a particular bond depends on its environment; so when we speak of bond energy we are usually referring to an average bond energy. However, you should note that some bond energies, such as H—H and Cl—Cl, can be precisely determined. Tables of bond energies enable us to calculate approximate enthalpies of reactions using the relation

$$\Delta H_{\text{rxn}} = \Sigma D(\text{bonds } broken) - \Sigma D(\text{bonds } formed)$$

▪ The above equation is another example of the application of Hess's law of heat summation. However, heats of reaction are seldom calculated in this manner since bond energies usually do not represent exact values, only an average for a particular bond in several environments.

▪ Bond energies represent thermodynamic quantities that can be measured or calculated. They also tell us about the strength of a bond: The larger the bond energy, the stronger the bond. For example, bond strength and bond energy increase in the following order of bond types between atoms:
single < double < triple.

An average **bond length** for bond types can be defined. These are shown in Table 8.5 in your text. You should note the following:

▪ There is a relationship between bond energy and bond length. As bond energy for a particular type of bond increases, the bond length shortens.

▪ Also note that a bond between two atoms grows shorter and stronger as the number of bonds between the two atoms increases.

Another quantity that can tell us something about the nature of a bond and its strength is the electronegativity difference between bonded atoms.

▪ **Electronegativity** (*EN*) is the relative ability of an atom in a molecule to attract electrons to itself.

▪ If electrons are shared *equally* between two bonded atoms, the atoms have equal values of electronegativity, and the bond is said to be **nonpolar.**

▪ When electrons are *unequally* shared between two covalently bonded atoms, the atoms have differing electronegativity values, and the bond is said to be **polar covalent.**

▪ A large difference in electronegativities between two bonded atoms leads to the formation of an ionic bond.

▪ Atoms with high electronegativity values usually have high ionization energies and large electron affinities (elements in the top, right side of the periodic table). Atoms with low electronegativity values usually have low ionization energies and low electron affinities (elements in the left side of the periodic table).

- Electronegativities of elements have been calculated and tabulated and are listed in Figure 8.7 of the text.

A polar bond results from unequal sharing of electrons between bonded atoms; thus, the bonded atoms become partially charged with respect to one another.

- The **polarity of a bond,** that is, the separation of charges, can be represented by the symbol $\leftrightarrow$. The tip of the arrow points toward the more electronegative atom (the one with a negative partial charge), and the other end points toward the less electronegative atom (the one with a positive partial charge).
- You can represent the polarity of the H   Cl bond in two ways:

$$\underset{\text{H—Cl}}{\xrightarrow{\hspace{1cm}}} \qquad \overset{\delta^+ \quad \delta^-}{\text{H—Cl}}$$

where δ represents a partial charge, a value less than one. The tip of the arrow points toward chlorine because Cl has an electronegativity value of 3.2, whereas that of H is 2.2.

EXERCISE 12   Estimate ΔH for the reaction

$$H_2(g) + Cl_2(g) \longrightarrow 2HCl(g)$$

using the following bond-energy values:

$$D(\text{H—H}) = 436 \text{ kJ/mol}$$
$$D(\text{Cl—Cl}) = 242 \text{ kJ/mol}$$
$$D(\text{H—Cl}) = 431 \text{ kJ/mol}.$$

Solution: You can calculate ΔH for the above reaction using the relation

$$\Delta H_{\text{rxn}} = \Sigma D(\text{bonds } broken) - \Sigma D(\text{bonds } formed)$$

or, for the reaction being considered:

$$\Delta H_{\text{rxn}} = D(\text{H—H}) + D(\text{Cl—Cl}) - 2D(\text{H—Cl})$$

The number 2 in front of the term $D(\text{H—Cl})$ is required because 2 mol of H—Cl bonds are formed.

$$\Delta H_{\text{rxn}} = (1 \text{ mol})(436 \text{ kJ/mol}) + (1 \text{ mol})(242 \text{ kJ/mol}) - (2 \text{ mol})(431 \text{ kJ/mol})$$
$$= -184 \text{ kJ}$$

EXERCISE 13   For each of the following pairs, which series shows the correct trend of electronegativity?
(a) $EN_B > EN_C > EN_N > EN_O > EN_F$ or $EN_F > EN_O > EN_N > EN_C > EN_B$
(b) $EN_C > EN_{Si} \approx EN_{Ge}$ or $EN_{Ge} \approx EN_{Si} > EN_C$
(c) $EN_{Cs} > EN_{Ge} > EN_F$ or $EN_F > EN_{Ge} > EN_{Cs}$.
   Solution: (a) $EN_F > EN_O > EN_N > EN_C > EN_B$. Electronegativity generally increases from left to right across a period. (b) $EN_C > EN_{Si} \approx EN_{Ge}$. The electronegativities of second-row elements are greater than those of elements in the third and fourth rows. (c) $EN_F > EN_{Ge} > EN_{Cs}$. The atoms of highest electronegativity are found in the

upper right-hand section of the periodic table; the next highest in electronegativity are found in the middle and the lower right-hand sections; and the lowest are found in the left-hand section of the periodic table.

EXERCISE 14 (a) For each of the following bonds, indicate with the symbol → its polarity: H—C; H—O; C—N; and B—F. (b) Which bond is the most polar?

Solution: (a) Using the data in Figure 8.7 of the text, determine the electronegativity of each atom. The more electronegative atom of a pair of bonded atoms is assigned the partial negative charge.

	$\overset{+\longrightarrow}{\text{H—C}}$	$\overset{+\longrightarrow}{\text{H—O}}$	$\overset{+\longrightarrow}{\text{C—N}}$	$\overset{+\longrightarrow}{\text{B—F}}$
Electronegativity values	↑ ↑ 2.2 2.5	↑ ↑ 2.2 3.4	↑ ↑ 2.5 3.0	↑ ↑ 1.8 4.0
Difference	0.3	1.2	0.5	2.2

(b) Calculate the difference in electronegativity values for each pair of atoms. The pair with the largest difference will be the most polar: B—F.

OXIDATION NUMBERS: NOMENCLATURE

Oxidation number reflects the charge assigned to an atom in a particular bonding situation. It is also a tool that helps you keep track of electrons in a chemical reaction.* The terms *oxidation number* and *oxidation state* are closely related. An atom in an oxidation state of +2 is said to have a +2 oxidation number. An atom may have several possible oxidation states, positive and negative, depending on the identity of the atoms bonded to it.

A few general rules have been developed to aid you in assigning oxidation numbers to atoms in compounds. *You must memorize these now.*

1. The oxidation number of an element in its elementary or uncombined state is zero. For example, in Cl_2 each chlorine atom has an oxidation number of zero.

2. In an ionic compound, the oxidation number of a monatomic ion is the same as its charge. For example, in KCl potassium has an oxidation number of +1 and the chloride ion has one of −1.

3. Certain elements almost always have the same oxidation number in their compounds. These elements are:

(a) Group 1A elements (Li, Na, K, Rb, Cs), with an oxidation number of +1;

(b) Group 2A elements (Be, Mg, Ca, Sr, Ba), +2;

(c) Group 3A elements (B, Al), +3;

(d) Fluorine, chlorine, bromine, iodine, −1 in binary compounds with metals (other states are possible when combined with nonmetals);

(e) Hydrogen: +1 (except for metallic hydrides, where its oxidation state is −1, as in CaH_2 and NaH);

---

*Caution Oxidation number may be related to the charge of an atom or it may have no relationship. For example, sulfur in $Na_2S_4O_6$ has an oxidation number of $+2\frac{1}{2}$, but an atom cannot have $\frac{1}{2}$ an electron, only an integral number.

(f) Oxygen: −2 (except in peroxide compounds, where the oxidation state is −1; in superoxide ions [O_2^-], where it is −$\frac{1}{2}$; and when bonded to F [+2 in OF_2].

4. In an A-B compound, the more electronegative element is assigned the negative oxidation number; the less electronegative one is assigned a positive oxidation number (for example, in SF_4, F is assigned −1 and thus S is +4).

5. In a neutral compound, the sum of oxidation numbers of all atoms is zero; in a compound with a charge, the sum is equal to the charge of that compound.

Oxidation numbers are also an aid for naming compounds. In this section, you review the rule first given in Section 2.6 in the text: If more than one oxidation state exists for an element, the oxidation state is given in Roman numerals after the element's name and is enclosed in parentheses. For example, $FeCl_3$ is iron(III) chloride. Note that with this method of nomenclature the number of atoms is not indicated for binary compounds.

EXERCISE 15   State the most common oxidation number(s) for each of the following elements and give an example of a compound in which an atom of the element has that oxidation number: **(a)** Li; **(b)** Sr; **(c)** Al; **(d)** Ag; **(e)** N; **(f)** O; **(g)** F; **(h)** Zn; **(i)** S; **(j)** H.
  Solution: **(a)** +1: LiF. All atoms in group 1A have a +1 oxidation number. **(b)** +2: SrO. All atoms in group 2A have a +2 oxidation number. **(c)** +3: $AlCl_3$. **(d)** +1: AgCl or Ag_2O. **(e)** −3: NH_3. +5: NO_3^-. **(f)** −2: CaO. **(g)** −1: HF. **(h)** +2: ZnS. **(i)** +6: SF_6. −2: H_2S. **(j)** +1: HCl. −1: NaH (only with metallic elements).

EXERCISE 16   Name the following compounds: **(a)** SF_6; **(b)** NO; **(c)** N_2O; **(d)** XeO_3; **(e)** P_2O_5.
  Solution: When an element has more than one possible oxidation state, its oxidation state is indicated by using a Roman numeral in parentheses after the name of the element: **(a)** sulfur(VI) fluoride; **(b)** nitrogen(II) oxide; **(c)** nitrogen(I) oxide; **(d)** xenon(VI) oxide; **(e)** phosphorus(V) oxide.

EXERCISE 17   Write chemical formulas for the following compounds: **(a)** nitrogen(III) fluoride **(b)** arsenic(III) oxide **(c)** cobalt(II) sulfate **(d)** manganese(IV) oxide.
  Solution: **(a)** NF_3; **(b)** As_2O_3. Note that oxygen has an oxidation number of −2; thus three oxygen atoms are required (3 × −2 = −6) to balance the total oxidation number of arsenic atoms (2 × +3 = +6). **(c)** $CoSO_4$. **(d)** MnO_2. Note that two oxygen atoms are required to balance the total oxidation number of manganese with an oxidation number of +4. By now you may have noted the following for binary compounds containing atom A with an oxidation number of +y and atom B with oxidation number of −x: The chemical formula of the compound can be generated by using the absolute value of the oxidation number as the subscript of the other atom, with integers reduced to their smallest ratio of whole numbers:

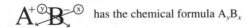   has the chemical formula A_yB_x

$Mn^{+4}O^{-2}$ becomes Mn_2O_4; you must reduce the subscripts to the simplest ratio of whole numbers—$Mn_{2/2}O_{4/2}$ or MnO_2.

SELF-TEST QUESTIONS

Key Terms

Having reviewed the key terms listed at the end of Chapter 8 and their applications, identify the following statements as true or false. If a statement is false, indicate why it is incorrect. Key terms are italicized in the statements.

8.1 The *Lewis structure* of N_2 is :N̈=N̈:

8.2 In the reaction

$$BF_3(g) + NH_3(g) \longrightarrow F_3B{:}NH_3(g)$$

the product formed contains a *single covalent bond*.

8.3 The bond between Ca and O in CaO is *covalent* in nature.

8.4 For the following hydrogen halides, *bond polarity* increases in the order

$$H{-}Cl < H{-}Br < H{-}I.$$

8.5 The ionic *valence* of each element in BaO is 2.

8.6 The *oxidation number* of iron in Fe_2O_3 is +3.

8.7 In $MgCl_2$, the Mg—Cl bond is *ionic*.

8.8 The *lattice energy* of MgO is less than that of NaCl.

8.9 In H_3CCH_3, ethane, there exists a *double bond*.

8.10 In HCCH, acetylene, there exists a *triple bond*.

8.11 The three ions O^{2-}, F^-, and Na^+ form an *isoelectronic series*.

8.12 The *valence electrons* of Fe are the two 4s electrons.

8.13 The fact that H and N have differing *electronegativity* values results in the N—H bond in NH_3 having a bond polarity.

8.14 When Cl_2 dissociates to form two chlorine atoms in the gas phase, an exothermic process is involved, and this energy is termed *bond energy*.

8.15 The bond in Cl_2 is an example of a *polar covalent bond*.

8.16 The reaction of calcium with oxygen to form CaO illustrates the *octet rule*.

8.17 The *oxidation state* of iodine in ICl is −1.

8.18 One of the three *resonance forms* of SO_3 is

8.19 A *Born-Haber cycle* used to calculate lattice energy will have in the cycle steps for ionization of a metal, electron affinity for a nonmetal, and the formation of gaseous atoms from elements.

8.20 The *formal charge* of an atom in an ionic compound is simply its ionic charge.

8.21 An *ionic bond* is characterized by a small difference in electronegativity values between the bonded atoms.

8.22 $FeCl_3$ contains a *metallic bond*.

8.23 The *electron-dot symbol* for silicon is S̈i·

8.24 CO_2 contains a *covalent bond*.

8.25 *Multiple bonding* occurs among the atoms carbon, nitrogen and oxygen.

8.26 The *bond-dissociation energy* of a C=N bond is larger than that for a C≡N bond.

8.27 Na_2 contains a *nonpolar covalent bond*.

Problems and Short-Answer Questions

8.28 When Li(s) reacts with $F_2(g)$, LiF(s) is produced, and 610 kJ/mol of heat is released. Since it requires energy to break the F—F bond in F_2 and to form Li^+ from Li(s), why should the formation of LiF(s) from Li(s) and $F_2(g)$ release energy?

8.29 Predict the formula of the solid ionic compound formed when each of the following pairs of atoms are combined:
    (a) Na and I   (c) Ca and Br
    (b) Ba and S   (d) Ca and H

8.30 Which element or ion in each of the following pairs is the larger?
    (a) Na or Na^+ (c) O^- or O^{2-}
    (b) Fe^{3+} or Fe^{2+} (d) K^+ or Cl^-

8.31 Draw the Lewis structures for:
    (a) $HCCCH_3$
    (b) ICl
    (c) $CaBr_2$
    (d) H_3O^+
    (e) CS_2 (S—C—S arrangement of atoms)

8.32 Name the following compounds:
    (a) SO_3 (f) $Fe(NO)_2$
    (b) Ag_2S (g) AuCl
    (c) $AsCl_3$ (h) $AuCl_3$
    (d) ICl   (i) $HgCl_2$
    (e) $Fe(NO_3)_3$

8.33 (a) Draw the Lewis structures for the following series of related compounds: N_2, HN_2H, and $H_2N_2H_2$.
    (b) How do the N—N bond distances and bond energies vary in this series?

8.34 Assign oxidation numbers to the elements in the following compounds:
    (a) HCl   (d) $CaCO_3$
    (b) H_2SO_4 (e) AlF_3
    (c) $Fe(NO_3)_2$ (f) $SnCl_6^{2-}$

8.35 (a) Arrange the following pairs of covalently bonded atoms in expected order of increasing bond polarity: C—N, P—Cl, I—Cl, and C—O.
    (b) For each pair, indicate the direction of the polarity of the bond, using the symbol ↔.

8.36 Estimate the standard enthalpy of formation of ammonia given the following average bond energies in kilojoules:

D(N—H), 391; D(N—N), 163; D(N=N), 418; D(N≡N), 941; D(H—H), 436.

8.37 Explain why the lattice energy of magnesium sulfide (3406 kJ/mol) is less than that of magnesium oxide (3850 kJ/mol).

Multiple-Choice Questions

8.38 What is the valence-electron configuration for Se^{2-}?
(a) $2s^2 2p^4$
(b) $2s^2 2p^6$
(c) $4s^2 4p^4$
(d) $4s^2 4p^6$
(e) $3s^2 3p^2$

8.39 Which of the following ionic crystals would you expect to have the largest lattice energy?
(a) LiCl
(b) LiBr
(c) CaO
(d) SrO
(e) $BaSO_4$

8.40 Which of the following is most likely to be an ionic compound?
(a) P_2O_5
(b) $SbCl_5$
(c) H_2Te
(d) $SrCl_2$
(e) (a), (b), and (c)

8.41 Why doesn't the ion Na^{2+} occur in ionic crystals?
(a) It would not be strongly attracted to an anion.
(b) The energy for the second ionization of sodium is too large.
(c) Insufficient lattice energy is released to compensate for the endothermic formation of gaseous Na^{2+} ions.
(d) (a) and (b)
(e) (b) and (c)

8.42 What is the valence-electron configuration for Ni^{2+}?
(a) $3d^6 4s^2$
(b) $3d^6 4s^1$
(c) $3d^8$
(d) $3d^5 4s^2$
(e) $3d^4 4s^2$

8.43 What is the correct order of increasing radii for the isoelectronic series Rb^+, Sr^{2+}, Se^{2-}, Br^-?
(a) $Rb^+ < Sr^{2+} < Se^{2-} < Br^-$
(b) $Br^- < Se^{2-} < Sr^{2+} < Rb^+$
(c) $Se^{2-} < Br^- < Rb^+ < Sr^{2+}$
(d) $Sr^{2+} < Rb^+ < Br^- < Se^{2-}$
(e) none of the above

8.44 Which of the following has the largest ionic radius?
(a) Be^{2+}
(b) Mg^{2+}
(c) Ca^{2+}
(d) Sr^{2+}
(e) Ba^{2+}

8.45 Which of the following compounds contains a double bond?
(a) N_2
(b) SO_2
(c) Cl_2
(d) NH_4^+
(e) SO_4^{2-}

8.46 Given the following average bond energies:

$$C—C = 348 \text{ kJ/mol}$$
$$C=C = 614 \text{ kJ/mol}$$
$$C≡C = 839 \text{ kJ/mol}$$

which of the following is most likely to contain a C—C bond that has a bond energy of 820 kJ/mol?
(a) C_2H_6
(b) C_3H_8
(c) C_4H_{10}
(d) C_2H_4
(e) C_2H_2

8.47 In which of the following compounds is the direction of the symbol ↔ *not* indicated correctly?
(a) HBr: H ↔ Br
(b) SO_2: S ↔ O
(c) BI_3: I ↔ B
(d) H_2S: H ↔ S
(e) none of the above

8.48 Which of the following is a valid Lewis structure for CO?
(a) :C̈—Ö:
(b) :C=O:
(c) :C̈=Ö:
(d) C≡O:
(e) :C≡O:

8.49 Which of the following is a valid Lewis structure for C_2Cl_4?

(a) 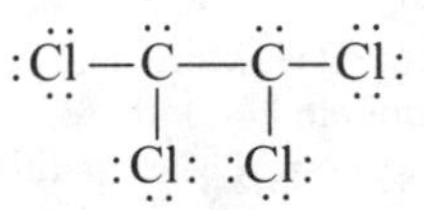 :C̈l—C—C—C̈l: with :Cl: :Cl: below

(b) :C̈l—C—C—Cl: with :Cl: :Cl: below

(c) :C̈l—C=C—C̈l: with :Cl: :Cl: below

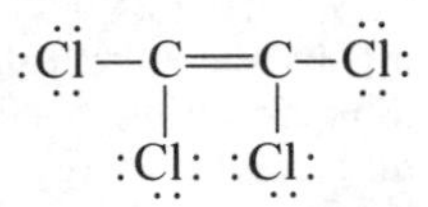

(d) 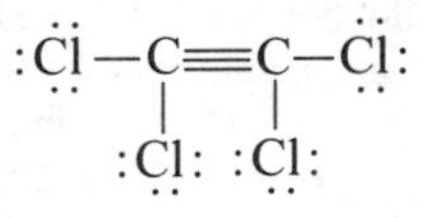 :C̈l—C≡C—C̈l: with :Cl: :Cl: below

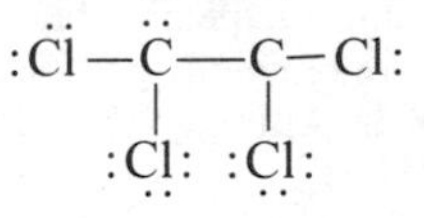

(e) none of the above

8.50 Which of the following are valid resonance structures for HN_3?
(a) H—N≡N—N̈
(b) H—N̈—N≡N:
(c) H—N̈=N=N̈:
(d) (a) and (b)
(e) (a), (b), and (c)

SELF-TEST SOLUTIONS

8.1 False. Each nitrogen atom contributes five valence electrons for a total of ten, or five pairs. The Lewis structure shown has six pairs and therefore is incorrect. The correct Lewis structure is N≡N **8.2** True. **8.3** False. Oxides of metals are ionic. **8.4** False. Bond polarity increases with increasing difference in electronegativity of the two bonded atoms. The order of electronegativity is: Cl > Br > I. Given that the electronegativity of hydrogen is 2.2, the most polar bond occurs when chlorine is bonded to hydrogen. **8.5** True. **8.6** True. **8.7** True. **8.8** False. Lattice energy is proportional to $Q_1 Q_2/d$. For MgO, $Q_1 Q_2$ is $2 \times 2 = 4$, and for NaCl it is $1 \times 1 = 1$. The lattice energy of MgO is larger because the product of MgO ion charges is four times greater than that of NaCl ion charges. **8.9** False. The Lewis structure of ethane is

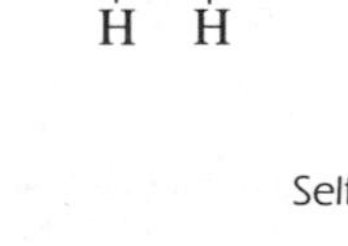

$$H-\underset{\underset{H}{|}}{\overset{\overset{H}{|}}{C}}-\underset{\underset{H}{|}}{\overset{\overset{H}{|}}{C}}-H$$

There is only a single bond between carbons, not a double bond. **8.10** True. The Lewis structure of acetylene is H—C≡C— H. **8.11** True. They all have the same electronic configuration, $1s^2 2s^2 2p^6$. **8.12** False. The valence electrons of iron also include the $3d^6$ electrons: $3d^6 4s^2$. **8.13** True. **8.14** False. Bond energy is an endothermic quantity (energy required). **8.15** False. The Cl—Cl bond is a nonpolar covalent bond because both atoms have identical electronegativity values. **8.16** True. In CaO both atoms achieve an octet of valence electrons. **8.17** False. Because chlorine is more electronegative than iodine, it is assigned a −1 oxidation state value; therefore iodine must have a +1 oxidation state so that the sum of oxidation states will equal zero. **8.18** True. **8.19** True. In addition, there will also be a term for the formation of the crystal from the gaseous ions. **8.20** True. Note-This will not be true for an atom contained within a polyatomic ion, but for simple salts such as $MgCl_2$ the statement is true. **8.21** False. An ionic bond is characterized by a large difference in electronegativity values. **8.22** False. $FeCl_3$ is an ionic compound. A metallic bond occurs when two metals are bonded together. **8.23** True. **8.24** True. **8.25** True. Most occurrences observed involve these three atoms. **8.26** False. More energy is required to break a triple bond than a double bond; therefore the bond-dissociation energy of a triple bond is larger than that of a double bond. **8.27** False. It contains a metallic bond. A nonpolar covalent bond involves two nonmetals possessing equivalent electronegativity value. **8.28** Although the two processes $F_2(g) \rightarrow 2F^-(g) + 2e^-$ and $Li(s) \rightarrow Li^+(g) + e^-$ require energy, the energy released in $Li^+(g) + F^-(g) \rightarrow LiF(s)$ more than compensates for this required energy. Thus, the formation of $LiF(s)$ from $Li(s)$ and $F_2(g)$ releases energy. **8.29** The sum of ion charges in a formula must equal zero for an ionic compound. **(a)** Na belongs to group 1A and thus has a 1+ charge; I has a 1− charge in ionic compounds. Therefore, they occur in a 1:1 ratio in an ionic compound because they have the same charge. The formula is NaI. **(b)** Ba belongs to group 2A and thus has a 2+ charge; and S, like O, normally has a 2− charge. The formula is BaS. **(c)** Ca, like Ba, has a 2+ charge; and Br, like I, has a 1− charge. Two bromine ions are needed to equal the 2+ charge of Ca. The formula is $CaBr_2$. **(d)** Ca has a 2+ charge. Since H cannot also have a positive charge, it must have a negative charge, in the form of a hydride ion (H^-). Two hydride ions are needed to balance the 2+ charge of Ca. The formula is CaH_2. **8.30 (a)** Na. the removal of an electron from Na to form Na^+ reduces repulsions between the remaining electrons in Na^+, allowing the electron cloud to contract. This makes Na^+ smaller than Na. **(b)** Fe^{2+}. The larger charge of Fe^{3+} causes the electron cloud to contract to a greater degree, making Fe^{3+} smaller than Fe^{2+}. **(c)** O^{2-}. The extra electron added to O^- to form O^{2-} expands the size of the O^- electron cloud; hence O^{2-} is larger than O^-. **(d)** Cl^-, K^+ and Cl^- are isoelectronic ions because they have the same electron configuration: [Ar]. Cl^- has fewer protons than K^- to attract its electrons. As a result, Cl^- is the larger ion. **8.31 (a)** The structural formula is

$$H-\overset{\displaystyle \underset{|}{H}}{\underset{|}{C}}$$

H—C—C—C—H with H above and below the third carbon

The number of valence electrons is (4 H atoms)(1 valence electron) + (3 C atoms)(4 valence electrons) = 16 electrons or 8 pairs. Note that the Lewis structure

requires 20 electrons; thus, multiple bonds are required and the correct Lewis structure is

H—C≡C—C—H (with H above and below the final carbon)

(b) For ICl, the number of valence electrons is (1 I atom)(7 valence electrons) + (1 Cl atom)(7 valence electrons) = 14 electrons or 7 pairs. The Lewis structure :Ï—Cl: fits this requirement. **(c)** $CaBr_2$ is an ionic compound; therefore, the ions must be separated. All of calcium's valence electrons are removed to form a 2+ ion. Each bromine atom adds one electron to achieve a rare gas electron configuration. The Lewis structure is $[:Br:]^- [Ca]^{2+} [:Br:]^-$. **(d)** In H_3O^+, all of the hydrogen atoms are bonded to oxygen. The number of valence electrons is (3 H atoms)(1 valence electron) + (1 O atom)(6 valence electrons) − 1 electron for positive charge = 8 valence electrons or 4 pairs. The Lewis structure is

$$\left[\begin{array}{c} H \\ | \\ \overset{..}{O} \\ H \quad H \end{array} \right]^+$$

(e) Carbon is the central element. The number of valence electrons is (1 C atom)(4 valence electrons) + (2 S atoms)(6 valence electrons) = 16 valence electrons, or 8 pairs. In order to achieve an octet of electrons about each atom, multiple bonds must be used and the Lewis structure is :S=C=S: **8.32 (a)** sulfur(VI) oxide or sulfur trioxide; **(b)** silver(I) sulfide; **(c)** arsenic(III) chloride or arsenic trichloride; **(d)** iodine(I) chloride or iodine monochloride; **(e)** iron(III) nitrate; **(f)** iron(II) nitrate; **(g)** gold(I) chloride; **(h)** gold(III) chloride; **(i)** mercury(II) chloride. **8.33 (a)** Using the procedures for determining the number of valence electrons and Lewis structures yields

:N≡N: H—N=N—H   N—N—N—H (with H below each of the central nitrogens)

(b) Bond strengths and bond energies increase with an increasing number of bonds between atoms. With increasing bond strength, bond distances shorten. Thus, for these three molecules N—N bond distances increase in the order N_2 < HN_2H < $H_2N_2H_2$. The N—N bond energies increase in the following order: $H_2N_2H_2$ < HN_2H < N_2 (that is, single bond < double bond < triple bond). **8.34** Using the rules given in this chapter yields the following: **(a)** HCl: H has an oxidation number of +1; Cl has an oxidation number of −1. The sum of the oxidation numbers is zero: $(1)(+1) + (1)(−1) = 0$. **(b)** H_2SO_4: H has an oxidation number of +1; S has an oxidation number of +6; O has an oxidation number of −2. The sum of the oxidation numbers is zero: $(2)(+1) + (1)(+6) + (4)(−2) = 0$. **(c)** $Fe(NO_3)_2$: Fe has an oxidation number of +2; N has an oxidation number of +5; O has an oxidation number of −2. The sum of the oxidation numbers is zero: $(1)(+2) + 2[(1)(+5) + (3)(−2)] = 0$. **(d)** $CaCO_3$: Ca has an oxidation number of +2; C has an oxidation number of +4; O has an oxidation number of −2. The sum of the oxidation numbers is zero: $(1)(+2) + (1)(+4) + (3)(−2) = 0$. **(e)** AlF_3: Al has an oxidation number of +3; F has an oxidation number of −1. The sum of the oxidation numbers is zero: $(1)(+3) + (3)(−1) = 0$. **(f)** $SnCl_6^{2-}$: Sn has an oxidation number of +4; Cl has an oxidation number of −1. Because the ion has a 2− charge, the sum of the oxidation numbers is −2: $(1)(+4) + (6)(−1) = −2$. **8.35 (a)** I—Cl ≈ C—N < C—O < P—Cl. The order is determined by differences in electronegativity values between the pairs. The greater the difference in electronegativity, the more polar the bond. Electronegativity values are as follows: I = 2.7, Cl = 3.2, difference = 0.5; C = 2.5, N = 3.0, difference = 0.5; C = 2.5 O = 3.4, difference = 0.9; P = 2.2, Cl = 3.2, difference = 1.0. **(b)** I → Cl; C → N; C → O; P → Cl. **8.36** The reaction for the formation of ammonia from its constituent elements is $N_2(g) + 3 H_2(g) → 2 NH_3(g)$. Thus, the standard enthalpy of formation is estimated by using the relationship ΔH_f = sum bond energies of reactants-sum bond energies of products = $D(N≡N) + 3D(H—H) − 2 × 3D(N—H) = 941$ kJ + 3(436 kJ) − 6(391 kJ) = −97 kJ. $D(N≡N)$ is used in the calculation because the nitrogen atoms in N_2 are bonded together by a triple bond. $2 ×$ 3D(N—H) is used in the calculation because there are two ammonia molecules and each ammonia molecule contains three N—H bonds; therefore there are a total of six N—H bonds. **8.37** Lattice energy is proportional to the product of ion charges, divided by the distance between the positive and negative ions. In this case, the charges of the positive ion, Mg^{2+} and the negative ions, O^{2-} and S^{2-} are a constant factor; thus the lattice energy difference between the two salts must result from a difference in distances between the ions. The oxide ion is a smaller ion than the sulfide ion; this should result in a Mg—O distance that is smaller than a Mg—S distance. Lattice energy increases in magnitude with decreasing distance between the ions; thus, the lattice energy of Mg—O should be greater than that of Mg—S. **8.38 (d)**. **8.39 (c)**. Lattice energy is proportional to Q_1Q_2/d. For CaO, $Q_1Q_2 = 2$ (same as for SrO and $BaSO_4$). And CaO has the smallest value of d. Thus it has the largest lattice energy. **8.40 (d)**. Ionic compounds usually contain a metal, and the electronegativity difference between bonded atoms is large. **8.41 (e)**. **8.42 (c)**. $4s$ electrons are lost before $3d$ electrons when transition metal ions are formed. **8.43 (d)**. In an isoelectronic series, size decreases with increasing atomic number. **8.44 (e)**. Size increases down a family. **8.45 (b)**. Draw the Lewis structures for all. Only SO_2 has a double bond

$$:\ddot{O}—\ddot{S}=\ddot{O}:$$

8.46 (e). Value of bond energy corresponds to that of a C≡C bond. Only C_2H_2 (H—C≡C—H) has a triple bond. C_2H_4 has a double bond, and all others have a single bond. **8.47 (c)**. **8.48 (e)**. The octet rule is obeyed, and the total number of valence electrons is correct. **8.49 (c)**. **8.50 (e)**.

9 | Molecular Geometry and Bonding Theories

OVERVIEW OF THE CHAPTER

9.1, 9.2 VSEPR MODEL: A TOOL FOR PREDICTING MOLECULAR STRUCTURE AND DIPOLE MOMENTS

Review: Lewis structures (8.1, 8.6); bond polarity (8.5).

Objectives: You should be able to:

1. Relate the number of electron pairs in the valence shell of an atom in a molecule to the geometrical arrangement around that atom.
2. Explain why unshared electron pairs exert a greater repulsive interaction on other pairs than do shared electron pairs.
3. Predict the geometrical structure of a molecule or ion from its Lewis structure.
4. Predict, from the molecular shape and the electronegativities of the atoms involved, whether a molecule can have a dipole moment.

9.3, 9.4 COVALENT BONDING, HYBRID ORBITALS AND MOLECULAR STRUCTURE

Review: Orbital diagrams (6.8); orbital types and shapes (6.6); concept of valence orbitals (6.8).

Objectives: You should be able to:

1. Explain the concept of hybridization and its relationship to geometrical structure.
2. Assign a hybridization to the valence orbitals of an atom in a molecule, knowing the number and geometrical arrangement of the atoms to which it is bonded.

9.5 HYBRIDIZATION IN MOLECULES CONTAINING BOTH SIGMA AND PI BONDS

Review: Lewis structures containing multiple bonds (8.4, 8.6, 8.7); resonance (8.6).

Objectives: You should be able to:

1. Formulate the bonding in a molecule in terms of σ bonds and π bonds, from its Lewis structure.
2. Explain the concept of delocalization in π bonds.

140

Objectives: You should be able to:

1. Explain the concept of orbital overlap and the reason why overlap may in some cases be zero because of symmetry.

2. Describe how molecular orbitals are formed by overlap of atomic orbitals.

3. Explain the relationship between bonding and antibonding molecular orbitals.

4. Construct the molecular-orbital energy-level diagram for a diatomic molecule or ion built from elements of the first or second row and predict the bond order and number of unpaired electrons.

TOPIC SUMMARIES AND EXERCISES

The three-dimensional array of atoms in a molecule, referred to as the geometrical structure of the molecule, depends on many factors. The valence-shell electron-pair **(VSEPR)** model contains concepts that will help you predict the geometrical shapes of simple molecules.

VSEPR MODEL: A TOOL FOR PREDICTING MOLECULAR STRUCTURE AND DIPOLE MOMENTS

▪ An essential idea of the VSEPR model is that valence-shell electron-pair repulsions about a central atom in a molecule are minimized, and that this minimization of repulsions is primarily responsible for the geometrical arrangement of bonded atoms about the central atom.

▪ The electrons responsible for the geometrical structure of a molecule are of two types: nonbonding pairs (also called unshared pairs or lone pairs) and bonding electron pairs.

▪ The geometrical structure of a molecule can be predicted using the following procedures:

1. Draw the Lewis structure of the molecule or ion, showing all bonding and nonbonding pairs of electrons.

2. Count the total number of nonbonding electron pairs and single-bonding electron pairs about the central atom. (*A multiple bond is counted as a single-bonding electron pair.*)

3. Use Table 9.1 (or Tables 9.1, 9.2, and 9.3 of the text) to determine the arrangement of atoms about the central atom. *Note that the structure of a molecule is described in terms of arrangement of atoms, not electron pairs.* For example, consider molecules where the total number of valence-electron pairs is the same: In the series CH_4, $:NH_3$, and $H\ddot{O}H$, each central atom is surrounded by four electron pairs. The geometrical arrangement of valence-electron pairs is the same for all in the series (tetrahedral). However, the structures of the molecules (in terms of arrangement of atoms) differ because of differing numbers of atoms about the central atom (CH_4, tetrahedral; NH_3, trigonal pyramidal; H_2O, nonlinear).

4. In a molecule or ion, atoms bonded to the central atom arrange themselves so as to minimize electron repulsions. The strength of this repulsion is in the following order:

nonbonding pairs-nonbonding pairs >
nonbonding pairs-bonding pairs >
bonding pairs-bonding pairs

These electron-pair repulsions cause distortions from the idealized geometry.

Table 9.1   Relationship between numbers of bonding and nonbonding electron pairs and molecular structure

Molecular type[d]	Total number of electron pairs about central atom	Number of single-bonding electron pairs	Number of unshared electron pairs (E)	Arrangment of atoms about central atom[a]	Example
AX_2	2	2	0	Linear	BeH_2
AX_3	3	3	0	Trigonal planar	BCl_3
AX_2E	3	2	1	Nonlinear	SO_2
AX_4	4	4	0	Tetrahedral	CH_4
AX_3E	4	3	1	Trigonal pyramidal	NH_3
AX_2E_2	4	2	2	Nonlinear	H_2O
AX_5	5	5	0	Trigonal bipyramidal	PF_5
AX_4E	5	4	1	Seesaw[b]	$TeCl_4$
AX_3E_2	5	3	2	T-shaped[b]	ClF_3
AX_2E_3	5	2	3	Linear[b]	XeF_2
AX_6	6	6	0	Octahedral	SF_6
AX_5E	6	5	1	Square pyramidal	BrF_5
AX_4E_2	6	4	2	Square planar[c]	BrF_4^-

[a]These structures are diagramed in Tables 9.1, 9.2, and 9.3 of the text.
[b]Nonbonding electron pairs are in the triangular plane.
[c]Nonbonding electron pairs are 180° apart.
[d]A represents the central atom. X represents the atoms bonded to the central atom. E represents the number of unshared electron pairs possessed by the central atom.

In Section 8.9 of the text you learned about electronegativity and polar bonds. You should have noted that bond polarity increases in magnitude with increasing electronegativity difference between the bonded atoms. A molecule can also be polar if its centers of negative and positive charge do not coincide.

▪ You should distinguish between a polar molecule and a polar bond. A polar bond occurs if the bonding electrons are shared unequally by the two bonded atoms. If a polyatomic molecule contains polar bonds, the molecule may be polar if the bond dipoles do not cancel.

▪ The magnitude of the polarity of a molecule is measured by its **dipole moment, μ.** The value of a dipole moment increases with increasing distance (r) between the centers of positive and negative charge and increasing magnitude of the positive and negative charges (Q): $\mu = Qr$.

▪ Dipole moments are generally reported in units of debye, D (1 debye = 3.33×10^{-30} coulombs-m).

The dipole moment of a molecule depends on both the magnitude of bond dipoles and the molecular geometry of the bond dipoles.

- If a polyatomic molecule contains equivalent polar bonds that are symmetrically arranged about the central atom, then the bond dipoles will cancel and the polyatomic molecule is not polar. Examples of such cases are CH_4 (tetrahedral), BF_3 (trigonal planar), and CO_2 (linear). A molecule is polar if the bond dipoles are not all equivalent such as in CH_3Br.

- Predicting whether or not a molecule is polar is not always straightforward. However, you can do this for simple molecules by first determining the geometrical shape of the molecule using the VSEPR model. Then, determine which bonds in the molecule are polar. Finally, analyze the geometrical arrangement of the bond dipoles to see if they cancel or not.

EXERCISE 1   For each of the following species, draw its Lewis structure: (a) NO_3^-; (b) XeO_3; (c) H_2S; (d) F_2SO_2; (e) $SnCl_2$; (f) PCl_3; (g) IF_5; and (h) SF_4. Then state the number of single bonding electron pairs and the number of electron pairs about the central atom for each structure. Finally, using Table 9.1, predict the structure about the central atom.

   Solution: Use the procedures developed in Sections 8.1 and 8.6 of the text to draw the Lewis structures. Your efforts should result in the following Lewis structures:

The answers to the other questions are summarized in the following table:

Compound	Total number of electron pairs about central atom	Total number of single-bonding electron pairs about central atom	Total number of unshared electron pairs about central atom	Structure
(a) NO_3^-	3[a]	3	0	Triangular
(b) XeO_3	4	3	1	Trigonal pyramidal
(c) H_2S	4	2	2	Nonlinear
(d) SO_2F_2	4	4	0	Tetrahedral
(e) $SnCl_2$	3	2	1	Nonlinear
(f) PCl_3	4	3	1	Trigonal pyramidal
(g) IF_5	6	5	1	Square pyramidal
(h) SF_4	5	4	1	Seesaw

[a]Remember that a multiple bond counts as one electron pair.

EXERCISE 2   Predict the structure of SO_2. Do you expect the OSO bond angle to expand or contract compared to the idealized geometry?

Solution: Each sulfur and oxygen atom contributes six valence-shell electrons to the total number of available bonding and unshared electrons. Thus, there are 18 valence electrons that are distributed about the sulfur and oxygen atoms. The Lewis structures for SO_2 are

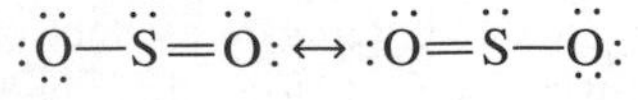

$$:\ddot{O}\!-\!S\!=\!\ddot{O}: \leftrightarrow :\ddot{O}\!=\!S\!-\!\ddot{O}:$$

Using Table 9.1, you should predict that the SO_2 molecule has the following nonlinear structure (V-shaped):

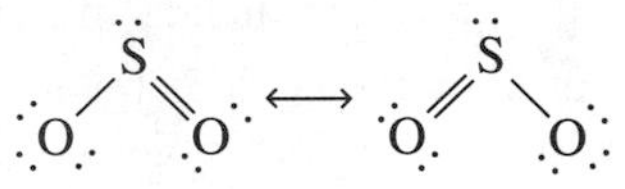

The repulsive forces existing between the nonbonding electron pair on sulfur and the bonding electron pairs on sulfur and oxygen atoms force the O-S-O bond angle to contract. The repulsive forces operating between a nonbonding electron pair and a bonding electron pair are greater than those between a bonding electron pair and another bonding electron pair. Thus, the O-S-O bond angle is not 120°, but less than this.

EXERCISE 3   Do both SO_2 and SO_3 possess dipole moments?
   Solution: First draw the Lewis structures of each and apply VSEPR rules to predict their geometrical structures. You determined the Lewis structure of SO_2 in Exercise 2. The Lewis structure of SO_3 differs from that of SO_2 by the addition of an oxygen atom

$$:\ddot{O} \longleftarrow :SO_2$$

to the unshared electron pair on sulfur in SO_2, yielding the following resonance forms:

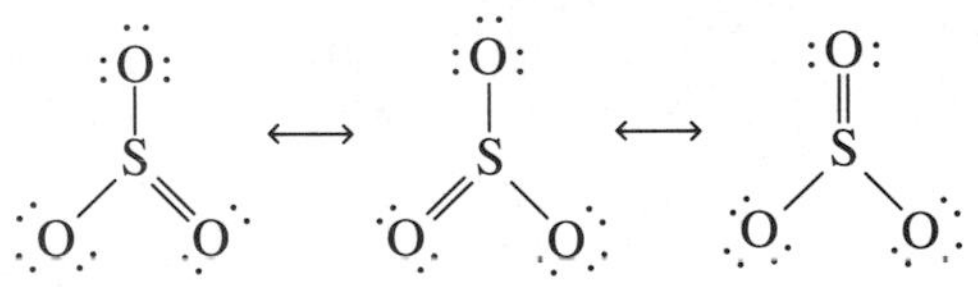

Around the sulfur atom in SO_2, there are two single-bonding electron pairs and one nonbonding electron pair; thus its structure is nonlinear (see Exercise 2). In SO_3, there are three single-bonding electron pairs about the sulfur atom, resulting in a trigonal-planar structure. The bond dipoles are shown below. Remember that the tip of the dipole arrow points toward the more electronegative atom.

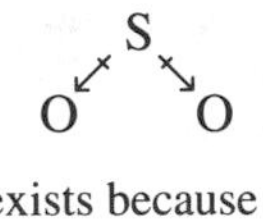

Dipole moment exists because centers of charge are not symmetrically arranged about sulfur.

There is no net dipole moment because all three bond dipoles are equivalent and symmetrically arranged.

COVALENT BONDING, HYBRID ORBITALS AND MOLECULAR STRUCTURE

Covalent bonding involves sharing of electrons between atoms to form a stable molecular structure. In section 9.3 of the text, the valence bond theory is introduced. The key idea is that a valence atomic orbital of one atom overlaps with a valence atomic orbital of another atom resulting in electron density shared between two nuclei. This build up of electron density holds the two positive nuclei together.

▪ Note in Figure 9.12 in the text the overlap of two $1s$ atomic orbitals to form an overlap region. This leads to a stable bond. In Figure 9.13 in the text you should observe that there is minimum in the potential energy diagram corresponding to the observed bond length for H_2. Atoms can not approach extremely close together because nuclei will begin to repel each other.

• Sigma (σ) bonds are shown in Figure 9.22 of the text. Note that electron density is concentrated between nuclei along the internuclear axis.

• Pi (π) bonds are shown in Figure 9.23 of the text. Note that the electrons density parallels the internuclear axis. A pi bond is weaker than a sigma bond. Also observe that a nodal plane exists in a pi bond. This is a region in which there is no electron density.

The geometrical structures of many molecules cannot be explained by considering the geometrical structure formed by the overlap of "pure" valence-shell atomic orbitals in bond formation. For example, consider the formation of a compound from beryllium and hydrogen atoms, BeH_2.

• Beryllium has no unpaired electrons available for bonding because the valence-shell electron configuration for beryllium is $2s^2$ and the shell is completely filled. Therefore, you might predict that beryllium does not form a compound with hydrogen.

• The compound BeH_2 is known to have a linear structure and equivalent beryllium-hydrogen bonds: H-Be-H.

• The existence of two linear Be-H bonds in BeH_2 implies that beryllium is using two new equivalent atomic orbitals, each with one electron, when it bonds with hydrogen.

• These new atomic orbitals are called **hybrid orbitals.** Hybrid orbitals are formed from linear combinations of "pure" atomic orbitals and have their own individual physical characteristics. They provide for better atomic orbital overlap, stronger bonds and more stable structures.

Given the structure of a covalent molecule, you can predict the type of hybrid orbitals used by the central atom in bonding to the other atoms. The type of hybrid orbitals used by an atom in its bond formations is related to the geometry of the single-bonding and nonbonding valence-shell electron pairs about the atom. Use the following procedure to determine the type of hybrid orbitals used by a central atom in a covalent compound.

1. Draw the Lewis structure for the compound.

2. Use the VSEPR model to predict the arrangement of valence-shell electron pairs about the atom of interest. (Remember that a multiple bond is treated as one electron pair).

3. Use Table 9.2 on the next page to predict the type of hybrid orbitals used by the central atom. Remember that nonbonding electron pairs can occupy hybrid orbitals.

The use of hybrid atomic orbitals to explain observed geometries of molecules provides an easy-to-use model at this stage of your learning of general chemistry. The question of whether or not an atom uses hybrid orbitals in bonding to other atoms is a more complex issue. For example, the bent angle in H_2S can also be explained by using pure atomic orbitals and then invoking electron pair repulsions to cause the angle to change. However, at this stage of your learning, use the hybrid atomic orbital model when molecular geometries cannot be explained by overlap of pure atomic orbitals.

Table 9.2 Types of hybrid orbitals

Number of electron pairs about the atom	Geometry of the electron pairs[a] (not atoms)	Types of hybrid orbitals used by the atom (types of atomic orbitals combined indicated in parentheses)	Example
2	Linear	sp $(s + p)$	BeH_2
3	Trigonal planar	sp^2 $(s + p + p)$	BI_3
4	Tetrahedral	sp^3 $(s + p + p + p)$	CH_4
4	Square planar	dsp^2 $(d + s + p + p)$	$PtCl_4^{2-}$
5	Trigonal bipyramidal	dsp^3 $(d + s + p + p + p)$	PF_5
6	Octahedral	d^2sp^3 $(d + d + s + p + p + p)$	SF_6

[a]This is also the geometry of hybrid orbitals about the central atom. Note that an unshared valence-electron pair can occupy a hybrid orbital.

EXERCISE 4 Boron trichloride has a trigonal-planar structure with the three bonds about the central atom at angles of 120°. Can its structure be explained using only the 2*s* and 2*p* valence-shell orbitals of boron? If not, suggest an alternative explanation using the concepts developed in Section 9.4 of the text.

Solution: The valence-shell orbital diagrams for boron and chlorine are

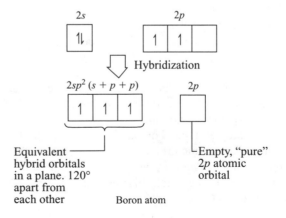

A covalent bond between two atoms can occur when a half-filled valence-shell orbital on one atom combines (overlaps) with an appropriate half-filled valence-shell orbital on the other. Boron has only one 2*p* atomic orbital with one electron that can overlap with a chlorine 2*p* atomic orbital containing one electron. Thus, you should predict the formula BCl, not BCl_3. To form three B—Cl bonds, the boron atom must use three hybrid orbitals with one electron in each. Also, these three hybrid orbitals of boron must be 120° apart in a trigonal plane. The nature of the three hybrid orbitals can be determined by visualizing the following process. First, a 2*s* electron in boron is promoted to obtain three unpaired electrons:

Second, the 2*s* and 2*p* atomic orbitals with one electron in each are hybridized, leaving one "pure" 2*p* atomic orbital in the boron atom:

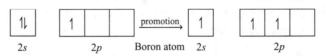

Three chlorine atoms, each with a half-filled $2p$ atomic orbital, combine with the three half-filled sp^2 hybrid orbitals on boron to yield three B—Cl bonds and the compound BCl_3.

EXERCISE 5 Predict the types of hybrid orbitals used by the central atom in the compounds in Exercises 1 and 2.

Solution: See Exercises 1 and 2 for the Lewis structures. In the following table, the central atom is underlined, and the total number of electron pairs about it are indicated:

Compound	Total number of electron pairs	Geometry of electron pairs	Hybrid orbitals used by central atom
$\underline{N}O_3^-$	3[a]	Trigonal planar	sp^2
$\underline{Xe}O_3$	4	Tetrahedral	sp^3
$H_2\underline{S}$	4	Tetrahedral	sp^3
$\underline{S}O_2F_2$	4	Tetrahedral	sp^3
$\underline{Sn}Cl_2$	3	Trigonal planar	sp^2
$\underline{P}Cl_3$	4	Tetrahedral	sp^3
$\underline{I}F_5$	6	Octahedral	d^2sp^3
$\underline{S}F_4$	5	Trigonal bipyramidal	dsp^3
$\underline{S}O_2$	3	Trigonal planar	sp^2

[a]Remember that a multiple bond counts as one electron pair.

Note: The geometry of electron pairs is different in some cases from the geometry of the molecule. H_2S is nonlinear, but the geometry of the electron pairs is approximately tetrahedral. Two of the sp^3 hybrids on sulfur are used in the bonds to the hydrogens. The other two sp^3 hybrids contain the unshared electron pairs. Also, you should realize that the experimentally observed structure of a molecule is determined by the locations of the atoms in the molecule and that we can only infer the locations of valence-shell unshared electron pairs from its structure.

The Lewis structures of some molecules show atoms sharing more than one pair of electrons; that is, multiple bonds exist between the atoms. For example, molecular nitrogen contains a triple bond:

$$: N \equiv N :$$

The nature and origin of multiple bonds in molecules is discussed in section 9.5 of the text. In the case of N_2, the mulitple bond consists of one sigma (σ) bond and two pi (π) bonds.

- A **sigma bond** is directed along the internuclear axis. Electron density is concentrated directly between the bonded atoms. Sigma bonds can form from:

    1. The overlap of two s atomic orbitals

    2. The overlap of an s orbital with a p orbital that is directed along the internuclear axis

    3. The overlap of two p orbitals, both directed along the internuclear axis; or the overlap of two hybrid orbitals, both directed along the internuclear axis. For example, the overlap of two sp hybrid orbitals to form a σ bond is shown in Figure 9.1 on page 148.

HYBRIDIZATION IN MOLECULES CONTAINING BOTH SIGMA AND PI BONDS

Figure 9.1 Two *sp* hybrid
orbitals overlap to form a σ
bond

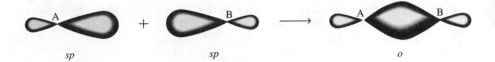

- **Pi bonds** occur with a σ bond if there are parallel *p* orbitals remaining on the atoms bonded together and if these *p* orbitals contain two electrons between them. A π bond consists of two distinct regions of electron density. One region is above the plane containing the internuclear axis, and the other is below the plane. An example of two $2p_x$ orbitals overlapping to form a π bond is pictured in Figure 9.2.

- In some molecules, the overlap of *p* orbitals among two or more neighboring atoms leads to a delocalized π bond. When such a condition exists, you write more than one Lewis structure to show resonance forms. The delocalization of π electrons in benzene is an important example discussed in the text.

EXERCISE 6 Describe the nature of bonding in the acetylide ion, C_2^{2-}.
  Solution: The Lewis structure for the acetylide ion is

$$[:C \equiv C:]^{2-}$$

All diatomic species are linear. The Lewis structure shows one σ and two π bonds between the bonded carbon atoms. The presence of only one σ bond in a plane containing the bonded atoms suggests that the two carbon atoms each use one of their two *sp* hybrid orbitals to form the σ bond. The two 2*p* orbitals remaining on each carbon atom are used to form two pi bonds, parallel to the plane of the sigma bond. The unshared electron pair on each carbon atom is contained within the nonbonding *sp* hybrid orbital.

EXERCISE 7 How many electrons occupy π orbitals in: (1) CO_2; **(b)** HCN?
  Solution: **(a)** The Lewis structure for CO_2 is:

$$\ddot{O} = C = \ddot{O}$$

It shows that in CO_2 there are two σ bonds (one between each C—O bond) and two π bonds (one between each C—O bond). Since each π bond contains two electrons, there is a total of four electrons occupying the π bonds. **(b)** The Lewis structure for HCN is

$$H - C \equiv N:$$

Since there are two π bonds between the carbon and nitrogen atoms, there is a total of four electrons occupying the π bonds.

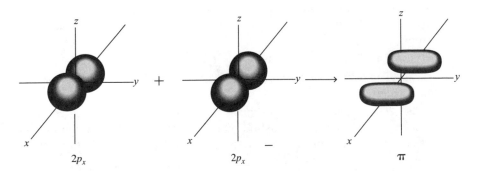

Figure 9.2 Two $2p_x$ hybrid
orbitals overlap to form a π
bond

EXERCISE 8 Show how the concept of delocalization of π electrons applies to NO_3^-.

Solution: The Lewis structures for NO_3^- are

The π electrons are not localized in any one nitrogen-oxygen bonded pair, but rather are associated with all three N—O bonds–that is, they are equally delocalized over the three N—O bonds. The following Lewis structure is one way of qualitatively representing the delocalization of the π electrons:

The dashed lines represent delocalized π electrons.

The Lewis model for covalent bonding, including the use of hybrid orbitals, provides us with a simple picture of bonding in which only valence-shell orbitals are used to construct localized bonds between atoms. However, this model does not always correctly predict the electronic structure of a molecule. For example, from the Lewis structure of O_2,

we would predict molecular oxygen to be **diamagnetic** (containing no unpaired electrons); however, it is observed experimentally to be **paramagnetic** (containing unpaired electrons).

The molecular orbital model for bonding is an alternative model. The molecular orbital model is contrasted to the Lewis model by the following:

- Electrons are not necessarily confined to bonds localized between two atoms or to lone pair orbitals.
- Atomic orbitals combine to form **molecular orbitals** that can spread or delocalize over an entire molecule.
- Atomic orbitals combine to form molecular orbitals that can have bonding or antibonding properties.

The essential principles of the molecular orbital model are:

- Atomic orbitals on two different atoms combine to form molecular orbitals.
- The number of molecular orbitals formed equals the number of atomic orbitals combined.
- When two atomic orbitals are linearly combined, a bonding and antibonding molecular orbital are formed.
- A **bonding molecular orbital** has electron density occurring primarily between the two nuclei, which holds the nuclei together.

• An **antibonding molecular orbital** has electron density not occurring primarily between the nuclei. A node occurs, thereby destabilizing the nuclei. The nuclei repel one another.

Molecular orbitals formed from two atomic orbitals are classified as sigma (σ) molecular orbitals or pi (π) molecular orbitals.

• **Sigma (σ) molecular orbitals** have electron density symmetrical about and along the internuclear bonding axis. A bonding molecular orbital concentrates electron density between the nuclei whereas an antibonding molecular orbital concentrates electron density away from the region between the nuclei. Be sure you know how σ_s and σ_p molecular orbitals are formed and their shapes.

• **Pi (π) molecular orbitals** have electron density above and below the internuclear bonding axis. Bonding π molecular orbitals concentrate electron density between nuclei whereas antibonding orbitals concentrate electron density away from the region between the nuclei. Note in Figure 9.33 of the text that π_{2p_x} and π_{2p_y} are equivalent, except that they are at 90° angles to one another.

You should memorize the energy-level diagram for molecular orbitals of second row diatomic elements as shown in Figure 9.34 of the text.

• Electrons enter molecular orbitals according to Hund's rule and Pauli's principle.
• The subscripts associated with σ and π molecular orbitals indicate the kind of atomic orbitals used to form a molecular orbital.
• Note that a bonding molecular orbital is of lower energy than either parent atomic orbital whereas antibonding molecular orbitals are of higher energy.

A useful concept for determining the stability of a diatomic molecule is **bond order.**

• Bond order is the *net* number of bonding electrons.
• Bond order = $\dfrac{\text{(number of bonding electrons} - \text{number of antibonding electrons)}}{2}$
• A bond order equal to zero means the combination of nuclei is not stable and the molecule will not permanently exist.
• The greater the bond order, the greater the bond strength.

EXERCISE 9   How many molecular orbitals are formed when the following numbers of atomic orbitals are combined: **(a)** two; **(b)** three?

Solution: The number of molecular orbitals formed is equal to the number of atomic orbitals used to form the molecular orbitals: **(a)** two; **(b)** three.

EXERCISE 10 Match the following labels with the correct molecular orbital:

$$\sigma_{2s}^*, \ \pi_{2p_x}^*, \ \pi_{2p_x}, \ \sigma_{1s}$$

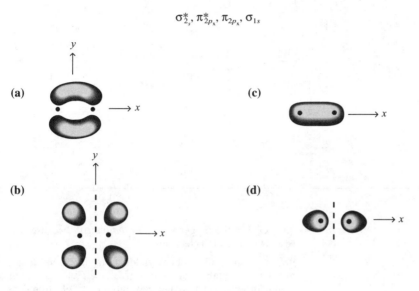

Solution: **(a)** π_{2p_x}; **(b)** $\pi_{2p_y}^*$; **(c)** σ_{1s}; **(d)** σ_{2s}^*.

EXERCISE 11 What are the bond orders for N_2, N_2^+, and N_2^-? Which one of the three species is the most stable?
Solution: First, determine the number of electrons each species has: N_2 has 14 electrons; N_2^+ has 13 electrons; and N_2^- has 15 electrons. Then, for each nitrogen species, place the electrons in molecular orbitals by using the energy-level diagrams given in Figure 9.34 in the text. Electrons are placed in molecular orbitals using Hund's rule and the Pauli exclusion principle: Electrons occupy lower energy levels before a higher one is filled, and if an energy level is degenerate (more than one molecular orbital of the same energy), all orbitals must contain one electron before pairing occurs. The electron configurations for N_2, N_2^+, and N_2^- are given in Table 9.3 on page 152. Next calculate the bond order for each:

$$\text{Bond order} = \frac{\left(\begin{array}{c}\text{no.}\\\text{bonding}\\\text{electrons}\end{array}\right) - \left(\begin{array}{c}\text{no.}\\\text{antibonding}\\\text{electrons}\end{array}\right)}{2}$$

$$N_2: \text{Bond order} = \frac{10-4}{2} = 3$$

$$N_2^+: \text{Bond order} = \frac{9-4}{2} = 2.5$$

$$N_2^+: \text{Bond order} = \frac{10-5}{2} = 2.5$$

N_2 is the most stable of the three species because it has the largest bond order. That is, N_2 has more net bonding electrons than does N_2^+ or N_2^-.

Table 9.3 Electron configurations of N_2, N_2^+, and N_2^-

	N_2 (14 electrons)		N_2^+ (13 electrons)		N_2^- (15 electrons)	
σ_{2p}^*	☐		☐		☐	
π_{2p}^*	☐	☐	☐	☐	↑	☐
σ_{2p}	↑↓		↑		↑↓	
π_{2p}	↑↓	↑↓	↑↓	↑↓	↑↓	↑↓
σ_{2s}^*	↑↓		↑↓		↑↓	
σ_{2s}	↑↓		↑↓		↑↓	
σ_{1s}^*	↑↓		↑↓		↑↓	
σ_{1s}	↑↓		↑↓		↑↓	

SELF-TEST QUESTIONS

Key Terms

Having reviewed the key terms listed at the end of Chapter 9 and their applications, identify the following statements as true or false. If a statement is false, indicate why it is incorrect. Key terms are italicized in the statements.

9.1 A *sigma bond* (σ) between carbon and oxygen atoms in CO could result from the overlap of carbon and oxygen $2p_z$ atomic orbitals, assuming that the bond axis is the x axis.

9.2 A *pi bond* (π) in CO would look like the following:

9.3 In the molecular energy levels for F_2, an *antibonding pi molecular orbital* lies lower in energy than a bonding pi molecular orbital.

9.4 A *sigma bonding molecular orbital* in He_2 formed from the overlap of $1s$ atomic orbitals would look like the following:

9.5 Graphite and diamond are *allotropic* forms of elemental carbon.

9.6 O_2 has the property of *paramagnetism* because it possesses two unpaired electrons.

9.7 A σ molecular orbital yields a stronger bond than a π molecular orbital because of a greater effective *overlap* of atomic orbitals and a greater concentration of electron density directly between bonded nuclei.

9.8 By application of a molecular-orbital energy-level diagram to Ne_2, we predict a stable molecule because it has a net *bond order* of zero.

9.9 According to the *VSEPR model,* the H_2S molecule should be linear.

9.10 *Hybridization* of $2s$ and $2p$ atomic orbitals of carbon is the reason why carbon can form four bonds to other atoms even though two bonds would normally be predicted from its valence-electron configuration.

9.11 A *molecular orbital* in H_2 is an allowed energy state for a proton.

9.12 Benzene is an example of a molecule in which electrons are *delocalized* over the $2s$ orbitals of carbon.

9.13 A *diamagnetic* substance is repelled by a magnetic field.

9.14 N_2 contains three *bonding pairs* of electrons between the nitrogen atoms.

9.15 A *molecular orbital diagram (energy-level diagram)* for homonuclear second-row elements as found in Figure 9.34 in the text shows the σ_{2p} molecular orbital lower in energy than a π_{2p} orbital.

9.16 The H—C—H *bond angle* in H_2CO equals the ideal bond angle of 120 degrees.

9.17 The *electron-pair geometry* in NH_3 is tetrahedral.

9.18 The *molecular geometry* of NH_3 is tetrahedral.

9.19 *Dipole moments* are vector quantities.

9.20 The tin atom in $SnCl_3^-$ possesses one *nonbonding pair* of electrons.

9.21 ICl is a *polar* molecule whereas I_2 is not.

9.22 A *bond dipole* exists when two bonded atoms have the same electronegativity.

9.23 According to *valence bond theory*, a chemical bond between two atoms occurs when a valence orbital on each atom overlaps and four electrons are shared.

9.24 The oxygen atom in water uses sp^3 *hybrid orbitals*.

9.25 According to *molecular orbital theory*, a sigma molecular orbital is never higher in energy than a pi molecular orbital.

9.26 A *bonding molecular orbital* concentrates electron density between two bonded atoms.

9.27 A *molecular orbital energy-level diagram* for F_2 shows the highest molecular orbital as empty of electrons.

9.28 A *pi molecular orbital* concentrates electron density along the bonding axis.

Problems and Short-Answer Questions

9.29 Draw the Lewis structures for the following molecules and predict their molecular structures and the types of hybrid orbitals used by the central atoms:

    (a) XeO_4 **(d)** ICl_2^-
    (b) PO_4^{3-} **(e)** SbF_5
    (c) TeF_6

9.30 Draw the Lewis structures for CO, CO_2, and CO_3^{2-} and predict the trend in C—O bond distances.

9.31 Why can't a p_x orbital on one atom overlap with a p_y orbital on another atom to form a bonding molecular orbital?

9.32 Write the molecular-orbital electron description for F_2 and F_2^+. Is either species paramagnetic?

9.33 In each of the following molecules, indicate the directions of the individual bond dipoles and the direction of the overall molecular dipole, if any:

    (a) ClCN   **(c)** BrCCBr
    (b) NO   **(d)** CF_4

9.34 Which types of hybrid orbitals on a central atom in a molecule give the following angles between atoms bonded to the central atom?

    (a) $180°$
    (b) $120°$
    (c) $90°$, planar structure
    (d) $90°$, octahedral structure

9.35 Give the formula of a molecule or ion in which
    (a) diatomic oxygen has a bond order of $1\frac{1}{2}$ (also construct the molecular-orbital energy levels for this oxygen species);
    (b) carbon uses sp^2 hybrid orbitals;
    (c) nitrogen forms three σ bonds;
    (d) phosphorus has five σ bonds.

9.36 Using NO_3^-, which contains delocalized π electrons, describe the hybrid atomic orbitals used by nitrogen. Would you expect NO_3^- to have a dipole moment?

9.37 Suggest a reason for the fact that at room temperature CO_2 is a gas but SiO_2 is a solid. (Hint: Silicon forms four σ bonds to oxygen, not two as might be indicated by the formula SiO_2.)

9.38 Would you expect N_2 or O_3 to be more reactive? Why?

Multiple-Choice Questions

9.39 Which of the following molecules possess(es) delocalized π electrons?
    (a) CH_2Cl_2 **(d)** **(a)** and **(b)**
    (b) OCN^- **(e)** **(b)** and **(c)**
    (c) N_2O

9.40 Which of the following is a property of a paramagnetic molecule?
    (a) unpaired electron(s) and attraction to a magnetic field;
    (b) unpaired electron(s) and repulsion from a magnetic field;
    (c) all electrons paired and attraction to a magnetic field;
    (d) all electrons paired and repulsion from a magnetic field;
    (e) both paired and unpaired electrons and repulsion from a magnetic field.

9.41 Which one of the following molecules possesses a trigonal-bipyramidal structure?
    (a) ICl_2 **(d)** PF_5
    (b) BrF_3 **(e)** PF_6^-
    (c) SF_4

9.42 Which of the following molecules has a trigonal-pyramidal structure?
    (a) BF_3 **(d)** **(a)** and **(b)**
    (b) NH_3 **(e)** **(b)** and **(c)**
    (c) $SOCl_2$

9.43 In which one of the following molecules would you expect the central atom to use sp^3d^2 hybrid orbitals?
    (a) PF_5 **(d)** SiO_2
    (b) BrF_5 **(e)** $PtCl_4^{2-}$ (square
    (c) CO_2 planar)

9.44 In a diatomic molecule of the second row, which molecular orbital has the highest energy?
    (a) σ_{1s},   **(d)** π_{2p}
    (b) σ_{2s}^* **(e)** π_{2p}^*
    (c) σ_{2p}

9.45 Which one of the following molecules has a bond order of 2?
    (a) Li_2 **(d)** N_2
    (b) B_2 **(e)** F_2
    (c) C_2

9.46 Which one of the following molecules does *not* have a dipole moment?
    (a) CO   **(d)** XeF_4
    (b) HBr   **(e)** NH_3
    (c) CH_3Cl

9.1 False. Only if the $2p_z$ orbitals overlap head on, along the z axis, would a σ bond form. **9.2** True. **9.3** False. A bonding molecular orbital is always more stable than an antibonding molecular orbital and thus is lower in energy. **9.4** False. This is a picture of a σ* orbital. In a σ molecular orbital, electron density is localized between the bonded nuclei, not outside as in a σ* orbital. **9.5** True. **9.6** True. **9.7** True. **9.8** False. A bond order of zero indicates that no stable bond exists between Ne atoms; therefore, Ne_2 does not form. **9.9** False. The Lewis structure for H_2S is

which is similar to that of H_2O. The repulsion between unshared valence electrons on sulfur and the bonding electrons between sulfur and hydrogen causes the H—S—H bond angle to contract from 180° to a smaller value. The molecule has a nonlinear structure. **9.10** True. **9.11** False. They are allowed energy states for an electron. **9.12** False. The electrons are delocalized over the 2p atomic orbitals of carbon. **9.13** True. **9.14** True. **9.15** False. The reverse is true. **9.16** False. The bond angle is less than the ideal bond angle of 120 degrees. The double bond of C=O in the molecule causes the H—C—H bond angle to contract because of electron-electron repulsions. **9.17** True. **9.18** False. The molecular geometry is trigonal pyramid. Note that the electron-pair geometry is not the same as the molecular geometry. **9.19** True. **9.20** True. **9.21** True. **9.22** False. A bond dipole exists when electronegativity values are different. **9.23** False. Two electrons are shared in a chemical bond. **9.24** True. **9.25** False. For certain elements, a $2s$ sigma molecular orbital is higher in energy than a $2p$ sigma molecular orbital. **9.26** True. **9.27** True. **9.28** False. A sigma molecular orbital concentrates electron density along the bonding axis. Electron density in a pi molecular orbital lies between the bonded atoms, but above and below the internuclear bonding axis. There is a node along this axis.

9.29 (a)

Tetrahedral—
sp^3 hybrids

(b)

Tetrahedral—
sp^3 hybrids

(c)

Octahedral—
d^2sp^3 hybrids

(d) [:Cl—I—Cl:]⁻ Linear, dsp^3 hybrids. The unshared electron pairs on iodine are in the radial (equatorial) plane of the trigonal bipyramidal arrangement of electron pairs.

(e)

Trigonal bipyramidal—
dsp^3 hybrids

9.30 The Lewis structures for CO, CO_2, and CO_3^{2-} respectively are: :C≡O:, Ö=C=Ö, and

The C—O bond types are: CO. a triple bond; CO_2, double bonds; and CO_3^{2-}, a bond that is between a single and a double bond. The greater the number of bonds between the C and O atoms, the shorter the bond distance will be. Thus we would predict that the C—O bond distances decrease in the order $CO_3^{2-} > CO_2 > CO$. **9.31** The p_x and p_y orbitals are oriented in space in such a manner that they are separated by an angle of 90°. Because their lobes of electron density are not oriented either directly toward one another or parallel to one another, there is insufficient orbital overlap to yield a stable and bonding molecular orbital. **9.32** F_2 (18 electrons): $(\sigma_{1s})^2 (\sigma_{1s}^*)^2 (\sigma_{2s})^2 (\sigma_{2s}^*)^2 (\pi_{2p})^2 (\pi_{2p})^2 (\sigma_{2p})^2 (\pi_{2p}^*)^2 (\pi_{2p}^*)^2$; F_2^+ (17 electrons): $(\sigma_{1s})^2 (\sigma_{1s}^*)^2 (\sigma_{2s})^2 (\sigma_{2s}^*)^2 (\pi_{2p})^2 (\pi_{2p})^2 (\sigma_{2p})^2 (\pi_{2p}^*)^2 (\pi_{2p}^*)^1$. F_2^+ is paramagnetic because one of its π_{2p}^* molecular orbitals has an unpaired electron. F_2 is diamagnetic because all of its electrons are paired.

9.33 (a) Cl ⇌ C ⇌ N
←———— +
            Molecular dipole

Since $EN_{Cl} > EN_N$, there is a net molecular dipole as indicated.

(b) N ⇌ O

The bond dipole and molecular dipole are coincident.

(c) Br ⇌ C — C ⇌ Br

No molecular dipole exists as the bond dipoles cancel because of their equivalence in magnitude and symmetrical arrangement.

(d)

CF_4 is a tetrahedral molecule. No net molecular dipole exists. **9.34 (a)** linear: sp; **(b)** trigonal planar: sp^2; **(c)** square planar: sp^2d; **(d)** octahedral: sp^3d^2. **9.35 (a)** O_2^-: The molecular-orbital energy levels are $(\sigma_{1s})^2 (\sigma_{1s}^*)^2 (\sigma_{2s})^2 (\sigma_{2s}^*)^2 (\pi_{2p})^2 (\pi_{2p})^2 (\sigma_{2p})^2 (\pi_{2p}^*)^2 (\pi_{2p}^*)^1$. **(b)** In $H_2C_2H_2$,

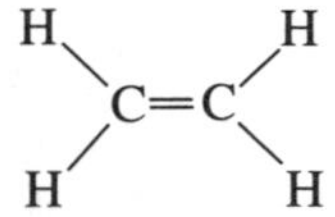

the arrangement of atoms around each carbon atom is trigonal planar. Therefore, each carbon is utilizing sp^2 hybrids. **(c)** NH_3 contains three single bonds that are σ bonds. **(d)** PCl_5 contains five single bonds that are σ bonds. **9.36** As explained in Exercise 8, all N—O bonds are equivalent because of π-electron delocalization:

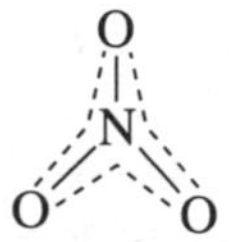

Since NO_3 has a trigonal-planar structure, the bond dipoles cancel, and there is no net molecular dipole. For a trigonal-planar structure to occur, nitrogen must use sp^2 hybrid orbitals for bonding to the three oxygen atoms. The remaining unhybridized p orbital on nitrogen is involved in the delocalized π bond. **9.37** CO_2 is a nonpolar molecular compound; thus electrostatic forces holding CO_2 molecules together are minimal, and it therefore exists as a gas. In SiO_2, a diamond-like lattice of Si—O bonds exists, with each silicon atom bonded to four oxygen atoms. Therefore, the total forces holding Si and O atoms together are more numerous than in the case of CO_2, and SiO_2 exists as a solid. **9.38** The Lewis structure of N_2 is

$$:N{\equiv}N:$$

and the Lewis structure of O_3 is

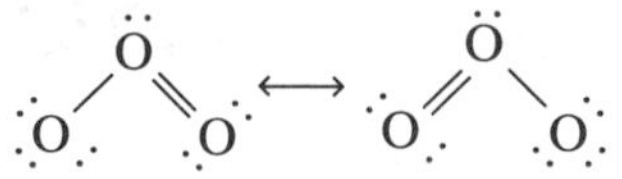

As shown by these structures, there is a triple bond between the nitrogen atoms, and the bond between oxygen atoms is between a single and a double bond. Since there are fewer net bonds between O atoms in O_3 than between N atoms in N_2, we expect the O—O bond in O_3 to be weaker than the N—N bond in N_2. Thus we would expect O_3 to be more reactive. This is verified by experimental evidence. **9.39 (e). 9.40 (a). 9.41 (d). 9.42 (e).** NH_3 and $SOCl_2$ are trigonal pyramidal; BF_3 is trigonal planar. **9.43 (b).** The unshared electron pair of Br occupies one of six equivalent sp^3d^2 hybrid orbitals (octahedral arrangement of electron pairs). **9.44 (e). 9.45 (c). 9.46 (d).** Square planar: All bond dipoles and nonbonding-electron-pair dipoles cancel because of the symmetrical structure.

Sectional Test 2

CHAPTERS 6–9

You should make this test as realistic as possible. **Tear out this test so that you do not have access to the information in the *Student's Guide*.** Use only the data provided in the questions and a periodic table. Do not check your answers until you are finished. If you answer questions incorrectly, review the section in the text indicated after each question.

Choose the best response for each question.

1. The energy required to break a nitrogen triple bond is 1,113 kJ/mol. Calculate the wavelength in meters of photons having sufficient energy to break this bond. ($h = 6.63 \times 10^{-34}$ J-s, $c = 3.00 = 10^8$ m/s.)

(a) $\dfrac{(6.63 \times 10^{-34} \text{ J-s})(3.00 \times 10^8 \text{ m/s})}{1,113 \text{ kJ}} = 1.79 \times 10^{-28}$ m

(b) $\dfrac{(6.63 \times 10^{-34} \text{ J-s})(3.00 \times 10^8 \text{ m/s})}{1,113 \times 10^6 \text{ J}} = 1.79 \times 10^{-31}$ m

(c) $\dfrac{(6.63 \times 10^{-34} \text{ J-s/molecule})(3.00 \times 10^8 \text{ m/s})}{(1,113 \text{ kJ/mol})(6.023 \times 10^{23} \text{ molecules/mole})}$

$= 5.92 \times 10^{-52}$ m

(d)

$\dfrac{\left(\dfrac{6.63 \times 10^{-34} \text{ J-s}}{\text{molecule}}\right)(3.00 \times 10^8 \text{ m/s})\left(6.023 \times 10^{23} \dfrac{\text{molecules}}{\text{mole}}\right)}{(1.113 \times 10^6 \text{ J/mol})}$

$= 1.07 \times 10^{-7}$ m

[6–1, 6–2]

2. What is the frequency of ultraviolet radiation with a wavelength of 15.3 nm?

(a) 1.96×10^{16}/s   **(b)** 1.96×10^7/s   **(c)** 4.59/s

(d) 4.59×10^{-9}/s

[6–1]

3. Which of the following statements is *not* correct?

(a) Photons of lower-frequency radiation have lower energies.

(b) Planck's study of black-body radiation led to the hypothesis that radiation is emitted in quantas of energy.

(c) Bohr's model of the hydrogen atom showed emission of energy by atoms occurs when an electron moves to a higher orbit.

(d) The lowest energy state of an electron in an atom is known as the ground state.

[6–1, 6–2, 6–3]

4. The energy of an electron in a Bohr orbit depends on the principal quantum number as follows:

(a) $E = R_H/n$ (b) $E = R_H/n^2$ (c) $E = -R_H/n$ (d) $E = -R_H/n^2$

[6–3]

5. As the principal quantum number of a Bohr orbit increases, the energy levels of a hydrogen atom

(a) become the same. (b) are closer together. (c) are further apart.

(d) approach an infinitely negative energy.

[6–3]

6. In the Bohr model of the hydrogen atom, which of the following statements is correct?

(a) When $n = \infty$, the electron is in its ground state.

(b) When $n = 1$, the electron is in an excited state.

(c) The transition $n = 2 \rightarrow n = 4$ represents emission of energy.

(d) The transition $n = 1 \rightarrow n = 3$ represents absorption of energy.

[6–3]

7. What is the ionization energy of the hydrogen atom in kJ/mol? ($R_H = 2.18 \times 10^{-18}$ J)

(a) $-R_H/1^2 = -2.18 \times 10^{-21}$ kJ/mol

(b) $R_H/1^2 = 2.18 \times 10^{-21}$ kJ/mol

(c) $(-R_H/1^2)(6.023 \times 10^{23}$ atoms/mol$) = -1.31 \times 10^3$ kJ/mol

(d) $(R_H/1^2)(6.023 \times 10^{23}$ atoms/mol$) = 1.31 \times 10^3$ kJ/mol

[6–3]

8. What is the initial principal quantum number for the transition to the $n = 1$ state if the frequency of the hydrogen line is 2.93×10^{15}/s? ($R_H = 2.18 \times 10^{-18}$ J, $h = 6.63 \times 10^{-34}$ J-s)

(a) 4 (b) 3 (c) 2 (d) 1

[6–3]

9. What are the possible values of m_l for $l = 3$?

(a) 3, –3 (b) 3, 0, 3 (c) 3, 2, 1, –1, –2, –3

(d) 3, 2, 1, 0, –1, –2, –3

[6–5]

10. Which set of quantum numbers for an electron in an atom is *not* allowed?

	n	l	m_l	m_s
(a)	2	1	0	–1/2
(b)	3	2	1	+1/2
(c)	3	3	2	+1/2
(d)	1	0	0	+1/2

[6–5]

11. Which electron has the label $3d$?

	n	l	m_l
(a)	3	2	1
(b)	3	1	0
(c)	3	0	0
(d)	3	3	2

[6–5]

12. Which hydrogen electron is *not* in an excited state?

    (a) $1s^1$ **(b)** $2s^1$ **(c)** $3p^1$ **(d)** $4s^1$

    [6–3, 6–9]

13. What is the total number of *orbitals* possible for $n = 3$?

    (a) $n = 3$ **(b)** $2n = 6$ **(c)** $n^2 = 9$ **(d)** $2n^2 = 18$

    [6–5, 6–7]

14. What is the maximum number of *electrons* that occupy the n = 4 energy level?

    (a) $n = 4$ **(b)** $2n = 8$ **(c)** $n^2 = 16$ **(d)** $2n^2 = 32$

    [6–5]

15. What is the value of l for a $5d$ electron?

    (a) $l = 0$ **(b)** $l = 1$ **(c)** $l = 2$ **(d)** $l = 3$

    [6–5, 6–7]

16. What are the possible m_l values for a $4p$ electron?

    (a) 4, 3, 2, 1, 0, –1, –2, –3, –4

    (b) 3, 2, 1, 0, –1, –2, –3

    (c) 2, 1, 0, –1, –2

    (d) 1, 0, –1

    [6–5, 6–7]

17. Which of these is *not* allowed?

    (a) $2s$ **(b)** $2f$ **(c)** $3p$ **(d)** $4d$

    [6–5, 6–7]

18. What is the correct label for the orbital

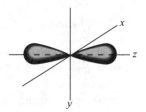

    (a) s_z **(b)** p_z **(c)** p_y **(d)** d_{z^2}

    [6–6]

19. What is the correct label for the orbital

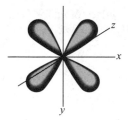

    (a) $d_{x^2-y^2}$ **(b)** d_{xy} **(c)** d_{xz} **(d)** d_{yz}

    [6–6]

20. Which order of energies of orbitals is correct in a many-electron atom?

    (a) $2s = 2p$ **(b)** $3s < 3p$ **(c)** $3d < 2s$ **(d)** $4s > 5s$

    [6–7]

21. Which statement is *not* true for many-electron atoms?

    (a) Outer electrons penetrate the electron clouds of inner electrons.

    (b) Outer electrons experience the full nuclear charge.

    (c) The $2s$ and $2p$ orbitals are of differing energies, whereas in a hydrogen atom they are the same.

(d) The $2p_x$ and $2p_y$ orbitals have the same energy in the absence of an applied magnetic field.

[6–5, 6–7]

22. Which of the following is the electron configuration of N (at. no. = 7) in its ground state?

(a) $1s^2 2s^2 2p^2$ (b) $1s^2 2s^2 2p^3$ (c) $1s^2 2s^2 2p^4$ (d) $1s^2 2p^5$

[6–8, 6–9]

23. Which of the following is the electronic configuration of Ni (at. no. = 28) in its ground state?

(a) $1s^2 2s^2 2p^6 3s^2 3p^6 3d^8 4s^2$ (b) $1s^2 2s^2 2p^6 3s^2 3p^6 3d^9 4s^1$

(c) $1s^2 2s^2 2p^6 3s^2 3p^6 4s^2 4p^6 5s^2$ (d) $1s^2 2s^2 2p^6 3s^2 3p^6 3d^6 4s^2 4p^2$

[6–8, 6–9]

24. Which element has the electron configuration $1s^2 2s^2 2p^6 3s^2 3p^2$?

(a) C   (b) Si   (c) N   (d) P

[6–8, 6–9]

25. The number of valence shell electrons for oxygen is

(a) 2   (b) 4   (c) 6   (d) 8

[6–8]

26. Which of the following are valence electrons for Sr?

(a) $5s^1$ (b) $5s^2$ (c) $5s^1 5p^1$ (d) $6s^2$

[6–8]

27. Which of the following valence-electron orbital diagrams is correct for the ground state of As?

(a) 4s: [↑↓]   4p: [↑][↑][↑]

(b) 4s: [↑↑]   4p: [↑][↑][↑]

(c) 4s: [↑↓]   4p: [↑↓][↑][ ]

(d) 4s: [↑↓]   4p: [↑][↓][↑]

[6–9, 6–10]

28. Which of the following atoms is the largest?

(a) C   (b) Si   (c) Ge   (d) Sn

[7–3]

29. Which of the following atoms is the largest?

(a) Na   (b) Mg   (c) O   (d) F

[7–3]

30. Which of the following arrangements of increasing ionization energy is correct?

(a) Na < Rb   (b) C < N   (c) Cl < S   (d) S < Te

[7–4]

31. Which element is expected to have the smallest first ionization energy?

(a) Be   (b) Rb   (c) F   (d) O

[7–4]

32. Which order of electron affinities is *not* correct?

(a) Cl > Br > I   (b) O > B > Li   (c) Cs > Ba > P

(d) S > Se > Te

[7–5]

33. Which of the following is the peroxide ion?
(a) O^{1-} (b) O^{2-} (c) O_2^{2-} (d) O_2^{-}
[7–8]

34. Which statement is true?

(a) The coinage metals (Cu, Ag, Au) are less reactive than the alkali metals.

(b) Beryllium shows properties very similar to those of magnesium.

(c) Group 2A elements are more reactive than group 1A elements.

(d) Group 1A elements form +2 ions.

[7–7]

35. KO_2 is

(a) potassium oxide (b) potassium peroxide
(c) potassium superoxide (d) potassium hydroxide

[7–7, 7–8]

36. An acceptable Lewis dot structure of the *ground* state of C is
(a) $\cdot\ddot{C}\cdot$ (b) $\cdot\ddot{C}\cdot$ (c) $\cdot\ddot{C}\cdot$ (d) $\cdot\dot{C}\cdot$
[8–1]

37. Which of the following statements is *false* for the reaction between $Rb(s)$ and $Cl_2(g)$?

(a) Rb is oxidized. (b) The product is RbCl.

(c) Cl^- ions are formed. (d) The product is covalent.

[8–2, 8–4]

38. Which of the following is *not* a typical property of an ionic solid?

(a) High melting point (b) Soft, moldable (c) Conductor of electricity when melted (d) Often soluble in water

[8–2]

39. Which of the following represents the correct order of lattice energies?

(a) $NaCl < MgCl_2$ (b) $MgO < Na_2O$ (c) $MgO = Al_2O_3$
(d) $NaF < CsBr$

[8–2]

40. Which of the following is *not* a correct chemical formula for a metallic oxide?

(a) Al_2O_3 (b) Na_2O (c) MgO (d) NO_2

[8–2]

41. Which of the following represents an isoelectronic series of ions, in order of *decreasing* ionic size?

(a) Na^+, K^+, Rb^+, Cs^+ (b) $O^{2-}, S^{2-}, Se^{2-}, Te^{2-}$
(c) $O^{2-}, F^-, Na^+, Mg^{2+}$ (d) $Ca^{2+}, K^+, Cl^-, S^{2+}$

[8–3]

42. Which order of ionic size is correct?

(a) $Na^+ > Mg^{2+}$ (b) $F^- > O^{2-}$ (c) $Rb^+ > K^+$ (d) $Ca^{2+} > Sr^{2+}$

[8–3]

43. Which of the following compounds is a covalent molecule?

(a) SO_2 (b) $CaCl_2$ (c) Al_2O_3 (d) Na_2S

[8–2, 8–4]

44. Which of the following is the correct Lewis dot structure for PCl_3?

(a) 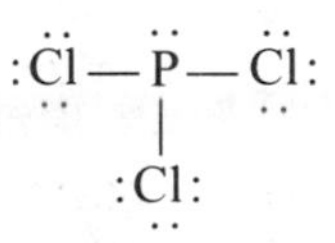 :Cl—P—Cl: with :Cl: below, P has lone pair on top

(c) 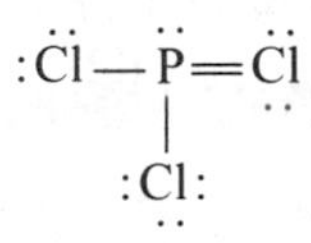 :Cl—P=Cl with :Cl: below

(b) 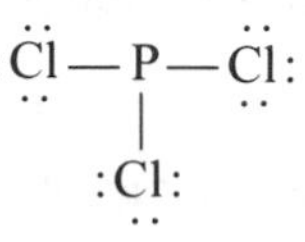 Cl—P—Cl: with :Cl: below

(d) 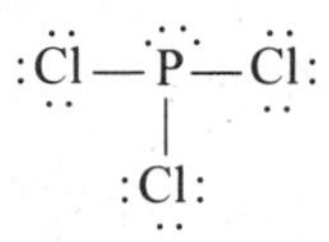 :Cl—P—Cl: with :Cl: below

[8–6]

45. Which of the following exhibits resonance?
  (a) H_2O (b) ClO_3^- (c) SO_2 (d) HCl
  [8–7]

46. How many lone pair *electrons* are on the Br atom in BrF_3?
  (a) 2 (b) 3 (c) 4 (d) 6
  [8–8]

47. Which of the following is the correct Lewis dot structure for KCN?
  (a) K—C≡N: (b) K—C≡N: (c) K^+:C≡N:$^-$
  (d) K^+:C=N:$^-$
  [8–6]

48. Which of the following statements concerning the bonding in the nitrite ion (NO_2^-) is *not* true?
  (a) It possesses 18 valence electrons.
  (b) The arrangement of bonds is O—O—N.
  (c) Two resonance structures can be written.
  (d) Its structure is trigonal planar.
  [8–6, 9–1]

49. Which of the following possesses a polar covalent bond?
  (a) NaCl(*s*) (b) O_2(*g*) (c) Al(*s*) (d) SO_2(*g*)
  [8–9]

50. Which is the most electronegative atom?
  (a) Cl (b) Se (c) Al (d) Ca
  [8–5]

51. Which species contains the strongest C—C bond?
  (a) CH_4 (b) C_2H_6 (c) C_2H_4 (d) C_2H_2
  [8–9]

Use the following responses for questions 52 through 56:
  (a) +7 (b) +6 (c) +5 (d) +4 (e) +2 (f) +1 (g) 0
  (h) −1 (i) −2 (j) −3 (k) −4 (l) −5
  [8–10]

52. What is the oxidation number of nitrogen in NO_3^-?
53. What is the oxidation number of chlorine in $NaClO_4$?
54. What is the oxidation number of iodine in I_2?
55. What is the oxidation number of nitrogen in N_2H_4?
56. What is the oxidation number of chromium in $K_2Cr_2O_7$?
57. Which name is *incorrect*?
  (a) CO_2, carbon dioxide (b) NH_4Cl, ammonium chloride
  (c) NO, nitrogen(I) oxide (d) HCl, hydrogen chloride
  [8–10]

58. Which name is correct?

(a) SF_6, sulfur pentafluoride   (b) $CaCl_2$, calcium chloride
(c) BaO, barium peroxide   (d) NCl_3, nitrogen chloride
[8–10]

59. Which geometry is *not* correct for the paired molecule?

(a) XeF_2, linear   (b) SF_6, octahedral   (c) BrF_5, square pyramid
(d) XeF_4, tetrahedral
[9–1]

60. Which of the following molecules has a dipole moment?

(a) CO_2 (b) PF_3 (c) NH_4^+ (d) SiH_4
[9–2]

61. Which of the following is *incorrect?*

(a) $SiCl_4$, regular tetrahedron, nonpolar   (b) HCN, bent, polar
(c) N_2, linear, nonpolar   (d) NH_3, trigonal pyramid, polar
[9–1, 9–2]

62. What is the expected Cl—C—Cl bond angle in C_2Cl_4?

(a) $60°$ (b) $109.5°$ (c) $120°$ (d) $180°$
[9–1, 9–4]

63. Which hybrid orbital does *not* agree with the expected geometry of the paired molecule?

(a) BF_3, sp^3 (b) $HgCl_2$, sp (c) PF_5, dsp^3 (d) SF_6, d^2sp^3
[9–5]

64. What is the type of hybridization used by carbon in C_2H_2 (acetylene)?

(a) sp (b) sp^2 (c) sp^3 (d) sp^2d
[9–5]

65. Which type of hybridization is associated with a square planar array of hybrid orbitals?

(a) sp (b) sp^3 (c) sp^2d (d) sp^3d^2
[9–5]

66. Which of the following describes the molecular orbital description for electrons in the $n = 2$ level of O_2^+?

(a) $\sigma_{2s}^2\sigma_{2s}^{*2}\pi_{2p}^4\sigma_{2p}^2\pi_{2p}^{*1}$ (b) $\sigma_{2s}^2\sigma_{2s}^{*2}\pi_{2p}^4\sigma_{2p}^2\pi_{2p}^{*1}$
(c) $\sigma_{2s}^2\sigma_{2s}^{*2}\sigma_{2p}^2\pi_{2p}^4\pi_{2p}^{*1}$ (d) $\sigma_{2s}^2\sigma_{2s}^{*2}\sigma_{2p}^2\pi_{2p}^4\sigma_{2p}^{*1}$
[9–7]

67. What is the bond order for He_2?

(a) 0 (b) $1/2$ (c) 1 (d) $1\ 1/2$
[9–6]

68. According to molecular orbital theory, which one of the following is paramagnetic?

(a) F_2 (b) H_2 (c) NO^+ (d) NO
[9–6, 9–7]

69. Which type of molecular orbital is

(a) π (b) π^* (c) σ_{2s}^* (d) σ_{2s}
[9–5]

70. Which of the following is paramagnetic based on its molecular orbital configuration?

(a) F_2 (b) N_2 (c) O_2^- (d) N_2^{2-}

[9-7]

ANSWERS

1. (d).   **2.** (a).   **3.** (c).   **4.** (d).   **5.** (b).   **6.** (d).   **7.** (d).   **8.** (b).   **9.** (d).
10. (c).   **11.** (a).   **12.** (a).   **13.** (c).   **14.** (d).   **15.** (c).   **16.** (d).   **17.** (b).   **18.** (b).
19. (b).   **20.** (b).   **21.** (b).   **22.** (b).   **23.** (a).   **24.** (b)   **25.** (c).   **26.** (b).   **27.** (a).
28. (d).   **29.** (a).   **30.** (b).   **31.** (b).   **32.** (c).   **33.** (c).   **34.** (a).   **35.** (c).   **36.** (b).
37. (d).   **38.** (b).   **39.** (a).   **40.** (d).   **41.** (c).   **42.** (c).   **43.** (a).   **44.** (a).   **45.** (c).
46. (c).   **47.** (c).   **48.** (b).   **49.** (d).   **50.** (a).   **51.** (d).   **52.** (c).   **53.** (a).   **54.** (g).
55. (i).   **56.** (b).   **57.** (c).   **58.** (b).   **59.** (d).   **60.** (b).   **61.** (b).   **62.** (c).   **63.** (a).
64. (a).   **65.** (c).   **66.** (b).   **67.** (a).   **68.** (d).   **69.** (b).   **70.** (c).

10 | Gases

OVERVIEW OF THE CHAPTER

10.1, 10.2
PROPERTIES AND
CHARACTERISTICS
OF GASES

Review: States of matter (1.1); dimensional analysis (1.5).

Objectives: You should be able to:

1. Describe the general characteristics of gases as compared to other states of matter, and list the ways in which gases are distinct.

2. Define atmosphere, torr and pascal, the most important units in which pressure is expressed. Also describe how a barometer and manometer work.

**10.3, 10.4, 10.5,
10.7** USING THE
GAS LAWS: SOLVING
PROBLEMS

Review: Density (1.3); temperature (1.3); mole (3.4); stoichiometry (3.1, 3.2, 3.6).

Objectives: You should be able to:

1. Describe how a gas responds to changes in pressure, volume, temperature, and quantity of gas.

2. Use the ideal-gas equation to solve for one variable (P, V, n, or T) given the other three variables or information from which they can be determined.

3. Use the gas laws, including the combined gas law, to calculate how one variable of a gas (P, V, n, or T) responds to changes in the one or more of the other variables.

4. Calculate the molar mass of a gas, given gas density under specified conditions of temperature and pressure. Also calculate gas density under stated conditions, knowing molar mass.

10.6 DALTON'S LAW
OF PARTIAL
PRESSURES:
MIXTURE OF GASES

Objectives: You should be able to:

1. Calculate the partial pressure of any gas in a mixture, given the composition of that mixture.

2. Calculate the mole fraction of a gas in a mixture, given its partial pressure and the total pressure of the system.

164

Review: Molecules (2.5); kinetic energy (4.1).

1. Describe how the distribution of speeds and the average speed of gas molecules changes with temperature.
2. Describe how the relative rates of effusion and diffusion of two gases depend on their molar masses (Graham's law).

Objectives: You should be able to:

1. Explain the origin of deviations shown by real gases from the relationship $PV/RT = 1$ for a mole of an ideal gas.
2. Cite the general conditions of P and T under which real gases most closely approximate ideal-gas behavior.
3. Explain the origins of the correction terms to P and V that appear in the van der Waals equation.

TOPIC SUMMARIES AND EXERCISES

Gases play an important role in the support of human life. If we lose contact with the atmosphere—which is a mixture of gases—for more than 4–5 min, permanent brain damage occurs; longer loss of contact eventually results in death. This is just one example of why we need to fully understand the physical properties and characteristics of gases and the physical laws that govern their behavior.

From observing the behavior of gases we find that:

- They expand to fill the container in which they are enclosed.
- They are compressible.
- They readily flow.
- They form homogenous mixtures with other gases.
- The volume of the gas molecules themselves is only a small portion of the total volume in which they are contained at room temperature and pressure.

The state (condition) of a gas is defined by giving the following properties:

- Temperature (in K) – T
- Volume (usually liters) – V
- Quantity (usually moles) – n
- Pressure (chemists commonly use atmosphere (atm) as a unit, but the SI unit is pascal (Pa)) – P

You need to remember the following about **gas pressure:**

- It is the force exerted by gas molecules per unit surface area.

10.8, 10.9 THE KINETIC-MOLECULAR THEORY OF GASES: MOLECULAR SPEEDS AND EFFUSION

10.10 DEPARTURES FROM THE IDEAL-GAS EQUATION

PROPERTIES AND CHARACTERISTICS OF GASES

• A **barometer** contains a column of mercury whose height is directly related to the pressure exerted by the atmosphere. Thus atmospheric pressure may be reported in units related to the height of the mercury column: mm Hg (or torr) or cm Hg.

• A **standard atmospheric pressure** represents 1 atm of pressure. This corresponds to the amount of pressure that supports a column of Hg 760 mm in height in a barometer at 0°C.

• A **manometer** is used to measure the pressure of enclosed gases, usually below atmospheric pressure. The pressure exerted is determined from the difference in heights of mercury levels in a U-tube.

• Units of pressure, and conversion factors you should be familiar with are:

$$1 \text{ atm} = 760 \text{ mm Hg} = 760 \text{ torr} = 101.3 \text{ kPa}$$
$$1 \text{ atm} = 29.9 \text{ in Hg} = 33.9 \text{ ft } H_2O = 10.3 \text{ m } H_2O$$

EXERCISE 1   **(a)** Suppose you were to construct a barometer using methanol as the liquid. Given that the density of Hg is 13.6 g/mL and that of methanol is 0.791 g/mL, and disregarding the effect of methanol gas molecules above methanol liquid, calculate the height of the methanol column at 1 atm of pressure at 0°C. **(b)** Why do we use Hg rather than methanol as a liquid in a barometer?

Solution: **(a)** At 1 atm of pressure and 0°C a mercury column has a height of 760 mm. Since the methanol is placed in a column of the same diameter, the force exerted by a column of mercury of 760 mm in length would be equivalent to a column of methanol of the same mass. The density of Hg is 17.2 times that of methanol per unit volume; so it will take 17.2 mm of methanol to equal the same mass of 1 mm of mercury. Therefore, the height of the methanol column is

$$(760 \text{ mm Hg}) \left(\frac{17.2 \text{ mm methanol}}{1 \text{ mm Hg}} \right) = 13{,}000 \text{ mm methanol}$$

(b) From this calculation you can see that mercury is an excellent choice for a liquid in a barometer because its column height at 1 atm is reasonable.

EXERCISE 2   A diver is 261 ft below sea level in pure water. What pressure, in atmospheres, is exerted upon the diver?

Solution: The pressure exerted upon the diver is the atmospheric pressure above the water plus the pressure exerted by the water above the diver. Using the equivalence relation 1 atm = 33.9 ft H_2O, the total pressure exerted upon the diver is calculated as follows:

Pressure = atmospheric pressure at sea level + pressure exerted by water

$$\text{Pressure} = 1 \text{ atm (at sea level)} + (261 \text{ ft}) \left(\frac{1 \text{ atm}}{33.9 \text{ ft}} \right)$$

$$\text{Pressure} = 1 \text{ atm} + 7.68 \text{ atm} = 8.68 \text{ atm}$$

USING THE GAS LAWS: SOLVING PROBLEMS

Problems involving application of the gas laws require an understanding of the quantitative relationships involving pressure (P), volume (V), quantity (n), and temperature (T in degrees Kelvin).

Gas laws discovered before the 20th Century that you need to remember are

• **Boyle's law:** For a given amount of a gas, volume is inversely proportional to pressure at constant temperature:

$$V = \frac{\text{constant}}{P} \qquad (T, \ n \text{ constant})$$

- **Charles's law:** For a given amount of a gas, volume is directly proportional to temperature at constant pressure.

$$V = \text{constant} \times T \qquad (P, \ n \text{ constant})$$

- **Avogadro's hypothesis:** Equal volumes of gases at the same temperature and pressure contain equal numbers of molecules.
- **Avogadro's law:** At constant pressure and temperature, the volume of a gas is directly proportional to the moles of gas:

$$V = \text{constant} \times n \qquad (P, \ T \text{ constant})$$

Eventually the four variables pressure (P), volume (V), moles of gas (n), and temperature (T) were tied together into one equation called the **ideal-gas equation:**

$$PV = nRT$$

- T must be expressed in degrees Kelvin.
- R is a constant, called the **gas constant,** whose value depends on the units chosen for pressure and temperature, $R = 0.0821$ L-atm/mol-K if pressure is measured in atmospheres and volume in liters.

Relationships that are useful for solving gas-law problems are in Table 10.1.

The following exercises demonstrate the methodology of solving several kinds of problems involving gases. When in doubt about how to solve such a problem, begin with the ideal-gas equation since it is the most general statement of the quantitative behavior of an ideal gas.

TABLE 10.1 Useful Relationships for Solving Gas Problems

Equation	Comment
Ideal-gas equation: $PV = nRT$	The most general description of the quantitative relationships among P, V, T, and n. It can be used as the starting point to solve most gas-law problems. Typically you use it when you are asked to calculate P, V, T, or n given the other three. See Exercise 4.
Same gas at two different states: $$\frac{P_1 V_1}{T_1} = \frac{P_2 V_2}{T_2} \ (n \text{ constant})$$	This equation is used when the same gas is compared under two different conditions (state 1 and state 2) with n constant. Typically you are asked to calculate P_2, V_2, or T_2 given two of the values and also P_1, V_1 and T_1. See Exercise 4.
Density of gas (g L): $$d = \frac{P(\mathcal{M} \text{ of gas})}{RT}$$	This is derived from $PV = nRT$. Rather than memorize the equation you should be able to derive it knowing that $$\frac{n}{V} = \frac{m/\mathcal{M}}{V} = \frac{g/V}{\mathcal{M}} = \frac{d}{\mathcal{M}} \quad \text{See Exercise 5.}$$
Molar mass of a gas ($\mathcal{M}$): $$(\mathcal{M}) = \frac{(\text{grams})(R)(T)}{PV}$$	This is also derived from $PV = nRT$ using the relationship $n = m/\mathcal{M}$.

EXERCISE 3   There is defined for gases a standard state, given the symbol STP, which is the standard reference condition for the properties of a gas. What is the standard state for a gas? What is the volume of one mole of an ideal gas at STP?

Solution: *The standard state for a gas corresponds to 0°C (273 K) and 1 atm of pressure.* The letters in STP stand for standard temperature and pressure. The volume of one mole of an ideal gas at STP is 22.4 L.

EXERCISE 4   A 5.00-L container is filled with $N_2(g)$ to a pressure of 3.00 atm at 250°C. What is the volume of a container that is used to store the same gas at STP? Solve this problem using each of the following approaches: (1) gas-law equation;

(2) $\dfrac{PV}{T}$ = constant at different conditions if number of moles are constant

(3) "common sense" and Charles's and Boyle's laws.

Solution: You are given that the initial conditions: $P = 3.00$ atm, $V = 5.00$ L, and $T = 523$ K (*remember:* when doing gas problems you must convert degrees Celsius to kelvin; $K = 250°C + 273 = 523$ K). The problem tells you that the final conditions are at STP (1.00 atm, and 273 K) and that you are to calculate the new volume of the gas.

Approach 1: The words *same gas* in the problem tell you that the number of moles of gas does not change; therefore n is constant. If you are not sure how to solve gas-law problems, always start with the ideal-gas law. Let's solve for n using the initial conditions and then use this value of n to solve for the new volume.

$$PV = nRT$$

or rearranging the equation to solve for n

$$n = \frac{PV}{RT} = \frac{(3.00 \text{ atm})(5.00 \text{ L})}{(0.0821 \text{ L-atm/K-mol})(523 \text{ K})} = 0.349 \text{ mol}$$

Use this value of n to solve for the new volume at STP:

$$V = \frac{nRT}{P} = \frac{(0.349 \text{ mol})(0.0821 \text{ L-atm/K-mol})(273 \text{ K})}{1.00 \text{ atm}} = 7.83 \text{ L}$$

Approach 2: Instead of using the previous approach you can shorten the number of steps involved in the calculation if you use the following equation: $P_1V_1/T_1 = P_2V_2/T_2$. It can be used whenever the moles of gas do not change.

Substituting the quantities $P_1 = 3.00$ atm, $V_1 = 5.00$ L, $T_1 = (250 + 273)$ K, $P_2 = 1.00$ atm, and $T_2 = 273$ K into the relation gives

$$\frac{(3.00 \text{ atm})(5.00 \text{ L})}{523 \text{ K}} = \frac{(1.00 \text{ atm})(V_2)}{273 \text{ K}}$$

Solving for V_a gives a volume of 7.83L.

Approach 3: When the moles of gas remain constant you can use a "common sense" method of solving for the new volume. Boyle's law tells you that volume varies inversely with pressure and Charles's law tells you that it varies directly with absolute temperature. Thus it makes sense to say

New volume = old volume × pressure correction ratio × temperature correction ratio

In this problem the pressure decreases from 3.00 atm to 1.00 atm. A decrease in the pressure increases the volume according to Boyle's law. Thus the pressure correction ratio must increase the volume of the gas. This ratio must be 3.00 atm/1.00 atm. The temperature decreases from 523 K to 273 K. A decrease in the temperature decreases the volume of a gas according to Charles's law. Thus the temperature correction ratio must be less than one. This ratio must be 273 K/523 K. You can now solve for the new volume

$$\text{New volume} = 5.00 \text{ L} \times \frac{3.00 \text{ atm}}{1.00 \text{ atm}} = \frac{273 \text{ K}}{523 \text{ K}} = 7.83 \text{ L}$$

This approach is no different from that in Approach 2 except that instead of using an equation you memorized, you used your knowledge of the behavior of gases to solve for an unknown quantity.

EXERCISE 5 Cyclopropane is used as a general anesthetic. It has a molar mass of 42.0g. What is the density of cyclopropane gas at 25°C and 1.02 atm?

Solution: In this problem you are given P, T, and the molar mass (M) of the gas. You are asked to calculate density ($d = m/V$). The ratio m/V for a gas can be related to its pressure, temperature, and (M) by using the ideal-gas equation:

$$PV = nRT = \left(\frac{m}{M}\right)RT$$

Rearranging the equation so the ratio m/V is on the left side of the equal sign yields:

$$\frac{m}{V} = \frac{(M)P}{RT} = d$$

The gas density of cyclopropane is therefore

$$d = \frac{(M)P}{RT}$$

$$= \frac{\left(42.0 \frac{\text{g}}{\text{mol}}\right)(1.02 \text{ atm})}{(0.0821 \text{ L-atm/K-mol})(298 \text{ K})}$$

$$= 1.75 \text{ g / L}$$

Converting the unit of density to g/cm³,

$$d = \left(1.75 \frac{\text{g}}{\text{L}}\right)\left(\frac{1 \text{ L}}{10^3 \text{ cm}}\right) = 1.75 \times 10^{-3} \text{ g/cm}^3$$

EXERCISE 6 What volume of $N_2(g)$ at 720 torr and at 23°C is required to react with 7.35 L of $H_2(g)$ at the same temperature and pressure?

$$N_2(g) + 3H_2(g) \longrightarrow 2NH_3(g)$$

Solution: Avogadro's hypothesis tells us that equal volumes of gases at the same temperature and pressure contain equal numbers of moles of gas. Therefore, the number of moles of gas and volumes of gas at the same temperature and pressure are directly proportional. This means that the volumes of gases, measured at the same temperature and pressure, are in the same ratio as the coefficients in the balanced chemical equation. From the balanced chemical equation,

$$1 \text{ mole } N_2 \eqsim 3 \text{ moles } H_2$$

or since the volumes of gases and the number of moles are directly related at the same temperature and pressure

$$1 \text{ liter } N_2 \eqsim 3 \text{ liters } H_2$$

$$\text{Volume } N_2 = 7.35 \text{ liters } H_2 \times \frac{1 \text{L } N_2}{3 \text{L } H_2} = 2.45 \text{ liters } N_2$$

DALTON'S LAW OF PARTIAL PRESSURES: MIXTURE OF GASES

The relationship describing the behavior of gases in a mixture was originally proposed by John Dalton in the nineteenth century. You need to know the following about **Dalton's Law of partial pressure** and its application:

- The total pressure (P_t) of a mixture of ideal gases is the sum of the individual pressures (partial pressures) each ideal gas would exert if it were the only gas in the container:

$$P_t = P_1 + P_3 + \ldots P_n \qquad (n \text{ gases in mixture})$$

- $P_i = \dfrac{n_i RT}{V}$ (P_i = partial pressure of an individual gas, n_i = moles of an individual gas, V = total volume of container)

- $P_t = (n_1 + n_2 + n_3 + \ldots n_n)\dfrac{RT}{V}$ (Use this equation to calculate P_t when you know the number of moles of each gas)

Dalton's law of partial pressure is commonly applied when measuring the pressure of a gas stored over water.

- A gas stored over water also contains water vapor. At a given temperature, the partial pressure of water vapor is known and is tabulated in tables.
- P_t = partial pressure of dry gas + partial pressure of water vapor.

$$P_t = P_{gas} + P_{H_2O}$$

- Therefore, the partial pressure of the dry gas is calculated by

$$P_{gas} = P_t - P_{H_2O}$$

Mole fraction and percentages of molecules of each gas present in a mixture are also used to describe the composition of gas mixtures. Note that mole fraction and percentage expressed as a decimal are the same numerically. Key concepts are

- **Mole fraction** (X) is the ratio of the number of moles of a component in a mixture to the total number of moles present.

$$X_i = \frac{\text{moles of component } i}{\text{total number of moles in mixture}}$$

- The partial pressure of gas A (P_A) in a mixture can be calculated from its mole fraction and the total gas pressure (P_t) of the mixture:

$$P_A = X_A P_t$$

- The volume percent of a gas in a mixture is equal to the mole fraction of that gas. Thus, a mixture of a gas containing 11.1% of oxygen has a mole fraction of oxygen of 0.111.

EXERCISE 7   Oxygen gas evolved from heating $KClO_3(s)$ is collected over water in a bottle. The total volume of gas in the bottle is 150.0 mL at 27.0°C and 810.0 torr.

Calculate the partial pressure of the collected $O_2(g)$. The partial pressure of pure water at 27.0°C is 22.4 torr.

Solution: A gas collected over water contains water vapor. Thus the measured pressure is a sum of the partial pressures of oxygen and water above the water:

$$P_t = P_{O_2} + P_{H_2O}$$

Solving for P_{O_2} gives:

$$P_{O_2} = P_t + P_{H_2O} = 810.0 \text{ torr} - 22.4 \text{ torr} = 787.6 \text{ torr}$$

EXERCISE 8 Answer the following questions concerning a gas mixture that is held in a 1.00-L container at 25.0°C and contains 0.0200 mole nitrogen and 0.0300 mol ammonia at a total pressure of 3.00 atm. **(a)** What is the mole fraction of ammonia gas? **(b)** What is the partial pressure of ammonia gas?

Solution: **(a)** The mole fraction is calculated from the relationship

$$X_{NH_3} = \frac{\text{moles of } NH_3}{\text{moles of } NH_3 + \text{moles } N_2}$$

$$= \frac{0.0300 \text{ moles } NH_3}{0.0300 \text{ moles } NH_3 + 0.0200 \text{ moles } O_2} = \frac{0.0300 \text{ moles}}{0.0500 \text{ moles}}$$

$$= 0.600 \quad \text{(note that there are no units)}$$

(b) Substitute the mole fraction of ammonia, X_{NH_3}, and the total pressure, P_t, into the equation

$$P_{NH_3} = X_{NH_3}P_t = (0.600)(3.00 \text{ atm}) = 1.80 \text{ atm}$$

The partial pressure of ammonia is 60% of the total gas pressure, 1.80 atm.

EXERCISE 9 A 0.50-L container holds 0.25 mol of N_2 and 0.15 mol of O_2 at 25°C. What is the total pressure exerted by the mixture of gases?

Solution: The total pressure equals the sum of the partial pressures exerted by the gases in the mixture. The partial pressure exerted by each gas can be calculated using the ideal-gas equation $PV = nRT$:

$$P_{O_2} = \frac{n_{O_2}RT}{V} \quad \text{and} \quad P_{N_2} = \frac{n_{N_2}RT}{V}$$

Rather than substitute the known values into these equations to calculate the partial pressures, let us see if there is a way to reduce the number of calculations. We know from Dalton's law of partial pressures that the total pressure is a sum of partial pressures:

$$P_t = P_{O_2} + P_{N_2} \quad \text{or} \quad P_t = \frac{n_{O_2}RT}{V} + \frac{n_{N_2}RT}{V}$$

Rearranging the right-hand side of the equation gives the following:

$$P_t = (n_{O_2} + n_{N_2})\frac{RT}{V}$$

and substituting for the known values gives:

$$P_t = \left[\frac{\left(0.0821\frac{\text{L-atm}}{\text{mol-K}}\right)(298 \text{ K})}{0.50 \text{ K}}\right] \times (0.25 \text{ mol} + 0.15 \text{ mol}) = 20 \text{ atm}$$

Note that the above can be generalized to:

$$P_t = (n_a + n_b + n_c + \cdots + n_z)\frac{RT}{V}$$

where n_a, n_b, etc. are the number of moles of each gas in the mixture.

THE KINETIC-MOLECULAR THEORY OF GASES: MOLECULAR SPEEDS AND EFFUSION

Application of the ideal-gas law in solving gas-law problems does not require any knowledge of the behavioral nature of gas molecules. By using some simple assumptions about the behavioral nature of gas molecules, scientists in the nineteenth century formulated the kinetic-molecular theory of gases, from which the ideal-gas law can be derived. In Section 10.8 of the text, you will find a detailed discussion of these assumptions and their implications for the molecular behavior of gases. You should read these carefully and understand them.

Perhaps the main idea to grasp from the kinetic-molecular theory of gases is that *at the same temperature, molecules of all gases have the same average translational kinetic energy*. Two types of molecular speeds are discussed in the text: root mean square (rms) and average.

- Rms speed μ is not the same as average speed. The rms speed is the square root of the average of the squared velocities of the gas molecules $[1/N(v_1^2 + v_2^2 + \cdots + v_N^2)]^{1/2}$ whereas the average speed is the average of the velocities $[1/N(v_1 + v_2 + \cdots + v_N)]$.
- The rms speed of gas molecules is slightly larger than the average speed.
- Note in Figure 10.13 of the text that speeds increases with temperature and also study the shape of the curve.
- The rms speed of a gas is *inversely* proportional to the molecular molar mass of the gas at a given temperature:

$$\mu = \sqrt{\frac{3\,RT}{\mathcal{M}}}$$

Thus at a particular temperature lighter molecules move faster than heavier molecules. For example, H_2 molecules move faster than N_2 molecules at 25°C. See Figure 10.14 in the text.

Effusion is the flow of gas molecules through a small pinhole or opening. **Graham's law of effusion** states for two gases escaping through the same pinhole at the same temperature

$$\cdot \frac{\text{rate of effusion of } A}{\text{rate of effusion of } B} = \sqrt{\frac{\mathcal{M}_B}{\mathcal{M}_A}}$$

- Note that the rate of effusion for substance A is inversely proportional to the mass of one mole of substance A.
- Usually time of effusion, rather than rate of effusion is measured. Measured times are *inversely* proportional to rates

$$\frac{\text{time}_B}{\text{time}_A} = \frac{\text{rate}_A}{\text{rate}_B} = \sqrt{\frac{\mathcal{M}_B}{\mathcal{M}_A}}$$

Lighter substances effuse faster than heavier substances.

- Do not confuse diffusion with effusion. Diffusion is a process in which a substance gradually mixes with another. An example of this is the fragrance of a perfume smelled after the bottle is opened.

Another concept discussed in this chapter is:

- **Mean free path:** the average distance gas molecules travel between collisions. In general, as the molecular size of a gas particle increases, and/or the number of particles per unit volume increases (more dense), the mean free path decreases.

EXERCISE 10 Some of the following essential parts of the kinetic-molecular theory of gases are stated incorrectly. Where this is the case, give the statement so it is correct. **(a)** A gas consists of molecules in ceaseless, ordered motion. **(b)** These molecules occupy a large percentage of the volume of the container. **(c)** The time between the collision of two gas molecules is long compared to the time that the molecules are in contact during their collision. **(d)** Attractive and repulsive forces act between the molecules. **(e)** The average kinetic energy of the molecules is proportional to the absolute temperature.

Solution: **(a)** A gas consists of molecules in ceaseless, *chaotic* (random) motion. **(b)** The gas molecules themselves occupy a *small, almost negligible* percentage of the volume of the container. **(c)** Correct. **(d)** The attractive or repulsive forces acting between the molecules, including gravitational forces, are negligible. **(e)** Correct.

EXERCISE 11 Under the same conditions of temperature and pressure, which gas effuses through a tiny hole in a box at a faster rate, $O_2(g)$ or $SO_2(g)$?

Solution: The ratio of effusion rates for two gases through the same hole is given by the relation

$$\frac{r_1}{r_2} = \sqrt{\frac{M_2}{M_1}}$$

Substituting the values for the molecular weights of $SO_2(g)$ and $O_2(g)$ into this equation yields

$$\frac{r_{SO_2}}{r_{O_2}} = \sqrt{\frac{32.0 \text{ g/mol}}{64.1 \text{ g/mol}}} = 0.707$$

The rate of effusion of $SO_2(g)$ molecules is 0.707 of the rate of effusion of $O_2(g)$ molecules. Thus O_2 molecules move through the hole at a faster rate.

EXERCISE 12 The average speed of $Ar(g)$ at 273.15 K is 380.8 m/s. How does the value of μ compare to 380.8 m/s?

Solution: μ is the root-mean-square speed and is calculated using the relationship:

$$\mu = \sqrt{\frac{3 RT}{M}}$$

Substituting for R, T, and M gives:

$$\mu = \sqrt{\frac{3(8.314 \text{ J K-mol})(273.15 \text{ K})}{39.95 \text{g mol}} \times \frac{10^3 \text{ g}}{1 \text{ kg}}}$$

The conversion of g to kg is necessary so that the units can be converted to m^2/s^2 ($1\ J = 1\ kg - m^2/s^2$).

$$\mu = \sqrt{1.705 \times 10^5\ \frac{J}{kg} \times \frac{1\ kg - m^2/s^2}{1\ J}} = 413.0\ m/s$$

This calculation shows that μ (root-mean-square speed) is greater than the average speed for Ar. This is also true for other gases.

DEPARTURES FROM THE IDEAL-GAS EQUATION

The behavior of gases departs from the ideal-gas equation primarily because of two factors:

- Gas molecules are attracted to one another at short distances. Therefore, gas molecules do not collide with a wall of a container as often as is expected for an ideal gas. This causes the observed (experimental) gas pressure to be less than the ideal gas pressure.
- The volume available for a gas to move randomly about in a container is less than the total volume of the container. Gas molecules possess finite volume and exclude part of the volume of the container from one another. Therefore, the experimental volume is greater than the ideal volume, the volume actually available to the gas molecules.

The **Van der Waals equation** accounts for the above two factors:

$$\left(P_{exp} + \frac{an^2}{V^2} \right)(V_{exp} - nb) = nRT$$

In $\left(P_{exp} + \dfrac{an^2}{V^2} \right)$ the term an^2/V^2 corrects for the intermolecular attractions that cause the observed pressure to be lower than what would be expected for an ideal gas. Typically, the more polar a substance is the larger the value of it will be. In $(V_{exp} - nb)$, the nb term corrects for the volume excluded to gas molecules because of their size. As expected, the value of b increases with increasing molecular size.

EXERCISE 13   The quantity PV/RT can be used to show whether 1 mol of a gas acts like an ideal gas as the pressure is varied over a wide range. **(a)** What is the value of PV/RT for 1 mol of an ideal gas? **(b)** When will PV/RT for a gas be greater than 1? **(c)** When will PV/RT for a gas be less than 1?

Solution: **(a)** $PV = nRT$ for an ideal gas. The ratio PV/RT equals n, which has the value of 1 for 1 mol of an ideal gas. **(b)** The observed value of PV is greater than the ideal value of PV for a gas when the observed volume—that is, the measured volume—is larger than the ideal volume. This happens when the gas molecules occupy a significant portion of the volume of the container. When this is the case, not all of the volume of the container is available to the gas molecules because the gas molecules exclude space from each other. Thus the gas molecules have an available ideal volume that is less than the observed volume. This behavior of gases occurs at very high pressures. **(c)** A gas shows this behavior if it is not an ideal gas and if the following relation exists:

$$PV_{experimental} < PV_{ideal}$$

The observed pressure will be less than the ideal pressure when gas molecules are attracted to each other and do not collide with the walls as frequently as in the ideal case.

EXERCISE 14 (a) Which term in the van der Waals equation accounts for the fact that gas molecules have a finite volume and exclude one another from space? (b) Under what conditions does a real gas show ideal-gas behavior?

Solution: (a) The term nb accounts for the excluded volume. The term $V - nb$ corresponds to the ideal volume. (b) When the volume becomes very large, the term n^2a/V^2 becomes very small compared to P, and the term $P + n^2a/V^2$ reduces to P. Also, the value of nb in the term $V - nb$ becomes very small compared to V, and the term reduces to V. The equation $(P + n^2a/V^2)(V - nb) = nRT$ then reduces to $PV = nRT$. Note that the volume becomes large when the pressure is low. Also, at high temperature the motional energies are larger relative to intermolecular forces between gas molecules, and this leads to more ideal behavior.

EXERCISE 15 What is the pressure exerted by one mole of Ar gas at a volume of 2.00 L at 300 K when it acts as an ideal gas and as a nonideal gas? $a = 1.34$ L^2-atm/mol^2 and $b = 0.0322$ L/mol for Ar(g). Explain any significant departure from ideality. Qualitatively predict if CO_2(g) would show greater or lesser departure from ideality. $a = 3.59$ L^2-atm/mol^2 and $b = 0.427$ L/mol for CO_2(g).

Solution: First, rearrange the van der Waals equation so that pressure is only on one side of the equal sign:

$$P + \frac{n^2a}{V^2} = \frac{nRT}{V - nb} \quad \text{or} \quad P = \frac{nRT}{V - nb} - \frac{n^2a}{V^2}$$

Substitute the following values into the van der Waals equation:
$n = 1.00$; $V = 2.00$ L; $T = 300$ K; $R = 0.0821$ L-atm/K-mol;
$n^2a = (1.00 \text{ mol})^2 (1.34 \text{ L}^2\text{-atm/mol}^2) = 1.34$ L^2-atm;
$nb = (1.00 \text{ mol})(0.0322 \text{ L/mol}) = 0.0322$ L

$$P = \frac{(1.00 \text{ mol})(0.0821 \text{ L-atm mol-K})(300 \text{ K})}{(2.00 \text{ L} - 0.0322 \text{ L})} - \frac{1.34 \text{ L}^2\text{-atm}}{(2.00 \text{ L})^2}$$

$$P = 12.5 \text{ atm} - 0.335 \text{ atm} = 12.2 \text{ atm}$$

The pressure calculated from the ideal-gas equation is

$$P = \frac{nRT}{V} = \frac{(1.00 \text{ mol})(0.0821 \text{ L-atm/mol-K})(300 \text{ K})}{2.00 \text{ L}}$$

$$P = 12.3 \text{ atm}$$

The calculated pressures are almost equal. The slight reduction in pressure from the ideal pressure is caused by the a term. If CO_2 had been chosen as the gas, the departure from ideality would have been greater because its a value is significantly larger.

SELF-TEST QUESTIONS

Key Terms

Having reviewed the key terms listed at the end of Chapter 10 and their applications, identify the following statements as true or false. If a statement is false, indicate why it is incorrect. Key terms are italicized in the statements.

10.1 The *pressure* exerted by a 10-cm column of mercury with a diameter of 1 cm is the same as that exerted by a 1-cm column of mercury with a diameter of 10 cm.

10.2 *Two standard atmospheres* of pressure correspond to 202.650 kPa of pressure.

10.3 *Standard temperature and pressure* correspond to 25°C and 1 atm pressure.

10.4 According to the *ideal-gas equation,* as a flexible balloon rises in altitude its size decreases, assuming the gas does not leak through the balloon's skin. (Atmospheric pressure decreases with increasing altitude.)

10.5 The following relationship is a statement of *Boyle's law: P = c/V.*

10.6 The following figure, which shows data plotted from experimental measurements, demonstrates the validity of *Charles's law:*

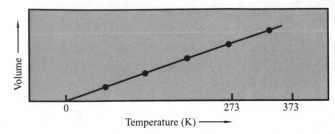

10.7 According to *Avogadro's hypothesis,* when one volume of nitrogen reacts with three volumes of hydrogen to form ammonia, four volumes of ammonia should form.

10.8 380 *torr* corresponds to 0.5 atmospheres of pressure.

10.9 The *mean free path* of O_2 in the atmosphere is expected to decrease with increasing altitude.

10.10 According to *Graham's law,* you should expect $NH_3(g)$ to effuse faster through a tiny hole than $CO_2(g)$.

10.11 In *van der Waals equation,*

$$\left(P + \frac{an^2}{V^2}\right)(V - nb) = nRT$$

the term an^2/V^2 corrects the ideal pressure, P, for attractive forces between molecules.

10.12 When two gases undergo *effusion* through the same hole at the same temperature, we can conclude that they have the same average kinetic energy.

10.13 When you measure the rate of NH_3 spreading throughout a long tube, you are measuring the rate of *diffusion* of NH_3.

10.14 According to the postulates of the *kinetic-molecular theory* of gases, gases have differing average kinetic energies at the same temperature.

10.15 In the ideal-gas equation, the *gas constant R* equals PV/nT.

10.16 In a mixture of two gases, A and B, A has a partial pressure of 20 mm Hg, and B has a partial pressure of 30 mm Hg at 25°C in a total volume of 1 L. According to *Dalton's law of partial pressures,* if these two gases were allowed to expand to 2 L at 25°C, the new total pressure of the gas mixture would be 100 torr, double the total pressure before expansion occurred.

10.17 The vapor pressure of ether is greater than that of ethanol. Therefore you should expect more *vapor* above ether than ethanol in a closed system.

10.18 A *Pascal* is a unit of pressure which equals 1 N/m^2.

10.19 The units *atmosphere* and torr are equivalent.

10.20 According to *Avogadro's law,* the pressure of a gas

maintained at constant temperature and volume is directly proportional to the number of moles of the gas.

10.21 You would expect a gas at high pressures to behave like an *ideal gas.*

10.22 If the mole fraction of a gas increases in a mixture of gases, the *partial pressure* of the gas increases.

10.23 The *mole fraction* of 0.02 mol HCl in 10 mol H_2O is 0.02/10.

10.24 The *root-mean-square speed* of a gas will increase as the molar mass of the gas increases.

Problems and Short-Answer Questions

10.25 A tire pressure gauge measures the pressure exerted by the air within the tire above atmospheric pressure. Given that atmospheric pressure at sea level is about 14.5 lb/in.^{2}, answer the following questions:
    (a) What is the total pressure of air in a tire when the gauge reads 28.0 lb/in.^{2}, at 25°C?
    (b) Suppose the temperature of the tire described in **(a)** rises to 35°C with no volume change; what is the new gauge reading?

10.26 A mixture of 2.00 g of $O_2(g)$ and 3.25 g of $SO_2(g)$ exerts how many atmospheres of pressure inside a 2.00-L container at 300 K?

10.27 How many liters of $O_2(g)$ at STP are evolved when 3.25 g of $KNO_3(s)$ decomposes to $KNO_2(s)$ and $O_2(g)$?

10.28 How many molecules of $CH_4(g)$ are contained in 1 mL at STP?

10.29 What volume of oxygen is required for the complete combustion of 10.0 L of $CH_4(g)$ if the gases are all at 600 K and 1.00 atm?

10.30 Calculate the rate of flow of $O_2(g)$ through the walls of a porous tube if $H_2(g)$ flows at the rate of 3.95×10^{-3} mL/sec under identical conditions.

10.31 Why do gases behave more like ideal gases at higher temperatures than at lower temperatures?

10.32 Helium is a gas at the boiling point of liquid nitrogen, 77 K. If 2.00 g of He is placed inside a 1.50-L container immersed in liquid nitrogen at 77 K, what is the pressure exerted by the helium gas?

10.33 A mixture of gases at 2.00 atm pressure and at 273 K contains 0.70 mol of N_2 and 0.30 mol of O_2. Calculate the partial pressure of each gas in the mixture.

10.34 How do the average kinetic energies and rms speeds of samples of N_2 and CO compare at 25°C?

10.35 A glass bulb was filled with the vapors of a volatile liquid at 100°C, and 750 torr pressure. The net weight of the sample in the bulb under these conditions was 0.124 g. When the bulb was filled with water at 25°C, the volume of the bulb was found to be 120.2 cm^3. Calculate the molar mass of the unknown gas.

10.36 A mixture containing 0.250 mol $N_2(g)$, 0.150 mol $CH_4(g)$ and 0.100 mol $O_2(g)$ is confined in a 1.00 L vessel at 35°C.

    (a) Calculate the partial pressure of $O_2(g)$ in the mixture.

    (b) Calculate the total pressure of the mixture.

    (c) If the temperature is raised to 50°C, what is the new partial pressure of $O_2(g)$, its mole fraction and its percentage in the gas mixture?

Multiple-Choice Questions

10.37 According to the ideal-gas equation, which of the following is true?

    (a) If gases are mixed, the partial pressure of each lowers the partial pressure of the others.

    (b) For Boyle's law to apply, a gas must be kept at constant pressure.

    (c) the volume of a gas is not changed if it is heated from 0°C to 100°C and at the same time the pressure is increased from 750 torr to 850 torr;

    (d) the volume of a gas doubles when the centigrade temperature doubles if all other variables are held constant;

    (e) The volume of a gas decreases by a factor of 2 when the pressure is doubled if all other variables are held constant.

10.38 Which of the following statements about the kinetic-molecular theory of gases is true?

    (a) It gives a satisfactory explanation of Boyle's law.

    (b) It is based on the concept that gas molecules occupy a finite portion of the container in which they are contained.

    (c) It can be applied to crystals.

    (d) It explains why small molecules exert a smaller pressure than do the same number of larger molecules at the same temperature and in the same volume.

    (e) It assumes that small and large molecules have the same speed at the same temperature.

10.39 Deviations in the behavior of gases from the ideal-gas equation: (1) occur because gas molecules occupy a finite volume in a container; (2) occur because attractions between gas molecules exist; (3) decrease with increasing pressure; (4) decrease with increasing temperature; (5) decrease with decreasing volume.

    (a) (1) and (2)

    (b) (1), (2), and (3)

    (c) (1), (2), and (4)

    (d) (1), (2), (4), and (5)

    (e) (1), (2), (3), (4), and (5)

10.40 A 15.0-L gas sample is heated from 29°C to 67°C at constant-pressure conditions. Under these conditions, which of the following statements is correct?

    (a) The number of moles of gas increases.

    (b) The number of moles of gas decreases.

    (c) The volume decreases.

    (d) The volume increases.

    (e) None of the above are correct.

10.41 If the density of a gas is 0.08987 g/L at STP conditions, what is its molar mass?

    (a) 0.08987 g   **(d)** 89.87 g

    (b) 2.01 g   **(e)** 100 g

    (c) 249 g

10.42 A sample of $CO_2(g)$ occupies 3.0 L at 35°C at 1.0 atm. What is its new volume if the temperature and pressure are changed to 48°C and 1.5 atm?

    (a) 3.0 L   **(d)** 4.3 L

    (b) 1.9 L   **(e)** 2.1 L

    (c) 4.7 L

10.43 What is the density in grams per liter of Ne gas at STP conditions?

    (a) 1.000 g/L

    (b) 0.9009 g/L

    (c) 1.0971 g/L

    (d) 0.0971 g/L

    (e) insufficient data provided

10.44 Several commercial drain cleaners contain $NaOH(s)$ and small amounts of $Al(s)$. When one of these cleaners is added to water, a reaction occurs resulting in the formation of H_2 bubbles:

$$2Al(s) + 2OH^-(aq) + 2H_2O \longrightarrow 3H_2(g) + 2AlO_2^-(aq)$$

The purpose of the H_2 bubbles is to agitate the solution and thereby increase the cleansing action. According to this equation, how many liters of $H_2(g)$ are released when 0.200 g of Al are dissolved in an excess of OH^- at 25°C and 1.00 atm pressure?

    (a) 1.00 L   **(d)** 2.72 L

    (b) 0.111 L   **(e)** 0.272 L

    (c) 0.0111 L

10.45 Which of the following statements about the properties of a gas are true when the temperature is increased at constant volume: (1) pressure increases; (2) pressure decreases; (3) average molecular speed increases; (4) average molecular speed decreases; (5) ideality of gas decreases.

    (a) (1), (2), (3), (4), and (5)

    (b) (2) and (4)

    (c) (2), (4), and (5)

    (d) (1) and (3)

    (e) (1), (3), and (5)

10.46 Which of the following statements about the density of a gas is correct?

    (a) It is independent of temperature.

    (b) It decreases with increasing temperature at constant pressure.

    (c) It is independent of pressure.

    (d) It decreases with increasing pressure at constant temperature.

    (e) It doubles when the volume of a container is doubled without a change in pressure or temperature.

10.47 What volume of oxygen will combine with 4 L of methane (CH_4) during its combustion reaction, all at STP?

    (a) 1 L   **(d)** 6 L

    (b) 2 L   **(e)** 8 L

    (c) 4 L

SELF-TEST SOLUTIONS

10.1 False. The force per unit area exerted by Hg in a column is related to the height of the mercury column and independent of its shape. The pressure exerted by the 10-cm column is greater. **10.2** True. 1 atm = 101.325 kPa, or 2 atm = 202.650 kPa. **10.3** False. Standard temperature is 0°C. **10.4** False. With increasing altitude, the value of pressure (P) decreases, and the value of V increases: $V = nRT/P$. **10.5** True. **10.6** True. **10.7** False. $N_2 + 3H_2 \rightarrow 2NH_3$; 1 volume + 3 volumes $\rightarrow$ 2 volumes. Two volumes of NH_3 form. **10.8** True. **10.9** False. Mean free path increases with decreasing pressure. Pressure decreases with increasing altitude. **10.10** True. $NH_3(g)$ is lighter than $CO_2(g)$. **10.11** False. It corrects the *observed* pressure (P in the equation). **10.12** True. **10.13** True. **10.14** False. At the same temperature, gases have the same average kinetic energies. **10.15** True. **10.16** False. Initially the total pressure is 50 mm Hg: $P = P_1 + P_2 = 20$ mm Hg + 30 mm Hg = 50 mm Hg. When the volume of the container is doubled at constant temperature the partial pressure of each gas must decrease by a factor of 2. Thus the new total pressure is 10 mm Hg + 15 mm Hg = 25 mm Hg. **10.17** True. **10.18** True. **10.19** False. A torr is equivalent to 1 mm Hg. **10.20** False. The volume of a gas maintained at constant temperature and pressure is .... **10.21** False. At high pressures, particles of gas are close together and thus experience strong interactions. At low pressures, they tend to behave as ideal gases. **10.22** True. **10.23** False. The mole fraction is 0.02/(0.02 + 10). The denominator is the sum of all moles of components in the mixture. **10.24** False. It will decrease with increasing molar mass. rms is proportional to the square root of the inverse of molar mass. **10.25** (a) $P_t = P_{gauge} + P_{atmosphere} = 28.0$ lb/in.^{2} + 14.5 lb/in.^{2} = 42.5 lb/in.^{2} (b) $T_1 = (25 + 273)K = 298$ K; $T_2 = (35 + 273)K = 308$ K. Since the number of moles of gas remains constant, we can use the relation of $P_1V_1/T_1 = P_2V_2/T_2$, and substitute into it the values of P_1, T_1, and T_2. V_1 and V_2 are the same; thus $P_1/T_1 = P_2/T_2$ or $P_2 = P_1(T_2/T_1)$: $P_2 = (42.5$ lb/in.$^2)(308$ K$)/(298$ K$) = 43.9$ lb/in.^{2}. The gauge pressure is 43.9 lb/in.^{2} − 14.5 lb/in.^{2} = 29.4 lb/in.^{2}. **10.26** The total pressure is the sum of partial pressures: $P_1 = P_{O_2} + P_{SO_2} = (n_{O_2} RT/V) + (n_{SO_2} RT/V) = RT/V (n_{O_2} + n_{SO_2})$. Since R, T, and V are known quantities, we need only calculate the number of moles of each gas in order to calculate P_1:

$$n_{O_2} = \frac{2.00 \text{ g}}{32.00 \text{ g mol}} = 0.0625 \text{ mol}$$

$$n_{SO_2} = \frac{3.25 \text{ g}}{64.00 \text{ g/mol}} = 0.0507 \text{ mol}$$

$$P_t = \left(0.08206 \frac{\text{L-atm}}{\text{K-mol}}\right)$$
$$\times \frac{(300\,\text{K})(0.0625 \text{ mol} + 0.0507 \text{ mol})}{2.00 \text{ L}}$$

10.27 The chemical equation describing the reaction is $2KNO_3(s) \rightarrow 2KNO_2(s) + O_2(g)$. The mole equivalence relations statement is 2 mol $KNO_3 \backsimeq$ 2 mol $KNO_2 \backsimeq$ 1 mol O_2.

At STP, 1 mol of an ideal gas has a volume of 22.4 L (see Exercise 3). The number of moles of KNO_3 are (3.25 g/(101.1 g/mol) = 0.0321 mol. The volume of O_2 evolved at STP is

$$V = \left(22.4 \frac{\text{L}}{\text{mol}}\right)(0.0321 \text{ mol KNO}_3)$$
$$\times \left(\frac{1 \text{ mol O}_2}{2 \text{ mol KNO}_3}\right) = 0.360 \text{ L}$$

10.28 One milliliter of $CH_4(g)$ at STP contains the following number of moles:

$$n = \frac{PV}{RT} = \frac{(1 \text{ atm})(0.00100 \text{ L})}{\left(0.08206 \frac{\text{L-atm}}{\text{K-mol}}\right)(273 \text{ K})}$$
$$= 4.46 \times 10^{-5} \text{ mol}$$

The number of molecules of $CH_4(g)$ is its number of moles times Avogadro's number:

$$\text{Molecules} = (4.46 \times 10^{-5} \text{ mol})$$
$$\times \left(6.022 \times 10^{23} \frac{\text{molecules}}{\text{mol}}\right)$$

$$\text{Molecules} = 2.69 \times 10^{19} \text{ molecules}$$

10.29 The balanced chemical equation describing the combustion of $CH_4(g)$ is $CH_4(g) + 2O_2(g) \rightarrow CO_2(g) + 2H_2O(l)$. Equal volumes of gases at the same temperature and pressure contain an equal number of moles. One mole of $CH_4(g)$ requires 2 mol of $O_2(g)$ to complete the combustion. Thus, 1 volume of $CH_4(g)$ requires 2 volumes of $O_2(g)$, or a volume ratio of 2:1. Therefore, $V_{O_2} = 2V_{CH_4} = (2)(10.0 \text{ L}) = 20.0 \text{ L}$. **10.30** The ratio of flow rates is

$$r_{O_2} / r_{H_2} = \sqrt{\mathcal{M}_{H_2} \big/ \mathcal{M}_{O_2}} \text{ where } r \text{ is the flow rate.}$$

Substituting the appropriate values into the equation gives

$$\frac{r_{O_2}}{3.95 \times 10^{-3} \text{ mL sec}} = \sqrt{\frac{2.02 \text{ g/mol}}{32.00 \text{ g/mol}}}$$

Solving for the rate of flow of $O_2(g)$ yields $r_{O_2} = 9.92 \times 10^{-4}$ mL/sec. **10.31** At higher temperatures the velocities of molecules significantly increase as compared with the forces of attraction between molecules. Because the motional energies of gas molecules cause a reduction in the effect of intermolecular attractions at higher temperatures, the molecules stick together less upon collision. This results in a decrease in the value of the an^2/V^2 term in the $(P + an^2/V)$ component of van der Waals equation. Thus at high temperatures the "experimental" pressure becomes more like the "ideal-gas" pressure.

10.32

$$PV = nRT$$

$$P = \frac{nRT}{V}$$

$$= \frac{(2.00 \text{ g He})\left(\dfrac{1 \text{ mol He}}{4.00 \text{ g He}}\right)\left(0.08206 \dfrac{\text{L-atm}}{\text{K-mol}}\right)(77 \text{ K})}{1.50 \text{ L}}$$

$$= 2.11 \text{ atm}$$

10.33 Let n_t equal total number of moles of gases. Then the unknown volume of the system is

$$P_t V_t = n_t RT \qquad V_t = \frac{n_t RT}{P_t}$$

The partial pressures are calculated as follows:

$$P_{N_2} = \frac{n_{N_2} RT}{V_t} = \frac{n_{N_2} RT}{\left(\dfrac{n_t RT}{P_t}\right)} = \left(\frac{n_{N_2}}{n_t}\right) P_t = X_{N_2} P_t$$

$$= \left(\frac{0.70 \text{ mol}}{0.70 \text{ mol} + 0.30 \text{ mol}}\right)(2.00 \text{ atm})$$

$$= 1.4 \text{ atm}$$

$$P_{O_2} = 2.00 \text{ atm} - 1.4 \text{ atm} = 0.6 \text{ atm}$$

10.34 The average kinetic energies of N_2 and CO are exactly the same at 25°C because all gaseous substances have the same average kinetic energies at the same temperature. To compare the rms speeds of the gas molecules at 25°C, use the following relation:

$$\frac{u_{N_2}}{u_{CO}} = \sqrt{\frac{\mathcal{M}_{CO}}{\mathcal{M}_{N_2}}} = \sqrt{\frac{18 \text{ g mol}}{28 \text{ g mol}}} = 0.80$$

Therefore $u_{N_2} = 0.80(u_{CO})$; N_2 molecules move with an average speed that is 80% of the value for CO. **10.35** $PV = nRT = (g/\mathcal{M}) RT$. Solving for $\mathcal{M}$ gives

$$\mathcal{M} = \frac{gRT}{PV}$$

$$= \frac{(0.124 \text{ g})\left(0.0821 \dfrac{\text{L-atm}}{\text{K-mol}}\right)(373 \text{ K})}{(750 \text{ torr})\left(\dfrac{1 \text{ atm}}{760 \text{ torr}}\right)(120.2 \text{ cm}^3)\left(\dfrac{1 \text{ L}}{10^3 \text{ cm}^3}\right)}$$

$$= 32.0 \text{ g/mol}$$

10.36 (a) $P_{O_2} = \dfrac{(\text{moles oxygen})RT}{V}$

$$= \frac{(0.100 \text{ mol})\left(\dfrac{\text{L-atm}}{\text{mol-K}}\right)(308 \text{ K})}{1.00 \text{ L}}$$

$$P_{O_2} = 2.53 \text{ atm}$$

(b) $P_{O_2} = X_{O_2} P_t$ or $P_t = P_{O_2} X_{O_2}$

$$= \frac{2.53 \text{ atm}}{\left(\dfrac{0.100 \text{ mol}}{(0.250 \text{ mol} + 0.150 \text{ mol} + 0.100 \text{ mol})}\right)}$$

$$P_t = 2.53 \text{ atm}/0.200 = 12.7 \text{ atm}$$

(c) Use $PV = nRT$ to calculate the new total pressure, where n is the total number of moles in the mixture. Then use $P_i = X_i P_t$ to calculate the new partial pressure.

$$P_t = \frac{nRT}{V} = \frac{(0.500 \text{ mol})\left(0.0821 \dfrac{\text{L-atm}}{\text{mol-K}}\right)(323 \text{ K})}{1.00 \text{ L}}$$

$$= 13.3 \text{ atm}$$

$$P_{O_2} = X_{O_2} P_t = \left(\frac{0.100 \text{ mol}}{0.500 \text{ mol}}\right)(13.3 \text{ atm}) = 2.66 \text{ atm}$$

The mole fraction does not change unless more material is added to the mixture. The percentage of oxygen in the mixture is the same as its mole fraction multiplied by 100: 20.0%. **10.37 (e). 10.38 (a). 10.39 (c). 10.40 (d).** $V = V_1 (T_2/T_1)$. Since $T_2/T_1 > 1$, volume increases. **10.41 (b).** At STP conditions, 1 mol of gas occupies 22.4 L. Therefore the volume of the gas in 22.4 L, its molecular weight, is $(0.08987 \text{ g/L})(22.4 \text{ L}) = 2.01 \text{ g}$. **10.42 (e).**

$$V_2 = \left(\frac{P_1}{P_2}\right)\left(\frac{T_2}{T_1}\right)(V_1)$$

$$= \left(\frac{1.0 \text{ atm}}{1.5 \text{ atm}}\right)\left(\frac{321 \text{ K}}{308 \text{ K}}\right)(3.0 \text{ L}) = 2.1 \text{ L}$$

10.43 (b). At STP conditions, 22.4 L contains 1 mol of Ne. Therefore

$$d = \frac{1 \text{ mol}}{22.4 \text{ L}} = \frac{20.18 \text{ g}}{22.4 \text{ L}} = 0.901 \text{ g/L}$$

10.44 (e). To calculate the number of moles of H_2 produced you need to make the conversion: grams Al $\rightarrow$ moles Al $\rightarrow$ moles H_2.

$$\text{Moles } H_2 = (0.200 \text{ g Al})\left(\frac{1 \text{ mol Al}}{26.98 \text{ g Al}}\right) \times \left(\frac{3 \text{ mol } H_2}{2 \text{ mol Al}}\right)$$

$$= 0.0111 \text{ mol } H_2$$

Using the ideal-gas equation, you can calculate the volume of gas produced:

$$V = \frac{nRT}{P}$$

$$= \frac{(0.0111 \text{ mol}) \left(0.08206 \, \dfrac{\text{L-atm}}{\text{K-mol}} \right) (298 \text{ K})}{1 \text{ atm}}$$

$$= 0.272 \text{ L or } 272 \text{ mL}$$

10.45 (d). 10.46 (b). 10.47 (e). $2O_2(g) + CH_4(g) \rightarrow CO_2(g) + 2H_2O(l)$. According to Avogadro's hypothesis, equal volumes of gases at the same T, P contain the same number of molecules (moles). Two moles of $O_2(g)$ are needed for each mole of $CH_4(g)$; thus 2 volumes × 4 L of $O_2(g)$ are required.

Intermolecular Forces, Liquids and Solids

11

OVERVIEW OF THE CHAPTER

Review: Enthalpy (5.3); kinetic-molecular theory of gases (10.8); nature of matter (1.1, 8.2).

Objectives: You should be able to: Employ the kinetic-molecular model to explain the differences in motion of particles in gases, liquids, and solids, and how these relate to their states.

11.1 KINETIC-MOLECULAR THEORY FOR LIQUIDS AND SOLIDS

Review: Electronegativity (8.5); bond dipole (8.5); dipole moment (8.5, 9.2).

Objectives: You should be able to: Describe the various types of intermolecular attractive forces, and state the kinds of intermolecular forces expected for a substance given its molecular structure.

11.2 INTERMOLECULAR FORCES: HYDROGEN BONDING

Review: Nature of gases (10.1); pressure exerted by gases (10.2); X radiation (6.1).

Objectives: You should be able to:

1. Explain the meaning of the terms viscosity, surface tension, critical temperature, and critical pressure, and account for the variations in these properties in terms of intermolecular forces and temperature.

2. Explain the way in which the vapor pressure of a substance changes with intermolecular forces and temperature.

3. Describe the relationship between the pressure on the surface of a liquid and the boiling point of the liquid.

4. Given the needed heat capacities and enthalpies for phase changes, calculate the heat absorbed or evolved when a given quantity of a substance changes from one condition to another.

11.3, 11.4 PHYSICAL PROPERTIES OF LIQUIDS AND SOLIDS: CHANGES OF STATE

Objectives: You should be able to: Draw a phase diagram of a substance given appropriate data, and use a phase diagram to predict which phases are present at any given temperature and pressure.

11.6 PHASE DIAGRAMS

11.7, 11.8 SOLIDS | *Review:* Density (1.3); covalent bonds (8.4); ionic bonds (8.2).

Objectives: You should be able to:

1. Distinguish between crystalline and amorphous solids.

2. Determine the net contents in a cubic unit cell, given a drawing or verbal description of the cell. Use this information, together with the atomic weights of the atoms in the cell and the cell dimensions, to calculate the density of the substance.

3. Describe the most efficient packing patterns of equal-sized spheres.

4. Predict the type of solid (molecular, covalent network, ionic, or metallic) formed by a substance, and predict its general properties.

5. Use the radii of the cations and anions in a substance to predict the coordination number of each ion.

TOPIC SUMMARIES AND EXERCISES

KINETIC-
MOLECULAR
THEORY FOR
LIQUIDS AND SOLIDS

According to the kinetic-molecular theory, a gas consists of a collection of molecules in constant, turbulent motion.

▪ Gas particles move about in the total volume of the container they occupy; however, the volume occupied by the gas particles themselves is small compared to the total volume.

▪ Gas particles remain largely separated from one another because the average energy of forces between gas molecules is considerably smaller than their average kinetic energy.

Why is a liquid formed when a gas is cooled at an appropriate temperature and pressure? The kinetic-molecular theory helps you understand the physical processes that are occurring.

▪ As the temperature of a gas is lowered, the average kinetic energy of gas particles decreases and the attractive forces between gas molecules begin to dominate. With sufficient cooling, gas molecules cluster together and form a liquid.

▪ The average kinetic energy of liquid molecules is considerably less than that of gas molecules.

▪ In the condensed state of a liquid, the molecules are in close proximity to one another and undergo frequent collisions.

Why is a solid formed when a liquid is sufficiently cooled?

▪ With additional lowering of temperature, attractive forces between particles become very large compared to the average kinetic energy associated with the movement of the particles and a solid forms.

- Particles in a solid have minimal or no translational (kinetic) energy, but they do have vibrational energy.

- Particles in a solid occupy certain positions relative to one another and this arrangement extends throughout the solid.

- When a solid and a liquid exist in a state of dynamic equilibrium, the rate of freezing equals the rate of melting; when a liquid and a gas exist in a state of dynamic equilibrium, the rate of condensation equals the rate of vaporization.

EXERCISE 1   Which of these physical properties is characteristic of a liquid, a gas, or both: **(a)** flows; **(b)** virtually incompressible; **(c)** slow diffusion of molecules; **(d)** always occupies entire volume of container?
Solution: **(a)** Both liquids and gases flow. **(b)** This is characteristic of a liquid; a gas is compressible. **(c)** This is characteristic of a liquid; molecules in a gas diffuse rapidly because of their high kinetic energies. **(d)** This is true of a gas; it occupies the entire volume in which it is contained.

EXERCISE 2   Using the kinetic theory of liquids, explain why the temperature of a volatile liquid decreases when it is poured into a shallow, insulated dish.
Solution: To answer this question you need to remember that at the surface of a volatile liquid only molecules with very high speeds have sufficient kinetic energy to escape from the liquid and thereby form gas molecules. Since a range of molecular speeds occurs for molecules within a liquid, not all molecules have sufficient kinetic energy to overcome attractive forces holding them to other nearby liquid molecules. Therefore, when molecules possessing high kinetic energy escape, the average kinetic energy of the molecules within the insulated liquid must decrease. Because the liquid is insulated, energy from the surroundings is not available to raise the average kinetic energy of the remaining liquid molecules. With a decrease in the average kinetic energy of the molecules in the liquid, the temperature of the liquid decreases.

INTERMOLECULAR FORCES: HYDROGEN BONDING

Attractive forces between neutral molecules are referred to as intermolecular forces. In Section 11.2 of the text three types of intermolecular forces are characterized and distinguished. The properties and characteristics of these intermolecular forces, and forces between ions, are summarized in Table 11.1 on page 184.

A special kind of electrostatic attraction between molecules occurs when hydrogen is bonded to nitrogen, oxygen, or fluorine atoms.

- A hydrogen atom attached to one of these three elements possesses a substantial partial positive charge. The positive end of each of the following bond dipoles (H atom)

$$\overset{\longleftarrow +}{N-H} \qquad \overset{\longleftarrow +}{O-H} \qquad \overset{\longleftarrow +}{F-H}$$

is capable of strongly interacting with an unshared electron pair possessed by a nitrogen, oxygen, or fluorine atom of an adjacent molecule. The intermolecular attraction that results is called a **hydrogen bond.** For example, the hydrogen bonds between HF molecules can be pictured as

$$---H—F---H—F---H—F---$$

The dashes between HF molecules represent hydrogen bonds.

Table 11.1   The Various Types of Intermolecular Forces

Type of intermolecular attraction	Characteristics of the interaction	Energy dependence on distance	Examples
Ion-dipole	Attractive force between an ion and an oppositely charged end of a permanent dipole possessed by a neutral molecule	$\dfrac{1}{r^2}$	KCl dissolved in water to give $K^+ \cdots OH_2$ and $Cl^- \cdots H_2O$
Dipole-dipole*	Positive end of permanent dipole on one molecule aligns itself with negative end of permanent dipole on another molecule Only significant in effect when molecules are close together	$\dfrac{1}{r^6}$	$\overset{\delta^+\delta^-}{HCl} \cdots \overset{\delta^+\delta^-}{HCl}$
London dispersion*	Short-range attractive forces between molecules resulting from momentary mutual distortion (polarization) of electron clouds Magnitude increases with increasing molecular volume and molecular mass	$\dfrac{1}{r^6}$	$Ne(g)$ $N_2(g)$ $C_8H_8(l)$

*Together known as van der Waals forces

- Hydrogen-bond energies are a small percentage of the energies of ordinary covalent bonds; however, they have important consequences for the properties of many chemical substances.

- You should learn to identify molecules that have the capacity to undergo hydrogen bonding with another like or differing molecule: Look for the presence of a N—H, O—H or F—H bond in a molecule.

EXERCISE 3   London dispersion forces allow nonpolar compounds to become liquefied. Assign the boiling points −162°C, −88°C, −42°C, and 0°C to the following nonpolar compounds: C_4H_{10}, CH_4, C_2H_6, and C_3H_8.

Solution: The magnitude of London dispersion forces between molecules is related to the extent to which electron clouds in molecules can be polarized. In general, the larger the size of an atom or molecule, the greater the magnitude of London dispersion forces. The boiling points of a series of related nonpolar compounds should increase with increasing molecular mass (increasing size of molecule), since the boiling points of such substances depend on the London dispersion forces between molecules.

Compound	Molecular mass (amu)	Number of electrons	b.p. (°C)
CH_4	16.05	10	−162
C_2H_6	30.08	18	−88
C_3H_8	44.11	26	−42
C_4H_{10}	58.14	34	0

EXERCISE 4   Explain why the replacement of a hydrogen atom in CH_4 by a chlorine atom to form CH_3Cl causes an increase in boiling point from −164°C to −24.2°C.

Solution: The difference in boiling-point temperatures exists because CH_4 molecules are nonpolar, whereas CH_3Cl molecules are polar. The additional dipole-dipole forces among CH_3Cl molecules as compared to only London dispersion forces among CH_4 molecules result in the increase in boiling point from $-164°C$ for CH_4 to $-24.2°C$ for CH_3Cl.

EXERCISE 5 Answer the following questions about hydrogen bonds: **(a)** How many hydrogen bonds can one water molecule form to other water molecules? **(b)** To what kinds of atoms must a hydrogen atom be bonded in a molecule in order for it to participate in hydrogen bonding? **(c)** Hydrogen-bond energies are on the order of 40 kJ/mol. Compare this number to the energies of covalent bonds found in Table 8.4 of the text.

Solution: **(a)** Four. For example, in ice the two unshared electron pairs on the oxygen atom of a water molecule each participate in a hydrogen bond to another water molecule. The two hydrogen atoms of the same water molecule each bond to an oxygen atom on another water molecule (see Figure 11.7 in the text). **(b)** It must be bonded to the highly electronegative atoms N, O, or F. **(c)** Hydrogen bonds are weaker than covalent bonds. Covalent-bond energies are on the order of 130 kJ/mol to 550 kJ/mol.

In Sections 11.3 –11.5 of the text, many new terms and concepts relating to the nature of liquids and solids are introduced. You should study these carefully.

PHYSICAL PROPERTIES OF LIQUIDS AND SOLIDS: CHANGES OF STATE

- The physical properties of a liquid are described by equilibrium vapor pressure, boiling point, melting point, viscosity, and surface tension.
- Table 11.2 summarizes some of the characteristics of these physical properties.

A relative measure of the forces holding particles of a substance together (intermolecular forces) is given by either critical temperature or critical pressure.

Table 11.2   Physical Properties of Liquids

Physical property	Definition	Characteristics
Vapor pressure	Pressure exerted by a gas (vapor) in equilibrium with its liquid	Increases nonlinearly with temperature Volatile liquids have high vapor pressures
Normal boiling point	The temperature at which the vapor pressure of a liquid equals *1 atm pressure*	The higher a liquid's vapor pressure, the lower is its normal boiling temperature
Boiling point	The temperature at which the vapor pressure of a liquid equals the external pressure	Boiling point increases with increasing external pressure
Viscosity	A measure of the resistance of liquids to flow under a standard applied force	The larger the viscosity of a liquid, the more slowly it flows Viscosity of a liquid decreases with increasing temperature The SI unit for viscosity is $N\text{-sec}/m^2$
Surface tension	The energy required to increase the surface of a liquid by a unit amount	Cohesive intermolecular forces cause a liquid surface to become as small as possible A liquid exhibits a meniscus in a glass tube when cohesive forces do not equal adhesive forces The SI unit for surface tension is J/m^2

- **Critical temperature** is the highest temperature at which a liquid can exist; at any higher temperature, the substance exists as a gas.
- The pressure required to liquify a gas at its critical temperature is termed **critical pressure.**
- The larger the values of critical temperature and critical pressure, the stronger the intermolecular forces holding the particles together.

A **phase change** occurs when matter is transformed from one physical state to another.

- The phase changes you should know are shown in Figure 11.1.
- During a phase change temperature remains constant.

Use the following information when you calculate enthalpy changes associated with phase changes.

- **Enthalpy of fusion,** ΔH_{fus}, is used to calculate the heat required to melt one mole of a substance at its melting point. Energy is required to disrupt the regularity of the solid state in order for particles in the liquid state to form.
- **Enthalpy of vaporization,** ΔH_{vap}, is the heat required to vaporize one mole of a substance at its boiling point. Energy must be supplied so that liquid particles gain sufficient kinetic energy to escape from one another at the surface of the liquid.
- To calculate the enthalpy change associated with a change in temperature of a substance in a particular phase at constant pressure, use specific heat capacity or molar heat capacity, C_p.

Molar heat capacity = specific heat capacity × mass of one mole.

A **dynamic equilibrium** exists when two or more forms of the same matter exist together and there is no apparent change in the quantity of matter, temperature or pressure.

- At the microscopic level, particles are passing from one phase or form to another and back again at equal rates.
- When a solid and a liquid exist in a state of dynamic equilibrium, the rate of freezing equals the rate of melting; when a liquid and a gas exist in a state of dynamic equilibrium, the rate of condensation equals the rate of vaporization.

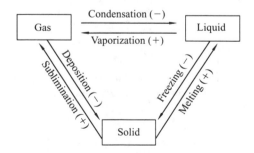

Figure 11.1 Phase changes. A plus sign indicates an endothermic process, and a minus sign indicates an exothermic process.

EXERCISE 6   The viscosity of a liquid decreases with increasing temperature. Why?

   Solution: **(a)** The viscosity of a liquid is a measure of its resistance to flow under some standard applied force. Increasing the temperature of a liquid causes the average kinetic energy of the molecules to increase. This increase in the average kinetic energy of molecules in the liquid overcomes some of the forces existing between liquid molecules, and thus the molecules are better able to move with respect to each other. That is, their resistance to motion or flow is decreased, resulting in decreased viscosity.

EXERCISE 7   Explain what happens to the boiling point of a liquid in an open container when the following occur: **(a)** The temperature of the surroundings increases without a change in atmospheric pressure. **(b)** The pressure above the liquid increases. **(c)** The volume of the liquid is doubled.

   Solution: **(a)** The boiling point does not change as long as the atmospheric pressure does not change. As the temperature of the surroundings increases, the rate of evaporation or boiling increases. **(b)** The increased pressure above the liquid requires an increase in the vapor pressure of the solution to cause boiling. Thus the boiling point increases with an increase in pressure. **(c)** The boiling point does not change because the vapor pressure has not changed. Vapor pressure does not depend on the volume of a liquid.

EXERCISE 8   Calculate the enthalpy change associated with converting 100 g of water at −10.0°C to steam at 110.0°C. $\Delta H_{fus} = 6.02$ kJ/mol, $\Delta H_{vap} = 40.67$ kJ/mol, C_p of ice = 37.78 J/mol −°C, C_p of $H_2O(l) = 75.3$ J/mol −°C, and C_p of steam = 33.1 J/ mol −°C.

   Solution: First, determine the changes that occur in the physical state of water:

$$\text{Ice}(-10.0°C) \xrightarrow{\ 1\ } \text{ice } (0°C) \xrightarrow{\ 2\ }$$
$$\text{liquid } (0°C) \xrightarrow{\ 3\ } \text{liquid } (100.0°C) \xrightarrow{\ 4\ }$$
$$\text{steam } (100.0°C) \xrightarrow{\ 5\ } \text{steam } (110.0°C)$$

Then calculate the enthalpy change associated with each physical change:

$$\Delta H_1 = n_{ice} C_{P_{ice}} \Delta T$$

$$= (100 \text{ g}) \left(\frac{1 \text{ mol}}{18.0 \text{ g}} \right) \left(\frac{37.78 \text{ J}}{\text{mol-}°C} \right) (0°C - (-10.0°C))$$

$$= 2100 \text{ J} = 2.10 \text{ kJ}$$

$$\Delta H_2 = n\Delta H_{fus} = (100 \text{ g}) \left(\frac{1 \text{ mol}}{18.0 \text{ g}} \right) \left(\frac{6.02 \text{ kJ}}{1 \text{ mol}} \right) = 33.4 \text{ kJ}$$

$$\Delta H_3 = n_{liquid} C_{P_{liquid}} \Delta T$$

$$= (100 \text{ g}) \left(\frac{1 \text{ mol}}{18.0 \text{ g}} \right) \left(\frac{75.3 \text{ J}}{\text{mol-}°C} \right) (100.0°C - 0°C)$$

$$= 41,800 \text{ J} = 41.8 \text{ kJ}$$

$$\Delta H_4 = n\Delta H_{vap} = (100 \text{ g}) \left(\frac{1 \text{ mol}}{18.0 \text{ g}} \right) \left(\frac{40.67 \text{ kJ}}{1 \text{ mol}} \right) = 226 \text{ kJ}$$

$$\Delta H_5 = n_{steam} C_{P_{steam}} \Delta T$$

$$= (100 \text{ g}) \left(\frac{1 \text{ mol}}{18.0 \text{ g}} \right) \left(\frac{33.1 \text{ J}}{\text{mol-}°C} \right) (110.0°C - 100°C)$$

$$= 1840 \text{ J} = 1.84 \text{ kJ}$$

The total enthalpy change (ΔH) is

$$\Delta H = \Delta H_1 + \Delta H_2 + \Delta H_3 + \Delta H_4 + \Delta H_5$$
$$= 2.10 \text{ kJ} + 33.4 \text{ kJ} + 41.8 \text{ kJ} + 226 \text{ kJ} + 1.84 \text{ kJ}$$
$$= 305 \text{ kJ}$$

PHASE DIAGRAMS

Depending on pressure and temperature, different phases of a particular substance may exist in equilibrium. A **phase diagram** concisely shows through a graphical presentation how the phases of a substance change with pressure and temperature.

- Lines are used to represent a condition of equilibrium between two phases.

- A single point connecting three lines on a phase diagram shows the pressure and temperature at which all three phases are in equilibrium. The condition at which all three phases of a substance are in equilibrium is called a **triple point.**

- Regions between lines represent conditions when a single phase exists. These ideas are briefly illustrated in Exercises 9 and 10.

EXERCISE 9   The following figure shows a phase diagram for CO_2. **(a)** What phases exist in the regions labeled 1, 2, and 3? **(b)** What conditions exist at the points labeled A and B?

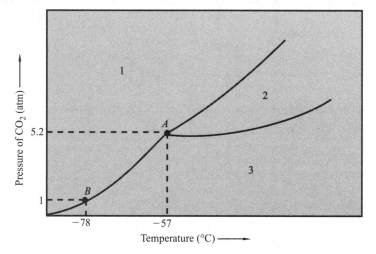

Solution: **(a)** The regions labeled 1, 2, and 3 correspond respectively to the pressure-temperature conditions at which the phases solid, liquid, and gas exist. **(b)** Point *A* corresponds to the triple point of CO_2, the pressure and temperature at which all three phases coexist. Point *B* corresponds to the temperature at 1 atm at which CO_2 sublimes —that is, an equilibrium exists between the solid and gas phases.

EXERCISE 10   Refer to the figure in Exercise 9 to answer the following questions. **(a)** What happens to the state of CO_2 when the temperature is increased at a *constant* pressure of 5.2 atm starting at −70°C? **(b)** Repeat question **(a)** when the pressure is reduced from 7 atm pressure to 1 atm pressure at a *constant* temperature of −60°C?
Solution: **(a)** At the given initial conditions, CO_2 is in region 1 and is a solid. With increasing temperature, CO_2 reaches the triple point at A and all three phases exist; then it becomes a liquid; next an equilibrium between liquid and gas phases occurs; and finally CO_2 becomes a gas in region 3. **(b)** Initially CO_2 is a solid. As the pressure is lowered it remains a solid until the pressure-temperature condition on the solid-gas phase line is reached and both phases occur. With a further lowering in pressure a gas forms.

Two types of solids are discussed: crystalline solids and amorphous solids.

- **Crystalline solids** consist of an ordered three-dimensional arrangement of particles. Section 11.5 emphasizes in its coverage this type of solid.
- **Amorphous solids** do not form a regular arrangement of particles.
- Crystalline solids can be classified in terms of the types of particles and the attractive forces between them. Study carefully Table 11.6 in the text. Note particularly the crystal classifications, the form of unit particles and the forces between particles. Use your knowledge of intermolecular forces to understand the characteristics of each type of crystal classification.

The three-dimensional arrangement of atoms or molecules in a crystal—called crystal lattice or space lattice—is often shown using a unit-cell representation.

- A **unit cell** is the smallest repeating unit of the three-dimensional crystal lattice. The entire space lattice can be considered an extension of the unit cell in three dimensions.
- You should look at Figures 11.30 through 11.33 in the text and be able to recognize or reproduce the three cubic unit cells (simple cubic, body-centered cubic, and face-centered cubic) and the NaCl unit cell shown.

You can also analyze crystal structures by considering how spherical particles in a crystal lattice are packed together. Close-packed structures contain the most efficient packing arrangements of spheres.

- **Hexagonal close-packed structures** result from an ABABAB . . . . arrangement of layers: atom A in a plane of atoms sits over itself in every other layer. See Figures 11.35 and 11.36 in the text.
- **Cubic close-packed structures** result from an *ABCABCABC* . . . . arrangement of layers: atom *A* in a plane of atoms sits over itself in every third layer below it.
- In both types of close-packed structure, each sphere has 12 nearest neighbors: The number of nearest neighbors is termed **coordination number.**

Solids can be classified according to the types of forces between particles. Four types of solids are presented in Table 11.6 in the text. You should focus on the types of forces between particles and the properties of the solids. Key relationships to note are:

- Intermolecular forces such as London forces, dipole-dipole and hydrogen bonds are associated with solids which are soft, have low to moderately high melting points and have poor thermal and electrical conduction.
- Atoms bonded by covalent bonds are usually very hard and have high melting points.
- Electrostatic attractions are dominant in ionic solids and give rise to brittle or hard materials and poor thermal or electrical conduction.
- Metallic elements are bonded by metallic bonds, which involve mobile

valence electrons. This produces solids which are soft to very hard, and have excellent thermal and electrical conduction.

EXERCISE 11 Identify the following unit cell as to type:

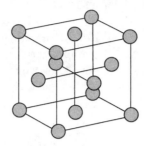

Solution: By referring to the figure, we see immediately that it belongs to a cubic unit cell group because eight particles are located at the corners of the cell and are spatially arranged to form a cube. Since particles are also located at the center of each face of the cube, this unit cell is known specifically as a face-centered cubic cell.

EXERCISE 12 The following figure represents the cubic unit cell for CsCl:

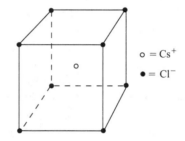

$o = Cs^+$
$\bullet = Cl^-$

Calculate the number of formula units per unit cell for CsCl.

Solution: There are eight Cl^- ions at the corners of the cube and one Cs^+ ion at the center. An ion on a corner is shared by eight other unit cells and contributes only $\frac{1}{8}$ of itself to the unit cell. An ion at the center of a cube contributes all of itself to the unit cell. Therefore:

$$\text{Number of } Cl^- \text{ per unit cell} = (8)(\tfrac{1}{8}) = 1$$
$$\text{Number of } Cs^+ \text{ per unit cell} = (1)(1) = 1$$

The ratio of Cs^+ to Cl^- is 1:1; thus the number of formula units per unit cell is 1.

EXERCISE 13 At room temperature iron crystallizes in a body-centered cubic close-packed structure. The length of an edge of the cubic cell is 287 pm. A body-centered cubic structure has 2 atoms per unit cell. Given that the density of Fe is 7.86 g/cm³, calculate the mass of an iron atom.

Solution: The length of an edge of the cubic unit cell of iron is 287 pm, or 287×10^{-12} m = 2.87×10^{-8} cm. From this length, the volume of the cubic unit cell of iron is calculated:

$$V = l \times l \times l = l^3 = (2.87 \times 10^{-8} \text{ cm})^3 = 2.36 \times 10^{-23} \text{ cm}^3$$

From the given density, and the calculated volume, the mass of the unit cell is

$$d = \frac{m}{V}$$

or

$$m = d \times V = \left(\frac{7.86 \text{ g}}{1 \text{ cm}^3} \right)(2.36 \times 10^{-23} \text{ cm}^3) = 1.85 \times 10^{-22} \text{ g}$$

Note that the unit cell contains two iron atoms; thus, if you divide the mass of the unit cell by two, you determine the mass of a single iron atom:

$$\text{mass Fe atom} = \left(\frac{1 \text{ unit cell}}{2 \text{ Fe atoms}} \right)\left(\frac{1.85 \times 10^{-22} \text{ g}}{1 \text{ unit cell}} \right)$$

$$= 9.25 \times 10^{-23} \frac{\text{g}}{\text{Fe atom}}$$

SELF-TEST QUESTIONS

Key Terms

Having reviewed the key terms listed at the end of Chapter 11 and their applications, identify the following statements as true or false. If a statement is false, indicate why it is incorrect. Key terms are italicized in the statements.

11.1 Liquid water is placed in a jar, which is then sealed. Unless water vapor is also added before the jar is sealed, a *dynamic equilibrium* between the liquid and vapor phases cannot be attained.

11.2 At an atmospheric pressure of 700 mm Hg, water boils at 97.7°C. This temperature is termed the *boiling point* of water at 700 mm Hg of pressure.

11.3 The *normal boiling point* of water is the temperature at which the vapor pressure of liquid water equals the atmospheric pressure.

11.4 As the temperature of a liquid decreases, its *vapor pressure* also decreases.

11.5 At low operating temperatures in an automobile engine, we want an oil that lets a piston slide easily against the sides of the cylinder in which it is contained. Thus at low temperatures we should use an oil that has a high *viscosity* rating.

11.6 At 20°C alcohol tends to "wet" or spread out on a piece of wax paper more readily than water. From this observation we can say that at 20°C alcohol has a lower *surface tension* than water.

11.7 A liquid is drawn up into a thin tube by *capillary action*.

11.8 At low temperatures, argon can be liquefied because *London dispersion* forces between Ar molecules are smaller than the average kinetic energy of argon particles.

11.9 H_2O has a lower normal boiling point than H_2S because of the existence of *hydrogen bonds* between H_2S molecules.

11.10 In general, the larger the *intermolecular forces* existing between molecules, the lower the temperature at which a liquid is converted to a solid at 1 atm of pressure.

11.11 We should expect KCl to form an *amorphous solid*.

11.12 $Ca(NO_3)_2$ forms a *crystalline solid* or simply a *crystal*.

From what we know about the periodic table, we should expect $Ba(NO_3)_2$ also to be a crystal in the solid state.

11.13 A crystalline solid can be thought of as built up from a small, repeating unit known as the *unit cell*.

11.14 The temperature at which the equilibrium

$$H_2O(s) \rightleftharpoons H_2O(l)$$

exists at 2 atm of pressure is called the *melting point*. The temperature at which this equilibrium exists at 1 atm of pressure is called the *normal melting point*.

11.15 $CO_2(s)$ will sublime completely to form $CO_2(g)$ at any temperature below its *triple point* providing the pressure above the solid phase is always below the equilibrium vapor pressure.

11.16 *Coordination number* of a cation refers to the number of cations that can surround a particular cation in a crystal.

11.17 In the two very similar close-packing forms of spheres, *cubic close-packing* and *hexagonal close-packing*, each sphere is touched by eight other spheres in the same plane.

11.18 We should expect the *critical temperature* of HCl to be higher than that of CH_4.

11.19 As critical temperature increases for a series of related substances, *critical pressure* decreases.

11.20 Figure 11.2 could represent the *phase diagram* for a pure substance.

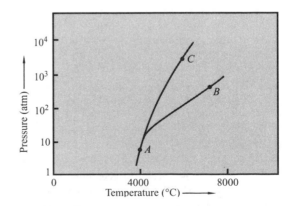

Figure 11.2 Phase diagram.

11.21 According to Figure 11.2, it is possible for the substance to undergo a *phase change* from solid to liquid at 9 atm of pressure (point *A*).

11.22 The *ion-dipole forces* between Ca^{2+} and water should be greater than those between Na^+ and water.

11.23 Significant *dipole-dipole forces* should exist between CCl_4 and SO_2.

11.24 The *polarizability* of CCl_4 should be larger than that of CH_4.

11.25 Water collected on a well waxed surface tends to form a spherical bead. This shows that the *surface tension* of water is high relative to the intermolecular forces between water and wax.

11.26 Oxygen freezes at a lower temperature than carbon dioxide. Thus, you should expect the *heat of fusion* of oxygen to be greater than that of carbon dioxide.

11.27 *Heat of vaporization* for a liquid is always smaller than its heat of fusion.

11.28 A *volatile* substance has a high heat of vaporization.

11.29 A crystal lattice represents a two-dimensional array of points describing a *crystalline solid*.

11.30 A *primitive cubic* unit cell shows lattice points only at the corners of a cube.

11.31 A *body-centered cubic* unit cell is a primitive cubic unit cell with the addition of a lattice point in the center of the cube.

11.32 A *face-centered cubic* unit cell is a body-centered unit cell with the addition of lattice points at the center of each side (face) of the cube.

11.33 *X-ray diffraction* techniques have shown that the spacing of layers of atoms in a solid crystal is usually about 2 to 20 angstroms.

11.34 *Molecular solids* tend to be rigid solids.

11.35 An example of a *covalent (network) solid* is diamond.

11.36 *Ionic solids* show good thermal conduction.

11.37 *Metallic solids* consist of arrays of atoms of metallic elements and show good thermal properties.

Problems and Short-Answer Questions

11.38 When a closed container containing liquid water is left for a time, two processes continually occur. What are these two processes, and how are their rates related?

11.39 Water boils at 71°C on top of Mt. Everest (8848 m above sea level) and at 86°C on top of Mt. Whitney (4418 m above sea level). Why does water boil at a higher temperature on Mt. Whitney than on Mt. Everest?

11.40 Should ionic salts in water increase the surface tension of aqueous solutions above the value for pure water?

11.41 What types of intermolecular forces must be overcome when a molecule of NH_3 escapes from liquid ammonia?

11.42 Explain why methanol (CH_3OH) has a higher normal boiling point and a higher normal melting point than methyl-

fluoride (CH_3F), given the following data:

Compound	Molecular weight amu	b.p. (°C)	m.p. (°C)
CH_3OH	32.04	65	−94
CH_3F	34.03	−78	−142

11.43 Which compound should have the higher melting point, butanol ($CH_3CH_2CH_2CH_2OH$) or KCl?

11.44 A heating curve shows the change in temperature of a substance as a function of the time it is heated. It is like a phase diagram, but instead of pressure as one of the axes, time is plotted. Figure 11.3 shows an idealized heating curve for a substance at a constant pressure of 1 atm beginning with a solid at a low temperature. Explain what physical changes in the substance are occurring in the regions from points A to B, B to C, C to D, D to E, and E to F. What phases are present at temperatures *x* and *y*?

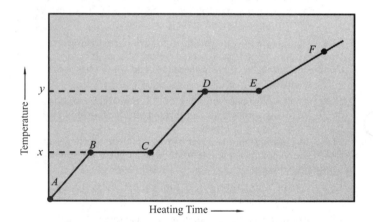

Figure 11.3 Heating curve.

11.45 The cooling curve of 1 mol of a substance is shown in Figure 11.4 on page 193.

(a) Why is the curve flat and horizontal between points *A* and *B*?

(b) The curve between points *C* and *D* results from a phenomenon known as supercooling. At points on this curve, a liquid has a temperature below its freezing point. Suggest a reason for the existence of this phenomenon.

(c) What are the phases of the substances at points *C*, *D*, and *E*?

11.46 The length of a CsCl cubic unit cell is 4.123 Å. Calculate the density of CsCl in g/cm³. See Exercise 12 for a diagram of the cubic unit cell.

11.47 Indicate the type of crystal (molecular, metallic, covalent-network, or ionic) each of the following would form upon solidification: **(a)** SO_3; **(b)** Ne; **(c)** Cs; **(d)** $MgCl_2$; **(e)** SiO_2.

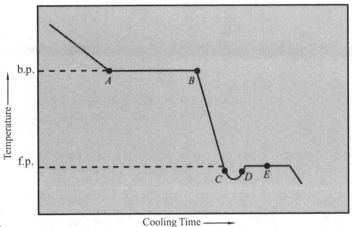

Figure 11.4   Cooling curve.

Multiple-Choice Questions

11.48   At 1 atm external pressure, water can exist at 100°C as
  (a)   only a solid;
  (b)   only a liquid;
  (c)   only a gas;
  (d)   both a solid and a liquid;
  (e)   both a liquid and a gas.

11.49   What type of cubic crystal is represented by the following diagram?
  (a)   Normal   (d)   body-centered
  (b)   cubic   (e)   simple
  (c)   face-centered

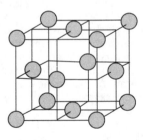

11.50   Which type of crystalline solid is SO_2 most likely to form?
  (a)   Ionic   (d)   metallic
  (b)   molecular   (e)   covalent
  (c)   atomic

11.51   Which of the following substances should have the highest normal boiling point?
  (a) CH_3Cl (d) CH_4
  (b) CH_3Br (e) CH_3OH
  (c) CH_3I

11.52   Which of the following intermolecular forces is important only at the shortest distances between interacting particles?
  (a)   Ion-ion;
  (b)   ion-dipole;

  (c)   ion-induced dipole;
  (d)   London dispersion forces;
  (e)   all equally insensitive to short distances.

11.53   Which of the following substances is most likely to exist as a crystalline solid at room temperature?
  (a)   HBr   (d) SO_2
  (b) Cl_2 (e) $ZnCl_2$
  (c) NF_3

11.54   In C_6H_5Cl, chlorobenzene, which of the following intermolecular forces operate in the liquid phase: (1) ion-ion; (2) ion-dipole; (3) dipole-dipole; (4) London dispersion?
  (a)   (1) and (2)   (d)   (2) and (4)
  (b)   (3) and (4)   (e)   all of them
  (c)   (1) and (3)

11.55   Which of the following substances crystallizes in a molecular-solid form?
  (a)   Ne   (d)   carbon
  (b)   Fe   (e) SO_2
  (c) KNO_3

11.56   Which of the following substances should have the highest melting point?
  (a)   Kr   (d) $CHCl_3$
  (b) SO_3 (e) H_2O
  (c) $Ca(NO_3)_2$

11.57   Which of the following statements is true about ice?
  (a)   It sinks in water.
  (b)   It is more dense than water.
  (c)   Hydrogen-bonding forces between H_2O molecules at 0°C are more extensive than those in water at 50°C.
  (d)   The melting of ice releases heat.
  (e)   All of the above are true.

11.58   Which of the following compounds should have the highest heat of sublimation?
  (a) C_6H_6 (benzene)   (d)   KF
  (b)   Xe   (e) C_8H_{18}
  (c) I_2

11.59   Which of the following represents the correct order of increasing boiling points for CCl_4, Cl_2, ClNO, N_2?
  (a) $CCl_4 < ClNO < Cl_2 < N_2$

(b) $N_2 < Cl_2 < ClNO < CCl_4$
(c) $Cl_2 < N_2 < CCl_4 < ClNO$
(d) $CCl_4 < N_2 < Cl_2 < ClNO$
(e) $ClNO < Cl_2 < N_2 < CCl_4$

SELF-TEST SOLUTIONS

11.1 False. A dynamic equilibrium will exist when the rate of evaporation equals the rate of condensation. **11.2** True. **11.3** False. Vapor pressure of liquid must equal 1 atm of pressure, not simply atmospheric pressure. **11.4** True. **11.5** False. We need an oil that flows easily; this is characterized by a low viscosity. **11.6** True. **11.7** True. **11.8** False. For liquefaction of a nonpolar substance to occur, London forces must be greater than kinetic-energy forces. **11.9** False. It is H_2O, not H_2S, that possesses the ability to form hydrogen bonds. Thus H_2O has the higher boiling point. **11.10** False. With large intermolecular forces, we usually observe higher freezing-point temperatures. **11.11** False, KCl, like NaCl, forms a crystalline solid. **11.12** True. **11.13** True. **11.14** True. **11.15** True. **11.16** False. It refers to the number of nearest-neighbor anions that surround the cation of interest. **11.17** False. Each is adjacent to six other spheres in the same plane. **11.18** True. Critical temperature increases with increasing intermolecular forces; HCl is polar, CH_4 is nonpolar. **11.19** False. Pressure is directly related to temperature; thus critical pressure tends to increase with increasing critical temperature. **11.20** True. **11.21** False. At 9 atm of pressure, only solid and gas exist; thus only a transformation from solid to gas is possible. **11.22** True. The higher charge of the calcium ion will more strongly interact with the negative end of the water dipole. **11.23** False. CCl_4 is a nonpolar molecule; therefore London forces will exist, but not dipole-dipole. **11.24** True. CCl_4 has more electrons and is a larger molecule. **11.25** True. **11.26** False. It takes less energy to melt oxygen; thus, it should have a lower heat of fusion. **11.27** False. Vaporization requires the particles to be separated, which requires more energy than simply moving particles from a rigid lattice to a liquid state. **11.28** False. A high heat of vaporization implies strong intermolecular forces in the liquid. This would result in a lower vapor pressure and a higher boiling point. **11.29** False. It is a three-dimensional array. **11.30** True. **11.31** True. **11.32** False. It is a primitive cubic cell with the addition of lattice points to the faces of the cube. **11.33** True. **11.34** False. They tend to be soft solids because intermolecular forces between particles are somewhat weak. **11.35** True. **11.36** False. They are poor conductors of heat. **11.37** True. **11.38** Vaporization of liquid and condensation of vapor continually occur at equal rates in a closed container that contains liquid water—that is, a state of dynamic equilibrium exists. **11.39** At higher altitudes, atmospheric pressure is less than at lower altitudes. Boiling occurs when the vapor pressure of the liquid equals the surrounding atmospheric pressure. Thus water has a lower vapor pressure and boiling point on Mt. Everest than on Mt. Whitney. **11.40** Ions in solution tend to pull water molecules into the interior of the solution because of ion-dipole attractions. The liquid surface in an aqueous ionic solution therefore contracts to a smaller area than is the case for pure water under identical conditions. More energy is required to increase the surface area of the

aqueous ionic solution than is required to increase that of pure water; thus the addition of ionic salts to water increases the surface tension of the aqueous solution above the value for pure water. Note that the increasing of the surface tension of water by addition of salts is of a far smaller magnitude than the increasing of surface tension by addition of detergent. **11.41** Several intermolecular forces must be overcome: London dispersion forces, dipole-dipole forces (NH_3 is polar), and hydrogen-bonding forces (NH_3 contains hydrogen atoms attached to the highly electronegative element N). **11.42** Both CH_3OH and CH_3F are polar and have similar molecular weights. If dipole-dipole forces and London dispersion forces were the only intermolecular forces, we would expect the compounds to have similar melting and boiling points. However, CH_3OH molecules are also attracted to each other by hydrogen-bonding forces, which must be overcome in the melting and boiling processes. The additional energy required to break hydrogen bonds between methanol molecules increases the melting and boiling points of CH_3OH compared to those of CH_3F. **11.43** In general, more energy is required to overcome coulombic forces in an ionic crystal than to overcome polar-covalent forces in a molecular crystal. Thus KCl, which forms an ionic solid, should have a higher melting point than butanol, which forms a molecular solid. **11.44** The physical changes occurring in the specified regions are as follows: A to B corresponds to heating the solid to its melting point; B to C corresponds to the melting of the solid at constant temperature; C to D corresponds to heating the liquid to its boiling point; D to E corresponds to the boiling of the liquid at constant temperature and pressure; E to F corresponds to heating the gas from the boiling point of the liquid to higher temperatures. Temperature x corresponds to the melting point of the solid. Thus both solid and liquid phases are present. Temperature y corresponds to the boiling point. At this temperature both liquid and gas phases may exist. **11.45 (a)** Between points A and B, heat is being removed from the substance at its boiling point with no change in the substance's temperature. During this time, condensation of the substance from gas to liquid occurs. The average kinetic energies of gas molecules are lowered to the point that attractive forces dominate and a liquid forms. As long as the kinetic energy removed is exactly equal to the energy released during the formation of intermolecular bonds between liquid particles, the temperature of the substance remains constant. This temperature is the boiling point of the liquid or condensation temperature of the gas. **(b)** Supercooling represents the time lag associated with the change from the random organization of liquid molecules to their regular and orderly arrangement in a crystal. This time lag often occurs when cooling is too rapid for a true dynamic equilibrium between liquid and solid to occur. Until the lowest point on the curve between C and D is reached, there are no crystal nuclei in the solution upon which other particles can be deposited to form larger crystals. **(c)** Point C corresponds to the supercooled liquid phase only. Point D corresponds to a mixture of tiny crystals and liquid. At this point crystals are forming rapidly. At point E the rate of crystal formation has "caught up" with the cooling rate, and the solid and liquid phases are in equilibrium. **11.46** CsCl has a cubic unit cell with one formula of CsCl per unit cell. The volume of the unit cell is $V = (4.123$ Å$)^3 = 70.08$ Å^3. The mass of the unit cell is the mass of one Cs^+ ion and one Cl^- ion:

$m = 132.90$ amu $+ 35.45$ amu $= 168.35$ amu. The density (m/V) of one unit cell is:

$$d = \left(\frac{168.35 \text{ amu}}{70.08 \text{ Å}^3} \right) \left(\frac{1 \text{ Å}}{10^{-8} \text{ cm}} \right)$$
$$= 2.402 \times 10^{24} \text{ amu/cm}^3$$

One gram is equivalent to 6.022×10^{23} amu. The density of CsCl expressed in g/cm^3 is

$$\left(2.402 \times 10^{24} \frac{\text{amu}}{\text{cm}^3} \right) \left(\frac{1 \text{ g}}{6.022 \times 10^{23} \text{ amu}} \right) = 3.989 \text{ g/cm}^3$$

11.47 (a) SO_3 is a polar covalent molecule; thus it is a molecular type. **(b)** Ne would possess London dispersion forces only between atoms; thus it is a molecular type. **(c)** Cs is a metal; thus it is a metallic type. **(d)** $MgCl_2$ consists of a metal and nonmetal; thus it is an ionic type. **(d)** SiO_2 contains a network of covalent Si—O linkages creating a three-dimensional array of interconnected silicon and oxygen atoms; thus it is a covalent-network type. **11.48 (e)**. **11.49 (c)**. **11.50 (b)**. Polar-covalent molecules such as SO_2 tend to crystallize in a molecular crystal lattice. **11.51 (c)**. Normally, the heaviest compound in a related series will have the highest boiling point because London dispersion forces increase with increasing weight. However, CH_3OH has the highest boiling point because hydrogen bonds also exist between CH_3OH molecules. **11.52 (d)**. **11.53 (e)**. A crystalline solid is likely to form when a compound is formed from a metallic element (Zn) and a nonmetal (Cl). **11.54 (b)**. **11.55 (e)**. Molecular solids usually form from polar or nonpolar *molecular* compounds. SO_2 is the only case listed that fits this requirement. **11.56 (c)**. In general, ionic substances have higher melting points than the atomic or molecular substances listed here. **11.57 (c)**. **11.58 (d)**. KF is the only polar substance listed. In general, a polar substance will have a higher heat of sublimation than a nonpolar substance. **11.59 (b)**. $N_2 < Cl_2 < CCl_4$ based on increasing London dispersion forces. ClNO needs to be placed. ClNO is polar and should have a higher boiling point than Cl_2. CCl_4 is *far* more massive than ClNO and thus should have a higher boiling point, even though CCl_4 is nonpolar.

12 | Modern Materials

OVERVIEW OF THE CHAPTER

12.1 LIQUID CRYSTALS

Review: Liquids (11.1); ionic materials (11.7, 11.8); solids (11.8).

Objectives: You should be able to:

1. Recount the ways in which a liquid crystalline phase differs from an ordinary (isotropic) liquid phase.

2. Distinguish among the major classes of liquid crystalline phases.

3. Describe how liquid-crystalline materials are employed in liquid crystal displays for electronics equipment.

12.2 POLYMERS

Review: Mass-chemical formula-stoichiometric relationships (3.5, 3.6); intermolecular forces (11.2).

Objectives: You should be able to:

1. Describe what is meant by the terms polymer and monomer, and give some examples.

2. Explain how a polymer is formed from monomers via (a) addition polymerization or (b) condensation polymerization.

3. Explain how processing affects the properties of synthetic polymer fibers and how it relates to molecular orientations in the fiber.

4. Explain what is meant by the term "crystallinity" in polymers, and indicate how polymer properties generally vary with degree of crystallinity.

5. Describe the process of cross-linking in polymers, and explain how it affects polymer properties.

6. Describe the vulcanization of rubber and its effect on the properties of the material.

Review: Covalent structures (11.7, 8.4, 8.6, 7.8).

Objectives: You should be able to:

1. Define the term "ceramic" and give several examples of ceramic materials.

2. Indicate the advantages and disadvantages of engineering ceramics as compared with other materials in various applications.

3. Describe the sol-gel process for forming ceramic materials.

4. Give examples of ceramic composites and indicate their potential advantages over non-composite ceramic materials.

5. Define the term superconductivity and give an example of a super conducting ceramic oxide.

Review: Van der Waals forces (11.2); metallic oxides (7.6, 7.7).

Objectives: You should be able to:

1. Explain what is meant by the term "thin film" and provide several illustrations of applications of thin films.

2. Explain how thin films are formed via vacuum deposition, sputtering and chemical vapor deposition.

TOPIC SUMMARIES AND EXERCISES

In this chapter you will learn about the properties and applications of several types of modern materials. You will need to learn many new definitions; use the key term section of the self-study exercises to check your knowledge of them. As you read the chapter, note the following about modern materials: Their properties; how they are formed or manufactured; and specific applications.

Liquid crystals are substances that are above their melting points and exhibit properties intermediate between those of the solid and liquid phases. The intermediate phase, or phases, is described as liquid crystalline.

- Typical liquid crystalline substance show a molecular shape that is long and rod-like, with some rigidity. These rod-like substance can maintain parallel ordering in the liquid crystal state.
- Another molecular shape is flat, similar to a pancake. These molecules stack on top of each other.
- Ionic substances and flexible molecules usually don't show this behavior.

Liquid crystalline phases are characterized by the arrangements of the molecules.

- In the **nematic** phase, the simplest type, molecules are aligned along their long axis, but there is no ordering with respect to the alignment of their ends.

- In the **smectic** phase, the rods are aligned parallel and their ends are aligned together. The ordering that persists is due primarily to London dispersion forces and dipole-dipole interactions.

- In the **cholesteric** phase, characteristic of derivatives of the cholesterol molecule, flat, disc-shaped molecules align themselves in a stacking arrangement. Twisting of the stacks may occur and this property may be useful in specific applications.

- Commercial applications of liquid crystals are numerous, with the most well-known in the area of electrically controlled liquid crystal displays (*LCD*).

- By application of an electrical potential in an appropriate direction in a liquid crystalline substance, the orientation of the liquid crystal molecules can be altered. Depending on the orientation of the molecules, polarized light may pass through the liquid crystal medium, or not.

- Some cholesteric substances change color as a function of temperature.

- Liquid crystal behavior is seen in certain types of biological structures, such as in tissue structures.

EXERCISE 1   Which of the following central linkages would you expect to find in liquid crystal chemistry? Which liquid crystalline phases would primarily involve these linkages?

$$—C—C—\qquad —CH=CH—\qquad —N—N—$$
$$—CH=N—\qquad —CH_2—NH—\qquad —N=N—$$

Solution: Compounds that form the nematic and smectic liquid crystal phase are rod-shaped and fairly rigid about the long axis of the molecule. The presence of pi bonds prevents rotation of bonded atoms about the bond, and a series of pi bonds near one another in the molecule will give it rigidity. Thus, the linkages that show pi bonds should form a central linkage in a liquid crystal system: $—CH=CH—$; $—CH=N—$; $—N=N—$.

EXERCISE 2   What properties should a liquid crystal possess so that it can be used in electrical applications?
Solution: The liquid crystal phase should be stable at room temperature, colorless, chemically and photochemically stable, and should show a broad liquid crystal range covering ambient temperatures.

POLYMERS

Polymers form the basis of a huge industry involving such things as plastics, resins, lacquers, and synthetics. **Polymers** are macromolecules, of high molecular weights, that are formed from smaller repeating units called **monomers.**

- The coupling of monomers to form polymers is termed **polymerization.** The text shows in the reaction before Equation (12.1) how polyethylene is formed by the opening of a carbon-carbon double bond to form carbon atoms that possess a single electron. These carbon atoms join together into long chains by **addition polymerization**

- A chemical formula such as $+\!\!\begin{smallmatrix}H\\|\\C\\|\\H\end{smallmatrix}\!-\!\begin{smallmatrix}H\\|\\C\\|\\Cl\end{smallmatrix}\!\!+_n$ shows the repeating unit in the

polymer. The subscript n is used to denote that a large number of these units are bonded in a chain. This type of formula is characteristic of addition polymers.

- **Condensation polymerization** is shown in Equation (12.2) in the text. In the reaction shown, a hydrogen from the end of a diamine reacts with a OH group of adipic acid to form water which splits out. Look at the product, nylon-6,6 and observe that a N—C bond (thickened bond below) has formed between the diamine and adipic acid. The nitrogen atom initially had a hydrogen atom attached and the carbon atom had a OH group. Water may be a product in a condensation reaction because of the reaction between a hydrogen atom and a hydroxyl group:

$$R-N-(H+HO)-C-R' \rightarrow R-N-C-R' + H_2O$$

Polymers are typically amorphous substances, with no definite crystalline order. However, there are local regions of ordering termed **crystallinity** that arise from interactions between one polymer chain and another.

- **Thermoplastic** materials can be reshaped under different conditions. Such materials consist of independent linear chains of flexible polymer molecules. If branching occurs, that is side chains occur off the main chain, then the chains may gain ordering in certain regions because of intermolecular interactions between branches.
- **Plasticizers** are low molecular weight materials that are added to polymers to minimize inter-chain interactions. This makes polymers more flexible.
- **Thermosetting plastics** can not be reshaped readily after being shaped. When a thermosetting polymer is heated, chemical bonds between polymer chains form, creating cross-links. This density of cross-links causes the material to be rigid, and extremely difficult to reshape. The text shows in Equation (12.4) the vulcanization of rubber as an example of cross-linking. Note how sulfur is used to cross-link polymer chains by reactions at carbon-carbon double bonds to form a thermosetting polymer.
- Rubber is an **elastomer** because upon stretching or bending, the original shape is restored when the applied stress is removed. The cross-linking of sulfur helps prevent the chains from slipping; thus, the original shape can be restored.
- **Crosslinking** is a process of forming chemical bonds between chains in polymers. When crosslinking occurs, the polymer becomes more dense and rigid. Thermosetting polymers are crosslinked to maintain their shapes. Vulcanization of natural rubber involves crosslinking of cis-polyisoprene with sulfur, as shown in Figure 12.18 in your text. Usually only a small amount of crosslinking is necessary to achieve the desired physical and mechanical properties.

EXERCISE 3   What are the desired qualities of a plasticizer? Give an example of a natural plasticizer that was used to soften materials before the plastic industry was developed.

Solution: Plasticizers are additives to materials that increase their workability, flexibility and distensibility. A plasticizer must be mixable with the substrate to form a mixture that yields the desired product upon processing. The cost of adding the plasticizer must be economical compared to the final cost of the product. Neat's foot oil and whale sperm oil are two materials that were added to leather to soften it and make it water repellent before the advent of the plastics industry.

EXERCISE 4   Write a reaction showing the formation of polystyrene, $-(CHCH_2)-_n$.

What type of polymerization reaction did you write?

Solution: To write the reaction describing the formation of polystyrene you first must determine the monomer. Rewrite the chemical formula to show the arrangement of bonds in the carbon backbone:

$$-\left[\begin{array}{cc} \overset{\displaystyle H}{\underset{\displaystyle C_6H_5}{C}} & \overset{\displaystyle H}{\underset{\displaystyle H}{C}} \end{array}\right]-$$

From this structure, the monomer can be deduced by recognizing that a carbon-carbon double bond in the monomer is typically opened up to permit carbon atoms of different monomers to chain together. Thus, the monomer unit used to form polystyrene should be $CH\!=\!CH_2$ Therefore, the reaction describing the formation of polystyrene is C_6H_5

$$n(CH_2\!=\!CH_2) \rightarrow -(CH\!-\!CH_2)-_n$$
$$\quad\quad\quad\quad C_6H_5 \quad\quad\quad\quad C_6H_5$$

This reaction is an example of addition polymerization because the polymer backbone is formed by carbon atoms adding together without another molecule splitting out.

EXERCISE 5   How would you classify the following polymerization reaction?

$$nBr(CH_2)_{10}Br + 2nNa \rightarrow -(CH_2CH_2)-_{5n} + 2nNaBr$$

Solution: This reaction can be classified, in a broad sense, as a condensation reaction. When the carbon chains are united, there is a splitting out of bromine atoms which react with sodium atoms to form the salt sodium bromide. Also, in a condensation reaction, the polymeric product has fewer atoms in the repeating unit than found in the original monomer.

CERAMICS

Inorganic materials that are high-melting, normally hard or brittle, and stable to very high temperatures, are termed **ceramics.** Bonding can involve a covalent network structure or ionic, or a combination of the two.

- Ceramic materials are based on a variety of chemical forms; the major ones are silicates (Si and O), oxides of metals, carbides (metals and carbon), nitrides (metals and nitrogen) and aluminates (Al_2O_3 with other metallic oxides).
- Ceramic materials are used to replace other engineering materials such as wood, metals or plastics, to provide heat and corrosion resistance, low defor-

mity properties, lighter weight materials, or lower cost materials. A major disadvantage of engineering materials is their brittleness, a tendency to shatter. Also ceramics are difficult to manufacture into specific components in systems; thus ceramics are often more costly.

The method of processing ceramics to form parts is an important focus for research. Unless the process can prevent random, extremely tiny micro cracks or voids in the structure from forming, the material is subject to stress during use.

- One method for improving fracture resistance is to produce extremely pure, uniform, sub-micron ceramic particles, and then scinter them at high temperature under pressure to create the desired material.

- The **sol-gel process** is used to make extremely fine ceramic particles of uniform size. A gel is produced from a sol (a suspension of a metallic hydroxide) whose pH is adjusted to cause a condensation reaction involving splitting water out from two of the reacting particles. The resulting gel is heated, driving off all liquid, and fine metallic oxide powder is produced. Refer to equations (12.5–12.7) in the text for an example.

- **Ceramic composites** are formed from mixing two or more ceramics. Usually a ceramic host material, a ceramic matrix, has another ceramic added to it so as to alter the original properties of the host.

- **Ceramic fibers** are widely used as an additive to a ceramic matrix. A fiber has a length that is at least 100 times greater that its diameter.

- The text describes the formation of *silicon carbide* (SiC) and its use as a fiber. Also, *boron nitride,* BN, has been used as a ceramic fiber. When these fibers are added to a host matrix, the matrix is strengthened because the fibers resist deformations with respect to applied stress along their long axis.

In Section 12.3, applications of ceramics to the electronic industry, space program and electrical industry are discussed.

- Silicon carbide is used in host matrices and is widely used as an abrasive.

- Some ceramics such as quartz (crystalline SiO_2) are **piezoelectric.** When subjected to a mechanical stress, they produce an electrical potential; thus they are used in electronic circuits.

- **Thermistors** are ceramic materials whose electrical conductivity increases with increasing temperature. They are used to measure or control temperature, as heating elements and as electrical switches.

- Ceramic tiles are used in the space shuttle vehicles to prevent heat build upon re-entry into the atmosphere of the Earth. The composition is primarily a mixture of silica fibers and aluminum borosilicate fibers.

- One of the most exciting newer applications of ceramics is the fast growing field of **superconductors.** *Superconductivity* is the property of exhibiting a "frictionless" flow of electrons within a material and the Meissner effect, in which magnetic fields are excluded from their volume. Materials that show this property do so only below a certain temperature, called the *superconducting transition temperature,* T_c. Superconductors showing a T_c above 77 K, which means inexpensive liquid nitrogen instead of liquid helium can be used to cool the material, are typically mixed metallic oxides such as

$YBa_2Cu_3O_x$. Copper and oxygen are the most important constituents. Yttrium and barium can be replaced readily without losing superconductivity properties.

EXERCISE 6   The preparation of materials in making a ceramic is very important in manufacturing. Diffusion rates are slow during the firing of ceramics; therefore the distance through which diffusion must occur for reactions must be minimized. How is this accomplished in the manufacturing of ceramics?

Solution: To minimize the distance through which diffusion must occur, the particles reacting must be very near one another. This can be accomplished if the particles of ingredients are extremely tiny and thoroughly mixed or blended. The tiny size permits the particles to come close to one another and thus the particles have sufficient density so that diffusion will lead to a reaction.

EXERCISE 7   Ceramics are strong in resisting compression, but are brittle in bending. Explain how the presence of pre-existing defects in a ceramic creates brittleness.

Solution: The term brittle means to be easily broken or shattered. Consider the following two pictures, a ceramic with no voids (defects) and one with numerous small voids:

Note in the second figure that there are regions in which the ceramic material is thin compared to the entire material. These regions are more susceptible to being broken when the ceramic is bent. As these regions within the ceramic shatter, the overall ceramic structure is weakened, and further bending is possible, with additional shattering of the structure. In the first structure, the entire structure is able to resist bending, without local regions shattering because of the uniformity of solid structure.

THIN FILMS

Thin films of metals, metal oxides or organic substances are found in many applications. The thickness of thin films is typically 0.1 to 0.2 μm or greater. However, coatings such as paints and enamels are not considered thin films because their thickness is significantly greater.

- A thin film must adhere properly to the underlying substrate if it is to perform properly. The bonding of a thin film to a substrate may be by a chemical reaction or simply by intermolecular interactions. The later form of bonding is less robust, and the thin film is less stable.
- The electronic industry uses thin films extensively, employing them as conductors, resistors, capacitors, optical coatings, and protective coatings.

Three techniques for forming thin films are discussed: Vacuum deposition; sputtering; and chemical vapor deposition.

- **Vacuum deposition** involves vaporizing a metal, metal alloy or inorganic compound in a high vacuum chamber at a pressure of 10^{-5} mm Hg or less. A thin film is deposited on the surface of a substrate in a cooler region of the vacuum apparatus.
- **Sputtering** involves applying a high voltage to the source material, which is then carried in a plasma gas phase to a substrate that is a negative electrode. Sputtered atoms have a lot of energy and when they reach a substrate, they often can penetrate the surface of the substrate to bind more strongly than simply at the surface.

• **Chemical vapor deposition** involves depositing a volatile, stable chemical compound on the surface of a substrate, followed by a chemical reaction that produces a stable, thin film. Equations (12.8–12.11) in the text show some reactions that demonstrate chemical vapor deposition. Note that the initial reactants are gases, but the desired final product is a solid.

EXERCISE 8   In the vacuum deposition method of forming thin films, the physical form of the target is important. Particle size cannot be too fine or too large. Why?

   Solution: If the particle size is too fine, the target material will leave the boat or crucible when energized by an electron beam. Particles gain sufficient kinetic energy to escape the boat. If the particle size is too large, then heat transfer between particles is poor when the material is energized. This results in uneven heating and melting.

EXERCISE 9   The effectiveness of adhesion of thin films of gold, aluminum and iron on glass increases in the following order: Au < Al <<< Fe. What does this data suggest about the nature of adhesive forces between the metals and the surface of glass?

   Solution: There are two principal types of forces that hold thin films to a substrate: physiabsorption and chemisorption. The first involves van der Waals and electrostatic forces and the second involves the formation of chemical bonds; physiabsorption adhesion is less effective than chemisorption adhesion. The data suggests that both Au and Al adhere to the surface of glass by primarily physiaborption processes whereas Fe binds chemically to the surface. Studies have shown that iron forms oxide bonds with the glass to give greater adhesion.

SELF-TEST QUESTIONS

Key Terms

Having reviewed the key terms listed at the end of Chapter 12 and their applications, identify the following statements as true or false. If a statement is false, indicate why it is incorrect. Key terms are italicized in the statements.

12.1   An example of an amorphous *ceramic* is glass.

12.2   A *ceramic fiber* resists loads applied along its width.

12.3   Silicon carbide has been used as a ceramic fiber in forming *composites*.

12.4   A *liquid crystal* contains a liquid-like phase that is isotropic.

12.5   A *nematic* liquid crystal phase contains molecules that are aligned end to end.

12.6   In a *smectic* liquid crystal phase, attractive interactions between molecules occur when the molecules align with their long axes parallel.

12.7   In the *cholesteric* liquid crystalline phase cholesterol-like molecules form long fibers that align through stacking in layers of fibers.

12.8   An example of a *polymer* is cellulose, which is formed by the polymerization of glucose molecules.

12.9   The monomer unit making up *polyethylene* is C_2H_2.

12.10   Once a *plastic* is formed, it retains its shape under the application of heat and pressure.

12.11   Polyethylene is an example of a *thermoplastic material*.

12.12   A *thermosetting plastic* is easily reshaped after it is originally formed.

12.13   A rubber band can be called an *elastomer*.

12.14   Characteristic of some *condensation polymerization* reactions is the formation of water, along with a uniting of two reactants.

12.15   Polymers are amorphous materials and show no regions of *crystallinity*.

12.16   It is often noticed that the windows of a car form a film on the inside after the car has been exposed to extensive sunlight. This film is a *plasticizer* that is lost from plastic materials in the dash due to evaporation.

12.17   In *natural rubber* the isopropene unit has the following configuration

$$\begin{array}{ccc} CH_3 & & CH_3 \\ & C-C & \\ CH_3 & & CH_3 \end{array}$$

12.18   Some advantages of *ceramics* compared to natural materials are high resistance to heat and corrosion, lower weight in some cases, and potentially lower cost to use.

12.19   A *sol* consists of large particles dispersed in a medium.

12.20   Jello is comprised of primarily protein molecules dispersed in water, and is an example of a *gel*.

12.21   In the *sol-gel* process a gel is heated to a high temperature to drive off the sol.

12.22 Vulcanization of rubber involves *cross-linking* of polymers using nitrogen.

12.23 A *piezoelectric* ceramic generates an electrical potential when it is subjected to mechanical stress.

12.24 *Thermistors* exhibit electrical conductivity over a very wide temperature range that decreases with increasing temperature.

12.25 One of the most widely studied *superconductors* is $YBa_2Cu_3O_x$ which shows a T_c near 95 K.

12.26 A *superconducting ceramic* shows the property of exhibiting zero electrical resistance above a certain temperature.

12.27 If a silver salt solution is slowly evaporated on a surface and then heated, the resulting reflective surface is a *thin film*.

12.28 The coating of an optical lens with MgF_2 by *vacuum deposition* involves vaporizing the salt in a low pressure environment and depositing it on a substrate.

12.29 In the *sputtering* process of forming a thin film, the substrate is made to be the cathode.

12.30 In choosing a material to form a thin film using the *chemical vapor deposition* process, the melting point temperature of the substrate must be lower than that of the thin film.

Problems and Short-Answer Questions

12.31 Explain why salts such as NaCl and $MgCl_2$ do not exhibit liquid crystal behavior.

12.32 Compare the three major classes of liquid crystalline phases in terms of the ordering of molecules within their structures.

12.33 Show how the polymer polychloroprene is formed from chloroprene, $CH_2\!=\!CH\!-\!C\!=\!CH_2$
$$\underset{\displaystyle Cl}{|}$$

12.34 Explain why substantial chain branching in a polymer leads to a low density type.

12.35 Explain why glass is considered a type of ceramic.

12.36 Which of the following would you expect to be higher on the moh's scale (see Table 12.5 in the text), and why: Graphite (a form of carbon; it has been used as a lubricant), tungsten carbide, and magnesium oxide.

12.37 Would you expect the following molecule to exhibit liquid crystal properties?

$CH_3\!-\!CH\!=\!CH\!-\!CH_2\!-\!CH_2\!-\!CH_2\!-\!CH_2\!-\!CH_2\!-\!CH\!=\!CH\!-\!CH_3$

12.38 Small impurities affect the pitch of helical cholesteric liquid crystalline structures. How could this property be exploited to detect trace-vapors for certain chemicals?

12.39 Why are cross-linkers used in processing of polymers that are to be elastomers?

12.40 There are certain ceramics that can deform plastically at high temperatures by generation and motion of dislocations within the structure. Why are high temperatures required if a ceramic is to undergo plastic-like distortions?

12.41 Could $SiCl_4$ be used to replace SiH_4 in the chemical vapor deposition of silicon nitride?

Multiple-Choice Questions

12.42 Which type of material is used in measuring temperature changes based on color?
  (a) Nemetic liquid crystal
  (b) Smectic liquid crystal
  (c) Cholesteric liquid crystal
  (d) Thermistor ceramic

12.43 Which of the following characteristics is *not* a desired benefit of adding a plasticizer?
  (a) less elasticity
  (b) greater flexibility
  (c) greater workability
  (d) less brittleness

12.44 Which of the following is one of the principal products formed using thermoplastic resins?
  (a) ceramics
  (b) Superconducting material
  (c) PVC
  (d) LCD

12.45 A fiber must have the following property if it is added to a polymer to form an improved product.
  (a) Low tensile strength
  (b) Small length, large breadth
  (c) Very high crystalline melting point
  (d) High cohesive energy with polymer substrate

12.46 Which statement is *not* true about most ceramics
  (a) strong
  (b) low cost
  (c) abrasion resistant
  (d) withstand wide temperature range

12.47 Which of the following is likely to be useful in preparing a superrefractory ceramic?
  (a) Al   (c) Zr
  (b) Mg   (d) Si

12.47 Which of the following is *not* an advantage of the vacuum deposition method for thin film formation?
  (a) high rate of deposition
  (b) simplicity of design
  (c) economy
  (d) efficiency

12.48 Which statement is true about the target in the sputtering process of thin film deposition?
  (a) It is the substrate
  (b) It is the cathode
  (c) It is the plasma
  (d) It is granular

12.49 What type of sol is typically formed in the sol-gel process?
  (a) alkoxide   (c) metal amide
  (b) metal   (d) metal hydroxide

12.50 Which type of ceramic generates an electrical potential when subjected to mechanical stress?
  (a) thermistor   (c) piezoelectric
  (b) LCD   (d) superconductor

12.51 What coolant used in applications of superconductors is readily available and cools below 77 K?
  (a) N_2 (c) H_2
  (b) He   (d) Alcohol

SELF-TEST SOLUTIONS

12.1 True. **12.2** False. Stress is resisted along the long axis. **12.3** True. **12.4** False. Isotropic means unordered. The molecules are aligned along the long axis. **12.5** False. The ends are not aligned as they are in the smectic phase. **12.6** True. **12.7** False. The molecules are disc-shaped and somewhat flat. **12.8** True. **12.9** False. The monomer unit is C_2H_4. **12.10** False. Many types of polymers can be reshaped, such as polyethylene. **12.11** True. **12.12** False. These polymers resist changes in shape after once formed and set. **12.13** True. **12.14** True. **12.15** False. Regions of crystallinity, or ordering, are known in polymers. **12.16** True. **12.17** False. The basic unit is *cis*-polyisoprene

12.18 True. **12.19** False. The particles are extremely tiny. **12.20** True. **12.21** False. Water, not sol, is driven off. **12.22** False. Sulfur is the cross-linker, not nitrogen. **12.23** True. **12.24** False. Thermistors exhibit a moderately narrow electrical conductivity range that increases with temperature. **12.25** True. **12.26** Superconductivity occurs *below* a certain temperature. **12.27** True. **12.28** True. **12.29** False. The substrate is made the anode. **12.30** False. The substrate must have a higher melting point than the thin film. **12.31** Liquid crystalline substances tend to have long rod-like shapes, and to be somewhat rigid. This enables them to interact along the parallel axis. Salts do not possess a long rod-like shape, but have crystalline shapes that involve ions at crystal points in specific arrangements. Thus, salts do not have the appropriate structures to interact along a long axis. **12.32** The least ordered arrangement of molecules occurs in the nematic phase; the molecules are aligned along their long axis, but there is no ordering with respect to the ends of the molecules. The next least ordered arrangement of molecules is in the smectic phase. This phase also shows ordering with respect to the ends of molecules. Layers of aligned molecules occur with their long axes parallel. The cholesteric phase is the most ordered phase, with disc-shaped molecules aligned through a stacking arrangement. There are specific repeating arrangements from one layer to another. **12.33** In the presence of a suitable reagent, the double bonds located at the ends of chloroprene molecules open up, with one electron centered on each end carbon atom, and the other two electrons pairing to form a double bond between the middle carbons. The single electrons on the end carbon atoms in chloroprene molecules couple with single electrons on other chloroprene molecules to build up a long chain of chlorprene units:

12.34 Substantial chain branching in a polymer will prevent molecules from closely approaching one another. The branching chains of the polymer sterically interact with one another and prevent a close approach. Because the polymer chains are not strongly associated together, the density is less than if the molecules were in closer proximity. **12.35** Ceramics are composed of inorganic materials possessing a covalent network structure or ionic bonding, or a combination of the two. These materials are usually crystalline. Glass is not crystalline, but amorphous. However, it is composed of silicon and oxygen that bind together through a covalent network. Glass is hard, and brittle in the sense that it will fracture if stress is applied. Depending on the additives added to the silicate materials when glass is formed, glass can be stable to reasonably high temperatures. **12.36** The moh's scale is based on the relative ability of a substance to scratch the surface of another substance; the higher the number, the greater the ability. Ceramic materials usually have higher moh values than metals or nonmetals. Graphite, a form of carbon, does not have the hardness of diamond, another form of carbon; it tends to be soft. Tungsten carbide should have a high moh value because silicon carbide does as shown in Table 12.3 of the text. Magnesium oxide should also have a relatively high moh value, although not as high as BeO in Table 12.3 because of the larger size of magnesium. Thus, graphite should have the lowest moh value; magnesium oxide and tungsten carbide should have similar moh values, although the value for magnesium might be expected to be slightly less. **12.37** No. Although there are pi bonds within the backbone, they are too far apart to give rigidity to the molecule. This molecule is too flexible to form a liquid crystal state. **12.38** Chlosteric liquid crystals are typically colored. Their color is sensitive to both temperature and structure. A small change in either changes the color of the substance. If a trace-vapor of a chemical was absorbed by the chlosteric liquid crystal, the helical structure would be slightly altered. This would result in a unique color change that could be used to detect the presence of the chemical. **12.39** In an elastomer the mobility of chain segments must be low so that the elastomer retains its original shape when stress is released. Restriction of mobility of chain segments is achieved by a network of primary-bonds that cross-link the backbones of polymer chains. **12.40** The generation and motion of dislocation within a ceramic structure requires chemical bonds to be broken and remade. This is difficult in either a covalent network or ionic structure. High temperatures are needed to provide sufficient energy for bond-breaking. **12.41** Silane, SiH_4 reacts with ammonia at 900-1100°C as follows:

$$3SiH_4(g) + 4NH_3(g) \longrightarrow Si_3N_4(s) + 12H_2(g)$$

The silicon nitride solid remains on the surface of the substrate after the reaction occurs. If $SiCl_4$ were used in place of silane, HCl would form instead of H_2. HCl is more reactive than H_2; thus, it has a greater potential to react with silicon nitride. **12.42** (c). **12.43** (a). **12.44** (c). **12.45** (d). **12.46** (b). **12.47** (a)—leads to irregular particle size, and pinholes. **12.48** (b). **12.49** (d). **12.50** (c). **12.51** (b).

13 | Properties of Solutions

OVERVIEW OF THE CHAPTER

13.1, 13.3, 13.4
SOLUTIONS

Review: Energy (5.1, 5.2); pressure (10.2); intermolecular forces (11.2)

Objectives: You should be able to:

1. Describe the energy changes that occur in the solution process in terms of the solute-solute, solvent-solvent, and solute-solvent attractive forces; describe the role of disorder in the solution process.

2. Rationalize the solubilities of substances in various solvents in terms of their molecular structures and intermolecular forces.

3. Describe the effects of pressure and temperature on solubilities.

13.2
CONCENTRATION
EXPRESSIONS

Review: Molarity (4.1); solution (4.1); dimensional analysis (1.5)

Objectives: You should be able to:

1. Define mass percentage, parts per million, mole fraction, molarity, molality, and calculate concentrations in any of these concentration units.

2. Convert concentration in one concentration unit into any other (given density of the solution where necessary).

13.5 COLLIGATIVE
PROPERTIES

Review: Boiling point (11.4); melting point (11.4); vapor pressure (11.5); gas-law constant (10.4); pressure units (10.2)

Objectives:

1. Determine the concentration and molar mass of a nonvolatile non-electrolyte from its effect on the colligative properties of a solution.

2. Explain the difference between the magnitude of changes in colligative properties caused by electrolytes compares to those caused by nonelectrolytes.

3. Describe the effects of solute concentration on the vapor pressure, boiling point, freezing point, and osmotic pressure of a solution, and calculate any of these properties give appropriate concentration data.

Objective: You should be able to describe how a colloid differs from a true solution.

TOPIC SUMMARIES AND EXERCISES

A solution is a homogeneous mixture consisting of the solvent and one or more solutes.

- Dissolving a solute in a solvent requires three distinct processes: (1) separating solvent particles from one another; (2) separating solute particles from one another; and (3) solvent particles surrounding solute particles.
- The process whereby separated solute particles are surrounded by solvent particles is referred to as **solvation.** This process is called **hydration** if water is the solvent.
- The solubility of a solute in a particular solvent depends on: (1) energy effects; (2) the disorder of solute and solvent particles in solution compared to their initial state; (3) temperature; and (4) pressure.

Spontaneous dissolving of a solute in a particular solvent is favored by two factors:

1. The energy of the solution is lower than the sum of energies of the pure solvent and pure solute. This is equivalent to saying that dissolving a solute in a solvent is usually spontaneous when the process is exothermic. This happens when solute-solvent interactions are more energetic than the sum of energies of solute-solute and solvent-solvent interactions. However, endothermic solution processes can occur spontaneously if the following second factor operates to a sufficient degree.

2. A more disordered arrangement of solute and solvent particles exists in the solution compared to the arrangements of particles found in the pure solute and solvent.

- In general, we find that a solute is most likely to dissolve to a great extent in a solvent when the solute and solvent particles have similar structures and similar intermolecular attractive forces (that is, "likes dissolve likes").

When a solid is placed in a solvent and begins to dissolve, two dynamic opposing processes exist: (1) dissolving of the solid; and (2) crystallization of solute particles. When the rates of these two opposing processes are equal, the concentration of the dissolved solid becomes constant and no more solid dissolves.

- A solution is **saturated** when the concentration of the dissolved solid is constant and the two opposing rates are equal.

• An **unsaturated** solution contains the solute in a concentration less than needed to form a saturated solution. No solid will be apparent.

• Sometimes a **supersaturated** solution can be prepared: It contains a solute in a concentration greater than that in a saturated solution. Such solutions are unstable, and under the right conditions, the solid will crystallize until a saturated solution occurs.

• **Solubility** is the quantity of solute that needs to be dissolved in a given quantity of solvent to form a saturated solution. Solubility changes with temperature.

Increasing the pressure of a gas over a solvent increases the solubility of the gas, whereas pressure has minimal effect on the solubility of solids and liquids.

• The relationship between pressure and solubility of a gas is expressed by **Henry's law:**

$$C_g = kP_g$$

where k is a proportionality constant known as the Henry's law constant, C_g is the solubility of the gas in the solvent, and P_g is the pressure of the gas over the solution.

• Note the linear relationship between solubility and pressure.

• The solubility of a gas in general decreases with increasing temperature.

EXERCISE 1   Hexane and heptane are nonpolar liquids. **(a)** In a solution containing only hexane and heptane, what kinds of forces exist between molecules? **(b)** Why does a solution of hexane and heptane have a greater disorder particles than do pure hexane and pure heptane alone? **(c)** Would you expect NaCl(*s*) to dissolve in a solution containing only heptane and hexane?

  Solution: **(a)** London dispersion forces—that is, instantaneous induced-dipole attractions (see Section 11.2)—exist between nonpolar molecules such as heptane and hexane. **(b)** A solution containing hexane and heptane has a larger volume than either the separate hexane or heptane liquids from which the solution is formed. Thus, in the solution state, the hexane and heptane molecules have more room to move about, and there exist more possible kinds of arrangements of molecules than in the separated hexane and heptane liquids. This results in a more disordered state for the solution particles compared to that for the separated liquids. **(c)** No. An ionic salt such as NaCl dissolves in polar solvents, not in a nonpolar liquid containing only heptane and hexane.

EXERCISE 2   The formation of HCl(*aq*) from the reaction of HCl(*g*) and H$_2$O(*l*) leads to a solution with a more ordered state than the reactants alone. Explain why the solution particles are in a more ordered state than reactant particles, and why, in spite of this condition, HCl(*g*) dissolves in H$_2$O(*l*).

  Solution: The reactants consist of two phases, gas and liquid. The gaseous state is a highly disordered state because of the large volume available for gas molecules to move about randomly. When HCl(*g*) and H$_2$O(*l*) form a solution, a homogeneous mixture results in which HCl molecules are considerably more restricted in their motions than in the gaseous state. Therefore, the solution possess a more ordered arrangement of particles than do the separated reactants. Nevertheless, HCl and H$_2$O readily form a solution because of the strong attractive forces between polar HCl and H$_2$O molecules. This results in a large exothermic heat of solution that enables the solution process to occur.

EXERCISE 3 For each solute listed in the left-hand column of the following table, select a solvent in the right-hand column that will dissolve it:

Solute	Solvent
Sodium	Water
Gasoline	Mercury
KCl	Benzene
$NH_3(g)$	

Solution: Solvents usually dissolve solutes that have intermolecular forces similar to those found in the solvent. These kinds of forces found in the pure state of each substance are listed in parentheses.

Solute	Appropriate solvent
Sodium (metallic)	Mercury (metallic)
Gasoline (van der Waals)	Benezene (van der Waals)
KCl (ionic)	Water (dipole-dipole, hydrogen bonds)
$NH_3(g)$ (dipole-dipole, hydrogen bonds)	Water (dipole-dipole, hydrogen bonds)

EXERCISE 4 When you open a bottle of carbonated soft drink, bubbles of carbon dioxide gas escape from the solution. Explain this phenomenon using Henry's law.

Solution: According to Henry's law, the mass of a gas dissolved in a definite volume of a liquid is directly proportional to pressure. During the bottling of the beverage, carbon dioxide is forced into the solution by application of pressure. The bottle is tightly capped to ensure the carbon dioxide stays dissolved. When you open the bottle, the pressure above the solution is decreased to atmospheric pressure and thus the solubility of the carbon dioxide decreases. This results in $CO_2(g)$ escaping from solution, a process known as effervescence.

Molarity and mole fraction are two ways of expressing concentration. In Section 13.2 of the text, you are introduced to the following new concentration expressions: parts per million (*ppm*); molality (*m*); and mass percentage. Concentration expressions are defined and expressed in equation form in Table 13.1 on page 210.

CONCENTRATION EXPRESSIONS

- Recall that a solvent is a component of a solution whose phase does not change when the solution is formed; all other components are solutes. If more than one component has a phase that does not change, then the component of greatest concentration is the solvent.

- Concentration expressions can be used as conversion factors when you do problems. For example, molality (*m*) is used to convert between the number of moles of solute and mass of the solvent in kilograms. You used density in a similar manner in Chapter 1.

EXERCISE 5 A 3.0 g sample of ground water contains 3.5 *ppm* of arsenic ion. How many grams of arsenic are in this sample?

Solution: Parts per million (*ppm*) is defined as

$$ppm \text{ of a component} = \frac{\text{mass of component in solution}}{\text{total mass of solution}} \times 10^6$$

Table 13.1 Units of Concentration

Concentration unit	Definition	Equation
Molarity (M)	Number of moles of solute in a liter of solution	$M = \dfrac{\text{moles solute}}{\text{volume of solution in liters}}$
Mass percentage (wt %)	The percentage of mass of a component of a solution in a given mass of the solution	$\text{Mass \%} = \dfrac{\text{mass of solution component}}{\text{total mass of solution}} \times 100$
Parts per million (ppm)	The grams of a solute in a million (10^6) grams of solution. This is equivalent to 1 mg of solute per kg of solution.	$\text{ppm} = \dfrac{\text{mass of solute}}{\text{mass of solution}} \times 10^6$
Mole fraction (X)	Ratio of the number of moles of a component in a mixture to total number of moles of all components in solution	$X = \dfrac{\text{moles of component}}{\text{total moles of all components}}$
Molality (m)	Number of moles of solute in a kilogram of *solvent*	$m = \dfrac{\text{moles of solute}}{\text{mass of \emph{solvent} in kilograms}}$

Let x equal the number of grams of arsenic in the solution. Substituting the given values into this equation yields:

$$2.2 \; ppm = \frac{x}{3.0 \text{ g}} \times 10^6$$

Solving for the number of grams, x, gives:

$$x = \frac{(2.2)\,(3.0 \text{ g})}{1 \times 10^6} = 6.6 \times 10^{-6} \text{ g}$$

EXERCISE 6 An antifreeze mixture is a 11.0 molal ethylene glycol solution. Water is the solvent and it has a mass of 800 g. The molar mass of ethylene glycol is 62.0 g. What is the mass of the solution?

Solution: This problem demonstrates how molality is used as a conversion factor. To calculate the mass of the solution you need the masses of water and ethylene glycol; the former is a given quantity. The mass of ethylene glycol can be determined if you know how many moles of it are present in the solution—molality enables you to make this calculation from the mass of water and the molality of the solution.

$$\text{Moles ethylene glycol} = \text{molality} \times \text{mass of solvent in kg}$$

$$= \left(\frac{11.0 \text{ mol ethylene glycol}}{1 \text{ kg H}_2\text{O}}\right) (800 \text{ g H}_2\text{O}) \left(\frac{1 \text{ kg H}_2\text{O}}{1000 \text{ g H}_2\text{O}}\right)$$

$$= 8.80 \text{ mol ethylene glycol}$$

$$\text{Mass ethylene glycol} = (8.80 \text{ mol ethylene glycol}) \left(\frac{62.0 \text{ g ethylene glycol}}{1 \text{ mol}}\right)$$

$$= 546 \text{ g ethylene glycol}$$

$$\text{Mass solution} = 800 \text{ g H}_2\text{O} + 546 \text{ g ethylene glycol} = 1346 \text{ g}$$

EXERCISE 7 A solution contains 10.0 g of KOH and 900.0 mL of water. The volume of the solution is 904.7 mL. **(a)** Which component of the solution is the solute? Which is the solvent? **(b)** What is the molarity of the solution? **(c)** What is the mass percentage of KOH? What is the weight percentage of H_2O? **(d)** What is the mole fraction of KOH? **(e)** What is the molality of the solution

    Solution: **(a)** KOH is the solute because it is the component of the solution that occurs in the smallest amount; water is the solvent. Also, water is the solvent because KOH(s) dissolves in it to form a *liquid* solution—the same phase as water. **(b)** The molarity (M) of the solution is calculated from the relation

$$\text{Molarity} = \frac{\text{moles KOH}}{\text{volume of solution in liters}}$$

Substituting the appropriate values into this equation and solving for molarity of the solution yields

$$\text{Molarity} = \frac{(10.0 \text{ g KOH})\left(\dfrac{1 \text{ mol KOH}}{56.1 \text{ g KOH}}\right)}{0.9047 \text{ L}} = \frac{0.178 \text{ mol}}{0.9047 \text{ L}} = 0.197 \text{ mol/L}$$

(c) The mass percentage of KOH is calculated from the relation

$$\text{Mass \% KOH} = \left(\frac{\text{mass KOH}}{\text{mass of solution}}\right)(100)$$

The mass of water is calculated from its density and volume:

$$\text{Mass } H_2O = (\text{volume})(\text{density}) = (900.0 \text{ mL})\left(1.00 \frac{\text{g}}{\text{mL}}\right) = 900 \text{ g}$$

The mass percentage of KOH is

$$\text{Mass \% KOH} = \left(\frac{10.0 \text{ g KOH}}{10.0 \text{ g KOH} + 900 \text{ g } H_2O}\right)(100) = 1.10\%$$

The mass percentage of water is

$$100\% - \text{mass \% KOH} = 100\% - 1.10\% = 98.90\%$$

because the sum of mass percentages of all components in solution equals 100 percent. **(d)** The mole fraction of KOH (X_{KOH}) is calculated as follows:

$$X_{KOH} = \frac{\text{moles KOH}}{\text{moles KOH} + \text{moles } H_2O}$$

Calculate the number of moles of KOH and H_2O:

$$\text{Moles KOH} = (10.0 \text{ g KOH})\left(\frac{1 \text{ mol KOH}}{56.1 \text{ g KOH}}\right) = 0.178 \text{ mol}$$

and

$$\text{Moles } H_2O = (900.0 \text{ g } H_2O)\left(\frac{1 \text{ mol } H_2O}{18.1 \text{ g } H_2O}\right) = 50.0 \text{ mol}$$

Then substitute into expression for X_{KOH}

$$X_{KOH} = \frac{0.178 \text{ mol KOH}}{0.178 \text{ mol KOH} + 50.0 \text{ mol } H_2O} = 3.55 \times 10^{-3}$$

(e) The molality of the solution is calculated as follows:

$$\text{Molality} = \frac{\text{moles KOH}}{\text{weight of } \textit{solvent} \text{ in kilograms}}$$

$$\text{Molality} = \frac{0.178 \text{ mol KOH}}{(900 \text{ g})\left(\dfrac{1 \text{ kg}}{1000 \text{ g}}\right)} = 0.198 \; m$$

EXERCISE 8   Calculate the molarity of a solution that contains 18.0 percent HCl by mass and has a density of 1.05 g/mL.

   Solution: To calculate molarity, you must determine the number of grams of HCl in 1.00 L of solution. The mass percentage of HCl and the density of HCl can be used to calculate this mass. Density is used to calculate the mass of 1.00 L of solution:

$$\text{Mass of solution} = \left(1.05 \; \frac{\text{g}}{\text{mL}}\right)\left(\frac{1000 \text{ mL}}{1.00 \text{ L}}\right) = 1050 \; \text{g/L}$$

The number of grams of HCl in this solution is calculated from the given mass percentage:

$$\text{Mass \% HCl} = \left(\frac{\text{Mass HCl}}{\text{Mass solution}}\right)(100)$$

Or, rearranging the equation, we have

$$\text{Mass HCl} = \frac{\text{Mass \% HCl}}{100} \, (\text{Mass of solution})$$

$$\text{Mass HCl} = \left(\frac{18.0}{100}\right)\left(1050\frac{\text{g}}{\text{L}}\right) = 189 \; \text{g/L}$$

The molarity can be calculated now because you know that 189 g of HCl is contained in 1.00 L of solution.

$$\text{Molarity} = \frac{\text{moles HCl}}{1.00 \text{ L}} = \frac{(189 \text{ g HCl})\left(\dfrac{1 \text{ mol HCl}}{36.5 \text{ g HCl}}\right)}{1.00 \text{ L}} = 5.18 \; M$$

COLLIGATIVE PROPERTIES

Those properties of a solution that depend on the *number* of solute particles are called **colligative properties.** In Section 13.5 of the text, the following colligative properties of a solution are discussed: vapor-pressure lowering, boiling-point elevation, freezing-point depression, and osmotic pressure. The text discussion is limited to the effect of nonvolatile solutes on the colligative properties of a solution, assuming ideal conditions.

   ▪ Useful relations for calculating colligative properties of solutions are given in Table 13.2 on page 213.

   ▪ Remember to account for all forms of the solute when calculating X, m, or M. For example, when calculating the osmotic pressure of a 0.10 M Na_2SO_4 solution, you must use $M = 3 \times 0.10 \; M = 0.30 \; M$ because Na_2SO_4 ionizes to form one sodium ion and two sulfate ions.

   ▪ Another useful relation is **Raoult's law,** which enables you to calculate the

Table 13.2 Relations for Calculating Colligative Properties of Solutions

Colligative property	Quantitative relation	Comments
Vapor-pressure lowering (VPL)	$VPL = X_B P_A^\circ$	A nonvolatile solute lowers the vapor pressure of a solvent by an amount VPL X_B = mole fraction of solute particles P_A° = vapor pressure of pure solvent
Boiling-point elevation (ΔT_b)	$\Delta T_b = K_b m$	A nonvolatile solute raises the boiling point of a solution by an amount ΔT_b K_b = molal boiling-point elevation constant, which is unique for a given solvent m = molality of solute
Freezing-point depression (ΔT_f)	$\Delta T_f = K_f m$	A nonvolatile solute lowers the freezing point of a solution by an amount ΔT_f K_f = molal freezing-point depression constant, which is unique for a given solvent m = molality of solute
Osmotic pressure (π)	$\pi = MRT$	Osmosis is the net movement of solvent particles across a semipermeable membrane from a more dilute solution into a more concentrated one; Osmotic pressure (π) of a solution is the applied pressure required to prevent osmosis from pure solvent into a solution. *Isotonic* solutions have the same osmotic pressure. M = total molarity of all ionized and unionized solute species R = ideal-gas constant T = absolute temperature

vapor pressure of a solvent when a solute is present:

$$P_A = X_A P_A^\circ$$

In this equation, P_A° is the vapor pressure of pure solvent; X_A is the mole fraction of solvent; and P_A is the vapor pressure of solvent with a solute present.

EXERCISE 9   Compare the osmotic pressures of 0.10 M sucrose and 0.10 M NaCl aqueous solutions at 25°C.

Solution: Sucrose is a nonelectrolyte in water; therefore the molarity of solute particles is the same as the molarity of the solution, 0.10 M. The equation for osmotic pressure is

$$\pi = MRT$$

Substituting the known values of M, R, and T gives

$$\pi = \left(\frac{0.10 \text{ mol}}{1 \text{ L}} \right) \left(0.082 \, \frac{\text{L-atm}}{\text{mol-K}} \right) (298 \text{ K}) = 2.4 \text{ atm}$$

NaCl is a strong electrolyte in water and forms sodium and chloride ions: therefore the molarity of all solute particles in the solution is

$$M = M_{Na^+} + M_{Cl^-} = 0.10 \, M + 0.10 \, M = 0.20 \, M$$

This is twice that for the sucrose solution. You can see that the osmotic pressure of the NaCl solution should be twice that of the sucrose solution: $\pi = 2 \times 2.4$ atm = 4.8 atm.

EXERCISE 10   A solution is formed by dissolving 10.0 g of KCl in 500.0 g of H_2O.
(a) What is the new vapor pressure of H_2O at 25°C if the vapor pressure of pure H_2O is

23.8 mm Hg at 25°C? **(b)** What is the boiling-point elevation? K_b for H_2O is 0.52°C/m. **(c)** What is the freezing-point depression? K_f for H_2O is 1.86°C/m. **(d)** What is the osmotic pressure of the solution? Assume that the volume of the solution is 0.500 L.

Solution: **(a)** The vapor pressure of a volatile component in a solution, P_A, is related to its vapor pressure as a pure liquid, P_A°, and its mole fraction, X_A, as expressed by Raoult's law:

$$P_A = X_A P_A^\circ$$

The mole fraction of water in the solution is

$$X_{H_2O} = \frac{\text{moles } H_2O}{\text{moles } H_2O + \text{moles } K^+ + \text{moles } Cl^-}$$

$$= \frac{\text{moles } H_2O}{\text{moles } H_2O + (2)(\text{moles KCl})}$$

Since P_A is a colligative property of the solution, the denominator in the X_{H_2O} expression must be the sum of the number of moles of all solvent and solute species. KCl is a strong electrolyte and ionizes in water to form K^+ and Cl^- ions. The mole fraction of water is calculated as follows:

$$\text{Moles } H_2O = (500.0 \text{ g } H_2O)\left(\frac{1 \text{ mol } H_2O}{18.0 \text{ g } H_2O}\right) = 27.8 \text{ mol}$$

$$\text{Moles KCl} = (10.0 \text{ g KCl})\left(\frac{1 \text{ mol KCl}}{74.55 \text{ g KCl}}\right) = 0.134 \text{ mol}$$

$$X_{H_2O} = \frac{27.8 \text{ mol } H_2O}{27.8 \text{ mol } H_2O + (2)(0.134 \text{ mol KCl})} = 0.990$$

The partial pressure of the water when 10.0 g of KCl is present is thus

$$P_{H_2O} = X_{H_2O}P_{H_2O}^\circ = (0.990)(23.8 \text{ mm Hg}) = 23.6 \text{ mm Hg}$$

(b) The boiling-point elevation is found using the equation $\Delta T_b = K_b m$, where m is the molality of the solute and K_b is the boiling-point elevation constant for the pure solvent:

$$m = \frac{\text{moles of solute}}{\text{kg of solvent}} = \frac{\text{moles } K^+ \text{ ions} + \text{moles } Cl^- \text{ ions}}{\text{kg of solvent}}$$

$$m = \frac{0.268 \text{ mol ions}}{(500.0 \text{ g})\left(\dfrac{1 \text{ kg}}{1000 \text{ g}}\right)} = 0.536 \text{ } m$$

Substituting the values of m and K_b into $\Delta T_b = K_b m$ gives

$$\Delta T_b = (0.52 \text{ °C/}m)(0.536 \text{ } m) = 0.28 \text{ °C}$$

(c) Similarly, the freezing-point depression is found by substituting the values of m and K_f into

$$\Delta T_f = K_f m$$
$$\Delta T_f = (1.86 \text{ °C/}m)(0.536 \text{ } m) = 1.0 \text{ °C}$$

(d) Osmotic pressure is calculated using the equation $\pi = MRT$. First calculate the value of M

$$M = \frac{\text{moles of all solute species}}{\text{volume in L}}$$

$$= \frac{(0.134 \text{ mol KCl})(2 \text{ mol ions/1 mol KCl})}{0.500 \text{ L}}$$

$$= 0.536 \text{ } M$$

Then calculate π:

$$\pi = (0.0536\ M)\left(0.082\ \frac{\text{L-atm}}{\text{mol-K}}\right)(298\ \text{K}) = 13\ \text{atm}$$

A **colloid** contains gaseous, liquid, or solid particles with diameters approximately 10 Å to 2000 Å dispersed in a solvent-like substance.

| COLLOIDS

- Colloidal dispersions (colloids) scatter light very effectively and thus appear cloudy or opaque. The scattering of a light beam that is passed through a colloid is known as the **Tyndall effect.**

- Colloid particles do not settle under the influence of gravity as do larger particles.

- A **hydrophilic colloid** contains colloidal particles that are attracted to water molecules and thereby form a single-phase suspension. Often such colloidal solutions show a gelatinlike character and are known as gels.

- A **hydrophobic colloid** contains colloidal particles that are not attracted to water molecules; they do not form gels.

- Colloidal particles can be removed from a dispersing medium by coagulation, which involves increasing the size of colloidal particles until they settle out of solution. Dialysis involves using a semipermeable membrane to separate colloidal particles from ions in solution.

EXERCISE 11   A colloidal suspension of As_2S_3 can be prepared by reacting As_2O_3 with H_2S in water. (a) How could you show that the As_2S_3 solution is colloidal? (b) The colloidal solution appears gelatinous. Is it a hydrophilic or hydrophobic colloid?

Solution: (a) The solution will appear cloudy, and the suspended particles will not settle under the influence of gravity. Also, the solution will scatter a beam of light (Tyndall effect). (b) Since the As_2S_3 particles do not separate from H_2O (as evidenced by the gelatinlike character of the solution), the colloid must be hydrophilic.

EXERCISE 12   Identify the phase of the dispersed substance, the phase of the dispersing substance, and the colloid type for each of the following: beer foam, clouds, milk, smoke, and paints. Refer to Table 13.7 of the text for descriptions of different kinds of colloids.
      Solution:

Example	Dispersed phase	Dispersing phase	Type
Beer foam	Gas	Liquid	Foam
Clouds	Liquid	Gas	Aerosol
Milk	Liquid	Liquid	Emulsion
Smoke	Solid	Gas	Aerosol
Paint	Solid	Liquid	Sol

SELF-TEST QUESTIONS

Key Terms

Having reviewed the key terms listed at the end of Chapter 13 and their applications, identify the following statements as true or false. If a statement is false, indicate why it is incorrect. Key terms are italicized in the statements.

13.1   The *molality* of a toluene solution containing 0.12 mol of toluene in 0.500 kg of benzene is 0.24.

13.2   The *percentage* of NaOH in a solution containing 10.0 g of NaOH and 190 g of H_2O is 5.00 percent.

13.3   If 1.1 mol of $CH_3(CH_2)_3OH$ dissolves in 1 L of water, then the *solubility* of $CH_3(CH_2)_3OH$ in 100 mL of water is 1.1 mol.

13.4 In a solution of NaCl, the formation of $Na^+(aq)$ from the interaction between Na^+ particles and water molecules is called *hydration*.

13.5 A *saturated solution* of $CdCO_3$ forms when $CdCO_3(s)$ and dissolved $CdCO_3$ are in a state of dynamic equilibrium.

13.6 The solubility of $CdCO_3$ in water at 25°C is 4.48×10^{-4} g/mL. A 1-L solution that contains 3.0×10^{-2} g of dissolved $CdCO_3$ is *unsaturated;* one that contains 6.0×10^{-1} g of dissolved $CdCO_3$ is *supersaturated;* and one that contains 4.48×10^{-1} g of dissolved $CdCO_3$ is *saturated*.

13.7 We would expect water and crystal particles that are around 10,000 Å in size to form a *colloid*.

13.8 The *Tyndall effect* could be utilized to detect the presence of smoke in a darkened room.

13.9 If two solutions are connected by a semipermeable membrane and are isotonic, the solutions have the same *osmotic pressure*.

13.10 Two aqueous solutions, 0.1 M NaCl and 0.2 M NaCl, are separated by a semipermeable membrane that allows only water to pass through it. *Osmosis* occurs as water flows out of the 0.2 M NaCl solution through the semipermeable membrane into the 0.1 M NaCl solution.

13.11 An example of a *colligative property* is the observation that a 0.25 m $LaCl_3$ solution has a freezing point that is lower than that of pure water by an amount $(K_f)(0.25)$.

13.12 According to *Raoult's law,* the vapor-pressure lowering of a solvent over a solution is given by the product of the molarity of the solution and the vapor pressure of pure solvent.

13.13 According to *Henry's law,* doubling the partial pressure of a gas over a solvent in which it dissolves doubles the solubility of the gas in the solvent.

13.14 It is found that the vapor pressure of a solvent is directly dependent on its mole fraction. This suggests that it is an *ideal solution*.

13.15 A solution contains 0.30 mg of sodium ion per liter of solution: This corresponds to a concentration of sodium ion of 3.0 *ppm*.

13.16 *Crystallization* occurs more readily when a seed crystal is present.

13.17 The *molal boiling-point-elevation constant* for carbon tetrachloride is 2.53 °C/m. This means that 1 m nonvolatile solute particles in carbon tetrachloride will cause the solution to boil 2.53 °C higher than pure carbon tetrachloride.

13.18 In general, *molal freezing-point-depression constants* for solvents are smaller in magnitude than their molal boiling-point-elevation constants.

13.19 The presence of *ion pairs* in a solution will cause an increase in osmotic pressure.

13.20 CH_3Cl should have *hydrophilic* properties.

13.21 A $CH_3(CH_2)_{20}$–group in a molecule should cause the molecule to be *hydrophobic*.

Problems and Short-Answer Questions

13.22 A 0.500-L solution that contains 20.0 g of glucose has an osmotic pressure of 5.43 atm at 25°C. What is the molar mass of glucose?

13.23 Calculate the concentrations asked for in each problem.
(a) The molality of 142 g of Na_2CO_3 in 2.00 kg of water at 0°C (a saturated solution).
(b) The molality of 24.50 g of codeine, $C_{18}H_{21}NO_3$ in 150.5 g of ethanol, C_2H_5OH.
(c) The mole fraction of 32.3 g of NaCl in 265.0 g of H_2O.
(d) The percent-by-mass of HCl in a tile cleaner containing 140 g of HCl and 800 g of H_2O.

13.24 What volume of concentrated hydrochloric acid with a density of 1.19 g/mL and containing 37.2% by mass of HCl contains 150 g of HCl?

13.25 The freezing-point depressions for 0.01 m solutions of $Co(NH_3)_6Cl_3$, $MgSO_4$, NH_4Cl, and CH_3COOH are 0.0643°C, 0.0308°C, 0.0358°C, and 0.0193°C, respectively. K_f for H_2O is 1.86°C/m. Which compounds are strong electrolytes and which are weak? Which compound forms the greatest number of ions in a 0.01 m solution?

13.26 A 0.157 M NaCl solution is to be used to replace lost blood. The average osmotic pressure of blood is 7.7 atm at 25°C. Is the NaCl solution isotonic with blood?

13.27 When 20.00 g of sucrose is dissolved in 100.00 g of water at 20°C, a vapor-pressure lowering of 0.185 torr Hg is observed. The vapor pressure of pure H_2O at 20 °C is 17.54 torr. Determine the mass of 1 mole of sucrose from the vapor-pressure-lowering data.

13.28 Hydrophilic colloids do not settle out of solution as do dispersions of larger, visible particles. Why?

13.29 In each of the following pairs of substances, which substance will be more soluble in liquid NH_3?
(a) $NaCl(s)$ or $H_2(g)$
(b) $CCl_4(l)$ or $CH_3OH(l)$

13.30 Indicate whether each of the following processes proceeds with an increase or decrease in randomness (disorder):
(a) sublimation of solid CO_2;
(b) freezing of liquid ammonia;
(c) increasing the total pressure of $N_2(g)$ stored over water in a closed container.

13.31 Proteins are substances found in living cells. They are long molecules that include both nonpolar hydrocarbon segments and polar subunits. Proteins are flexible and can fold about themselves. If a protein is placed in water, in what way would you expect the protein to fold itself? (In other words, in what way are the polar and nonpolar segments arranged with respect to one another in the folded protein structure?)

13.32 A saturated solution of KBr in H_2O at 0°C has a molality of 0.4499. What is the solubility of KBr in H_2O in units of g KBr/100 mL of H_2O at 0°C? The density of H_2O at 0°C is 0.99988 g/mL.

Multiple-Choice Questions

13.33 What is the molality of ethylene glycol, $C_2H_4(OH)_2$, in a solution prepared by mixing 5.00 g of ethylene glycol in 125 g of water?
(a) 0.644   (d) 0.000619
(b) 0.000644   (e) none of these
(c) 0.619

13.34 What is the mole fraction of water in a solution prepared by mixing 12.5 g of H_2O with 220 g of acetone, C_3H_6O?

    (a) 0.817
    (d) 0.155
    (b) 0.845
    (e) none of these
    (c) 0.183

13.35 In which of the following solvents would you expect the solubility of $CaCl_2$ to be greatest?

    (a) CH_3OH
    (c) CCl_4
    (b) C_6H_6 (benzene)
    (d) H_2O
    (e) insufficient information to answer question

13.36 Given that $K_b = 2.53°C/m$ for benzene, which mass of acetone (CH_3COCH_3) must be dissolved in 200 g of benzene to raise the boiling point of benzene by 3.00 °C?

    (a) 7.26 g
    (d) 0.138 g
    (b) 0.0726 g
    (e) 13.8 g
    (c) 2.56 g

13.37 The presence of a nonvolatile solute in a volatile solvent will result in which of the following?

    (a) It will raise the freezing point and lower the vapor pressure and boiling point.
    (b) It will lower the freezing point and raise the vapor pressure and boiling point.
    (c) It will raise the freezing point, vapor pressure, and boiling point.
    (d) It will lower the freezing point and vapor pressure and raise the boiling point.
    (e) It will lower the boiling point and raise the freezing point and vapor pressure.

13.38 Which of the following substances might stabilize a colloidal suspension of oil in water?

    (a) octane, C_8H_{18}
    (b) sodium bicarbonate, $NaHCO_3$
    (c) sodium stearate, $NaCO_2(CH_2)_{16}CH_3$
    (d) HCl
    (e) $CaCl_2$

13.39 Which of the following solutions has the largest osmotic pressure?

    (a) 0.15 M NaCl
    (d) 0.05 M $Al(NO_3)_3$
    (b) 0.10 M $CaCl_2$
    (e) 0.20 M NH_3
    (c) 0.05 M $Ba(NO_3)_2$

13.40 When 0.200 g of a high-molecular-weight compound is dissolved in water to form 12.5 mL of solution at 25 °C, the osmotic pressure of the solution is found to be 1.10×10^{-3} atm. What is the molar mass of the compound?

    (a) 3.56×10^5 g
    (d) 2.98×10^3 g
    (b) 3.56×10^4 g
    (e) 3.00×10^4 g
    (c) 2.98×10^4 g

SELF-TEST SOLUTIONS

13.1 True. $m = 0.12$ mol/0.500 kg = 0.24. **13.2** True. Mass % = (10.0 g)/(10.0 g + 190 g)(100) = 5.00 %. **13.3** False. In 1000 mL of water, 1.1 mol of $CH_3(CH_2)_3OH$ dissolves. Thus in 100 mL of water, 0.11 mol of $CH_3(CH_2)_3OH$ dissolves. **13.4** True. **13.5** True. **13.6** True. **13.7** False. A particle size of $10^{4\text{–}3}$ is too large for a colloidal solution to form. **13.8** True. **13.9** True. **13.10** False. Water flows from the more dilute solution to the more concentrated solution until the osmotic pressures become equal. **13.11** False. Since $LaCl_3$ ionizes to form four particles, the total molality of the solution is 4×0.25 m = 1.00 m, not simply 0.25 m. Therefore, $\Delta T = K_f(1.00\ m)$. **13.12** False. Vapor-pressure lowering is directly related to the mole fraction of solvent and vapor pressure of pure solvent. **13.13** True. **13.14** True. An ideal solution obeys Raoult's law which states that the vapor pressure of a solvent is directly proportional to the product of its mole fraction and vapor pressure as a pure liquid. Since the vapor pressure is directly related to the mole fraction of the solvent, this suggests that the solution is ideal. **13.15** False. 1 *ppm* corresponds to 1 mg of solute per liter of solution. Thus the concentration should be 0.30 ppm. **13.16** True. **13.17** True. **13.18** False. Inspect Table 13.5 in the text. You will see the opposite is true. This results from the slope of the solid—gas equilibrium line in a phase diagram being less than that of the liquid—gas equilibrium line. **13.19** False. Ion-pairs reduce the total number of solute species in solution; therefore, osmotic pressure decreases. **13.20** False. A hydrophilic species will have a highly charged region or a number of very polar regions. This is not the case for the given molecule. **13.21** True. A hydrophobic group will be non-polar and molecular hydrophobic properties are enhanced by a large non-polar group.

13.22

$$\pi = MRT$$

$$= \frac{g}{(\mathcal{M})(V)}\,RT = \frac{gRT}{(\mathcal{M})(V)}$$

$$\mathcal{M} = \frac{gRT}{\pi V}$$

$$= \frac{(20.0\ \text{g})(0.082\ \text{L-atm/mol-K})(298\ \text{K})}{(5.43\ \text{atm})(0.500\ \text{L})}$$

$$= 180\ \text{g/mol}$$

13.23

(a)

$$m = \frac{\text{moles Na}_2\text{CO}_3}{\text{kg of solvent}}$$

$$= \frac{(142\ \text{g Na}_2\text{CO}_3)\left(\dfrac{1\ \text{mol}}{106.0\ \text{g}}\right)}{2.00\ \text{kg H}_2\text{O}}$$

$$= 0.670\ \text{molal}$$

(b)

$$m = \frac{\text{moles codeine}}{\text{kg of solvent}}$$

$$= \frac{(24.50\ \text{g codeine})\left(\dfrac{1\ \text{mol}}{299.4\ \text{g}}\right)}{(150.5\ \text{g ethanol})\left(\dfrac{1\ \text{kg}}{1000\ \text{g}}\right)}$$

$$= 0.5437\ \text{molal}$$

(c)

$$X(\text{NaCl}) = \frac{\text{mol NaCl}}{\text{mol NaCl} + \text{mol H}_2\text{O}}$$

$$= \frac{(32.3 \text{ g NaCl})\left(\dfrac{1 \text{ mol}}{58.4 \text{ g}}\right)}{\left[(32.3 \text{ g NaCl})\left(\dfrac{1 \text{ mol}}{58.5 \text{ g}}\right) + (265.0 \text{ g H}_2\text{O})\left(\dfrac{1 \text{ mol}}{18.01 \text{ g}}\right)\right]}$$

$$= \frac{0.533 \text{ mol}}{0.552 \text{ mol} + 14.71 \text{ mol}} = 0.0362$$

(d)

$$\% \text{ HCl} = \frac{140 \text{ g HCl}}{140 \text{ g HCl} + 800 \text{ g H}_2\text{O}} \times 100$$

$$= 14.9\% \text{ HCl}$$

13.24 This problem requires the following conversions: mass of HCl → mass of solution → volume of solution. The first conversion is accomplished by using 37.2% HCl-by-mass because there are 37.2 g of HCl per 100.0 g of solution

$$(150 \text{ g HCl})\left(\frac{100.0 \text{ g solution}}{37.2 \text{ g HCl}}\right) = 403 \text{ g solution}$$

Density is used to convert from mass of solution to volume of solution.

$$(403 \text{ g solution})\left(\frac{1 \text{ mL solution}}{1.19 \text{ g solution}}\right) = 339 \text{ mL solution}$$

13.25 The ΔT lowering for a 0.01 m weak electrolyte solution is approximately $\Delta T_f = K_f m = (1.86 \text{ °C}/m)(0.01 \ m) = 0.0186$ °C. Only the 0.01 m CH$_3$COOH solution has a ΔT_f approximately equal to 0.0186 °C; it is the only weak electrolyte. The other salt solutions produce freezing-point depressions significantly greater than that of acetic acid; thus they are all strong electrolytes. Since the 0.01 m solution of Co(NH$_3$)$_6$Cl$_2$ produces the greatest ΔT_f lowering, the molality of all species in the solution is the largest; it forms the greatest number of ions in solution. **13.26** The solutions are isotonic if they have the same osmotic pressure. The osmotic pressure of 0.157 M NaCl is calculated using the expression $\pi = MRT$. The value of M is 2(0.157 M) because there are two ions per NaCl formula:

$$\pi = (0.314 \ M)\left(0.082 \ \frac{\text{L-atm}}{\text{K-mol}}\right)(298 \text{ K}) = 7.7 \text{ atm}$$

The solutions are isotonic. **13.27** VPL = $X_{\text{sucrose}} P^\circ_{\text{H}_2\text{O}}$; $X_{\text{sucrose}} = $ mol sucrose/(mol sucrose + mol H$_2$O). For very dilute solutions, such as this one in which the number of moles of H$_2$O is far greater in magnitude than number of moles of sucrose, we can assume that (mol sucrose + mol H$_2$O) ≈ mol H$_2$O.

$$\text{VLP} = X_{\text{sucrose}} P^\circ_{\text{H}_2\text{O}} = \left(\frac{\text{mol sucrose}}{\text{mol H}_2\text{O}}\right) P^\circ_{\text{H}_2\text{O}}$$

$$\text{VLP} = \left[\frac{(\text{grams sucrose})\left(\dfrac{1 \text{ mol sucrose}}{\mathcal{M} \text{ sucrose}}\right)}{(\text{grams H}_2\text{O})\left(\dfrac{1 \text{ mol H}_2\text{O}}{\mathcal{M} \text{ H}_2\text{O}}\right)}\right] P^\circ_{\text{H}_2\text{O}}$$

$$\mathcal{M} \text{ sucrose} = \left(\frac{\text{grams sucrose}}{\text{grams H}_2\text{O}}\right)\left(\frac{1 \text{ mol sucrose}}{1 \text{ mol H}_2\text{O}}\right)$$

$$\times \left(\frac{P^\circ_{\text{H}_2\text{O}}}{\text{VPL}}\right)(\mathcal{M} \text{ H}_2\text{O})$$

$$= \left(\frac{20.00 \text{ g}}{100.00 \text{ g}}\right)\left(\frac{1 \text{ mol}}{1 \text{ mol}}\right)$$

$$\times \left(\frac{17.54 \text{ mm Hg}}{0.185 \text{ mm Hg}}\right)\left(\frac{18.0 \text{ g}}{1 \text{ mol}}\right)$$

$$= 341 \text{ g}$$

13.28 There are two primary reasons why hydrophilic colloidal particles do not settle out of a solution. The charge on all colloidal solution particles is the same, and repulsions between the similarly charged particles keep them separated. Also, their small size results in a high kinetic energy, which counteracts gravitational forces. **13.29** NH$_3$(l) is a polar solvent. The most polar substance of each pair will be the most soluble. **(a)** NaCl(s) is more soluble because it contains ions; H$_2$(g) is nonpolar. **(b)** CH$_3$OH(l) is more soluble because it is polar; CCl$_4$(l) is nonpolar. **13.30 (a)** Sublimation of solid CO$_2$ involves an increase in randomness because of change from organized crystal lattice to random nature of particles in gaseous state. **(b)** The freezing of liquid ammonia results in a decrease in randomness because a solid has a more organized arrangement of particles than a liquid. **(c)** With an increase in pressure, more N$_2$(g) molecules dissolve. These dissolved particles are more confined and have less random movement than when they were in the gas phase. Therefore, there is a decrease in disorder. **13.31** A protein folds itself in such a way that the polar units are exposed to polar water and the nonpolar units are tucked away inside the structure. The nonpolar units are in this way prevented from interacting with polar water molecules. This is another application of the "likes dissolve likes" concept. **13.32** To calculate the solubility of KBr in units of grams per 100 mL of H$_2$O, use the definitions of density and molality. From the density of H$_2$O, you calculate the mass of 100 mL of H$_2$O to be 0.099987 kg. From the definition of molality, the number of moles of KBr in 100 mL of water is calculated as follows:

Moles KBr in 100 mL H$_2$O

$$= (\text{molality})\left(\frac{0.099987 \text{ kg}}{100 \text{ mL H}_2\text{O}}\right)$$

$$= \left(\frac{0.4499 \text{ mol}}{\text{kg}}\right)\left(\frac{0.099987 \text{ kg}}{100 \text{ mL H}_2\text{O}}\right)$$

$$= 0.04498 \text{ mol}/100 \text{ mL H}_2\text{O}$$

The solubility of KBr in units of grams per 100 mL of H$_2$O is calculated as follows:

$$\left(\frac{0.04498 \text{ mol KBr}}{100 \text{ mL H}_2\text{O}}\right)\left(\frac{119.01 \text{ g KBr}}{1 \text{ mol KBr}}\right)$$

$$= 5.354 \text{ g KBr}/100 \text{ mL H}_2\text{O}$$

13.33

(a) $m = \dfrac{(5.00 \text{ g C}_2\text{H}_4(\text{OH})_2)\left(\dfrac{1 \text{ mol C}_2\text{H}_4(\text{OH})_2}{62.07 \text{ g C}_2\text{H}_4(\text{OH})_2}\right)}{(125 \text{ g H}_2\text{O})\left(\dfrac{1 \text{ kg}}{1000 \text{ g}}\right)}$

$= 0.644$

13.34

(d) $X_{\text{H}_2\text{O}} = \dfrac{(12.5 \text{ g H}_2\text{O})\left(\dfrac{1 \text{ mol H}_2\text{O}}{18.01 \text{ g H}_2\text{O}}\right)}{\left[(12.5 \text{ g H}_2\text{O})\left(\dfrac{1 \text{ mol H}_2\text{O}}{18.01 \text{ g H}_2\text{O}}\right) + (220 \text{ g C}_3\text{H}_6\text{O})\left(\dfrac{1 \text{ mol C}_3\text{H}_6\text{O}}{58.09 \text{ g C}_3\text{H}_6\text{O}}\right)\right]}$

$= 0.155$

13.35 (d). $CaCl_2$ is an ionic substance and will dissolve best in the most polar solvent, water.

13.36 (e).

$\Delta T = K_b m = K_b \left(\dfrac{\text{moles acetone}}{1 \text{ kg benzene}}\right)$

$\text{Moles acetone} = \left(\dfrac{\Delta T}{K_b}\right)(1 \text{ kg benzene})$

$= \left(\dfrac{3.00 \text{ °C}}{2.53 \text{ °C}/m}\right)(1 \text{ kg benzene})$

$= \left(\dfrac{1.19 \text{ mol acetone}}{1 \text{ kg benzene}}\right)(1 \text{ kg benzene})$

$= 1.19 \text{ mol}$

The number of moles of acetone in 200 g of benzene is calculated as follows:

$\dfrac{1.19 \text{ mol acetone}}{\text{moles acetone}} = \dfrac{100 \text{ g benzene}}{200 \text{ g benzene}}$

$\text{Moles acetone} = 0.238 \text{ mol}$

$\text{Mass acetone} = (0.238 \text{ mol acetone})$

$\times \left(\dfrac{58.09 \text{ g acetone}}{1 \text{ mol acetone}}\right)$

$= 13.8 \text{ g}$

13.37 (d). **13.38 (c)**. **13.39 (b)**. The solution with the highest molarity of particles has largest osmotic pressure.

13.40 (a). $M = \dfrac{\pi}{RT} = \dfrac{1.10 \times 10^{-3} \text{ atm}}{\left(0.082 \dfrac{\text{L} \cdot \text{atm}}{\text{K} \cdot \text{mol}}\right)(298 \text{ K})}$

$= 4.50 \times 10^{-5} \text{ mol/L}$

The number of moles in 0.200 g of the compound in 12.5 mL is:

$\text{Moles} = \left(4.50 \times 10^{-5} \dfrac{\text{mol}}{\text{L}}\right)\left(\dfrac{1 \text{ mol}}{1000 \text{ mL}}\right)(12.5 \text{ mL})$

$= 5.62 \times 10^{-7} \text{ mol}$

The molar mass of the compound is calculated as follows:

$\dfrac{0.200 \text{ g}}{\text{mass of 1 mol}} = \dfrac{5.62 \times 10^{-7} \text{ mol}}{1 \text{ mol}}$

$\text{Molar mass} = (0.200 \text{ g})\left(\dfrac{1 \text{ mol}}{5.62 \times 10^{-7} \text{ mol}}\right)$

$= 3.56 \times 10^5 \text{ g}$

Sectional Test 3

CHAPTERS 10–13

You should make this test as realistic as possible. **Tear out this test so that you do not have access to the information in the** *Student's Guide*. Use only data provided in the questions and a periodic table. Do not check your answers until you are finished. If you answer questions incorrectly, review the section in the text indicated after each question.

Choose the best response for each question.

1. Which of the following is not a property of a gas under normal conditions?

(a) Flows easily (b) Compressible (c) Completely fills its container (d) High density

[10–1]

2. The pressure of a gas is measured using a U-shaped manometer. The height of the mercury in the manometer is 13 cm on the side connected to the sample and 26 cm on the side connected to the atmosphere. Atmospheric pressure is 752 torr. What is the gas pressure of the sample?

(a) 882 torr (b) 765 torr (c) 752 torr (d) 622 torr

[10–2]

3. Which of the following is a statement of Boyle's Law

(a) $V = kn$ (P, T constant) (b) $P = kT$ (V, n constant)
(c) $PV = k$ (n, T constant) (d) $V = kT$ (n, P constant)

[10–3]

4. A gas at 30°C and 1 atm pressure has a volume of 3.50 L. What volume does the gas occupy at 40°C and 1 atm pressure?

(a) $\left(\dfrac{30°C}{40°C}\right)(3.50\text{ L}) = 2.63$ L   (b) $\left(\dfrac{40°C}{30°C}\right)(3.50\text{ L}) = 4.67$ L

(c) $\left(\dfrac{303\text{ K}}{313\text{ K}}\right)(3.50\text{ L}) = 3.39$ L   (d) $\left(\dfrac{313\text{ K}}{303\text{ K}}\right)(3.50\text{ L}) = 3.62$ L

[10–3]

5. How many moles of hydrogen gas are in a sample of H_2 gas with a volume of 9.00 L at a temperature of 100°C and at a pressure of 2.00 atm. (R = 0.08206 L-atm/K-mol)

(a) 2.19 moles (b) 0.588 moles (c) 0.429 moles (d) 0.0289 moles
[10–4]

6. A sample of gas at 750 torr and at a temperature of −50°C and a volume of 3.00 L is allowed to change so that the temperature is 200°C and gas pressure is 845 torr. What is the new volume?

(a) $\dfrac{T_2}{T_1} \times \dfrac{P_1}{P_2} \times V_1 = 5.65 \text{ L}$ (b) $\dfrac{T_1}{T_2} \times \dfrac{P_1}{P_2} \times V_2 = 1.26 \text{ L}$

(c) $\dfrac{T_2}{T_1} \times \dfrac{P_2}{P_1} \times V_1 = 7.17 \text{ L}$ (d) $\dfrac{T_2}{P_2} \times \dfrac{T_1}{P_1} \times V_1 = 0.499 \text{ L}$

[10–4]

7. A scuba diving tank is filled with 42 L of O_2 at 1.00 atm and 10 L of He at 1.00 atm and 27°C. The tank has a total volume of 6.0 L. What is the total pressure in the scuba tank at 25°C?

(a) 5.3 atm (b) 8.6 atm (c) 10.3 atm (d) 12.6 atm
[10–4]

8. What is the gas density of C_2H_2 at STP in units of g/L?

(a) 2.12 g/L (b) 1.83 g/L (c) 1.16 g/L (d) 0.850 g/L
[10–5]

9. A sample of $KClO_3$ is heated and decomposed as follows:

$$2KClO_3(s) \longrightarrow 2KCl(s) + 3O_2(g)$$

If 3.00 g of $KClO_3$ (molar mass = 122.55 g) is totally decomposed and the evolved O_2 collected in a 1.00 L vessel at 22°C, what pressure will the O_2 exert?

(a) 0.245 atm (b) 0.389 atm (c) 0.811 atm (d) 0.889 atm
[10–7]

10. Which of the following statements is true according to the kinetic molecular theory?

(a) Gravitational forces act upon gas particles.

(b) Average kinetic energy of gas particles is proportional to kelvin temperature.

(c) Gas particles exert forces upon each other.

(d) Gas particles have measurable volumes.

[10–8]

11. How many times faster (or slower) will H_2 gas pass through a pinhole than HF(*g*)?

(a) 0.101 (b) 0.318 (c) 3.15 (d) 9.91
[10–9]

12. A real gas typically exhibits behavior that is closest to an ideal gas at

(a) low pressure and high temperature

(b) high pressure and high temperature

(c) low pressure and low temperature

(d) high pressure and low temperature

[10–10]

13. 43.2 g of a gas occupies 22.4 L at 200°C and 2.00 atm. What is its molar mass?

(a) 49.9 g/mol (b) 37.6 g/mol (c) 21.6 g/mol (d) 13.2 g/mol
[10–5]

14. In what state of matter is molecular motion slowest?
    (a) Solid (b) Liquid (c) Aqueous (d) Gas
    [11–1]

15. What is the direct conversion of solid to gas termed?
    (a) Condensation (b) Sublimation (c) Evaporation (d) Freezing
    [11–6]

16. What is the enthalpy change for the conversion of 3.0 mol of ice at 0°C to water vapor at 130°C? Data: ΔH_f = 6.02 kJ/mol; C_p (H_2O)(l)) = 75.3 J/°C-mol; ΔH_v = 40.67 kJ/mol; C_p ($H_2O(g)$) = 33.1 J/°C-mol. (Note that C_p is in units of J, not kJ.)
    (a) 49.8 kJ (b) 55.2 kJ (c) 166 kJ (d) 237 kJ
    [11–4]

17. Which of these statements is true?
    (a) The vapor pressure of a liquid increases with decreasing temperature.
    (b) The boiling point of a liquid is independent of atmospheric pressure.
    (c) Vapor pressure varies directly with volume.
    (d) The higher a boiling point of a liquid at 1 atm atmospheric pressure, the greater the internal cohesive forces of the liquid.
    [11–4, 11–5]

18. The viscosity of a liquid
    (a) increases with increasing temperature
    (b) decreases with increasing temperature
    (c) increases with increasing quantity
    (d) decreases with decreasing quantity
    [11–3]

19. Which of these substances has the lowest boiling point?
    (a) NaCl (b) HF (c) H_2O (d) H_2
    [11–2]

20. Which of these substances has the highest melting point?
    (a) Cl_2 (b) H_2O (c) KCl (d) $CaCl_2$
    [11–2]

21. Which of these substances has the highest boiling point?
    (a) H_2O (b) H_2Se (c) H_2S (d) H_2Te
    [11.2]

22. How many KF formula units are there in a unit cell of KF? KF has a NaCl structure.
    (a) 1 (b) 2 (c) 4 (d) 6
    [11–8]

23. What is the density of KF(s) if it has a NaCl structure and a unit cell length of 5.35×10^{-10} m?
    (a) 1.38 g/cm^3 (b) 2.47 g/cm^3 (c) 3.92 g/cm^3 (d) 4.65 g/cm^3
    [11–8]

24. Which solvent should dissolve CCl_4 best?
    (a) H_2O(l) (b) CH_3OH(l) (c) HCl(aq) (d) benzene(l)
    [11–2]

25. Consider the phase diagram for water

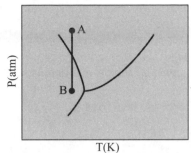

If the external pressure at point B is raised to the value of *A* at constant temperature, which of the following happens?

(a) Water sublimes (b) Ice changes to liquid water

(c) Ice changes to water vapor (d) A triple point forms

[11–6]

26. Which of the following should exhibit hydrogen bonding?

(a) CH_3OH (b) CH_4 (c) PH_3 (d) LiH

[11–2]

27. Which of the following is characteristic of a liquid crystal?

(a) Molecules are arranged in an ordered way.

(b) Molecules are arranged in a partial ordering.

(c) Low viscosity. (d) Low density.

[12–1]

28. Which of the following structural characteristics is likely to be present in a liquid crystal?

(a) Highly branched structure (b) Ionic bonds (c) Metallic bonds

(d) Long, axial structural backbone.

[12–1]

29. In which liquid crystalline phase are diskshaped molecules stacked in layers?

(a) chloesteric (b) smectic (c) nematic (d) polymeric

[12–1]

30. Which of the following bonding sites in a monomer would likely be the reactive site in forming a polymer during polymerization?

(a) C=O (b) C=C (c) C—H (d) C—Cl

[12–2]

31. Which of the following types of polymers can not be reshaped?

(a) Thermoplastic (b) Thermosetting plastic (c) Elastomer

(d) Fibers

[12–2]

32. Which statement about polymers is true?

(a) Mechanical strength of a polymer increases with increased crystallinity.

(b) Low density polymers often exhibit extensive chain branching.

(c) The greater the density of crosslinks in a polymer, the softer the material.

(d) Plasticizers increase the rigidity of polymers.

[12–2]

33. Glass is a

(a) crystalline ceramic. (b) noncrystalline ceramic. (c) an ionic compound. (d) polymer.

[12–3]

34. Which of the following is *not* used to form a ceramic?

(a) Alumina (b) Zirconia (c) Silicon carbide (d) Kevlar

[12–3]

35. A primary goal of the sol-gel process is

(a) to form a polymer.

(b) to activate a structural site in a monomer.

(c) to form extremely fine particles of uniform size.

(d) to process a polymer.

[12–3]

36. Which is true about ceramic fibers?

(a) They undergo condensation polymerization with ceramics.

(b) They exhibit great strength when loads are applied along the long axis.

(c) Silicon carbide is an example of one.

(d) They weakly resist deformations when imbedded in a ceramic matrix.

[12–3]

37. Which statement is true about superconducting ceramics?

(a) New superconducting ceramics have simple oxide structures.

(b) They exhibit superconductivity only above a superconducting transition temperature.

(c) They include within their volume all magnetic fields.

(d) The copper-oxygen content of new superconductors is of great significance.

[12–3]

38. For a thin film to be useful, it must possess several properties, one of which is

(a) Chemically stable in the environment.

(b) Shows low adhesiveness.

(c) Shows high density of dislocations.

(d) Exhibits diversity of thickness.

[12–4]

39. Which of the following processes requires a high voltage source?

(a) Vacuum deposition (b) Sputtering (c) Chemical vapor deposition (d) All of the above

[12–4]

40. Which statement is *not* true?

(a) A solute takes on the phase of the solvent.

(b) A solute is in lesser quantity than a solvent.

(c) A solute and solvent may be gases.

(d) A polar solvent dissolves a nonpolar solute.

[13–1]

41. What is the molarity of a solution that contains 32.0 g of HCl (molar mass = 36.46 g) in 2.50 L of solution?

(a) 0.00351 M (b) 0.0128 M (c) 0.351 M (d) 12.8 M

[13–2]

42. How many moles of H_2SO_4 are present in 250 mL of a 3.00 M H_2SO_4 solution?

   (a) 0.150 moles (b) 0.750 moles (c) 1.50 moles (d) 750 moles
   [13–2]

43. What is the molality of a solution of 100.0 g of methanol, CH_3OH (molar mass = 32.05 g), in 250 mL of water? The density of water is 1.00 g/mL.

   (a) 0.00400 molal (b) 0.0125 molal (c) 8.91 molal (d) 12.5 molal
   [13–2]

44. What is the mole fraction of NaCl (molar mass = 58.44 g) in a solution containing 20.2 g of NaCl and 55.0 g of water (molar mass = 18.02 g)?

   (a) 0.898 (b) 0.102 (c) 0.113 (d) 0.268
   [13–2]

45. When NaCl dissolves in water

   (a) energy is released when NaCl bonds are broken.
   (b) Na^+ interacts with the positive dipole of water.
   (c) Cl^- interacts with the positive dipole of water.
   (d) Na^+ ions remain independent of water molecules.
   [13–2, 13–3]

46. The solubility of a particular salt in water is 9.8 g/mL at 25°C. If 10.3 g is completely dissolved, the solution is

   (a) supersaturated (b) saturated (c) unsaturated (d) dilute
   [13–3]

47. An aqueous solution strongly conducts electrical current. The solution contains a(n)

   (a) nonelectrolyte (b) weak electrolyte (c) strong electrolyte
   (d) nonpolar solute
   [13–3]

48. The solubility of $CO_2(g)$ in water

   (a) increases with increasing temperature.
   (b) increases with decreasing pressure.
   (c) increases with increasing pressure.
   (d) is not affected by temperature or pressure.
   [13–4]

49. Which of the following is not a colligative property?

   (a) Vapor pressure lowering (b) Boiling point elevation
   (c) Osmotic pressure (d) Density
   [13–5]

50. What is the osmotic pressure in atmospheres of a solution containing 36.5 g of NaCl dissolved in water to form 3.50 L of solution at 25°C? ($R = 0.08206$ L-atm/mol-K.)

   (a) 8.73 atm (b) 4.36 atm (c) 1.28 atm (d) 2.57 atm
   [13–5]

51. What mass in grams of CH_3OH must be added to 500 g of water to produce a solution boiling at 102.35°C? ($K_b = 0.52$°C/m for water.)

   (a) 3.55 g (b) 4.52 g (c) 36.2 g (d) 72.4 g
   [13–5]

52. 1.08 g of a protein is dissolved in 50.0 mL of water and the osmotic pressure of the solution is found to be 5.86 torr at 25°C. What is the molar mass of the protein?

(a) 1.52×10^{-5} g

(b) 90.3 g

(c) 5.76×10^3 g

(d) 6.85×10^4 g

[13–5]

53. What is the typical particle range, in nm, for a colloid?

(a) 1–10

(b) 100–200

(c) 1–200

(d) 50–300

[13–6]

54. Which statement is true about colloidal suspensions?

(a) They settle under the influence of gravity in a reasonable time period.

(b) Colloidal particles have a high ratio of surface area to volume.

(c) They are not stable in light.

(d) Light passes directly through the suspension without interference.

[13–6]

55. Lava is a foam colloid. The dispersed phase and dispersion medium are respectively:

(a) gas, solid

(b) gas, liquid

(c) liquid, gas

(d) solid, gas

[13–6]

ANSWERS

1. (d). **2.** (a). **3.** (c). **4.** (d). **5.** (b). **6.** (a). **7.** (b). **8.** (c). **9.** (d).
10. (b). **11.** (c). **12.** (a). **13.** (b). **14.** (a). **15.** (b). **16.** (c). **17.** (d). **18.** (b).
19. (d). **20.** (d). **21.** (a). **22.** (c). **23.** (b). **24.** (d). **25.** (b). **26.** (a). **27.** (b).
28. (d). **29.** (a). **30.** (b). **31.** (b). **32.** (b). **33.** (b). **34.** (d). **35.** (c). **36.** (b).
37. (d). **38.** (a). **39.** (b). **40.** (d). **41.** (c). **42.** (b). **43.** (d). **44.** (b). **45.** (c).
46. (a). **47.** (c). **48.** (c). **49.** (d). **50.** (a). **51.** (d). **52.** (d). **53.** (c). **54.** (b).
55. (a).

Chemical Kinetics 14

OVERVIEW OF THE CHAPTER

14.1, 14.2 REACTION RATES: GENERAL CONSIDERATIONS

Review: Concentration units (13.2); graphing techniques (see Appendix in the text).

Objectives: You should be able to:

1. Express the rate of a given reaction in terms of the variation in concentration of a reactant or product substance with time.
2. Calculate the average rate over an interval of time, given the concentrations of a reactant or product at the beginning and end of that interval.
3. Calculate instantaneous rates from a graph of reactant or product concentrations as a function of time.
4. Explain the meaning of the term *rate constant* and state the units associated with rate constants.
5. Calculate rate, rate constants, or reactant concentration, given two of these together with the rate law.
6. Determine the rate law from experimental results that show how concentration affects rate.

14.3 REACTION RATES: FIRST AND SECOND ORDER

Review: Calculation of the slope of a straight line (see Appendix in the text); use of logarithms (see Appendix in the text).

Objectives: You should be able to:

1. Use the equations

$$\ln\left(\frac{[A]_0}{[A]_t}\right) = kt$$

$$\ln [A]_t = -kt + \ln [A]_0$$

to determine (a) the concentration of a reactant or product at any time after a reaction has started, (b) the time required for a given fraction of sample to react, or (c) the time required for a reactant concentration to reach a certain level.

2. Explain the concept of reaction half-life and describe the relationship between half-life and rate constant for a first-order reaction.

3. Use the equations

$$\ln [A]_t = -kt + \ln [A]_0$$

$$\frac{1}{[A]_t} = \frac{1}{[A]_0} + kt$$

to determine graphically whether the rate law for a reaction is first or second order.

4. Determine the rate law for a given higher-order reaction from the appropriate data.

14.4 FACTORS INFLUENCING REACTION RATES: TEMPERATURE AND ACTIVATION ENERGY

Review: Energy (5.1, 5.2); endothermic and exothermic processes (5.2).

Objectives: You should be able to:

1. Explain the concept of activation energy and how it relates to the variation of reaction rate with temperature.

2. Determine the activation energy for a reaction from a knowledge of how the rate constant varies with temperatures (the Arrhenius equation).

3. Use the collision model of chemical reactions to explain how reactions occur at the molecular level.

14.5 REACTION MECHANISMS

Objectives: You should be able to:

1. Explain what is meant by the mechanism of a reaction using the terms elementary steps, rate-determining step, and intermediate.

2. Derive the rate law for a reaction that has a rate-determining step, given the elementary steps and their relative speeds; or, conversely, choose a plausible mechanism for a reaction given the rate law.

14.6 CATALYSIS

Objectives: You should be able to:

1. Describe the effect of a catalyst on the energy requirements for a reaction.

2. Relate the factors that are important in determining the activity of a heterogeneous catalyst.

3. Explain how enzymes act as biological catalysts using the lock-and-key model.

TOPIC SUMMARIES AND EXERCISES

The study of chemical kinetics is concerned with (1) the velocity of chemical reactions—that is, the rate, or speed, at which reactants form products; (2) what factors influence the rates of reactions; and (3) how reactions occur. To describe the **rate** (or speed) **of a chemical reaction,** we use the ratio of the change in concentration of a reactant or product to the time interval required for the observed concentration change.

- For example, the rate at which N_2O_5 disappears in the chemical reaction

$$2N_2O_5(g) \longrightarrow 4NO_2 + O_2(g)$$

is written as

$$\text{Rate} = -\frac{\Delta[N_2O_5]}{\Delta t}$$

$$= -\left(\frac{\begin{array}{c}\text{concentration of } N_2O_5 \text{ at } t_{\text{final}} \\ -\text{concentration of } N_2O_5 \text{ at } t_{\text{initial}}\end{array}}{t_{\text{final}} - t_{\text{initial}}} \right)$$

- The Greek symbol Δ means "change in" or "difference in." Thus, $\Delta[N_2O_5]$ means the change in concentration of N_2O_5 at time t_{final} minus the concentration of N_2O_5 at time t_{initial}.
- Rates are assigned a positive quantity. The negative sign in front of $\Delta[N_2O_5]/\Delta t$ is necessary because $\Delta[N_2O_5]$ is a negative number (the concentration of N_2O_5, a reactant, decreases with time).

You need to differentiate between average rates and instantaneous rates of reactions.

- An **average rate** is calculated by taking the ratio of the change in concentration to change in time for a particular set of data points chosen from rate data. Note in Table 14.1 of the text that average rates vary depending on the pair of data chosen.
- An **instantaneous rate** is the slope of a tangent drawn at any point along a graph of concentration versus time. See Figure 14.1 on page 231. Note that instantaneous rates are rates at a particular time and thus vary with time. Unless otherwise stated, the rate of a reaction will refer to an instantaneous rate.
- **Initial rates** are instantaneous rates at time = 0.

Another way of describing the velocity of a chemical reaction is by a rate law.

- A **rate law** is an *experimentally* derived expression that relates the rate of a chemical reaction to the concentration of one or more of the species in the chemical reaction. For example, the rate law for the previously described decomposition of $N_2O_5(g)$ is

$$\text{Rate} = k[N_2O_5]$$

This expression tells you that the velocity of the reaction is directly proportional to the concentration of N_2O_5. If you double the concentration of N_2O_5, the rate of the reaction doubles.

▪ The constant k, a proportionality constant called the **rate constant,** is determined experimentally.

▪ A general form of the rate law for simple kinds of reactions is

$$\text{Rate} = k[\text{reactant 1}]^n[\text{reactant 2}]^m \ldots$$

The exponent associated with the concentration of a species in the rate law is called the **reaction order** for that species. If you add all the exponents ($n + m \ldots$), you obtain what is called the overall reaction order. Reaction orders do *not* have to be integral.

You need to use units with all rate calculations: Concentrations, time, rates and rate constants.

▪ Units of concentration are molarity for solutions and atmospheres of pressure for gases.

▪ Units of time are seconds, minutes, hours or any appropriate time period. Typically seconds or minutes are used.

▪ Units of rate are [concentration]/time; for example, Ms^{-1}. Note that the unit expression Ms^{-1} is equivalent to the expression M/s. Ms^{-1} can also be expressed as moles $L^{-1}s^{-1}$ because molarity, M, is equivalent to moles/L or moles L^{-1}.

▪ Units of the rate constant depend on the rate law. To solve for the units of the rate constant rearrange a rate law so that the rate constant, k, is on one side alone. Then substitute units into the other side and solve for the units of the rate constant.

EXERCISE 1　　(a) Write three rates for the reaction

$$2N_2O(g) \longrightarrow 2N_2(g) + O_2(g)$$

using the changes in the concentrations of $N_2O(g)$, $N_2(g)$, and $O_2(g)$ per unit time. (b) Relate the three to each other.

　　Solution:　　(a) The rate of the change of concentration of $N_2O(g)$ with time will be derived first. Assume that the $N_2O(g)$ concentration at time $t_{initial}$ is $[N_2O]_{initial}$ and that at time t_{final} it is $[N_2O]_{final}$. The decrease in the concentration of $N_2O(g)$ with a change in time is

$$\text{Rate}_1 = -\frac{[N_2O]_{final} - [N_2O]_{initial}}{t_{final} - t_{initial}} = \frac{\Delta[N_2O]}{\Delta t}$$

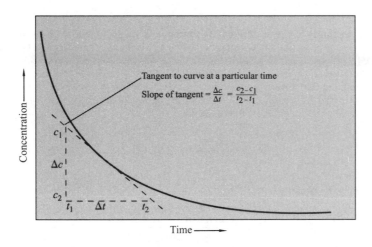

Slope of tangent $= \dfrac{\Delta c}{\Delta t} = \dfrac{c_2 - c_1}{t_2 - t_1}$

Figure 14.1 Graphing technique to determine an instantaneous rate.

The negative sign is required because $\Delta[N_2O]$ is a negative number; the N_2O concentration is decreasing with time, and the measured rate is by convention a positive quantity. The other rates are

$$\text{Rate}_2 = \frac{\Delta[N_2]}{\Delta t} = \frac{[N_2]_{\text{final}} - [N_2]_{\text{initial}}}{t_{\text{final}} - t_{\text{initial}}}$$

$$\text{Rate}_3 = \frac{\Delta[O_2]}{\Delta t} = \frac{[O_2]_{\text{final}} - [O_2]_{\text{initial}}}{t_{\text{final}} - t_{\text{initial}}}$$

The latter two rates do not require a negative sign, because the concentrations of O_2 and N_2 increase with time. **(b)** The three rates are related to each other as follows:

$$\text{Rate} = -\frac{1}{2}\frac{\Delta[N_2O]}{\Delta t} = \frac{1}{2}\frac{[N_2]}{\Delta t} = \frac{[O_2]}{\Delta t}$$

The factor of 1/2 before $\dfrac{\Delta[N_2O]}{\Delta t}$ and $\dfrac{[N_2]}{\Delta t}$ is needed because the rate of disappearance $N_2O(g)$ and the rate of appearance of $N_2(g)$ are each twice the rate of appearance of $O_2(g)$ (that is, for every two molecules of $N_2O(g)$ that react, two molecules of $N_2(g)$ and one molecule of O_2 appear). Thus, $N_2O(g)$ must disappear more rapidly than $O_2(g)$ appears. The factor of 1/2 simply causes all rates to be mathematically equal.

EXERCISE 2   Why are initial rates commonly used to determine rate laws?

Solution: An initial rate is the slope of the tangent, measured at time zero, to a curve such as shown in Figure 14.1. Take a ruler and use it as a tangent to the curve. Move the ruler along the curve until you reach time = 0. At this point, you should observe that the slope of the curve has its greatest value. During early changes in a reaction, the rates of greatest change are measured. Experimental errors associated with rate data are less when the measured rates are large.

EXERCISE 3   For the reaction

$$2NO(g) + 2H_2(g) \longrightarrow N_2(g) + 2H_2O(g)$$

the rate law is

$$\text{Rate} = k[NO]^2[H_2]$$

(a) Identify the rate constant in the rate law and indicate the overall reaction order. (b) What are the units of k if the unit of concentration is atmospheres? (c) How does the initial rate change if the initial concentration of $NO(g)$ is doubled and that of $H_2(g)$ remains constant?

Solution: (a) The rate constant is k. The overall order of the reaction is calculated by adding the powers to which each term is raised in the rate law. In this problem, the overall order of the reaction is $2 + 1$, or 3. (b) To solve for the units of k, you must relate k to all of the other terms in the rate law:

$$\text{Rate} = -\frac{\Delta[NO]}{\Delta t} = k[NO]^2[H_2]$$

Rearranging the equation:

$$k = \left(-\frac{\Delta[NO]}{\Delta t}\right)\left(\frac{1}{[NO]^2[H_2]}\right)$$

Substituting the appropriate units for each term in the expression yields the following:

$$k = \left(\frac{\text{atm}}{\text{s}}\right)\left(\frac{1}{(\text{atm})^2(\text{atm})}\right) = \frac{1}{(\text{s})(\text{atm})^2} = \text{atm}^{-2}\text{s}^{-1}$$

(c) The rate is proportional to the square of [NO] if the concentration of H_2 is held constant. Thus, doubling the concentration of NO causes the rate to increase by a factor of four.

EXERCISE 4 The rate law for the reaction

$$O_3(g) + 2NO(g) \rightarrow N_2O_5(g) + O_2(g)$$

contains $NO(g)$ and $O_3(g)$ with orders of $\frac{2}{3}$ each. Write the rate law.

Solution: The order with respect to a reactant is the power to which the concentration of that reactant is raised. The rate law is

$$-\frac{\Delta[O_3]}{\Delta t} = k[NO]^{2/3}[O_3]^{2/3}$$

EXERCISE 5 For the reaction

$$2NO(g) + 2H_2(g) \rightarrow N_2(g) + 2H_2O(g)$$

the rate law is

$$\frac{\Delta[N_2]}{\Delta t} = k[NO]^2[H_2]$$

Is there any relation between the balancing coefficients in the stoichiometric equation and the exponents in the rate law?

Solution: No. The exponent of $[H_2]$ in the rate law is 1; yet its balancing coefficient in the stoichiometric equation is 2. The rate law for any reaction must be experimentally determined. In general, it cannot be determined from inspection of the balanced equation.

EXERCISE 6 In three different experiments, the following data were obtained for the reaction $A + B \rightarrow C$:

Experiment number	Initial concentration of A, M	Initial concentration of B, M	Initial rate, Ms^{-1}
1	0.20	1.00	5.0×10^{-2}
2	0.20	2.00	2.0×10^{-1}
3	0.40	2.00	4.0×10^{-1}

Determine the rate law for the reaction.

Solution: The rate law may have the form

$$\text{Rate} = k[A]^n[B]^m$$

where n and m may be zero or some other number. One approach to solving this type of problem is to evaluate the value of n at two different rates when [B] is constant and then to evaluate the value of m when [A] is constant. To solve for the value of n, (Experiments 2 and 3) use the ratio of rates for two differing initial concentrations of A but with the same initial concentrations of B:

$$\frac{\text{Rate}_2}{\text{Rate}_3} = \frac{k[A]^n_{\text{expt2}}[B]^m_{\text{expt2}}}{k[A]^n_{\text{expt3}}[B]^m_{\text{expt3}}}$$

Since $[B]^m_{\text{expt2}} = [B]^m_{\text{expt3}}$, you can write

$$\frac{\text{Rate}_2}{\text{Rate}_3} = \frac{k[A]^n_{\text{expt2}}[B]^m_{\text{expt2}}}{k[A]^n_{\text{expt3}}[B]^m_{\text{expt3}}} = \frac{[A]^n_{\text{expt2}}}{[A]^n_{\text{expt3}}} = \left(\frac{[A]_{\text{expt2}}}{[A]_{\text{expt3}}}\right)^n$$

From the experimental data,

$$\frac{\text{Rate}_2}{\text{Rate}_3} = \frac{2.0 \times 10^{-1}\,Ms^{-1}}{4.0 \times 10^{-1}\,Ms^{-1}} = \left(\frac{0.20\,M}{0.40\,M}\right)^m$$

or

$$0.50 = (0.50)^n$$

The latter relation is valid only if the value of n is equal to 1. Repeating this procedure to evaluate the value of m when the concentration of A is kept constant (Experiments 1 and 2):

$$\frac{\text{Rate}_1}{\text{Rate}_2} = \left(\frac{[B]_{\text{expt1}}}{[B]_{\text{expt2}}}\right)^m$$

or

$$\frac{5.0 \times 10^{-2}\,Ms^{-1}}{2.0 \times 10^{-1}\,Ms^{-1}} = \left(\frac{1.00\,M}{2.00\,M}\right)^m$$

or

$$0.25 = (0.500)^m$$

For this latter relationship to be valid, the value of m must be 2. Therefore, the rate law is

$$\text{Rate} = -\frac{\Delta[A]}{\Delta t} = k[A][B]^2$$

REACTION RATES: FIRST AND SECOND ORDER

The rate law for the first-order reaction

$$C_2H_6(g) \longrightarrow C_2H_4(g) + H_2(g)$$

is

$$\text{Rate} = -\frac{\Delta [C_2H_6]}{\Delta t} = k[C_2H_6]$$

- This is a **first-order reaction** because there is only one reactant concentration in the rate law and its concentration is raised to the first power—that is, it has an exponent of one. Only a few reactions are first order; most reactions have more complex rate laws.

- An important relationship that first-order reactions obey is

$$\ln [A]_t = -kt + \ln [A]_0$$

where $[A]_t$ is the concentration at time t of species A and $[A]_0$ is the initial concentration of species A at the start of the reaction ($t = 0$). It is important to remember that the two concentrations $[A]_t$ and $[A]_0$ must have the same concentration units.

- An important characteristic of a first-order reaction is its **half-life.** The half-life, $t_{1/2}$, is the time required for a concentration of a reactant to decrease to one-half of its original value—that is, $[A]_{t_{1/2}} = \frac{1}{2}[A]_0$. For a first-order reaction, the half-life depends only on the rate constant and is given by the relationship

$$t_{1\,2} = \frac{0.693}{k}$$

- Note that the value of $t_{1/2}$ for a first-order reaction does not depend on concentrations. This is an important property of first-order reactions.

Most reactions have rate laws with overall orders of two or more. The rate law of a higher-order reaction is often determined by proposing a rate law and then determining whether the experimental data fit it.

- A simple **second-order rate law** is a rate law with one reactant that has a reaction order of two. The general rate law for this type of second-order reaction is

$$\text{Rate} = k[A]^2$$

- A second order reaction of this type obeys the following relationship

$$\frac{1}{[A]_t} = \frac{1}{[A]_0} + kt$$

▪ The half-life for a second-order reaction is

$$t_{1/2} = \frac{1}{k[A]_0}$$

Note that the half-life is concentration dependent.

EXERCISE 7   What type of curve results when ln (concentration at time t) is plotted against t for a first-order reaction?
     Solution: The equation for a first-order reaction is

$$\ln[A]_t = -kt + \ln[A]_0$$

where $[A]_t$ is the concentration at time t and $[A]_0$ is the concentration at the start of the reaction ($t = 0$). This is the equation of a straight line, with ln $[A]_t$ as one variable and t as the other. The slope of the line is $-k$. See Figure 14.2 in the text and note the linear relationship

EXERCISE 8   The following data were obtained for the decomposition of N_2O_5 in CCl_4 at 45°C:

Initial concentration of [N_2O_5], M	Initial rate, Ms^{-1}
0.50	1.55×10^{-4}
1.0	3.10×10^{-4}
1.5	4.65×10^{-4}

(a) Show that the reaction is first order. (b) What is the value of k? (c) What is the value of the half-life for the reaction?
     Solution: (a) If it is a first-order reaction, the rate law can be written as

$$-\frac{\Delta[N_2O_5]}{\Delta t} = k[N_2O_5]$$

The initial rate should double if the concentration of N_2O_5 is doubled. Doubling the concentration of N_2O_5 from 0.50 M to 1.0 M doubles the rate from 1.55×10^{-4} M to 3.10×10^{-4} M/sec; thus the reaction is first order. (b) The value of k is calculated by averaging the k values associated with each set of data. Rearranging the rate law results in the following:

$$\text{Initial rate} = k[N_2O_5]_0 \quad \text{or} \quad k = \frac{\text{initial rate}}{[N_2O_5]_0}$$

where $[N_2O_5]_0$ is the initial concentration of N_2O_5. The calculated values of k are

$$k = \frac{1.55 \times 10^{-4}\ Ms^{-1}}{0.50\ M} = 3.1 \times 10^{-4}\ s^{-1}$$

$$k = \frac{3.10 \times 10^{-4}\ Ms^{-1}}{1.0\ M} = 3.1 \times 10^{-4}\ s^{-1}$$

$$k = \frac{4.65 \times 10^{-4}\ Ms^{-1}}{1.5\ M} = 3.1 \times 10^{-4}\ s^{-1}$$

Obviously, the average value of k is 3.1×10^{-4} s^{-1}. **(c)** The half-life of the reaction is calculated using the relation

$$t_{1/2} = \frac{0.693}{k} = \frac{0.693}{3.1 \times 10^{-4} \text{ s}^{-1}} = 2.2 \times 10^3 \text{ s}$$

EXERCISE 9   The half-life of a first-order reaction is 6.00×10^{-2} s. What is the rate constant for the reaction?
   Solution: The rate constant is calculated by using the relation $k = 0.693/t_{1/2}$. Substituting 6.00×10^{-2} s for $t_{1/2}$ yields

$$k = \frac{0.693}{6.00 \times 10^{-2} \text{ s}} = 11.6 \text{ s}^{-1}$$

EXERCISE 10   A particular reaction is either first-order or second-order. How can you determine which one it is?
   Solution: Measure the half-life of the reaction. If the reaction is first-order, its half-life will be independent of concentration; if its half-life decreases with increasing concentration, it is second-order. Alternately, you can use a graphing procedure to distinguish between the two. To determine if the rate-law expression is a first- or second-order reaction, first graph both ln [A]$_t$ and 1/[A]$_t$ against t. If the reaction is first-order, the graph of ln [A]$_t$ versus t will be a straight line, with a slope of $-k$ and an intercept of ln [A]$_0$ (see Exercise 7). On the other hand, if it is a second-order reaction, the graph of 1/[A]$_t$ against t will be a straight line with a slope of k and an intercept of 1/[A]$_0$.

FACTORS INFLUENCING REACTION RATES: TEMPERATURE AND ACTIVATION ENERGY

For both exothermic and endothermic reactions to occur, energy must be available to the reactants so that their chemical bonds can undergo rearrangements to form products. There will be no reaction unless a minimum amount of energy, referred to as **activation energy** (E_a), is provided to the reactants to overcome the energy barrier between reactants and products.

• Reactant molecules gain activation energy from their collisions with other molecules. During collisions between molecules, kinetic energy is transferred to potential energy in chemical bonds. If sufficient kinetic energy is transferred to the chemical bonds of reactant molecules–that is, an amount of potential energy equal to or greater than the activation energy for the reaction—products may form.

• Even if the reactant molecules have gained sufficient activation energy, there may be no reaction unless the molecules are properly aligned during their collisions.

• If a collision between reacting molecules is successful in initiating the reaction, the reacting molecules form a transitory intermediate termed an **activated complex** or **transition state**. An activated complex has the maximum energy of any species formed during the reaction pathway of reactants going to products.

• The rate at which an activated complex forms (which determines the rate of the reaction) depends on both the activation energy for the reaction and the number of collisions that are effective in producing products.

Svante Arrhenius determined a relationship between the rate constant for a reaction and the activation energy required for the reaction to occur. Most reaction-rate data obey the **Arrhenius equation:**

$$\ln k = \ln A - \frac{E_a}{RT}$$

where A is a constant termed the frequency factor, R is the gas constant, T is absolute temperature, k is the rate constant, and E_a is the activation energy.

- Note that as the value of activation energy increases, the value of $\ln k$ decreases, and therefore, the value of k decreases. A lower value of k corresponds to a smaller rate of reaction.

- Also, because E_a is a positive quantity, the effect of increasing temperature is an increasing value of the rate constant; consequently the rate of reaction increases.

- Another convenient form of the Arrhenius equation that relates two rate constants at two different temperatures, T_1 and T_2, is

$$\ln \frac{k_1}{k_2} = \frac{E_a}{R}\left(\frac{1}{T_2} - \frac{1}{T_1}\right)$$

This is the most useful form when working problems.

Note: E_a and R must have the same units—convert E_a to joules because $R = 8.314$ J/K-mol.

EXERCISE 11 Calculate the Arrhenius activation energy for the decomposition of N_2O_5 from the following data: at 35°C, $k_1 = 6.60 \times 10^{-5}$ s^{-1}, and at 65°C, $k_2 = 2.40 \times 10^{-3}$ s^{-1}.

Solution: Since the rate-constant data are at two different temperatures, you can use the following equation to solve for the activation energy:

$$\ln \frac{k_1}{k_2} = \frac{E_a}{R}\left(\frac{1}{T_2} - \frac{1}{T_1}\right)$$

The given information is

$$k_1 = 6.60 \times 10^{-5} \text{ s}^{-1} \qquad T_1 = 308 \text{ K}$$
$$k_2 = 2.40 \times 10^{-3} \text{ s}^{-1} \qquad T_2 = 338 \text{ K}$$
$$R = 8.314 \text{ J/K-mol}$$

Substituting this data into the Arrhenius equation yields

$$\ln \frac{6.60 \times 10^{-5} \text{s}^{-1}}{2.40 \times 10^{-3} \text{s}^{-1}} = \frac{E_a}{8.314 \text{ J/K-mol}}\left(\frac{1}{338 \text{ K}} - \frac{1}{308 \text{ K}}\right)$$

Solving for E_a gives

$$E_a = \frac{(8.314 \text{ J K-mol})(\ln 2.75 \times 10^{-2})}{\left(\dfrac{1}{338 \text{ K}} - \dfrac{1}{308 \text{ K}}\right)}$$

$$= 1.04 \times 10^5 \text{ J/mol} = 104 \text{ kJ/mol}$$

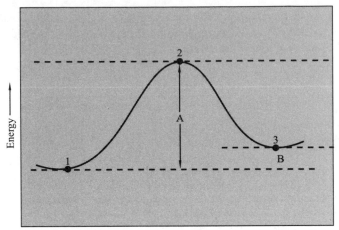

Figure 14.2 Graph of energy change during a chemical reaction.

Progress of reaction

EXERCISE 12   Figure 14.2 represents energy changes that might accompany a particular chemical reaction. Referring to this figure, answer the following questions: **(a)** Points 1, 2, and 3 correspond to the energy of which species in the reaction? **(b)** To what energy changes do A and B correspond? **(c)** Is the reaction exothermic or endothermic?

   Solution: **(a)** Point 1 corresponds to the energy of the reactants. Point 2 corresponds to the energy of the activated complex with an activation energy equal to E_a. Point 3 corresponds to the energy of the products of the reaction. **(b)** Energy change A corresponds to the activation energy, E_a. Energy change B corresponds to the net energy absorbed during the reaction. **(c)** Since energy is absorbed (as shown by the products being at a higher energy level than the reactants), this particular reaction is endothermic.

REACTION MECHANISMS

The rate law of a reaction does not depend on the stoichiometry of the reaction, but rather on the reaction mechanism. A **reaction mechanism** is a proposed sequence of elementary reactions which, when added together, result in the overall chemical equation that describes the reaction. An example of a proposed reaction mechanism containing two elementary reactions is that for the conversion of ozone into oxygen:

$$O_3(g) \longrightarrow O_2(g) + O(g) \tag{1}$$

$$O_3(g) + O(g) \longrightarrow 2O_2(g) \tag{2}$$

$$\overline{2O_3(g) \longrightarrow 3O_2(g)} \tag{3}$$

▪ Step 1 is a **unimolecular process** because only a single reactant is involved.

▪ Step 2 is **bimolecular** because the elementary reaction involves two reactant molecules.

▪ $O(g)$ is an **intermediate** because it is neither a reactant nor a product of the overall reaction but is consumed and formed in an elementary reaction.

▪ Note that when reactions (1) and (2) are added, the overall reaction (3) is obtained.

EXERCISE 13   What is the molecularity of each of the following elementary processes? Write the rate law for each.

(a) $CO(g) + Cl_2 \longrightarrow COCl_2(g)$

(b) $HClI(g) \longrightarrow HCl(g) + I(g)$

 Solution: **(a)** The molecularity of an elementary process is determined by the number of molecules that participate as reactants in an elementary step (process). In this case it is two. The rate law will be second order because it is a bimolecular process. It will also be first order in each reactant: A rate law for an elementary step is dependent only on the reactants because it is their collisions that determines the rate of the reaction

$$Rate = k[CO][Cl_2]$$

(b) Only one reactant molecule appears in the elementary process, therefore the reaction is first order in the reactant and the overall reaction order is one. The rate law is

$$Rate = k[HClI]$$

EXERCISE 14 For the reaction

$$2H_2(g) + 2NO(g) \longrightarrow N_2(g) + 2H_2O(g)$$

show that the proposed mechanism

(1) $$2NO(g) \underset{k_{-1}}{\overset{k_1}{\rightleftharpoons}} N_2O_2(g) \qquad \text{(fast)}$$

(2) $$N_2O_2(g) + H_2 \overset{k_2}{\longrightarrow} H_2O_2(g) + N_2(g) \qquad \text{(slow, rate determining)}$$

(3) $$H_2(g) + H_2O_2(g) \underset{k_{-3}}{\overset{k_3}{\rightleftharpoons}} 2H_2O \qquad \text{(fast)}$$

is consistent with the experimental rate law

(4) $$Rate = k[H_2][NO]^2$$

 Solution: If you write the rate law based only on the rate-determining step, the result is

(5) $$Rate = k_2[N_2O_2][H_2]$$

This rate law is not acceptable because N_2O_2 is an intermediate—it does not appear in the overall reaction. Therefore you need to look at one of the fast steps in the mechanism that will enable you to represent the concentration of N_2O_2 in terms of an original reactant. If you assume that the forward and reverse rates in step 1 are both fast and are approximately equal in value, then you can write

$$Rate_1 = k_1 [NO]^2 = rate_{-1} = k_{-1} [N_2O_2]$$

Solving for $[N_2O_2]$ yields

(6) $$[N_2O_2] = \frac{k_1[NO]^2}{k_{-1}}$$

Substituting the expression for $[N_2O_2]$ in equation (6) into equation (5) gives

$$Rate = k_2 \left(\frac{k_1[NO]^2}{k_{-1}} \right)[H_2] = \frac{k_2k_1}{k_{-1}} [H_2][NO]^2$$

If you define

$$k = \frac{k_2k_1}{k_{-1}}$$

then you have the experimentally observed rate law

$$\text{Rate} = k[H_2][NO]^2$$

CATALYSTS

Wilhelm Ostwald was the first to define a **catalyst** in terms of its influence on the rate of a chemical reaction: "A catalyst is a substance that changes the velocity of a chemical reaction without itself appearing in the end products."

- A catalyst increases the rate of a chemical reaction by providing a new reaction path with a lower activation energy.
- The concentration of a catalyst may appear in the rate law for a reaction.
- **Homogeneous catalysts** have the same phase as the reaction medium.
- A catalyst in a different phase than the reaction mixture is a **heterogeneous catalyst.** Heterogeneous catalysis usually involves the adsorption of a reactant molecule onto the surface of the heterogeneous catalyst at an active site.

EXERCISE 15   The primary process for the manufacturing of H_2SO_4 is the contact process. Sulfur is oxidized to $SO_2(g)$ in the presence of $O_2(g)$. $SO_2(g)$ is then oxidized to $SO_3(g)$ in the presence of $O_2(g)$ and the catalyst $V_2O_5(s)$ at 400°C. The $SO_3(g)$ is then reacted with $H_2O(l)$ to form $H_2SO_4(l)$. Is the oxidation of SO_2 to SO_3 an example of homogeneous or heterogeneous catalysis?

Solution: In homogeneous catalysis, the catalyzed reaction occurs in one phase; in heterogeneous catalysis, the catalyzed reaction occurs at a phase interface. The catalyzed reaction in this problem involves $SO_2(g)$, the solid catalyst V_2O_5. The catalysis occurs at the surface of the solid V_2O_5 and is thus an example of heterogeneous catalysis.

EXERCISE 16   The hydrolysis of sucrose (table sugar) in an acid solution is described by the chemical equation

$$\underset{\text{Sucrose}}{C_{12}H_{22}O_{11}(aq)} + H_2O(l) \longrightarrow \underset{\text{Glucose}}{C_6H_{12}O_6(aq)} + \underset{\text{Fructose}}{C_6H_{12}O_6(aq)}$$

The rate law for the reaction is

$$-\frac{\Delta[C_{12}H_{22}O_{11}]}{\Delta t} = k[C_{12}H_{22}O_{11}][H^+]$$

Why does $[H^+]$ appear in the rate law?

Solution: The H^+ ion must be involved in some step in the reaction mechanism because it appears in the rate law. The H^+ ion is a catalyst because it affects the rate but does not appear in the stoichiometric equation.

SELF-TEST QUESTIONS

Key Terms

Having reviewed the key terms listed at the end of Chapter 14 and their applications, identify the following statements as true or false. If a statement is false, indicate why it is incorrect. Key terms are italicized in the statements.

14.1   The expression

$$\text{Rate} = k[Br_2]$$

for the gaseous reaction

$$Br_2 \longrightarrow 2Br$$

is called a *rate law.*

14.2   From the expression

$$\text{Rate} = k[Br_2]$$

we can determine that for the reaction

$$Br_2 \longrightarrow 2Br$$

the *reaction order* is two.

14.3 The term k in the expression

$$Rate = k[Br_2]$$

is a *rate constant*.

14.4 The *reaction rate* for the gaseous reaction $Br_2 \rightarrow 2Br$ can be written as

$$-\frac{\Delta[Br_2]}{\Delta t}$$

14.5 For the gaseous reaction

$$2N_2O_5 \longrightarrow 4NO_2 + O_2$$

the rate law is

$$Rate = k[N_2O_5]$$

This reaction is an example of a *first-order reaction*.

14.6 If the rate constant for a first-order reaction is 1×10^{-3}/sec, the *half-life* of the reaction is 1.44×10^{-3}/sec. The half-life of any reaction is also independent of initial concentrations.

14.7 The *activated complex* for the reaction between $NO(g)$ and $O_3(g)$ might be of the form

It would form at the point of minimum energy in the rate-determining step.

14.8 A reaction that has a very large rate constant associated with its rate law would be expected to have a low *activation energy*.

14.9 According to the *Arrhenius equation,* as the temperature of a reaction increases, the value of the rate constant also increases.

14.10 The following is a proposed *reaction mechanism* for the gaseous reaction $2NO + 2H_2 \longrightarrow N_2 + H_2O$:

$$2NO + H2 \longrightarrow N_2 + H_2O_2 \quad \text{(slow)}$$
$$H_2O_2 + H_2 \longrightarrow 2H_2O \quad \text{(fast)}$$

Since this mechanism can be used to explain the experimentally observed rate law, this is sufficient proof of the validity of the proposed mechanism.

14.11 The first step in the proposed mechanism in question 14.10 is called the *rate-determining step* because it is the slowest step in the proposed mechanism.

14.12 A substance that increases the activation energy of a reaction and is not permanently changed in the process is called a *catalyst*.

14.13 Carbonic anhydrase catalyzes the decomposition of HCO_3^- to CO_2 in the body. Thus it is called an *enzyme*.

14.14 $HCl(aq)$ catalyzes the hydrolysis of aqueous sucrose. Therefore, HCl is an example of a *heterogeneous catalyst*.

14.15 Solid MnO_2 causes the rate of decomposition of molten $KClO_3$ to increase. MnO_2 is an example of a *homogeneous catalyst*.

14.16 The rate of decomposition of $NO(g)$ to yield $N_2O(g)$ and $NO_2(g)$ at 100 atm of pressure follows a third-order rate law:

$$Rate = k[NO]^3$$

From the rate law we can conclude that the decomposition of $NO(g)$ requires a *termolecular* reaction.

14.17 A plausible reaction mechanism for the decomposition of $N_2O(g)$ to $N_2(g)$ and $O_2(g)$ is

$$N_2O(g) \longrightarrow N_2(g) + O(g) \quad \textbf{(a)}$$
$$N_2O(g) + O(g) \longrightarrow N_2(g) + O_2(g) \quad \textbf{(b)}$$

$N_2(g)$ is an *intermediate* in the reaction.

14.18 Reaction **(a)** shown in problem 14.17 is an example of an *elementary* reaction.

14.19 Reaction **(a)** shown in problem 14.17 is a *bimolecular* reaction.

14.20 Neither reaction **(a)** nor reaction **(b)** in problem 14.17 is an example of a *unimolecular* reaction.

14.21 The *chemical kinetics* of a reaction gives us information about the spontaneous direction of a reaction.

14.22 For a first-order reaction, the *instantaneous rate* increases with time of reaction.

14.23 The rate law for the reaction of NH_4^+ with NO_2^- is first order in each reactant. Therefore, the *overall order* is two.

14.24 The magnitude of the *frequency factor* in the Arrhenius equation is related to the probablity that collisions are favorably oriented for reaction.

14.25 According to the *collision model,* reaction rates do not vary significantly with changes in the number of reactant molecules.

14.26 The *transition state* for a chemical reaction is the energy barrier between the starting molecule and the highest energy along the reaction pathway.

14.27 The *molecularity* for the following elementary step in a reaction mechanism

$$NO_2(g) + NO_2(g) \longrightarrow NO_3(g) + NO(g)$$

is four, the number of molecules reacting and formed.

14.28 *Adsorption* refers to the uptake of molecules into the interior of another substance.

14.29 A particular catalyst with a large number of *active sites* will be more effective than one with fewer.

14.30 A *substrate* may change the structural features around an active site when it binds.

14.31 In the *lock-and-key model* for the specificity of enzymes, the substrate is the key.

Problems and Short-Answer Questions

14.32 The following reaction occurs in mildly acidic solution:

$$2HCrO_4^-(aq) + 8H^+(aq) + 3As(OH)_3(s) \rightleftharpoons$$
$$2Cr^{3+}(aq) + 3H_3AsO_4(aq) + 5H_2O(l)$$

A study of reaction-rate data for this reaction shows that when the concentration of hydrogen ions is doubled the rate of the reaction increases by a factor of 8. When the concentration of either $HCrO_4^-$ or $As(OH)_3$ is doubled, the rate doubles. Write the rate law for the above reaction.

14.33 The rate law for the reaction

$$2NO(g) + H_2(g) \longrightarrow N_2(g) + 2H_2O(g)$$

is

$$Rate = \frac{\Delta[N_2]}{\Delta t} = k[NO]^2[H_2]$$

How is the initial rate affected by each of the following changes:

(a) the concentration of $NO(g)$ is doubled;
(b) a catalyst is present;
(c) the temperature is decreased;
(d) the volume of the reaction container is doubled?

14.34 Given the following kinetic data for the reaction

$$2A + B + C \longrightarrow 2D$$

$[A]_{initial}(M)$	$[B]_{initial}(M)$	$[C]_{initial}(M)$	Initial rate $(M\ s^{-1})$
0.1	0.1	0.1	x
0.2	0.1	0.1	$4x$
0.2	0.2	0.1	$32x$
0.2	0.1	0.2	$8x$

(a) Determine the rate law as a function of the concentrations of A, B, and C.
(b) What is the overall order of the reaction?

14.35 Given the following data for the reaction

$$CF_4(g) + H_2(g) \longrightarrow CHF_3(g) + HF(g)$$

determine the rate law for the reaction.

$[CF_4]_{initial}(M)$	$[H_2]_{initial}(M)$	Rate $(M\ s^{-1})$
0.10	0.10	45
0.15	0.10	67.5
0.20	0.20	180
0.20	0.30	270

14.36 An alteration in the structure of a certain virus follows first-order kinetics with an activation energy of 586 kJ/mol. The half-life of the reaction at 29.6°C is 1.62×10^4 s (1 yr = 3.154×10^7 s). What are the rate constant at 29.6°C and the half-life at 32.0°C for the alteration of structure of the virus?

14.37 The nuclear decay of $^{131}_{53}I$ to $^{131}_{54}Xe$ follows first-order kinetics and has a half-life of 8 days. Calculate the time required for 90 percent of $^{131}_{53}I$ to decay.

14.38 The decomposition of $CrCl^{2+}$ in water occurs as follows:

$$CrCl^{2+} \longrightarrow Cr^{3+} + Cl^-$$

The process is first order, and the rate constant for the reaction is 2.85×10^{-7} s^{-1}. If Cl^- is removed from the reaction as it forms so that it cannot recombine with Cr^{3+} and if 10 g of $CrCl^{2+}$ is present initially in 1 L of solution, how many grams of $CrCl^{2+}$ are left after 10 hr?

14.39 The rate law for the reaction

$$2NO(g) + O_2(g) \longrightarrow 2NO_2(g)$$

is

$$-\frac{\Delta[O_2]}{\Delta t} = k[NO]^2[O_2]$$

The proposed mechanism is

$$NO(g) + O_2(g) \longrightarrow NO_3(g)$$
$$NO(g) + NO_3(g) \longrightarrow 2NO_2(g)$$

(a) The reaction mechanism does not include a collision among three molecules. Since the reaction order is three, shouldn't the mechanism include such a collision?
(b) Which step in the proposed mechanism is the slowest step?
(c) What is the molecularity of each step?

14.40 The proposed mechanism for the catalytic oxidation of V^{3+} in the presence of Cu^{2+} and Fe^{3+} is

$$V^{3+} + Cu^{2+} \longrightarrow V^{4+} + Cu^+ \quad \text{(slow)}$$
$$Cu^+ + Fe^{3+} \longrightarrow Cu^{2+} + Fe^{2+} \quad \text{(fast)}$$

(a) Write the overall reaction equation.
(b) Which ion acts as a catalyst?
(c) What is the rate law for the overall reaction?

14.41 The decomposition of $HI(g)$ to $H_2(g)$ and $I_2(g)$ is catalyzed by platinum. The observed rate equation is

$$Rate = k[HI][\text{surface area of Pt}]$$

(a) Is this an example of homogeneous or heterogeneous catalysis?
(b) Why is the surface area of Pt in the rate law?

Multiple-Choice Questions

14.42 Given that a reaction is exothermic and has an activation energy of 50 kJ/mol, which of the following statements are correct: (1) the reverse reaction has an activation energy equal to 50 kJ/mol; (2) the reverse reaction has an activation energy less than 50 kJ/mol; (3) the reverse reaction has an activation energy greater than 50 kJ/mol; (4) the reaction rate increases with temperature; (5) the reaction rate decreases with temperature?

(a) (1) and (4) (d) (2) and (5)
(b) (2) and (4) (e) (3) and (5)
(c) (3) and (4)

14.43 A catalyst increases the rate of a reaction by doing which of the following? (a) Increasing reactant concentrations; (b) increasing temperature; (c) decreasing temperature; (d) increasing activation energy of reaction; (e) decreasing activation energy of reaction.

14.44 Given the following mechanism for a reaction:

$$Cl_2(g) \longrightarrow 2Cl(g)$$
$$2NO(g) + 2Cl(g) \longrightarrow N_2(g) + 2ClO(g)$$
$$2ClO(g) \longrightarrow Cl_2(g) + O_2(g)$$

which of the following is a catalyst in the reaction?

(a) Cl_2 (d) ClO
(b) N_2O_5 (e) O_2
(c) N_2

14.45 What can we say about catalysts, given the following information?

$$CO(g) + 3H_2(g) \xrightarrow{\text{Ni catalyst}} CH_4(g) + H_2O(g)$$
$$CO(g) + 2H_2(g) \xrightarrow[\text{catalyst}]{\text{ZnO Cr}_2O_3} CH_3OH(g)$$

(a) Catalysts are nonspecific in their activity.
(b) Catalysts are highly specific in their activity.
(c) Ni is a better catalyst than ZnO/Cr_2O_3.
(d) ZnO/Cr_2O_3 is a better catalyst than Ni.
(e) Most metals can act as catalysts.

14.46 What are the units of k for the rate law

$$\text{Rate} = k[A][B]^2$$

when the concentration unit is mol/L?

(a) s^{-1} (d) $L^2 \, mol^{-2} \, s^{-1}$
(b) s (e) $L^2\text{-}s^2 \, mol^{-2}$
(c) $L \, mol^{-1} \, s^{-1}$

14.47 The burning of paper is an exothermic process. However, for paper to burn, it must be brought up to a temperature equal to or greater than its kindling temperature. Once the required activation energy has been provided, which of the following will cause the paper to continue burning?

(a) Energy absorbed from the atmosphere;
(b) the heat released during the reaction;
(c) paper molecules colliding together;
(d) low activation energy of paper;
(e) none of the above.

14.48 A proposed mechanism for a reaction is

$$CH_3OH + HCl \longrightarrow CH_3^+ + H_2O + Cl^-$$
$$CH_3^+ + Cl^- \longrightarrow CH_3Cl$$

What is the overall reaction conforming to this proposed mechanism?

(a) $CH_3Cl + H_2O \longrightarrow CH_3OH + HCl$
(b) $CH_3OH + HCl \longrightarrow CH_3Cl + H_2O$
(c) $CH_3Cl \longrightarrow CH_3^+ + Cl^-$
(d) $CH_3^+ + Cl^- \longrightarrow CH_3Cl$
(e) $HCl \longrightarrow H^+ + Cl^-$

14.49 Given the energy-profile diagram in Figure 14.3, which of the following phrases could correctly describe the kind of reaction shown: (1) an exothermic reaction; (2) an endothermic reaction; (3) a reaction with two steps in its reaction mechanism; (4) a reaction with three steps in its reaction mechanism; (5) a catalyzed reaction involving two successive steps?

(a) (1) and (3)
(b) (2) and (4)
(c) (1), (3), and (5)
(d) (2), (4), and (5)
(e) (1), (2), (3), (4), and (5)

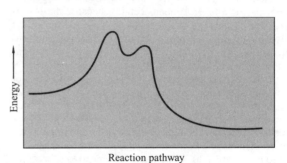

Reaction pathway

Figure 14.3 An energy-profile diagram.

14.50 Figure 14.4 is an energy-profile diagram for two reactions, A and B. Which of the following statements about the reactions are correct: (1) reaction B is slower than reaction A; (2) reaction A is slower than reaction B; (3) the reverse reaction of B has a higher activation energy than that for the reverse reaction of A; (4) the reverse reaction of A has a higher activation energy than that for the reverse reaction of B; (5) the energy-profile diagram for reaction B could represent a catalyzed reaction A?

(a) (2) and (3)   (d) (1), (4), and (5)
(b) (2) and (4)   (e) (2), (3), and (5)
(c) (1) and (3)

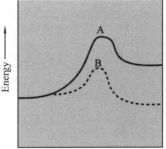

Reaction pathway

Figure 14.4 An energy-profile diagram for two reactions.

14.1 True. **14.2** False. The reaction order is one because the exponent associated with $[Br_2]$ is one. **14.3** True. **14.4** True. **14.5** True. **14.6** False. $t_{1/2} = 0.693/k = 0.693/(1 \times 10^{-3}$ s$^{-1}) =$ 693 s. The second part of the statement is true. **14.7** False. An activated complex forms at the *maximum* energy point in a rate-determining step. **14.8** True. **14.9** True. For elementary reactions that we are studying, E_a is always positive, and thus the statement is true. No reaction has an associated negative E_a value. A few free radical reactions appear to have $E_a = 0$. **14.10** False. A proposed mechanism must be proved by observation of intermediates to show that they indeed form. **14.11** True. **14.12** False. A catalyst *lowers* the activation energy for a reaction. **14.13** True. **14.14** False. It is a homogeneous catalyst because the phases of HCl and sucrose are the same. **14.15** False. It is a heterogeneous catalyst because the phases of MnO_2 and $KClO_3$ are different. **14.16** False. A third-order rate law by itself does not prove that the reaction is termolecular. In fact, reaction among three NO molecules in the gas phase is highly unlikely. **14.17** False. An intermediate is a species in the reaction mechanism that is neither a reactant nor a product of the overall reaction. Thus $N_2(g)$ is not an intermediate because it is a product of the overall reaction. $O(g)$ is an example of an intermediate. **14.18** True. **14.19** False. It is a unimolecular reaction because in the reaction only one molecule of reactant is shown to form products. **14.20** False. See answer to problem 14.20. **14.21** False. Chemical kinetics gives us information about the speeds or rates of chemical reactions, not the most likely direction. **14.22** False. Instantaneous rate is greatest at the start of a chemical reaction. **14.23** True. **14.24** True. **14.25** False. Reaction rates increase with increasing number of reactant molecules. **14.26** False. Transition state refers to the chemical species that forms at the highest entry along the reaction pathway. **14.27** False. It is two, the number of molecules reacting in the elementary step. **14.28** False. It refers to the binding of molecules to the surface of another species. **14.29** True. **14.30** True. **14.31** True. **14.32** The given data state that the rate of the reaction depends on the concentrations of H^+, $HCrO_4^-$, and $As(OH)_3$. Thus, the rate law will have the form Rate = $k[H^+]^n[HCrO_4^-]^m[As(OH)_3]^l$. $[HCrO_4^-]^m$ and $[As(OH)_3)]^l$ have the same exponential values because doubling each concentration doubles the rate—that is, $m = l = 1$ since 2(concentration) = 2(rate). The exponent value for $[H^+]$ must reflect the fact that $(2 \times$ concentration$)^n = 2^n($concentration$)^n = 2^n$ (rate) = 8(rate)—that is, $2^n = 8$, or $n = 3$. Thus the rate law is Rate = $k[H^+]^3[HCrO_4^-][As(OH)_3]$ **14.33** **(a)** The rate is 4 times greater because the concentration of NO is squared. **(b)** The rate is increased. **(c)** The rate decreases. **(d)** The reduction in volume decreases the concentration of NO and H_2; thus the rate is reduced. **14.34 (a)** Doubling the concentration of A increases the rate by 4; thus, in the rate law A must be squared. Doubling the concentration of B increases the rate by a factor of 8; thus, B must have a power of 3 $(2^3 = 8)$ in the rate law. Doubling the concentration of C doubles the rate; thus, C must have a power of 1 in the rate law. The rate law is $-\Delta[B]/\Delta t = k[A]^2[B]^3[C]$. **(b)** The overall order of the reaction is 6(2 + 3 + 1). **14.35** Assuming that the rate law has the form Rate = $k[CF_4]^n[H_2]^m$, we will use the approach in Exercise 6 to solve this problem:

$$\frac{\text{Rate}_1}{\text{Rate}_2} = \frac{k[0.10]^n[0.10]^m}{k[0.15]^n[0.10]^m} = \left(\frac{0.10}{0.15}\right)^n$$

$$\frac{45\ M\ \text{s}^{-1}}{67.5\ M\ \text{s}^{-1}} = 0.667 = (0.667)^m$$

Therefore $n = 1$.

$$\frac{\text{Rate}_3}{\text{Rate}_4} = \frac{k[0.20][0.20]^m}{k[0.20][0.30]^m}$$

$$= \frac{180\ M\ \text{s}^{-1}}{270\ M\ \text{s}^{-1}} = 0.667 = (0.667)^m$$

Therefore $m = 1$. The rate law is Rate = $k[CF_4][H_2]$. **14.36** The rate constant at 29.6°C is calculated using the relation $k_1 = 0.693/t_{1/2} = 0.693/1.62 \times 10^4$ s $= 4.28 \times 10^{-5}$ s^{-1}. The half-life at 32.0°C is calculated from the rate constant at 32.0°C, which can be calculated using the Arrhenius equation. E is given as 587 kJ/mol, but A is not given. However, k_2 at 32.0°C can be calculated without knowledge of the value of A using the equation

$$\ln\left(\frac{k_1}{k_2}\right) = \frac{E_a}{R}\left(\frac{1}{T_2} - \frac{1}{T_1}\right)$$

$$\ln\left(\frac{4.28 \times 10^{-5}/\text{sec}}{k_2}\right) = \frac{(587\ \text{kJ/mol})(1000\ \text{J/kJ})}{(8.314\ \text{J/K-mol})}$$

$$\times\left(\frac{1}{305.0\ \text{K}} - \frac{1}{302.6\ \text{K}}\right)$$

Solving for k_2 gives $k_2 = 2.68 \times 10^{-4}$ s^{-1}. Thus, the half-life for the alteration of the virus at 32.0°C is $t_{1/2} = 0.693/k_2 = 0.693/(2.68 \times 10^{-4}$ s$^{-1}) = 2.58 \times 10^3$ s. **14.37** The time required for 90 percent of ^{131}I to decay can be calculated using the equation for a first-order process: $\ln [A]_t = -kt + \ln [A]_0$ where $[A]_t$ is the concentration at time t, and $[A]_0$ is the concentration at $t = 0$. Thus

$$k = \frac{0.693}{t_{1/2}}$$

$$= \frac{0.693}{(8\ \text{days})(24\ \text{hr/day})(3600\ \text{sec/hr})}$$

$$= 1.00 \times 10^{-6}\text{s}^{-1}$$

When 90 percent of ^{131}I has decayed, there is 10 percent of the original amount left, or $[A]_t = 0.10\ [A]_0$. And $\ln [A]_t - \ln [A]_0 = -kt$ or $\ln ([A]_t/[A]_0) = -kt$ or $\ln (0.10[A]_0/[A]_0) = \ln 0.10 = -kt$. Solving for t: $t = $ s or 26.5 days. **14.38** In the answer to the previous problem, we derived the relation $\ln ([A]_t/[A]_0) = -kt$. In this problem we know $[A]_0$, k, and t, are asked to solve for $[A]_t$. Since the ln expression contains a ratio of concentrations, you can use a ratio of gram quantities because the volume and molar mass of $CrCl^{2+}$ cancel.

$$\ln\left(\frac{[A_t]}{10.00 \text{ g}}\right) = (-2.85 \times 10^{-7} \text{s}^{-1})(10 \text{ hr})$$

$$\times \left(\frac{60 \text{ min}}{1 \text{ hr}}\right)\left(\frac{60 \text{ sec}}{1 \text{ min}}\right)$$

$$= -0.0103$$

Solving for $[A_t]$ yields $[A_t]$ = 9.90 g. After 10 hr, only 0.10 g of $CrCl^{2+}$ has decomposed. **14.39 (a)** No, A collision among three molecules is not likely. The $[NO]^2$ term results because the slowest step in the mechanism depends on [NO] and some other concentration term that is directly related to $[NO][O_2]$. **(b)** If the first elementary step were the rate determining step, the rate law would be $k[NO][O_2]$. This does not agree with the observed rate law. Therefore, the second elementary step must be the rate determining step. You can show this by using the approach in Exercise 14. **(c)** Both steps are bimolecular. **14.40 (a)** Adding the two steps in the mechanism results in the overall equation $V^{3+} + Fe^{3+} \rightarrow V^{4+} + Fe^{2+}$. **(b)** Cu^{2+} is the catalyst, because it is used in the slow step and reformed in the fast step. **(c)** The rate law is Rate = $k[V^{3+}][Cu^{2+}]$ from the stoi-chiometry of the slow step. **14.41 (a)** It is an example of het-erogeneous catalysis, because the catalysis of HI(g) occurs at a solid-gas interface—the surface of solid platinum and the gases. **(b)** A surface Pt atom absorbs an HI molecule, enabling the formation of $H_2 + I_2$. The rate of the reaction depends on the number of surface Pt atoms, that is, the surface area of Pt available for adsorbing HI molecules. Thus, the surface area of Pt is involved in the slowest step and occurs in the rate law. **14.42 (c). 14.43 (e). 14.44 (a).** A catalyst is regenerated during a reaction. $Cl_2(g)$ catalyzes the decomposition of NO(g). **14.45 (b).** Changing the catalyst changed the nature of products; thus catalysts are specific in the kinds of products they help form.

14.46 (d). $k = \dfrac{(M \text{ s}^{-1})}{(M)(M)^2} = \dfrac{1}{(M)^2(s)}$

$$= \frac{L^2}{(\text{mol})^2(s)} = L^2\text{mol}^{-2}\text{s}^{-1}$$

14.47 (b). The energy released causes other parts of the paper to reach kindling temperature. **14.48 (b). 14.49 (c). 14.50 (a).**

15 | Chemical Equilibrium

OVERVIEW OF THE CHAPTER

15.1, 15.2, 15.3, 15.4 THE EQUILIBRIUM STATE: K_p AND K_C

Review: Gas laws (10.3, 10.4); gas-law constant (10.4); stoichiometric equivalences (3.8); concept of dynamic equilibrium (11.4).

Objectives: You should be able to:

1. Write the equilibrium expression for a balanced chemical equation, whether heterogeneous or homogeneous.

2. Numerically evaluate K_c (or K_p) from a knowledge of the equilibrium concentrations (or pressures) of reactants or products, or from the initial concentrations and the equilibrium concentration of at least one substance.

3. Interconvert K_c and K_p.

15.3, 15.4, 15.5 CALCULATING EQUILIBRIUM CONCENTRATIONS

Review: Stoichiometric relationships (3.6).

Objectives: You should be able to use the equilibrium constant to calculate equilibrium concentrations.

15.6 REACTION QUOTIENT AND LE CHATELIER'S PRINCIPLE: DIRECTION OF A CHEMICAL REACTION

Review: Enthalpy (5.3, 5.4).

Objectives: You should be able to:

1. Calculate the reaction quotient, Q, and by comparison with the value of K_c or K_p determine whether a reaction is a equilibrium. If it is not at equilibrium, you should be able to predict in which direction it will shift to reach equilibrium.

2. Explain how the relative equilibrium quantities of reactants and products are shifted by changes in temperature, pressure, or the concentrations of substances in the equilibrium reaction.

3. Explain how the change in equilibrium constant with change in temperature is related to the enthalpy change in the reaction.

Review: Catalysis (14.6); Rates of chemical reactions (14.1, 14.2).

15.6 CATALYSIS
AND EQUILIBRIUM

Objective: You should be able to describe the effect of a catalyst on a system as it approaches equilibrium.

TOPIC SUMMARIES AND EXERCISES

At constant temperature, a mixture of $H_2(g)$ and $I_2(g)$ in a closed container reacts to form $HI(g)$ until the partial pressures of $H_2(g)$, $I_2(g)$, and $HI(g)$ are constant with time. If the container originally contains $HI(g)$ instead of $H_2(g)$ and $I_2(g)$, $HI(g)$ decomposes until the partial pressures of $H_2(g)$, $I_2(g)$, and $HI(g)$ are constant with time. This reaction is an example of a reversible reaction that leads to an equilibrium state.

**THE EQUILIBRIUM
STATE: K_p AND K_c**

▪ A state of **chemical equilibrium** exists when both forward and reverse reactions are proceeding at equal rates and concentrations of reactants and products, pressure, volume, and temperature do not change with time.

▪ Double arrows between reactants and products in a chemical reaction show that an equilibrium state exists.

Another statement of the equilibrium condition is the **law of mass action.**

▪ This law states that if an equilibrium condition exists for a reaction then there also exists for the reaction a constant K, called the **equilibrium constant.** Its value depends on a ratio concentrations of products to concentrations of reactants at equilibrium. More specifically, the equilibrium constant for the general reaction

$$jA + kB \rightleftharpoons pR + qS$$

has the form

Stoichiometric coefficient
in chemical equation

$$\text{Equilibrium constant} = K = \frac{[R]^p[S]^q}{[A]^j[B]^k} \quad \begin{array}{l} \leftarrow \text{products} \\ \leftarrow \text{reactants} \end{array}$$

▪ In the equilibrium-constant expression, the concentration of each substance is raised to a power equal to its coefficient in the balanced chemical equation.

The phase change of solid water to liquid water at constant temperature is an example of **heterogeneous equilibrium**—that is, a reversible reaction involving substances in two or more phases.

▪ At constant temperature, the value of the equilibrium constant for a heterogeneous system depends only on the equilibrium concentrations of solutes and gases. The concentrations of pure liquids and pure solids are constant and do not affect the position of equilibrium.

• *Thus, the concentration of a pure solid or liquid is not shown in the equilibrium-constant expression for a heterogeneous equilibrium.* For example, the equilibrium-constant expression for

$$PbBr_2(s) \rightleftharpoons Pb^{2+}(aq) + 2Br^-(aq)$$

is

$$K = [Pb^{2+}][Br^-]^2$$

Two kinds of concentration units are encountered when measuring quantities of gases: Molarity and atmospheres of pressure.

• When all the concentration units are expressed as molarity and used to evaluate an equilibrium-constant value, the equilibrium constant is labeled K_c.

• When all the concentration units are expressed as atmospheres of pressure, we write K_p.

• In general, the values of K_p and K_c for the same gaseous equilibrium are different. A general relationship between the two is

$$K_p = K_c(RT)^{\Delta n}$$

where Δn equals the number of moles of gas products minus the number of moles of gas reactants for the reaction being considered. The number of moles of liquid and solid present in the equilibrium are not included in the calculation of Δn. Only in the case in which Δn equals zero does K_p equal K_c.

EXERCISE 1   Write equilibrium-constant expressions for the following reversible reactions:

(a) $H_2(g) + I_2(g) \rightleftharpoons 2HI(g)$
(b) $2Cl_2(g) + 2H_2O(g) \rightleftharpoons 4HCl(g) + O_2(g)$
Solution:

(a) $K = \dfrac{[HI]^2}{[H_2][I_2]}$

Remember that K involves a ratio of the concentrations of products to reactants. Also, the concentrations of substances may be raised to a power greater than one. The concentration of HI is raised to the second power because there is a two in front of HI(g) in the balanced equation. Similarly, [H$_2$] and [I$_2$] both have exponents of one.

(b) $K = \dfrac{[HCl]^4[O_2]}{[Cl_2]^2[H_2O]^2}$

Again, the exponent of each substance in the equilibrium expression is raised to the same coefficient in front of the substance in the balanced equation.

EXERCISE 2   How is the equilibrium-constant expression for the reaction

$$2HI(g) \rightleftharpoons H_2(g) + I_2(g)$$

related to the one for the reverse reaction:

$$H_2(g) + I_2(g) \rightleftharpoons 2HI(g)$$

Solution: For the reaction

$$2HI(g) \rightleftharpoons H_2(g) + I_2(g)$$

the associated equilibrium constant is

$$K' = \frac{[H_2][I_2]}{[HI]^2}$$

The equilibrium constant for the reverse reaction is

$$K = \frac{[HI]^2}{[H_2][I_2]} = \frac{1}{K'}$$

Note that the first equilibrium-constant expression is the reciprocal (inverted form) of the second one. The equilibrium-constant expression for any reversed reaction is simply the reciprocal of the equilibrium-constant expression for the original reaction. There fore the numerical values are also the reciprocal of one another.

EXERCISE 3 Write the equilibrium-constant expression for each of the following heterogeneous equilibria:

    (a) $H_2(g) + I_2(s) \rightleftharpoons 2HI(g)$
    (b) $H_2(g) + Br_2(l) \rightleftharpoons 2HBr(g)$
    (c) $(CH_3)_4Sn(s) \rightleftharpoons (CH_3)_4Sn(g)$
    (d) $Ag(CN)_2^-(aq) + AgI(s) \rightleftharpoons I^-(aq) + 2AgCN(s)$
    Solution: Remember that only gases and solutes are shown in the equilibrium-constant expression.

(a) $K = \dfrac{[HI]^2}{[H_2]}$

The concentration of the pure solid I_2 is included in the value of K.

(b) $K = \dfrac{[HBr]^2}{[H_2]}$

Similarly, the concentration of liquid Br_2 is included in the value of K.

(c) $K_p = P_{(CH_3)_4Sn}$

The equilibrium constant K_p equals the partial pressure of gaseous $(CH_3)_4Sn$.

(d) $K = \dfrac{[I^-]}{\left[Ag(CN)_2^-\right]}$

The concentrations of the two solids involved in this equilibrium are incorporated into the equilibrium constant.

EXERCISE 4   A 1.000-L container contains an equilibrium mixture of $H_2(g)$, $I_2(g)$, and $HI(g)$ at 721 K. The number of moles of each substance present is 1.843×10^{-3}, 1.843×10^{-3}, and 1.310×10^{-2}, respectively. Calculate the values of K_c and K_p for the reaction

$$H_2(g) + I_2(g) \rightleftharpoons 2HI(g)$$

Solution: Since the container has a volume of 1.000 L, the molarity of each species is equivalent to the number of moles of each species present. Thus the molarity of H_2 is 1.843×10^{-3}; the molarity of I_2 is 1.843×10^{-3}; and the molarity of HI is 1.310×10^{-2}. Substituting these values into the following equilibrium-constant expression yields the value of K_c:

$$K_c = \frac{[HI]^2}{[H_2][I_2]}$$

$$= \frac{\left(1.310 \times 10^{-2}\right)^2}{\left(1.843 \times 10^{-3}\right)\left(1.843 \times 10^{-3}\right)}$$

$$= 50.52$$

The value of K_p is calculated by using the relation $K_p = K_c(RT)^{\Delta n}$. In this problem Δn equals the number of moles of gas products minus the number of moles of gas reactants: $2 - (1 + 1) = 0$. Therefore

$$K_p = K_c(RT)^0 = K_c(1) = K_c = 50.52$$

EXERCISE 5 N_2O_4, a colorless gas, and NO_2, a brown gas, exist in equilibrium as follows:

$$2NO_2(g) \rightleftharpoons N_2O_4(g)$$

A closed container at 25°C is charged with N_2O_4 at a partial pressure of 0.51 atm and NO_2 at a partial pressure of 0.56 atm. At equilibrium the partial pressure of N_2O_4 is found to be 0.54 atm. What is K_p for the equilibrium?

Solution: First, write the equilibrium expression for the problem:

$$K_p = \frac{P_{N_2O_4}}{P_{NO_2}^2}$$

To solve for the equilibrium constant, you need to determine the equilibrium partial pressures of both gases. You are given the initial partial pressures and the equilibrium partial pressure of N_2O_4. You can determine the equilibrium partial pressures from the stoichiometry of the equilibrium as follows: (1) Write the equilibrium and underneath it develop a table of concentrations that shows (2) initial conditions, (3) the change that occurs when the reaction proceeds to equilibrium, and (4) the equilibrium concentrations determined from either the given information or from the changes in partial pressures.

		$2NO_2$	$\rightleftharpoons$	N_2O_4
(1)				
(2)	Initial pressure (atm):	0.56		0.51
(3)	Change in pressure (atm):	?		?
(4)	Equilibrium pressure (atm):	?		0.54

At this point, you do not know the information for the second line and the equilibrium partial pressure of NO_2. However, you can determine the change in pressure from the

N_2O_4 data. You can calculate the change in N_2O_4 pressure from the difference in equilibrium pressure and initial pressure: 0.54 atm − 0.51 atm = +0.03 atm. The change in NO_2 pressure is a negative quantity because N_2O_4 gained pressure; it is calculated from this change and the stoichiometry:

$$\text{Change in pressure of } NO_2 = -\left(0.03 \text{ atm } N_2O_4\right)\left(\frac{2 \text{ mol } NO_2}{1 \text{ mol } N_2O_4}\right) = -0.06 \text{ atm}$$

Thus, the equilibrium partial pressure of NO_2 is the sum of the initial partial pressure and the change in partial pressure, 0.56 atm + (−0.06 atm) = 0.50 atm. The table of concentrations now becomes:

	$2NO_2 \rightleftharpoons N_2O_4$	
Initial pressure (atm)	0.56	0.51
Change in pressure (atm)	−0.06	+0.03
Equilibrium pressure (atm)	0.50	0.54

You can now solve for K_p by substituting the equilibrium partial pressures in the third line of the previous table into the equilibrium constant expression:

$$K_p = \frac{P_{N_2O_4}}{P_{NO_2}^2} = \frac{0.54}{(0.50)^2} = 2.2$$

Note that the value of K does not have units; therefore the unit of pressure is not shown in the calculation.

The large majority of chemical equilibrium problems require calculating equilibrium concentrations. Equilibrium concentrations are calculated from the relationship between K and equilibrium concentrations. The next series of sample exercises explore how to solve for equilibrium concentrations. At the end of this section is a summary of the steps that were used in solving the sample exercises.

CALCULATING EQUILIBRIUM CONCENTRATIONS

EXERCISE 6   A 1.0-L container initially holds 0.015 mol of H_2 and 0.020 mole of I_2 at 721 K. What are the concentrations of H_2, I_2, and HI after the system has achieved a state of equilibrium? The value of K_c is 50.5 for the reaction $H_2(g) + I_2(g) \rightleftharpoons 2HI(g)$.

   Solution: You are given only the starting concentrations of H_2 and I_2, which are 0.015 M and 0.020 M, respectively, since the volume is 1.0-L. You are asked to find the equilibrium concentration of all substances in the equilibrium. To solve for the equilibrium concentration, use an approach similar to that used to solve Sample Exercise 15.11 in the text. Since the system contains H_2 and I_2 and no HI, the concentrations of H_2 and I_2 must decrease when they react to form HI at equilibrium. According to the stoichiometry of the balanced chemical equation, if you let x equal the molarity of H_2 that reacts, then x also equals the molarity of I_2 that reacts, and $2x$ equals the molarity of HI that forms. The last equivalence is derived by the following conversions:

$$1 \text{ mol } H_2 \simeq 1 \text{ mol } I_2 \simeq 2 \text{ mol HI}$$

$$[HI] = \left(x \frac{\text{mol } H_2}{L}\right)\left(\frac{2 \text{ mol HI}}{1 \text{ mol } H_2}\right) = 2x \text{ mol HI/L}$$

All of this information is summarized by the following table of concentrations:

	$H_2(g)$	+	$I_2(g)$	$\rightleftharpoons$ 2HI(g)
Initial (M):	0.015		0.020	0
Change (M):	$-x$		$-x$	$+2x$
Equilibrium (M):	$(0.015 - x)$		$(0.020 - x)$	$2x$

Substituting the equilibrium concentrations into the K_c expression for the reaction $H_2(g) + I_2(g) \rightleftharpoons 2HI(g)$ yields

$$K_c = \frac{[HI]^2}{[H_2][I_2]} = \frac{(2x)^2}{(0.015 - x)(0.020 - x)} = \frac{4x^2}{3.0 \times 10^{-4} - 0.035x + x^2} = 50.5$$

Rearranging this equation into quadratic form ($ax^2 + bx + c = 0$), gives

$$46.5x^2 - 1.77x + 0.0152 = 0$$

Dividing through by 46.5 yields

$$x^2 - 0.0381x + 3.27 \times 10^{-4} = 0$$

To solve this equation for the value of x, we must use the quadratic formula (see Appendix A.3 in the text):

$$x = \frac{-b \pm \sqrt{(b^2) - 4ac}}{2a}$$

$$= \frac{-(-0.0381)}{2(1)} \pm \frac{\sqrt{(-0.0381)^2 - 4(1)(3.27 \times 10^{-4})}}{2(1)}$$

$$= \frac{0.0381 \pm 0.0119}{2}$$

$$= 0.025 \text{ and } 0.0131$$

One solution, 0.025, is not valid because it is larger in value than either initial concentration. Thus $x = 0.0131$ is the only valid (physically significant) answer. Therefore

$$[H_2] = 0.015 - x = 0.015 - 0.013 = 0.002 \ M$$
$$[I_2] = 0.020 - x = 0.020 - 0.013 = 0.007 \ M$$
$$[HI] = 2x = 2(0.013) = 0.026 \ M$$

EXERCISE 7   A container initially holds 1.00 atm of NO_2 and 2.00 atm of NO. The equilibrium that is established is

$$2NO_2(g) \rightleftharpoons 2NO(g) + O_2(g)$$

All of the species are gases. K_p is $6.76. \times 10^{-5}$. What is the equilibrium partial pressure of O_2?

   Solution: You can develop a concentration table as you did in Exercise 6. The given quantities are the initial concentrations of NO_2 and NO; the question asks for the equilibrium partial pressure of O_2. Because there is no initial O_2, the reaction shifts to the right to produce O_2: The partial pressure of NO_2 decreases and the partial pressure of NO and O_2 increases. Let x equal the equilibrium partial pressure of O_2.

	$2NO_2(g)$	$\rightleftharpoons$	$2NO(g)$	$+ O_2(g)$
Initial (atm):	1.00		2.00	0
Change (atm):	$-2x$		$+2x$	$+x$
Equilibrium (atm):	$(1.00 - 2x)$		$(2.00 - 2x)$	x

The change in pressure of NO is $+2x$ because of the stoichiometry of the reaction—two moles of NO form for every mole of O_2 that forms. Using a similar analysis, the change in pressure of NO_2 is $-2x$. For NO and O_2 to form, NO_2 must dissociate; two moles of NO_2 dissociated for every mole of O_2 that forms. The number of moles of gas and atmospheres of pressure are directly related at constant temperature and volume because of the ideal-gas law, $PV = nRT$.

Note that the value of K_p is small; therefore you should expect the amount of NO_2 that dissociates, x, to be very small, particularly when compared to 2.00 and 1.00. Thus you can approximate the equilibrium concentration of NO_2 and NO as follows: $(1.00 - 2x)$ atm $\simeq 1.00$ atm and $(2.00 - 2x)$ atm $\simeq 2.00$ atm.*

$$K_p = \frac{[NO]^2[O_2]}{[NO_2]^2} \quad \text{or} \quad 6.76 \times 10^{-5} = \frac{(2.00)^2(x)}{(1.00)}$$

Solving for x, the equilibrium partial pressure of O_2, gives $x = 1.69 \times 10^{-5}$ atm.

*Note: The value of x is indeed much smaller than 2.00 or 1.00. For example, the percent error caused by the approximation 2.00 atm $- x \simeq 2.00$ atm is $x/2.00 \times 100 = 1.69 \times 10^{-5}/2 \times 100 = 8.5 \times 10^{-4}$ %. Any time this percent is less than 5%, the approximation will be considered valid.

SUMMARY OF STEPS INVOLVED IN CALCULATING EQUILIBRIUM CONCENTRATIONS

1. Write a balanced chemical equation describing the chemical changes that occur.

2. Identify knowns and the unknown asked for in the question.

3. Assign a variable, such as x, to the unknown. This variable usually represents a change in concentration.

4. Prepare a table of concentrations, using the substances involved in the chemical equilibrium:

      Row 1   chemical reaction describing equilibrium

      Row 2   initial concentrations

      Row 3   changes in concentration (x usually first appears here)

      Row 4   equilibrium concentrations (add rows 2 + 3)

5. Write the equilibrium constant expression.

6. Solve for the unknown variable by using K and the equilibrium concentrations in row 4. Determine if any approximations can be made to simplify the problem.

7. Answer the question.

8. Check the approximation for validity: For expressions of type $C + x$ or $C - x$, x can be neglected if

$$\frac{x}{C} \times 100 < 5\%$$

Up to a 5% error caused by the approximation is acceptable.

REACTION QUOTIENT AND LE CHATELIER'S PRINCIPLE: DIRECTION OF A CHEMICAL REACTION

Reactions that do not go to completion achieve an equilibrium state. The extent to which a reaction proceeds in approaching an equilibrium state is indicated by the value of K.

- $K \gg 1$: The reaction tends to proceed far to the right; mostly product forms at equilibrium.
- $K \ll 1$: The reaction lies far to the left; very little product forms at equilibrium.
- $K \approx 1$: Both the products and the reactants exist with neither highly favored versus the other.

A reaction that is not at equilibrium proceeds to a state of equilibrium. The direction of change in approaching equilibrium is determined by the **reaction quotient, Q.**

- A reaction-quotient expression for a given reaction has the same form as the reaction's equilibrium-constant expression, but it is numerically evaluated using the given reactant and product concentrations.
- The relationships between numerical values of K and Q and the direction of a reaction are summarized in Table 15.1.

Table 15.1   Relationships Among Numerical Values of K and Q

Relationship	Interpretation	Direction of change
$Q = K$	System is at equilibrium	none
$Q > K$	System is not at equilibrium; reaction proceeds to form more reactants until $Q = K$.	$\leftarrow$
$Q < K$	System is not at equilibrium; reaction proceeds to form more products until $Q = K$.	$\rightarrow$

The composition of a reaction at equilibrium may be altered by changing the concentration of a product or reactant, or by changing the pressure, volume, or temperature of the system. The question of what effect a change in one of these variables has on the extent of a reaction can be qualitatively answered by using **LeChatelier's Principle:**

- *When a stress is applied to a system at equilibrium, the system will change, if possible, in the direction of the reaction that will relieve the applied stress.*
- Alternatively, the principle may also be stated as follows: *When a system at equilibrium is disturbed, the system adjusts so as to restore the equilibrium state.*
- The effects of changing concentration, pressure, or temperature on a system at equilibrium are summarized in Table 15.2 on page 255. Note that changes in concentration and pressure do not change the value of K for a reaction.

Table 15.2 Effect of Changing Concentration, Pressure, or Temperature on a System at Equilibrium

Change	Effect
Concentration	
(1) Increasing the concentration of a substance	Reaction shifts to *remove* some of the added substance
(2) Decreasing the concentration a substance	Reaction shifts to produce *more* of the substance
Pressure	
(1) Increasing total gas pressure	Reaction shifts in a direction that causes the total number of moles of *gases* to *decrease*
(2) Decreasing total gas pressure	Reaction shifts in a direction that causes the total number of moles of *gases* to *increase*
Temperature	
(1) Increase	*Exothermic reaction:* K decreases—reaction shifts toward reactants
	Endothermic reaction: K increases—reaction shifts toward products
(2) Decrease	*Exothermic reaction:* K increases—reaction shifts toward products
	Endothermic reaction: K decreases—reaction shifts toward reactants

Note: In Table 15.2 that the value of K changes when temperature changes. Thus, when heat is added or removed from a system at equilibrium (i.e., a temperature change), the position of equilibrium is modified because the value of K is altered.

EXERCISE 8 What changes in the equilibrium composition of the reaction

$$N_2(g) + O_2(g) \rightleftharpoons 2NO(g)$$

will occur at constant temperature if: **(a)** the pressure of $N_2(g)$ is increased; **(b)** the pressure of $NO(g)$ is decreased; **(c)** the total pressure of the system is increased; **(d)** the total volume of the system is increased?

   Solution: **(a)** Increasing the concentration (partial pressure) of $N_2(g)$ shifts the equilibrium to the right, removing added $N_2(g)$ and also some $O_2(g)$, and thereby forming more $NO(g)$. **(b)** Decreasing the concentration (partial pressure) of $NO(g)$ shifts the equilibrium to the right, forming more $NO(g)$ and consuming $N_2(g)$ and $O_2(g)$. **(c)** The reaction proceeds in a direction that reduces the total number of moles of gas. Since the sum of moles of N_2 and O_2, 2, equals the number of moles of $NO(g)$ formed, 2, there is no way for the system to change to reduce the number of moles of gas. Thus, the composition of the equilibrium mixture does not change. **(d)** Pressure is inversely proportional to volume at constant temperature. Although the total pressure is decreased, no change in the equilibrium mixture occurs as explained in **(c)**.

EXERCISE 9 What change in the equilibrium composition of the reaction

$$2CO(g) + O_2(g) \rightleftharpoons 2CO_2(g)$$

occurs if the pressure of the system is increased?

Solution: The system responds to an increased pressure by reducing the total number of moles of gas until the equilibrium state is again established. There are three moles of gaseous reactants and two moles of gaseous product in the balanced equation. Formation of more product, $CO_2(g)$, and consumption of reactants, $CO(g)$ and $O_2(g)$, result in a net decrease in the total number of moles of gas. Thus the reaction shifts to the right.

EXERCISE 10   In which direction do the following reactions at equilibrium proceed when temperature is increased?

(a) $CH_4(g) + 2O_2(g) \rightleftharpoons CO_2(g) + 2H_2O(l)$ (exothermic)

(b) $H_2(g) + I_2(g) \rightleftharpoons 2HI(g)$ (endothermic)

Solution: Increasing temperature is caused by adding heat. **(a)** Since the reaction is exothermic, the chemical equation may be rewritten so heat is included as a product:

$$CH_4(g) + 2O_2(g) \rightleftharpoons CO_2(g) + 2H_2O(l) + \text{heat}$$

According to LeChatelier's principle, the position of equilibrium shifts to the left in order to remove the added heat. This results in a smaller value of K for the equilibrium. **(b)** The reaction is endothermic; heat may be considered as a reactant:

$$\text{Heat} + H_2(g) + I_2(g) \rightleftharpoons 2HI(g)$$

According to LeChatelier's principle, the position of equilibrium shifts to the right in order to remove the added heat. This results in a larger value of K for the equilibrium.

EXERCISE 11   The value of K_c for the reaction

$$H_2(g) + I_2(g) \rightleftharpoons 2HI(g)$$

is 50.5. In what direction will the reaction proceed if the initial concentrations are as follows: $[H_2] = 0.015\ M$; $[I_2] = 0.0012\ M$; $[HI] = 0.025\ M$?
Solution: First write the reaction-quotient expression for the reaction and then substitute into it the given concentrations (C). Remember that the reaction-quotient expression has the same form as the equilibrium-constant expression.

$$Q = \frac{C_{HI}^2}{C_{H_2} C_{I_2}} = \frac{(0.025)^2}{(0.015)(0.0012)} = 35$$

Next, compare the values of Q and K_c ($K_c = 50.5$). In this problem, $Q < K_c$. Therefore, the reaction proceeds to form more products until the value of Q increases to equal the value of K_c.

EXERCISE 12   According to LeChatelier's principle, what changes in the composition of the equilibrium system

$$H_2(g) + Br_2(l) \rightleftharpoons 2HBr(g)$$

occur when: **(a)** $Br_2(l)$ is added; **(b)** the partial pressure of $H_2(g)$ is increased; **(c)** the volume of the container that contains the equilibrium system is increased at constant temperature?
Solution: **(a)** No change occurs because the equilibrium composition is independent of the total quantity of *liquid* Br_2. **(b)** The increased partial pressure of $H_2(g)$ is partly offset by reaction of $H_2(g)$ with $Br_2(l)$ to form more $HBr(g)$. **(c)** The increase in volume is accompanied by a decrease in the total pressure of the system. (Remember that

$P \propto 1/V$.) The system reacts to a decrease in the total pressure by forming more moles of gas. The partial pressure of a pure liquid or solid in a system at equilibrium is constant and thus does not change so long as the temperature is unchanged. The consumption of $H_2(g)$ and $Br_2(l)$ to form $HBr(g)$ produces more moles of gas. Thus the reaction changes by forming more product.

EXERCISE 13   At 25°C the equilibrium constant for the reaction

$$\alpha\text{-D-glucose}(aq) \rightleftharpoons \beta\text{-D-glucose}(aq)$$

is 1.75. What are their equilibrium concentrations if initial concentrations are 1.00 mol/liter for both?

Solution: The first step in this problem is to determine the direction of the reaction when it changes from its initial state to the equilibrium state by using the reaction quotient. You need this information because you do not know if β-D-glucose has its concentration increased or decreased at equilibrium.

$$Q = \frac{C_{\beta\text{-D-glucose}}}{C_{\alpha\text{-D-glucose}}} = \frac{1.00}{1.00} = 1.00$$

Because $Q < K$, the reaction shifts to form more product. Therefore, let x equal the concentration of β-D-glucose formed during the change.

	$\alpha\text{-D-glucose}(aq) \rightleftharpoons \beta\text{-D-glucose}(aq)$	
Initial (*M*):	1.00	1.00
Change (*M*):	$-x$	$+x$
Equilibrium (*M*):	$(1.00 - x)$	$(1.00 + x)$

The value of K is large; thus, the *no* approximations should be made.

$$K = \frac{[\beta\text{-D-glucose}]}{[\alpha\text{-D-glucose}]} = 1.75$$

$$\frac{(1.00 + x)}{(1.00 - x)} = 1.75$$

Rearranging the above equation gives

$$1.00 + x = 1.75(1.00 - x) = 1.75 - 1.75x$$

or

$$2.75x = 0.75$$

$$x = \frac{0.75}{2.75} = 0.27 \ M$$

The equilibrium concentrations are therefore

$$[\beta\text{-D-glucose}] = (1.00 + x) \ M = 1.27 \ M$$
$$[\alpha\text{-D-glucose}] = (1.00 - x) \ M = 0.73 \ M$$

A reaction at equilibrium is dynamic at the molecular level. At equilibrium, the forward reaction is occurring at the same rate as the reverse reaction. Although the concentrations of reactant and product molecules remain constant, they are continuously forming and disappearing. By speeding up both forward and reverse reac-

CATALYSIS AND EQUILIBRIUM

tions, a **catalyst** decreases the time required for a reaction to reach equilibrium. *A catalyst does not change the value of the equilibrium constant.*

EXERCISE 14 For the reaction

$$HCN(aq) \rightleftharpoons H^+(aq) + CN^-(aq)$$

the rate of forward reaction equals $k_f[HCN]$ and the rate of reverse reaction equals $k_r[H^+][CN^-]$. Derive the equilibrium constant for the reaction, using the rate expressions. Assume that the reaction occurs by a single-step mechanism.

Solution: The net reaction rate at any time equals the rate of forward reaction minus the rate of reverse reaction. At equilibrium, the net reaction rate is zero. Thus, for the given reaction

$$\text{Net reaction rate} = \text{forward rate} - \text{reverse rate}$$

$$k_f[HCN] - k_r\left[H^+\right]\left[CN^-\right] = 0$$

$$k_f[HCN] = k_r\left[H^+\right]\left[CN^-\right]$$

$$\frac{k_f}{k_r} = \frac{\left[H^+\right]\left[CN^-\right]}{[HCN]} = K$$

As you can see from this equation, the equilibrium constant (K) is just the ratio of the rate constants for the forward and reverse reactions (k_f/k_r).

Note: This is valid if the reaction occurs by a single-step mechanism. The analysis for multistep mechanisms is usually more complicated.

EXERCISE 15 The equilibrium constant for the reaction

$$2SO_2(g) + O_2(g) \rightleftharpoons 2SO_3(g)$$

is 279. **(a)** Is the reaction fast or slow? **(b)** Nitrogen oxide catalyzes the reaction. How does the value of the equilibrium constant change when nitrogen oxide is added?

Solution: **(a)** The equilibrium constant provides no information about the rate of reaction. **(b)** The equilibrium constant does not change in value because a catalyst affects the forward and reverse reaction rates equally.

EXERCISE 16 Does a catalyst affect the enthalpy change in a reaction?

Solution: The enthalpy change for a reaction is independent of the way products are formed because it is a state function (see Section 5.2 in the text). However, a catalyst, by speeding up the reaction, causes the rate of heat absorption or evolution to increase. For example, in an exothermic reaction the catalyst causes all the heat to be released in a short time as compared to a much longer time when the catalyst is absent. However, the position of equilibrium is unchanged, and the total quantity of heat evolved is the same in each case.

SELF-TEST QUESTIONS

Key Terms

Having reviewed the key terms listed at the end of Chapter 15 and their applications, identify the following statements as

true or false. If a statement is false, indicate why it is incorrect. Key terms are italicized in the statements.

15.1 An example of *heterogeneous equilibrium* is the ionization of the slightly soluble salt $BaSO_4$:

$$BaSO_4(s) \overset{H_2O}{\rightleftharpoons} Ba^{2+}(aq) + SO_4^{2-}(aq)$$

The constant K for this equilibrium is given by the expression

$$K_c = \frac{[Ba^{2+}][SO_4^{2-}]}{[BaSO_4]}$$

15.2 When a closed system contains constant concentrations of $BaSO_4(s)$, $Ba^{2+}(aq)$, and $SO_4^{2-}(aq)$ at constant temperature, it has reached a state of *chemical equilibrium.*

15.3 The formation of $NH_3(g)$ from $N_2(g)$ and $H_2(g)$ by means of the *Haber process* is enhanced by increasing temperature.

15.4 According to the *law of mass action,* the equilibrium constant for the reaction

$$2CO_2(g) \rightleftharpoons 2CO(g) + O_2(g)$$

is

$$K_c = \frac{[CO][O_2]}{[CO_2]}$$

15.5 When the acid HCN exists in water, it slightly ionizes (dissociates) as follows:

$$HCN(aq) \rightleftharpoons H^+(aq) + CN^-(aq)$$

When the forward and reverse rates for this ionization are equal, we can state that a condition of *homogeneous equilibrium* exists.

15.6 An equilibrium exists between $CO_2(g)$ in human lungs and $CO_2(aq)$ on the blood system at the interface between the lung and blood tissues. According to *LeChatelier's principle,* an increase in the CO_2 concentration in the air a person breathes should cause a reduction in the CO_2 concentration in the person's blood system.

15.7 Suppose that during the Haber process we find $[NH_3] = 1.20\ M$, $[N_2] = 0.80\ M$, and $[H_2] = 0.50\ M$. By comparing the value of the *reaction quotient, Q,* to the value of K_c, which is 0.105 for the Haber reaction, we know that the system is at equilibrium.

15.8 The *equilibrium expression* for the reaction $N_2(g) + O_2(g) \rightleftharpoons 2NO(g)$ is

$$K = \frac{[N_2][O_2]}{[NO]^2}$$

15.9 The *equilibrium constant* for a particular reaction, K, is equal to K^{-1} for the reverse reaction.

Problems and Short-Answer Questions

15.10 Write K_c equilibrium-constant expressions for the following reactions:
  (a) $Br_2(g) \rightleftharpoons Br_2(l)$
  (b) $Ag_3PO_4(s) \rightleftharpoons 3Ag^+(aq) + PO_4^{3-}(aq)$
  (c) $C_2H_4(g) + H_2(g) \rightleftharpoons C_2H_6(g)$
  (d) $2C_2H_6(g) + 7O_2(g) \rightleftharpoons 4CO_2(g) + 6H_2O(l)$

15.11 Given K_c for the following three reactions, what is the value of K_p?
  (a) $I_2(g) + Cl_2(g) \rightleftharpoons 2ICl(g)$
  $K_c = 2.0 \times 10^5$ at 25°C
  (b) $N_2O_4(g) \rightleftharpoons 2NO_2(g)$
  $K_c = 0.90$ at 120°C
  (c) $CaCO_3(s) \rightleftharpoons CaO(s) + CO_2(g)$
  $K_c = 6.0 \times 10^{-3}$ at 740°C

15.12 2.00 mol of NO and some $O_2(g)$ are placed in a 1.00-L container at 460°C. When the system reaches equilibrium, we find 0.00156 mol of O_2 and 0.500 mol of $NO_2(g)$. Calculate K_c for the reaction

$$2NO(g) + O_2(g) \rightleftharpoons 2NO_2(g)$$

15.13 A 1.00-L container initially contains 0.476 mol of $SO_3(g)$ at 1105 K. What is K_c for the reaction

$$2SO_3(g) \rightleftharpoons O_2(g) + 2SO_2(g)$$

if 50.0 percent of the $SO_3(g)$ has decomposed at equilibrium?

15.14 At 2400°C, K_c equals 3.5×10^{-3} for the reaction

$$N_2(g) + O_2(g) \rightleftharpoons 2NO(g)$$

Determine whether the following systems are at equilibrium and, if not, in which direction each reaction is expected to proceed:
  (a) $[N_2] = 0.030\ M$: $[O_2] = 0.060\ M$: $[NO] = 0.020\ M$
  (b) $[N_2] = 0.050\ M$: $[O_2] = 0.070\ M$: $[NO] = 3.5 \times 10^{-3}\ M$
  (c) $[N_2] = 0.020\ M$: $[O_2] = 0.050\ M$: $[NO] = 1 \times 10^{-4}\ M$

15.15 At 400°C, the equilibrium-constant value for the reaction

$$2Ag_2O(s) \rightleftharpoons 4Ag(s) + O_2(g)$$

is 0.145. The partial pressure of $O_2(g)$ in air is 0.21 atm. At 400°C and in the presence of air, will $Ag_2O(s)$ decompose, or will finely divided $Ag(s)$ react with $O_2(g)$?

15.16 K_c for the reaction

$$NiO(s) + CO(g) \rightleftharpoons Ni(s) + CO_2(g)$$

is 4.54×10^3 at 936 K. (a) Is a system containing $NiO(s)$, $Ni(s)$, $CO(g)$ with a partial pressure of 2×10^{-3} atm, and $CO_2(g)$ with a partial pressure of 2.26×10^{-1} atm at equilibrium? (b) Can a system initially containing only $NiO(s)$ and $CO(g)$ achieve an equilibrium state? (c) Can a system initially containing only $NiO(s)$ and $CO_2(g)$ achieve an equilibrium state?

15.17 At 603 K, K_c for the reaction

$$NH_4Cl(g) \rightleftharpoons NH_3(g) + HCl(g)$$

is 1.1×10^{-2}. What is the equilibrium concentration of $NH_3(g)$ if the initial concentration of $NH_4Cl(g)$ is 0.500 M?

15.18 What changes in the equilibrium composition of the reaction

$$N_2O_4(g) \rightleftharpoons 2NO_2(g) \qquad \Delta H° = 58 \text{ kJ/mol}$$

occur if:

    (a) the partial pressure of $NO_2(g)$ is increased;
    (b) the total pressure of the system is increased;
    (c) the temperature of the system is decreased?

15.19 Oxygen is supplied to the body by hemoglobin (Hb). Each hemoglobin molecule carries up to four oxygen molecules. Carbon monoxide also reacts with Hb. The equilibrium constant for the reaction of Hb and CO is more than 200 times the equilibrium constant for the reaction of Hb and O_2. Using this information, explain why carbon monoxide is a poison.

15.20 For the equilibrium

$$C(s) + CO_2(g) \rightleftharpoons 2CO(g)$$

the value of K_p is 167.5 at 1000°C. If a 5-L container initially contains 2.00 atm pressure of $CO_2(g)$ and $C(s)$ at 1000°C, what are the partial pressures of $CO_2(g)$ and $CO(g)$ at equilibrium?

Multiple-Choice Questions

15.21 Which of the following factors affect equilibrium concentrations: (1) magnitude of equilibrium constant: (2) catalyst; (3) initial concentrations of products and reactants; (4) temperature, (5) overall reaction rate?

    (a) (1), (2), (3)   **(d)** (1), (3), (4), (5)
    (b) (1), (3), (4)   **(e)** all of them
    (c) (1), (2), (3), (4)

15.22 For a system to be at equilibrium, the relationship between reaction quotient, Q, and equilibrium constant is which of the following?

    (a) $Q = K = 1$ **(d)** $Q < K$
    (b) $Q = K = 0$ **(e)** $Q > K$
    (c) $Q = K$

15.23 Which of the following is a general expression relating K_p and K_c?

    (a) $K_p = K_c(RT)^{-\Delta n}$ **(d)** $K_p = K_c(RT)^{\Delta n}$
    (b) $K_c = K_p(RT)^{\Delta n}$ **(e)** $K_p = K_c + (RT)^{\Delta n}$
    (c) $K_p = K_c$

15.24 For the reaction

$$Cl_2(g) \rightleftharpoons 2Cl(g)$$

if molarity is used instead of partial pressure for the concentration unit at 1000°C to calculate the value of K_c, what will be the relationship of the numerical value of K_c to that of Kp?

    (a) larger than K_p
    (b) smaller than K_p
    (c) equal to K_p
    (d) nearly identical to K_p
    (e) impossible to compare with K_p without further information

15.25 The position of equilibrium for the reaction

$$ZnO(s) + H_2(g) \rightleftharpoons Zn(s) + H_2O(g)$$

does *not* depend upon which of the following: (1) concentra-

tion of $ZnO(s)$; (2) concentration of $H_2(g)$; (3) concentration of $Zn(s)$; (4) concentration of $H_2O(g)$; (5) the value of K_c?

    (a) (1), (2), (5)   **(d)** (2), (4)
    (b) (2), (3), (5)   **(e)** (5)
    (c) (1), (3)

15.26 Which of the following reactions, initially at equilibrium, will form more product when the total pressure is increased at constant temperature?

    (1) $2CO_2(g) \rightleftharpoons 2CO(g) + O_2(g)$
    (2) $N_2(g) + 3H_2(g) \rightleftharpoons 2NH_3(g)$
    (3) $SO_2(g) \rightleftharpoons S(s) + O_2(g)$
    (a) (1)   **(d)** all of them
    (b) (2)   **(e)** none of them
    (c) (3)

15.27 Which of the following reactions, initially at equilibrium, will shift to the left when the temperature is decreased at constant pressure?

    (1) $2C_2H_2(g) + 5O_2(g) \rightleftharpoons$
$$4CO_2(g) + 2H_2O(l) \quad \Delta H = -1297 \text{ kJ}$$
    (2) $CO_2(g) \rightleftharpoons O_2(g) + C(s)$
$$\Delta H = 393 \text{ kJ}$$
    (3) $4Fe(s) + 3O_2(g) \rightleftharpoons 2Fe_2O_3(s)$
$$\Delta H = -1644 \text{ kJ}$$
    (a) (1)   **(d)** all of them
    (b) (2)   **(e)** none of them
    (c) (3)

15.28 For the reaction $A + B \rightleftharpoons C$ to be at equilibrium, the relationship between forward (r_f) and reverse (r_r) reaction rates is

    (a) $r_f = r_r$ **(d)** $r_f = r_r = 1$
    (b) $r_f < r_r$ **(e)** $r_f > 1, r_r < 1$
    (c) $r_f > r_r$

15.29 Which of the following is *not* affected when a catalyst influences a gaseous chemical reaction?

    (a) Forward and reverse reaction rates;
    (b) initial reaction rate;
    (c) value of activation energy;
    (d) reaction mechanism pathway
    (e) value of equilibrium constant.

15.30 At 240°C, the value of K_c is 20 for the reaction

$$PCl_3(g) + Cl_2(g) \rightleftharpoons PCl_5(g)$$

If a 1.0-L container contains 0.25 mol of PCl_5 at 240°C, how many moles of it dissociate to form PCl_3 and Cl_2? (You'll need to use the quadratic formula to solve this problem.)

    (a) 0.0081 mol   **(d)** 0.11 mol
    (b) 0.090 mol   **(e)** 0.25 mol
    (c) 0.013 mol

SELF-TEST SOLUTIONS

15.1 False. It is an example of heterogeneous equilibrium. The constant K_c for the equilibrium is: $K_c = [Ba^{2+}][SO_4^{2-}]$. **15.2** True. **15.3** False. The value of K_c for the Haber process decreases with increasing temperature because the reaction is exothermic. **15.4** False. $K_c = [CO]^2[O_2]/[CO_2]^2$. **15.5** True.

15.6 False. CO_2(lung) $\rightleftarrows$ CO_2(blood). According to LeChatelier's principle, increasing [CO_2 (lung)] causes more CO_2 (blood) to form. **15.7** False. The reaction is $N_2 + 3H_2 \rightleftarrows 2NH_3$. Consequently: $Q = [NH_3]^2/[N_2]^2[H_2]^3 = (1.20)^2/(0.80)^2(0.50)^3 = 18$. Since $Q > K$, the reaction is not at equilibrium and reacts to form more reactants until $Q = K$. **15.8** False. The equilibrium expression is related to the concentration of products/reactants. The given equilibrium expression is reactants/products. The correct equilibrium expression is the reciprocal of the one shown. **15.9** True. **15.10 (a)** $K_c = 1/[Br_2]$. $Br_2(l)$ does not appear in the K_c expression because it is a pure liquid. **(b)** $K_c = [Ag^+]^3[PO_4^{3-}]$. $Ag_3PO_4(s)$ does not appear in the K_c expression because it is a pure solid. **(c)** $K_c = [C_2H_6]/[C_2H_4][H_2]$. **(d)** $K_c = [CO_2]^4/[C_2H_6]^2[O_2]^7$. H_2O does not appear in the K_c expression because it is a pure liquid. **15.11** The general relationship between K_p and K_c is $K_p = K_c(RT)^{\Delta n}$ where Δn equals the number of moles of gaseous products minus the number of moles of gaseous reactants, and the value of R is 0.0821 L-atm/mol-K. Remember that T is in degrees Kelvin. **(a)** $T = 25 + 273 = 298$ K; $\Delta n = 2 - (1 + 1) = 0$; $K_p = (2.0 \times 10^5)(0.0821 \times 298)^0 = (2.0 \times 10^5)(1) = 2.0 \times 10^5 = K_c$. Whenever the value of Δn equals 0, K_p equals K_c. **(b)** $T = 120 + 273 = 393$ K; $\Delta n = 2 - 1 = 1$; $K_p = (0.90)(0.0821 \times 393)^1 = 29$. **(c)** $T = 740 + 273 = 1013$ K; $\Delta n = 1 - 0 = 1$. (In calculating Δn, the moles of solid or liquid are not included.) $K_p = K_c(RT)^1 = K_c(0.0821 \times 1013)^1 = 0.50$. **15.12** The K_c expression for $2NO(g) + O_2(g) \rightleftarrows 2NO_2(g)$ is $K_c = [NO_2]^2/[NO]^2[O_2]$. In this problem we are given [O_2] and [NO_2] at equilibrium, but we are not given [NO]. However, we can calculate [NO] because we know its original concentration and how much NO_2 has formed. From the stoichiometry of the reaction, we see that for every 1 mol of NO that reacts, 1 mol of NO_2 forms. Thus the amount of NO_2 present at equilibrium represents the amount of NO that has reacted: [NO] = $2.00 - 0.50 = 1.50M$. Substituting these equilibrium concentrations into the K_c expression yields $K_c = (0.50)^2/(1.50)^2(0.00156) = 71$. **15.13** According to the stoichiometry of the reaction, when 50.0 percent of 0.476 M SO_3 decomposes at equilibrium: 0.238 M SO_2 and ($\frac{1}{2}$)(0.238 M) O_2 form. The equilibrium concentration of SO_3 is 0.476 M − 0.238 M = 0.238 M. Solving for K_c: $K_c = [SO_2]^2[O_2]/[SO_3]^2 = (0.238)^2 (0.119)/(0.238)^2 = 0.119$. **15.14** The reaction-quotient expression has the same form as the K_c expression but is evaluated at the given conditions. For the reaction $N_2(g) + O_2(g) \rightleftarrows 2NO(g)$, $Q = [NO]^2/[N_2][O_2]$. Evaluate the magnitude of the reaction quotient for each set of concentrations and compare it to the value for K_c: **(a)** $Q = (0.020)^2/(0.030)(0.060) = 0.222$. Since $Q > K_c(3.5 \times 10^{-3})$, the reaction proceeds to the reactant side until $Q = K_c$. **(b)** $Q = (3.5 \times 10^{-3})^2/(0.050)(0.070) = 3.5 \times 10^{-3}$. Since $Q = K_c$, the system is at equilibrium. **(c)** $Q = (1 \times 10^{-4})^2/(0.020)(0.050) = 1 \times 10^{-5}$. Since $Q < K_c$ (3.5 × 10^{-3}), the reaction proceeds to produce more products until $Q = K_c$. **15.15** The equilibrium-constant expression for the reaction is $K_p = P_{O_2} = 0.145$. $Ag_2O(s)$ decomposes as long as the partial pressure of $O_2(g)$ is less than 0.145 atm. Since the partial pressure of $O_2(g)$ is 0.21 atm, $Ag(s)$ reacts with $O_2(g)$ until the partial pressure of $O_2(g)$ is 0.145 atm. **15.16 (a)** The equilibrium-constant expression for the reaction is $K_c = [CO_2]/[CO] = 4.54 \times 10^3$. For the system to be at equilibrium, the reaction-quotient ratio [CO_2]/[CO] must equal 4.54×10^3. The value of the reaction quotient, calculated using the given

data, is $(2.26 \times 10^{-1})/(2 \times 10^{-3})$, which equals 1.13×10^2. The ratio does not equal the value of K_c; thus the system is not at equilibrium. **(b)** Yes. They will react to form Ni(s) and $CO_2(g)$ until equilibrium is established. **(c)** No. Either CO(g) or Ni(s) also must be present. **15.17** Let x be the molar concentration of $NH_3(g)$ formed. This gives us the following:

$$NH_4Cl(g) \rightleftharpoons NH_3(g) + HCl(g)$$

	$NH_4Cl(g)$	$NH_3(g)$	$HCl(g)$
Initial (M):	0.500	0	0
Change (M):	$-x$	$+x$	$+x$
Equilibrium (M):	$(0.500 - x)$	x	x

Thus, $K_c = [NH_3][HCl]/[NH_4Cl] = (x)(x)/(0.500 - x) = 1.1 \times 10^{-2}$. Solving for x using the quadratic formula yields $x = [NH_3] = 6.9 \times 10^{-2}$ mol/L. **15.18 (a)** Increasing the partial pressure of $NO_2(g)$ causes the equilibrium to shift so that the partial pressure of $NO_2(g)$ is decreased toward its initial value. This results in the formation of more $N_2O_4(g)$. **(b)** Increasing the total pressure causes the equilibrium to shift so that the total pressure is decreased toward its initial value. The total pressure is decreased by the formation of fewer moles of gas: $NO_2(g)$ forms $N_2O_4(g)$. **(c)** Decreasing the temperature of the system causes the equilibrium to shift so that heat is evolved. Heat is evolved when $NO_2(g)$ molecules combine to form $N_2O_4(g)$. Thus the concentration of $N_2O_4(g)$ increases and the concentration of $NO_2(g)$ decreases. **15.19** The reaction Hb + CO $\rightarrow$ Hb $\cdot$ CO occurs to a greater extent than the reaction Hb + O_2 $\rightarrow$ Hb $\cdot$ O_2. A high CO concentration in the blood prevents the formation of Hb $\cdot$ O_2 because the Hb molecules are primarily in the form Hb $\cdot$ CO. Thus little O_2 is transported through the body. **15.20** Using the stoichiometry of the equilibrium $C(s) + CO_2(g) \rightleftarrows 2CO(g)$, we can tabulate concentrations of CO_2 and CO. We need not consider $C(s)$ because its concentration value is included in the value of K_c.

	CO_2	CO
Initial (atm):	2.00	0
Change (atm):	$-x$	$+2x$
Equilibrium (atm):	$(2.00 - x)$	$2x$

(The expression $2x$ is required for CO because for every 1 mol of CO_2 that reacts 2 mol of CO forms.) Thus, $K_p = [CO]^2/[CO_2] = (2x)^2/(2.00 - x) = 167.5$ or $4x^2 = (2.00 - x)(167.5) = 335 - 167.5x$ or $4x^2 + 167.5x - 335 = 0$. Using the quadratic formula to solve for x yields

$$x = \frac{-(167.5) \pm \sqrt{(167.5)^2 - 4(4)(-335)}}{(2)(4)} = 1.91$$

Thus [CO_2] = $(2.00 - x)$ atm = $(2.00 - 1.91)$ atm = 0.09 atm and [CO] = $2x$ = 3.82 atm. **15.21 (b)**. **15.22 (c)**. **15.23 (d)**. **15.24 (b)**. $K_c = K_p/(RT)^{\Delta n} = K_p/(RT)^1$. Since K_p is divided by RT, $K_c < K_p$. **15.25 (c)**. Pure solids do not affect the position of equilibrium. **15.26 (b)**. The system restores equilibrium by changing so that fewer total moles of gas exist, that is, so there is less total pressure. **15.27 (b)**. Since in an endothermic reaction heat can be considered as a reactant, the removal of heat by lowering the temperature will cause the equilibrium to shift to the left to produce more heat. **15.28 (a)**. **15.29 (e)**. **15.30 (b)**. $K_c = [PCl_5]/[PCl_3][Cl_2] = (0.25 - x)/(x)(x) = 20$; $x = 0.090$ M or 0.090 mol in 1.0 L.

16 Acid-Base Equilibria

OVERVIEW OF THE CHAPTER

16.2 ACIDS AND BASES IN WATER: BRONSTED-LOWRY THEORY

Review: Acids and bases (4.3); hydrogen bond (11.2); Lewis structures (8.6).

Objectives: You should be able to:

1. List the general properties that characterize acidic and basic solutions and identify the ions responsible for these properties.
2. Define Bronsted acid, Bronsted base, and conjugate acid-base pair; identify the conjugate base associated with a given Bronsted acid and the conjugate acid associated with a given Bronsted base.

16.1, 16.3 ACIDS AND BASES; THE ROLE OF WATER; THE pH SCALE

Review: Concept of equilibrium and equilibrium constant (15.1, 15.2); acids and bases (4.3).

Objectives: You should be able to:

1. Explain what is meant by the autoionization of water, and write the ion-product constant expression for this process.
2. Define pH; calculate pH from a knowledge of [H$^+$] or [OH$^-$], and perform the reverse operation.

16.3, 16.4 STRONG ACID AND BASE SOLUTIONS: pH CALCULATIONS

Review: Concentration units (13.1); strong electrolytes (4.2).

Objective: You should be able to identify the common strong acids and bases and calculate the pHs of their aqueous solutions given their concentrations.

16.5, 16.6 WEAK ACID AND BASE SOLUTIONS: pH CALCULATIONS

Review: Use of quadratic formula (see Appendix in the text); weak electrolytes (4.2); equilibrium calculations (15.5).

Objectives: You should be able to:

1. Calculate the pH for a weak acid solution in water, given the acid concentration and K_a; calculate K_a given the acid concentration and pH.

2. Calculate the pH for a weak base solution in water, given the base concentration and K_b; calculate K_b given the base concentration and pH.

3. Calculate the percent ionization for an acid or base, knowing its concentration in solution and the value of K_a or K_b.

Review: Ions (2.5); hydration (13.1).

Objectives: You should be able to:

1. Determine the relationship between the strength of an acid and that of its conjugate base; calculate K_b from a knowledge of K_a, and vice-versa.
2. Predict whether a particular salt solution will be acidic, basic or neutral.

16.7, 16.8 SALT SOLUTIONS: HYDROLYSIS

Review: Bond strength (8.9); electronegativity (8.5); polarization (8.5); Lewis structures (8.6, 8.8).

Objectives: You should be able to:

1. Explain how acid strength relates to the polarity and strength of the H—X bond.
2. Predict the relative acid strengths of oxyacids.

16.9 FACTORS AFFECTING ACID-BASE STRENGTH

Review: Hydration (13.2); Lewis structures (8.6, 8.8).

Objectives: You should be able to:

1. Define an acid or base in terms of the Lewis concept.
2. Predict the relative acidities of solutions of metal salts from a knowledge of metal-ion charges and ionic radii.

16.10 THE LEWIS MODEL OF ACIDS AND BASES

TOPIC SUMMARIES AND EXERCISES

In Section 4.3 of the text, the acid-base properties of substances in water are characterized.

ACIDS AND BASES IN WATER: BRONSTED-LOWRY THEORY

▪ Acidic solutions are characterized by the presence of hydrated protons, $H^+(aq)$. An acid transfers a proton to water to form a hydrogen-bonded species. This species, sometimes called the **hydronium ion,** is often written as H_3O^+.
▪ Basic solutions are characterized by the presence of $OH^-(aq)$ in greater quantity than $H^+(aq)$.

J. Bronsted and T. M. Lowry proposed a model for characterizing the reactions of acids and bases in terms of proton (H^+) transfer.

- An **acid** is defined as any substance able to donate a proton in a chemical reaction, and a **base** is defined as any substance capable of accepting a proton in a chemical reaction.
- For example, look at the following reaction:

$$\underset{\text{Base}}{H_2O(l)} + \underset{\text{Acid}}{\textcircled{H}\, Cl(aq)} \longrightarrow H_3O^+(aq) + Cl^-(aq)$$

Proton transfer

Water is a base in this reaction because it accepts a proton from the acid HCl to form a hydrated proton, $H_3O^+(aq)$. For the remainder of this study guide, the hydrated proton will be represented as $H^+(aq)$. The above reaction becomes

$$HCl(aq) \longrightarrow H^+(aq) + Cl^-(aq)$$

- Acids such as HCl, HNO_3, or H_2SO_4 ionize in water to form a hydrated proton and a species called the **conjugate base** of the acid:

$$\underset{\text{Acid}}{HCl(aq)} \longrightarrow H^+(aq) + \underset{\substack{\uparrow \\ \text{Conjugate} \\ \text{base of HCl}}}{Cl^-(aq)}$$

Note that the conjugate base of HCl results from the *loss of one proton*.

- Bases such as NH_3 react with water to form a hydrated hydroxide ion and a species called the **conjugate acid** of the base.

$$\underset{\text{Base}}{NH_3(aq)} + H_2O(l) \rightleftharpoons \underset{\substack{\uparrow \\ \text{Conjugate acid} \\ \text{of } NH_3}}{NH_4^+(aq)} + OH^-(aq)$$

Note that the conjugate acid of NH_3 results from the *gain of one proton*.

- A **conjugate acid-base pair** is therefore an acid and a base, such as HF and F^-, that differ only in the presence or absence of *one* proton.
- When water reacts with ammonia it acts as an acid. When water reacts with HCl it acts as a base. A substance capable of acting both as an acid and a base is called **amphoteric.** NH_3 is another example of an amphotric substance: It can act as a base forming NH_4^+, or as an acid when it loses a proton to form NH_2^-.

The strength of an acid is measured by its tendency to donate a proton, and that of a base by its tendency to form hydroxide ions or accept a proton.

- **Strong acids** react completely with water to form $H^+(aq)$ and their conjugate bases. Strong acids are said to undergo 100 percent dissociation in terms of proton donation to water.

- **Strong bases** ionize completely in water—that is, all ionizable hydroxide units in the base come off and are hydrated.

- The ionization reactions of strong acids and bases with water are written with a single arrow between products and reactants. For example, the ionization reactions of HBr and KOH are represented as follows:

$$\text{Strong acid: HBr}(aq) \longrightarrow \text{H}^+(aq) + \text{Br}^-(aq)$$
$$\text{Strong base: KOH}(aq) \longrightarrow \text{K}^+(aq) + \text{OH}^-(aq)$$

- **Weak acids** and **bases** only partly ionize in water. The reactions of weak acids and bases with water are written with double arrows between products and reactants. We do this to show that an equilibrium condition exists: that is, a condition of no apparent change in the concentrations of all species with time. For example, the ionization of the weak acid HF is represented as follows:

$$\text{HF}(aq) \rightleftharpoons \text{H}^+(aq) + \text{F}^-(aq)$$

- *It is important to realize that the stronger the acid, the weaker its conjugate base and the stronger the base, the weaker its conjugate acid.* For example, the weak acid HF has a conjugate base F^- that has a strong tendency to attract a proton to form HF. You need to understand that the conjugate base of a weak acid is a weak base, not a strong base in water. Similarly, the conjugate acid of a weak base is a weak acid, not a strong acid.

EXERCISE 1 (a) Write a chemical equation describing the reaction of water with each of the following acids and bases: $\text{HCl}(aq)$, strong acid; $\text{HBr}(aq)$, strong acid; $\text{HNO}_2(aq)$, weak acid; $\text{NH}_3(aq)$, weak base; $\text{NaOH}(s)$, strong base. (b) Which of the following would you expect to have a significant capacity to react with a hydrated proton: Cl^-, Br^-, NO_2^-, NH_4^+?

Solution: (a) Remember that one arrow between products and reactants is used with a strong acid or base and double arrows are used with weak acids and bases:

$$\text{HCl}(aq) \longrightarrow \text{H}^+(aq) + \text{Cl}^-(aq)$$
$$\text{HBr}(aq) \longrightarrow \text{H}^+(aq) + \text{Br}^-(aq)$$
$$\text{HNO}_2(aq) \rightleftharpoons \text{H}^+(aq) + \text{NO}_2^-(aq)$$
$$\text{NH}_3(aq) + \text{H}_2\text{O}(l) \rightleftharpoons \text{NH}_4^+(aq) + \text{OH}^-(aq)$$
$$\text{NaOH}(s) \xrightarrow{\text{H}_2\text{O}} \text{Na}^+(aq) + \text{OH}^-(aq)$$

(b) Only those anions that are derived from *weak* acids will react significantly with hydrated protons. NO_2^- (from the weak acid HNO_2) reacts with $\text{H}^+(aq)$ to form HNO_2.

EXERCISE 2 Indicate which reactant is the acid and which is the base in each of the following reactions:
  (a) $\text{HC}_2\text{H}_3\text{O}_2(aq) + \text{NaOH}(aq) \longrightarrow \text{NaC}_2\text{H}_3\text{O}_2(aq) + \text{H}_2\text{O}(l)$
  (b) $\text{ZnS}(s) + 2\text{HCl}(aq) \longrightarrow \text{H}_2\text{S}(g) + \text{ZnCl}_2(aq)$
  (c) $\text{KCN}(aq) + \text{H}_2\text{O}(l) \longrightarrow \text{KOH}(aq) + \text{HCN}(aq)$
Solution: An acid donates a proton, and a base accepts the proton. (a) $\text{HC}_2\text{H}_3\text{O}_2$ is the acid, and NaOH is the base. (The OH^- ion accepts the proton.) (b) HCl is the acid and ZnS is the base. (The S^{2-} ion accepts the protons to form H_2S.) (c) H_2O is the acid and KCN is the base. (The CN^- ion accepts the proton.)

EXERCISE 3   Write the conjugate acid for each of the following species: **(a)** H_2O; **(b)** OH^-; **(c)** CO_3^{2-}; **(d)** NH_3.

　　Solution: *The conjugate acid is formed by the addition of **one** proton:* **(a)** $H_3O^+(aq)$ or $H^+(aq)$; **(b)** H_2O; **(c)** HCO_3^-; **(d)** NH_4^+.

EXERCISE 4   Write the conjugate base for each of the following species: **(a)** H_2O; **(b)** OH^-; **(c)** H_2SO_4; **(d)** NH_3.

　　Solution: *The conjugate base is formed by the removal of **one** proton:* **(a)** OH^-; **(b)** O^{2-}; **(c)** HSO_4^-; **(d)** NH_2^-.

EXERCISE 5 CN^- reacts with water as follows:

$$CN^-(aq) + H_2O(l) \rightleftharpoons HCN(aq) + OH^-(aq)$$

The Cl^- ion does *not* react with water to form HCl. Which acid is stronger, $HCl(aq)$ or $HCN(aq)$? Why?

　　Solution: The information provided in the question tells us that Cl^- is a weaker base than CN^-: CN^- associates with a proton from H_2O to form HCN. We also know that HCl and HCN are the conjugate acids of the bases Cl^- and CN^-. Using the principle that the weaker a base, the stronger its conjugate acid, we can conclude that HCl is a stronger acid than HCN. Cl^- essentially has no affinity for a hydrated proton.

ACIDS AND BASES; THE ROLE OF WATER; THE pH SCALE

Water undergoes **autoionization,** as shown in the chemical equation

$$H_2O(l) \rightleftharpoons H^+(aq) + OH^-(aq)$$

▪ The equilibrium-constant expression for the above is called the **ion-product constant** for water, K_w. At 25°C,*

$$K_w = [H^+][OH^-] = 1.0 \times 10^{-14}$$

▪ In pure water at 25°C, $[H^+] = [OH^-] = 10^{-7}M$.

▪ You should recognize that because of the autoionization of water, both H^+ and OH^- ions exist in any acid or base aqueous solution. The relationship $K_w = [H^+][OH^-] = 1.0 \times 10^{-14}$ enables you to calculate $[H^+]$ or $[OH^-]$ if you know the value of the other quantity:

$$\left[OH^-\right] = \frac{1.0 \times 10^{-14}}{\left[H^+\right]} \quad \text{or} \quad \left[H^+\right] = \frac{1.0 \times 10^{-14}}{\left[OH^-\right]}$$

For convenience, the hydrogen-ion concentration of a solution is commonly expressed in terms of **pH,** where pH is defined as:

$$pH = -\log[H^+] = \log\left(\frac{1}{[H^+]}\right)$$

---

*You should note that the value for the concentration of pure water, $[H_2O]$, does not appear in the K_w expression. The true equilibrium expression is $K_{eq} = [H^+][OH^-]/[H_2O]^2$. The value of $[H_2O]$ is essentially constant in solutions of dilute acids and bases, 55.5 M.

- When you are given pH and asked to solve for $[H^+]$, you can use the 10^x function of a scientific calculator as follows:

$$[H^+] = 10^{-pH}$$

where $x = -pH$

- You can describe acidity and basicity in terms of either $[H^+]$ or pH (see Table 16.1).

- Most solutions have a pH range of 0-14, but pH values greater than 14 and less than zero are possible

Other useful relationships in doing problems are:

$$pOH = -\log[OH^-]$$
$$pK_w = -\log K_w = 14$$
$$pK_w = pH + pOH = 14$$

Table 16.1   Relationship of [H⁺] and pH to acidity and basicity

Kind of solution	$[H^+], [OH^-]$ relationship	pH requirement
Neutral	$[H^+] = [OH^-]$	pH = 7
Acidic	$[H^+] > [OH^-]$	pH < 7
Basic	$[OH^-] > [H^+]$	pH > 7

EXERCISE 6   A solution has a hydroxide-ion concentration of 1.5×10^{-5} M. **(a)** What is the concentration of hydronium ions in this solution? **(b)** What is the pH of this solution? **(c)** Is the solution acidic, basic, or neutral?
    Solution: **(a)** You can calculate the hydronium-ion concentration by use of the relation $K_w = [H^+][OH^-] = 1.0 \times 10^{-14}$:

$$[H^+] = \frac{K_w}{[OH^-]} = \frac{1.0 \times 10^{-14}}{1.5 \times 10^{-5}} = 6.7 \times 10^{-10} \, M$$

(b) The pH of the solution is calculated by using the relation pH = $-\log [H^+]$ and substituting into it the value of $[H^+]$:

$$pH = -\log [H^+] = -\log (6.7 \times 10^{-10}) = -(-9.17) = 9.17$$

With a scientific calculator; this calculation is easily made by using the $\log_{10}$ function (*not* $\ln_e$). **(c)** The solution is basic because $[OH^-] > [H^+]$. Also, the pH is higher than 7, which is a requirement for a basic solution.

EXERCISE 7   A solution has a pH of 5.60. Calculate the hydrogen-ion concentration.
    Solution: To solve for $[H^+]$ in this solution, you will need to use the definition of pH:

$$pH = -\log [H^+]$$

Rearrange it so log $[H^+]$ is on the left-hand side of the equation:

$$-\log [H^+] = pH$$

or

$$\log [H^+] = -pH$$

Substituting for the value of pH gives

$$\log [H^+] = -5.60$$

The number -5.60 is a sum of a logarithmic characteristic and mantissa. If you do not have a scientific calculator with a 10^x or y^x function, you will need to use a log table to calculate $[H^+]$. If you have a calculator with a 10^x or y^x function, find the value of 10^{-pH}, which equals $[H^+]$.

$$[H^+] = 10^{-5.60} = 2.5 \times 10^{-6}$$

EXERCISE 8   Calculate $[H^+]$ for a solution with a pOH of 4.75.
   Solution: pOH is defined as $-\log [OH^-]$. The pH and the pOH of a solution are related to each other by the relation $pH + pOH = 14$. The pH of a solution with a pOH of 4.75 is

$$pH = 14 - pOH = 14 - 4.75 = 9.25$$

Using the procedure applied in Exercise 7 to calculate $[H^+]$ gives

$$\log [H^+] = -pH = -9.25 \qquad [H^+] = 5.6 \times 10^{-10}$$

STRONG ACID AND BASE SOLUTIONS: pH CALCULATIONS

You should know (memorize) these common strong acids: HCl, HBr, HI, HNO_3, $HClO_4$, and H_2SO_4 and strong bases: $LiOH$, $NaOH$, KOH, $RbOH$, $CsOH$, $Mg(OH)_2$, $Ca(OH)_2$, $Ba(OH)_2$, and $Sr(OH)_2$.

- All dissolved strong acids and bases completely give up their ionizable protons or hydroxide units to water.

- In the case of H_2SO_4, only one proton is completely donated to water, forming HSO_4^-. HSO_4^- is a weak acid.

- The hydrogen-ion concentration of a strong acid solution is calculated by assuming that the ionizable protons are dissociated 100 percent. A similar assumption is made for ionizable hydroxide units in strong bases.

EXERCISE 9   Calculate the pH of a $0.10\ M\ HNO_3$ solution.
   Solution: The first step in solving the problem is to write the ionization reaction:

$$HNO_3(aq) \longrightarrow H^+(aq) + NO_3^-(aq)$$

The second step is to write the initial concentrations and then the concentrations of all species after HNO_3 ionizes 100 percent. It is useful to tabulate the concentrations in solving problems of this type:

	$HNO_3 \xrightarrow{100\%}$	H^+ +	NO_3^-
Initial (*M*):	0.10	0	0
Ionization (*M*):	−0.10	+0.10	+0.10
Final (*M*):	0	0.10	0.10

The final HNO_3 concentration is zero because all of it has ionized to form equal molar amounts of H^+ and NO_3^-.

$$pH = -\log [H^+] = -\log(0.10) = -(-1.00) = 1.00$$

EXERCISE 10 Calculate the pOH and *p*H of an 0.02 *M* KOH solution.
Solution: The problem-solving approach is the same as in Exercise 9:

	KOH $\longrightarrow$ K$^+$	+	OH$^-$
Initial (*M*):	0.02	0	0
Ionization (*M*):	-0.02	$+0.02$	$+0.02$
Final (*M*):	0	0.02	0.02

The KOH ionizes 100 percent, and 0 *M* KOH remains after it ionizes.

$$pOH = -\log[OH^-] = -\log(0.02) = 1.7$$
$$pH = pK_w - pOH = 14 - 1.7 = 12.3$$

The equilibrium constants associated with weak acid and base ionizations are called the **acid-dissociation constant, K_a**, and the **base-dissociation constant, K_b**.

Examples of dissociation constants are shown for the following reactions of HF and NH$_3$ with water:

$$HF(aq) \rightleftharpoons H^+(aq) + F^-(aq) \qquad K_a = \frac{[H^+][F^-]}{[HF]}$$

$$NH_3(aq) + H_2O(l) \rightleftharpoons NH_4^+(aq) + OH^-(aq) \qquad K_b = \frac{[NH_4^+][OH^-]}{[NH_3]}$$

- In all K_a and K_b expressions, the concentration of water, which is a constant relative to any changes in concentrations of solute species, is incorporated into the values of K_a and K_b. Thus, the term [H$_2$O] is not shown in any K_a or K_b expression.
- You should know common weak acids (for example, NH$_4^+$, HCN, HF, HC$_2$H$_3$O$_2$) and weak bases (NH$_3$ and amines and basic anions.)

H$_2$CO$_3$, H$_2$S, and H$_3$PO$_4$ are examples of **polyprotic acids;** they have more than one ionizable proton.

- When describing the reactions of weak polyprotic acids with water, you can treat the ionization of each proton in a stepwise manner. Each ionization equilibrium produces one hydrated proton, and each ionization has an associated K_a value.
- For example, the stepwise ionization reactions of H$_2$S are described as follows:

$$H_2S(aq) \rightleftharpoons H^+(aq) + HS^-(aq) \qquad K_{a1} = \frac{[H^+][HS^-]}{[H_2S]}$$

$$HS^-(aq) \rightleftharpoons H^+(aq) + S^{2-}(aq) \qquad K_{a2} = \frac{[H^+][S^{2-}]}{[HS^-]}$$

Weak bases are generally of two kinds.

WEAK ACID AND BASE SOLUTIONS: pH CALCULATIONS

- The first includes ammonia (NH_3) and compounds related to NH_3 that are called amines. (Amines are formed when one or more N-H bonds in ammonia are converted to an N-C bond.) Hydroxylamine, NH_2OH, contains an N-OH bond instead of an N-C bond.

- The second includes anions of weak acids, such as $C_2H_3O_2^-$, F^-, CN^-, and S^{2-}. Conjugate bases of weak acids react with water in a manner that is illustrated by the reaction of F^- with H_2O:

$$F^-(aq) + H_2O(l) \rightleftharpoons HF(aq) + OH^-(aq) \quad K_b = \frac{[HF][OH^-]}{[F^-]}$$

Note that OH^- ion is produced in all reactions of a weak base with water.

- Solutions that are basic because of the presence of an anion are normally formed from salts containing a basic anion and a group 1A or group 2A cation. These cations have no significant acidic properties. Examples of basic salts are NaF, $Ca(CN)_2$, and $KC_2H_3O_2$.

Problem-solving techniques used for calculating pH and percent ionization of weak acid and base solutions are shown in the following exercises and in the next topic section. Also, review the summary of steps involved in calculating equilibrium concentrations in Chapter 15 of the *Student's Guide*.

EXERCISE 11   Calculate the percentage of hydroxylamine ionized in a 0.020 *M* hydroxylamine solution. K_b is 1.1×10^{-8} for hydroxylamine (NH_2OH).

Solution: The percentage of hydroxylamine ionized is the amount of its conjugate acid formed divided by the initial concentration of hydroxylamine times 100. The amount of hydroxylamine's conjugate acid formed by the reaction of hydroxylamine with water is calculated using the problem-solving technique shown in the Sample Exercises 16.10, 16.11, and 16.12 in the text. First write the chemical equation for the reaction and then the K_b expression:

$$H_2NOH(aq) + H_2O(l) \rightleftharpoons H_3NOH^+(aq) + OH^-(aq)$$

$$K_b = \frac{[H_3NOH^+][OH^-]}{[H_2NOH]} = 1.1 \times 10^{-8}$$

Tabulate the initial and equilibrium concentrations of all species involved in the equilibrium-constant expression. Let x equal the molarity of $H_3NOH^+(aq)$ ions formed.

	H_2NOH	$\rightleftharpoons$	H_3NOH^+	+	OH^-
Initial (*M*):	0.020		0		0
Reaction (*M*):	$-x$		$+x$		$+x$
Equilibrium (*M*):	$(0.020 - x)$		x		x

Note that for every molecule of H_2NOH reacting with water, one $OH^-(aq)$ ion and one $H_3NOH^+(aq)$ ion form. Thus, if x moles per liter of $H_3NOH^+(aq)$ are formed at equilibrium, then x moles per liter of OH^- must also have formed, and x moles per liter of H_2NOH must have reacted with water. Substitute the equilibrium concentrations into the equilibrium-constant expression:

$$K_b = \frac{[H_3NOH^+][OH^-]}{[H_2NOH]} = \frac{(x)(x)}{0.020 - x} = 1.1 \times 10^{-8}$$

The small size of K_b tells you that very little hydroxylamine reacts with water; thus the value of x is much smaller than 0.020 M. Making the approximation that the value of x is much smaller than 0.020 M gives:

$$K_b = \frac{(x)(x)}{0.020 - x} \simeq \frac{(x)(x)}{0.020} = 1.1 \times 10^{-8}$$

Solving for x:

$$x^2 = (0.020)(1.1 \times 10^{-8}) = 2.2 \times 10^{-10}$$
$$x = \sqrt{2.2 \times 10^{-10}} = 1.5 \times 10^{-5}\ M = [H_3NOH^+] = [OH^-]$$

The approximation used in solving for x is considered valid if x is no more than 5 percent of the quantity it is subtracted from. In this case, x is $(1.5 \times 10^{-5}/0.020)(100) = 7.5 \times 10^{-2}\%$ of the quantity it is subtracted from; thus the assumption is valid. The percentage of hydroxylamine ionized is as follows:

$$\%\ \text{ionized}\ =\ \left(\frac{\text{amount of species ionized}}{\text{original concentration}} \right)(100)$$

$$\%\ \text{ionized}\ =\ \left(\frac{x}{0.020} \right)(100)\ =\ \left(\frac{1.5 \times 10^{-5}}{0.020} \right)(100)\ =\ 0.075\%$$

EXERCISE 12   Calculate the pH of a 0.015 M acetic acid solution. The value of K_a is 1.8×10^{-5} for acetic acid.

   Solution: You can calculate the pH of a 0.015 M $HC_2H_3O_2$ solution by using a problem-solving procedure similar to the one used in the previous exercise. First, write the equilibrium and the K_a expression:

$$HC_2H_3O_2(aq) \rightleftharpoons H^+(aq) + C_2H_3O_2^-(aq)$$

$$K_a = \frac{[H^+][C_2H_3O_2^-]}{[HC_2H_3O_2]} = 1.8 \times 10^{-5}$$

Next, develop a concentration table, letting x equal the molarity of H^+ ions formed at equilibrium. Note that x must also equal the equilibrium concentration of $C_2H_3O_2^-$ because $C_2H_3O_2^-$ ions are formed in the same molar amount as H^+ ions.

	$HC_2H_3O_2$	$\rightleftharpoons$	H^+	$+$	$C_2H_3O_2^-$
Initial (*M*):	0.015		0		0
Reaction (*M*):	$-x$		$+x$		$+x$
Equilibrium (*M*):	$(0.15 - x)$		x		x

Again, assume that x is smaller in value than the number from which it is subtracted: $0.015 - x \simeq 0.015$.

$$K_a = \frac{(x)(x)}{0.015} = \frac{x^2}{0.015} = 1.8 \times 10^{-5}$$

Solving for x we have

$$x^2 = (0.015)(1.8 \times 10^{-5}) = 2.7 \times 10^{-7}$$

$$x = \sqrt{2.7 \times 10^{-7}} = 5.2 \times 10^{-4} = [H^+]$$

and

$$pH = -\log[H^+] = -\log(5.2 \times 10^{-4}) = 3.29$$

EXERCISE 13 Explain for a triprotic acid why the following trend in equilibrium constants is observed:

$$K_{a1} > K_{a2} > K_{a3}$$

Solution: The first ionization step involves dissociation of one hydrogen ion from a neutral molecule. This requires less energy than the second ionization step, which involves dissociation of a hydrogen ion from a species with a 1− charge. The hydrogen ion and the species with a 1− charge experience coulombic electrostatic attraction for each other. Similarly, the third ionization step involves dissociation of a hydrogen ion from a species with a 2− charge, which requires even more energy for ion separation than the previous two ionizations. The greater the energy required to remove a proton from a weak acid, the smaller the value of K_a.

EXERCISE 14 **(a)** Explain why the hydrogen-ion concentration for a 0.10 *M* H_3PO_4 solution can be determined from its first ionization step. The successive equilibrium constants are: $K_{a1} = 7.5 \times 10^{-3}$, $K_{a2} = 6.2 \times 10^{-8}$, and $K_{a3} = 4.8 \times 10^{-13}$. **(b)** Calculate the pH of this solution.

Solution: **(a)** The three equilibria involved are:

$$H_3PO_4(aq) \rightleftarrows H^+(aq) + H_2PO_4^-(aq) \qquad K_{a1} = 7.5 \times 10^{-3}$$

$$H_2PO_4^-(aq) \rightleftarrows H^+(aq) + HPO_4^{2-}(aq) \qquad K_{a2} = 6.2 \times 10^{-8}$$

$$HPO_4^{2-}(aq) \rightleftarrows H^+(aq) + PO_4^{3-}(aq) \qquad K_{a3} = 4.8 \times 10^{-13}$$

The amount of hydrogen ion produced by the second and third ionizations is not appreciable compared to the amount produced by the first ionization: the values of K_{a2} and K_{a3} are considerably smaller than the value of K_{a1}—and the concentrations of $H_2PO_4^-$ and HPO_4^{2-} are much lower than the concentration of H_3PO_4. Therefore, you can assume that the amount of hydrogen ions produced approximately equals that formed from the first ionization. The total amount of $H_2PO_4^-(aq)$ produced from the first ionization is also effectively the same as the hydrogen-ion concentration because the amount of $H_2PO_4^-$ dissociated is small, as indicated by the small value of K_{a2}. **(b)** Using the first equilibrium to solve for the pH, you can construct the following:

	H_3PO_4 $\rightleftarrows$	H^+	+ $H_2PO_4^-$
Initial (*M*):	0.10	0 *M*	0 *M*
Reaction (*M*):	−x	+x	+x
Equilibrium (*M*):	(0.10 − x)	x	x

Since the value of K_{a1}, 7.5×10^{-3}, is moderately large, you will not be able to assume that 0.10 is much larger than the value of *x*:

$$K_{a1} = \frac{\left[H^+\right]\left[H_2PO_4^-\right]}{\left[H_3PO_4\right]} = \frac{(x)^2}{0.10 - x} = 7.3 \times 10^{-3}$$

Rearranging the above into a quadratic form gives

$$x^2 + 7.3 \times 10^{-3}\, x - 7.3 \times 10^{-4} = 0$$

Solving for *x* using the quadratic formula yields

$$x = \frac{-\left(7.3 \times 10^{-3}\right) \pm \sqrt{\left(7.3 \times 10^{-3}\right)^2 - 4(1)\left(-7.3 \times 10^{-4}\right)}}{2}$$

$$= 0.024\ M = \left[H^+\right]$$

Thus,

$$pH = -\log\left[H^+\right] = -\log(0.024) = 1.62$$

0.1 M solution of NH_4NO_3, $Fe(NO_3)_2$, or $Al(NO_3)_3$ are acidic. Examples of basic salt solutions are 0.1 M solutions of $NaC_2H_3O_2$, KCN, and CaF_2. The basicity or acidity of these solutions is caused by the reaction of an ion with water to produce $H^+(aq)$ or $OH^-(aq)$. Such a reaction is often referred to as **hydrolysis.**

- *Anions that are conjugate bases of weak acids undergo hydrolysis to form basic solutions.* An illustration of this is the reaction of the weak base, the acetate ion, with water:

$$C_2H_3O_2^-(aq) + H_2O(l) \rightleftharpoons HC_2H_3O_2(aq) + OH^-(aq)$$

- *Anions of strong acids, such as NO_3^-, Cl^-, Br^-, and ClO_4^- do **not** undergo hydrolysis.*

- *Some common cations that react with water to form acidic solutions are NH_4^+, Fe^{3+}, Fe^{2+}, Cu^{2+}, Al^{3+}, Zn^{2+}, and Cr^{3+}.* The chemical equations describing the reactions of cations other than NH_4^+ with water will be discussed in the Lewis acid-base theory section.

- *Alkali metal ions (Na^+, Li^+, Cs^+, and Rb^+) and heavier alkaline earth metal ions (Mg^{2+}, Ca^{2+}, Ba^{2+}, and Sr^{2+}) do not form acidic solutions.*

Sometimes K_b values for basic anions or K_a values for acidic cations derived from ammonia or amines are not available in tables.

- Fortunately, there is a relationship (see derivation in Section 16.7 of the text) that enables you to calculate one of these values if the other is known:

$$K_a \text{ (of acid)} \times K_b \text{ (conjugate base of acid)} = K_w$$

- Therefore,

$$K_a = \frac{K_w}{K_b} \quad \text{or} \quad K_b = \frac{K_w}{K_a}$$

EXERCISE 15   Classify the following aqueous solutions as neutral, basic, or acidic: **(a)** KCl; **(b)** NH_4NO_3; **(c)** Na_2CO_3; **(d)** RbF.

Solution: Determine if the cation is acidic or the anion is basic, using the principles given in the summary. **(a)** Neutral. K^+ is a group IA metal and therefore is not acidic. Cl^- is the conjugate base of the strong acid HCl and thus is not basic. **(b)** Acidic. NH_4^+ is the conjugate acid of the weak base NH_3, and therefore is acidic. NO_3^- is the conjugate base of the strong acid HNO_3 and thus is not basic. **(c)** Basic. Na^+, like K^+, is not acidic. CO_3^{2-} is the conjugate base of the weak acid HCO_3^- and is therefore basic. **(d)** Basic. Rb^+, like Na^+ and K^+, is not acidic. F^- is derived from the weak acid HF and is therefore a basic ion.

EXERCISE 16   Calculate the pH of a 0.25 M sodium formate ($NaCHO_2$) solution, given that the value of K_a of formic acid ($HCHO_2$) is 1.8×10^{-4}.

Solution: Proceeding as in Exercise 11, we write appropriate chemical reactions and an equilibrium-constant expression. Note that sodium salts ionize 100 percent in water. Therefore the initial concentration of CHO_2^- is 0.25 M, the same concentration as for $NaCHO_2$. Since CHO_2^- is derived from a weak acid, $HCHO_2$, it will react with water to form a *basic* solution and its conjugate acid, $HCHO_2$:

$$NaCHO_2(aq) \xrightarrow{100\%} Na^+(aq) + CHO_2^-(aq)$$

$$CHO_2^-(aq) + H_2O \rightleftharpoons HCHO_2(aq) + OH^-(aq)$$

K_b for CHO_2^- is not given, but K_a for $HCHO_2$ is. Therefore, K_b is calculated from the value of K_a:

$$K_b = \frac{[HCHO_2][OH^-]}{[CHO_2^-]} = \frac{K_w}{K_a} = \frac{1.0 \times 10^{-14}}{1.8 \times 10^{-4}} = 5.6 \times 10^{-11}$$

In the previous equation, the relation $K_b \times K_a = K_w$ is used to solve for the unknown K_b value. The value of K_a is given in the question. Next, you can construct a concentration table to solve for the hydroxide ion concentration.

	CHO_2^- + H$_2$O $\rightleftharpoons$ HCHO$_2$ +	OH$^-$
Initial (*M*):	0.25	0 0
Reaction (*M*):	$-x$	$+x$ $+x$
Equilibrium (*M*):	$(0.25 - x)$	x x

Since the value of K_b is small, you can assume $x \ll 0.25$, resulting in $0.25 - x \approx 0.25$. Substituting equilibrium concentration values into the K_b expression gives

$$K_b = \frac{(x)(x)}{0.25} 5.6 \times 10^{-11}$$

Solving for x, pOH, and pH yields

$$x = \sqrt{(0.25)(5.6 \times 10^{-11})} = 3.7 \times 10^{-6} M$$

$$pOH = -\log[OH^-] = -\log(x) = 5.43$$

$$pH = 14 - pOH = 8.57$$

FACTORS AFFECTING ACID-BASE STRENGTH

The acids discussed in Chapter 16 of the text are primarily of two types: (a) binary acids, such as HCl, HF, and H_2S; and (b) oxyacids, such as H_3PO_4, H_2SO_4, and $C_2H_3O_2$. Binary acids of the type H_nX show acidic properties when the strength of the H—X bond is low, and X is a highly electronegative atom (for example, F, Br or Cl), or the X^- ion is very stable.

- For a series of H_nX in the same family of nonmetallic elements, the acidity tends to increase with increasing atomic number of X.
- In the case of the hydrogen halides, hydrofluoric acid is a weak acid as contrasted to the strong acidic characteristics of HCl, HBr, and HI. The H—F bond is extremely strong and thus H—F does not easily ionize.
- Metallic hydrides are either basic or show very limited acid-base behavior.

Oxyacids contain a nonmetal central atom that is bonded to one or more oxygen atoms and to one or more hydroxide groups. For example, the structural formulas of H_2SO_4 and $C_2H_4O_2$ are

$$\overset{\displaystyle :\overset{..}{O}H}{\underset{\displaystyle :\overset{..}{O}H}{\overset{..}{O}=\overset{..}{S}=\overset{..}{O}}} \qquad \overset{\displaystyle H \quad :\overset{..}{O}:}{\underset{\displaystyle H}{H-\overset{|}{\underset{|}{C}}-\overset{\|}{C}-\overset{..}{O}H}}$$

sulfuric acid   acetic acid

A general formula representation of an oxyacid is $O_xY(OH)_y$, where Y is the central element.

- The relative acidity of oxyacids, $O_xY(OH)_y$, that have the same number of oxygen atoms and hydroxide units (that is, the same values of x and y) but differing central atoms increases with increasing electronegativity of the central atom.

- The relative acidity of oxyacids that have different numbers of oxygen atoms and hydroxide units (that is, different values of x and y), but the same central atom increases with increasing oxidation number of the central atom. The oxidation number of the central atom will increase with an increasing number of oxygen atoms attached to it.

- The carboxyl group is found in a group of organic acids known as carboxylic acids. The general formula of a carboxylic acid is RCOOH, where R represents all groups attached to the —COOH carboxyl group.

Bases generally involve compounds with ionizable hydroxide groups, or with a nitrogen atom that has an available electron pair.

- Base strengths of amines (R—NH$_2$, for example) decrease with increasing electronegativity of attached R groups.

- Ionizable hydroxide units are found with the more electopositive elements, such as Na$^+$, K$^+$, Ca^{2+}, and Ba^{2+}.

EXERCISE 17   Although NaOH and HOCl have the same formula type, Y-OH, they have different acid-base properties. Explain why NaOH is a base and HOCl is an acid.

Solution: H—O—Cl is a polar covalent compound. Chlorine is a highly electronegative atom; it draws the bonding electrons in the polar covalent O—H bond toward the polar covalent O—Cl bond. This further polarizes the O—H bond and weakens it sufficiently so that the proton can ionize in water to form H$^+$ (*aq*). Sodium is an electropositive atom and forms ionic compounds. Thus, NaOH is an ionic solid with discrete Na$^+$ and OH$^-$ ions, and it dissolves in water to give Na$^+$(*aq*) and OH$^-$(*aq*) ions.

EXERCISE 18   Which acid is the stronger in water, sulfurous acid, whose Lewis structure is

$$\underset{\displaystyle :\overset{..}{\underset{..}{O}}:}{H-\overset{..}{\underset{..}{O}}-\overset{..}{\underset{|}{O}}-\overset{..}{\underset{..}{O}}-H}$$

or sulfuric acid, whose Lewis structure is

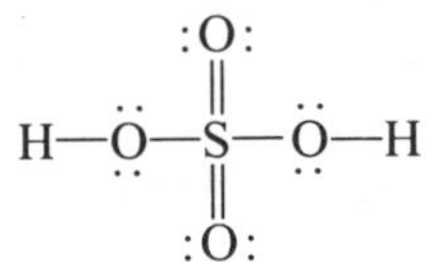

$$H-\overset{\cdot\cdot}{\underset{\cdot\cdot}{O}}-\overset{\overset{\displaystyle :O:}{\|}}{\underset{\underset{\displaystyle :O:}{\|}}{S}}-\overset{\cdot\cdot}{\underset{\cdot\cdot}{O}}-H$$

Solution: Sulfurous and sulfuric acid are oxyacids with different numbers of oxygen atoms and hydroxide groups, but with the same central atom. The sulfur atom in sulfuric acid has a +6 oxidation number; in sulfurous acid, the sulfur atom has a +4 oxidation number. Sulfuric acid is a stronger acid than sulfurous acid because the sulfur atom, with a +6 oxidation number, strongly polarizes the bonding electrons in the O—H bond toward the oxygen atom so that the proton is more easily removed by a base. In sulfurous acid, the sulfur atom has a smaller oxidation number of +4, resulting in a less polarized O-H bond and thus a lesser acidity.

THE LEWIS MODEL OF ACIDS AND BASES

The Bronsted-Lowry definition of an acid and a base is limited to substances that are able to donate or accept protons. A more general concept of acids and bases was proposed by G. N. Lewis.

- A **Lewis acid** is any substance that accepts an electron pair.
- A **Lewis base** is any substance that donates an electron pair.
- For example, BCl_3 is a Lewis acid because boron has an empty orbital that can accept an electron pair from a Lewis base like NH_3, as shown in the following reaction:

$$BCl_3 + :NH_3 \longrightarrow Cl_3B{:}NH_3$$

- Note that all Bronsted-Lowry acids and bases are Lewis acids and bases, but the opposite is not necessarily true. BCl_3 is a Lewis acid but not a Bronsted-Lowry acid because it has no ionizable protons.

The Lewis model enables you to understand the mechanisms of many acid-base reactions. Of particular interest in Section 16.10 of the text is the hydration of metal ions and the hydrolysis of metal ions.

- Dissolved metal ions are hydrated in water. For example, the aluminum ion occurs as $Al(H_2O)_6^{3+}$. It forms when the Lewis acid Al^{3+} reacts with the Lewis base water as follows:

$$Al^{3+} + 6 :\overset{\cdot\cdot}{O}H_2 \longrightarrow Al(:\overset{\cdot\cdot}{O}H_2)_6^{3+}$$

This is an example of a hydration reaction.

- Protons of hydrated water molecules bonded to certain metal ions may dissociate to form $H^+(aq)$. For example, a proton in one of the water molecules bonded to Al^{3+} is easily lost to the bulk solvent:

$$Al(H_2O)_6^{3+} \rightleftharpoons Al(H_2O)_5(OH)^{2+} + H^+(aq)$$

This is another example of a hydrolysis reaction.

- Metal ions with a high charge/ionic radius ration, such as Fe^{3+}, Al^{3+}, Cu^{2+}, Zn^{2+}, and Cr^{3+}, hydrolyze to the greatest extent, and thus form acidic solutions. Large ions with a +1 charge, such as Na^+, K^+, Cs^+, exhibit no acidic behavior.

EXERCISE 19 Using as an example the reaction of H^+ with OH^-, show that the Lewis definition of an acid is consistent with the Brønsted-Lowry definition.

Solution: H^+ is a Lewis acid because it acts as an electron-pair acceptor from electron-donor bases such as OH^-. This is shown by the following Lewis diagram:

$$H^+ + :\ddot{O}\!-\!H^- \longrightarrow [H^+ + \ddot{O}\!-\!H^-] \longrightarrow H\!-\!\ddot{O}\!-\!H$$

Therefore, any proton-donating substance (a Brønsted acid) is also a Lewis acid.

EXERCISE 20 Which reactant molecule or ion in the reaction

$$Ag^+(aq) + 2NH_3(aq) \longrightarrow Ag(NH_3)_2^+(aq)$$

is the Lewis acid? Which is the Lewis base?

Solution: The Lewis acid is $Ag^+(aq)$ because it accepts electron density from the Lewis base NH_3 to form the Lewis acid-base compound $Ag(NH_3)_2^+$ as follows:

$$H_3N: \longrightarrow Ag^+ \longleftarrow :NH_3$$

EXERCISE 21 Write the hydrolysis reaction for $Zn(H_2O)_4^{2+}$.

Solution: The bulk water acts as a base and removes a proton from one of the hydrated water molecules:

$$Zn(H_2O)_4^{2+} \rightleftharpoons Zn(H_2O)_3(OH)^+ + H^+(aq)$$

SELF-TEST QUESTIONS

Key Terms

Having reviewed the key terms listed at the end of Chapter 16 and their applications, identify the following statements as true or false. If a statement is false, indicate why it is incorrect. Key terms are italicized in the statements.

16.1 The *hydrolysis* reaction of $S^{2-}(aq)$ yields an acidic solution.

16.2 $K_a = \dfrac{[H^+][CN^-]}{[H_2O][HCN]}$

is the *acid-dissociation constant* expression for the ionization of HCN in water.

16.3 $K_b = \dfrac{[NH_4^+][OH^-]}{[NH_3]}$

is the *base-dissociation constant* for the reaction of $NH_3(aq)$ with $H_2O(l)$.

16.4 Donation of a proton from one water molecule to another to form $H^+(aq)$ and $OH^-(aq)$ is an example of an *autoionization* process.

16.5 The expression

$$K_w = \frac{[H^+][OH^-]}{[H_2O]^2}$$

is called the *ion-product constant* for water.

16.6 The *pH* of a 0.01 *M* NaOH solution is 2.

16.7 In the reaction

$$SO_3(g) + H_2O(l) \longrightarrow H_2SO_4(aq)$$

the *Lewis acid* is SO_3.

16.8 In the reaction in question 16.7, H_2O is a *Lewis base* because it donates a proton to SO_3 to form the acid H_2SO_4.

16.9 H_3PO_4 is a weaker *oxyacid* than H_3PO_3.

16.10 NH_3 acts as a *Bronsted-Lowry acid* in the reaction

$$NH_3(aq) + H_2O(l) \rightleftharpoons NH_4^+(aq) + OH^-(aq)$$

16.11 A *Bronsted-Lowry base must* contain a OH^- ionizable unit.

16.12 HSO_4^- is the *conjugate base* of H_2SO_4.

16.13 A *conjugate acid* is a substance formed by the addition of one proton to a Bronsted base.

16.14 CH_4 in water is an example of a *polyprotic acid*.

16.15 In the *autoionization* of water, equal molar quantities of hydrogen and hydroxide ions are formed.

16.16 The *hydronium ion* is a hydrated hydrogen ion and carries a positive charge.

16.17 H_2O and O^{2-} form a *conjugate acid-base pair*.

16.18 The hydroxide ion cannot act as an *amphoteric* substance.

16.19 For a substance to exist as an *amine* it must have a nitrogen atom with a lone pair of electrons.

16.20 A *carboxylic acid* contains the —COOH group, and this group contains a hydrogen attached to a carbon atom.

Problems and Short-Answer Questions

16.21 Identify the following substances as strong or weak acids or bases in water:
- (a) HCl;
- (b) H_2S;
- (c) NH_3;
- (d) KOH;
- (e) $HClO_4$;
- (f) HF;
- (g) H_3PO_4.

16.22 Give the conjugate acid of each of the following bases:
- (a) NH_2^-
- (b) CO_3^{2-}
- (c) HS^-
- (d) OH^-.

16.23 Give the conjugate base of each of the following acids:
- (a) HCl
- (b) $H_2PO_4^-$
- (c) NH_4^+
- (d) H_2CO_3.

16.24 Identify the acid and base in each of the following reactions:
- (a) $NH_4^+(aq) + H_2O(l) \rightleftharpoons NH_3(aq) + H^+(aq)$
- (b) $HC_2H_3O_2(aq) + H_2O(l) \rightleftharpoons H^+(aq) + C_2H_3O_2^-(aq)$
- (c) $HF(aq) + NaOH(aq) \longrightarrow NaF(aq) + H_2O(l)$

16.25 (a) Classify the following reactions as to type: Bronsted-Lowry, Lewis, or both.
- (1) $H_2O + NH_2^- \rightleftharpoons NH_3 + OH^-$
- (2) $BF_3 + F^- \rightleftharpoons BF_4^-$
- (3) $SO_3 + H_2O \rightleftharpoons H_2SO_4$

(b) For (3), write a mechanism using Lewis structures to show how H_2SO_4 is formed.

16.26 Calculate the pH and pOH of the following solutions:
- (a) 0.15 M HBr;
- (b) 0.0025 M $HClO_4$;
- (c) 0.075 M NaOH;
- (d) 0.0036 M $Ba(OH)_2$.

16.27 The pH of a 0.10 M nitrous acid solution is 2.17. Calculate K_a for a nitrous acid using this information.

16.28 What must be the molarity of an acetic acid solution so that its pH is 3.50?

16.29 What is the pH of a 0.20 M HClO solution? K_a for HClO is 3.2×10^{-8}.

16.30 Calculate the pH of a 0.35 M NH_4Cl solution. $K_b = 1.8 \times 10^{-5}$ for NH_3.

16.31 Calculate the pH of a solution labeled as 0.055 M NaCN. $K_a = 4.9 \times 10^{-10}$ for HCN.

16.32 Are the following salt solutions acidic, basic, or neutral? Write a net ionic equation that shows any reaction that occurs:
- (a) $KCl(aq)$;
- (b) $Fe(NO_3)_3(aq)$;
- (c) $NaF(aq)$.

16.33 Predict whether the salt $KHCO_3$ will form an acidic or basic solution in water, given the following data:

$$HCO_3^-(aq) \rightleftharpoons H^+(aq) + CO_3^{2-}(aq)$$
$$K_{a2} = 5.6 \times 10^{-11}$$

$$H_2CO_3(aq) \rightleftharpoons H^+(aq) + HCO_3^-(aq)$$
$$K_{a1} = 4.3 \times 10^{-7}$$

16.34 Ethanol, CH_3CH_2OH, is only faintly acidic in water. Why?

16.35 Which member of the following pairs would you expect to produce the more acidic solution:
- (a) NaCl or Na_2S;
- (b) $FeCl_2$ or $FeCl_3$;
- (c) $Cr(NO_3)_3$ or $Ca(NO_3)_2$;
- (d) NaH or HBr?

16.36 Ascorbic acid, $H_2C_6H_6O_6$, is a polyprotic acid containing two ionizable hydrogens. K_{a1} is 8.0×10^{-5} and K_{a2} is 1.6×10^{-12}. Calculate the pH of a 2.0×10^{-3} M solution of ascorbic acid.

Multiple-Choice Questions

16.37 Which of the following salts would form a basic solution in water?
- (a) NH_4NO_3
- (b) KNO_3
- (c) $CaCl_2$
- (d) $FeBr_3$
- (e) $CaCO_3$

16.38 Which of the following salts would form an acidic solution in water: (1) NH_4Cl; (2) $Fe(NO_3)_3$; (3) KCl; (4) NaF; (5) RbS?
- (a) (1) and (2)
- (b) (3) and (4)
- (c) (4) and (5)
- (d) all of them
- (e) none of them

16.39 What is the pH of a solution containing 2.5 g of NaOH dissolved in 100 mL of water?
- (a) 0.2
- (b) 13.8
- (c) 1.2
- (d) 12.8
- (e) 3.2

16.40 Which of the following 0.10 M solutions has the largest pH?
- (a) KNO_3
- (b) KF
- (c) KCl
- (d) K_2CO_3
- (e) K_2S

16.41 The pOH of a solution of NaOH is 11.30. What is the $[H^+]$ for this solution?

(a) 2.0×10^{-3} (d) 4.0×10^{-12}

(b) 2.5×10^{-3} (e) 6.2×10^{-8}

(c) 5.0×10^{-12}

(b) 2.04×10^{-9} (e) 6.92×10^{-3}

(c) 6.25×10^{-8}

16.42 Which one of the following binary compounds would you expect to be the most acidic?

   (a) NaH   (d) H_2O

   (b) CH_4 (e) H_2S

   (c) SnH_4

16.43 Which one the following oxyacids would you expect to be the most acidic?

   (a) HClO   (d) acetic acid

   (b) HBrO   (e) all are basic, not

   (c) HIO   acidic

16.44 Which one of the following is a Lewis acid, but *not* a Brønsted acid?

   (a) HCl   (d) KOH

   (b) BBr_3 (e) CH_4

   (c) NH_3

16.45 Which one of the following metal ions would you expect to show the greatest acidic properties in water?

   (a) K^+ (d) Cu^{2+}

   (b) Cs^+ (e) Al^{3+}

   (c) Ca^{2+}

16.46 Which of the following binary hydrides is the most basic?

   (a) HCl   (d) SiH_4

   (b) H_2S (e) NaH

   (c) PH_3

16.47 What is the percent dissociation of a 0.15 M HCN solution? K_a (HCN) = 4.9×10^{-10}

   (a) 7.35×10^{-11} (d) 5.71×10^{-3}

   (b) 8.57×10^{-3} (e) 5.71×10^{-5}

   (c) 8.57×10^{-6}

16.48 How many moles of $HC_3H_3O_2$ in a 200 mL solution are required to produce a solution with pH = 2.90? K_a ($HC_2H_3O_2$) = 1.8×10^{-5}

   (a) 1.79×10^{-2} (d) 1.62×10^{-3}

   (b) 8.82×10^{-2} (e) 1.79×10^{-3}

   (c) 8.95×10^{-2}

16.49 Which of the following 0.10 M solutions has the lowest pH?

   (a) $NaNO_2$ (d) NH_4Cl

   (b) $KClO_4$ (e) NH_3

   (c) K_2S

16.50 What is the pH of a 0.030 M $Ba(OH)_2$ solution?

   (a) 1.52   (d) 12.78

   (b) 12.48   (e) 0.03

   (c) 1.22

16.51 What is the pH of a 0.25 M weak acid that is 2.2% ionized?

   (a) 0.60   (d) 2.00

   (b) 1.00   (e) 2.26

   (c) 1.60

16.52 The pH of a 0.15 M weak base B is 9.25. What is K_b for B?

   (a) 1.38×10^{-10} (d) 2.39×10^{-7}

SELF-TEST SOLUTIONS

16.1 False. The hydrolysis reaction is $S^{2-}(aq) + H_2O \rightleftarrows$ $HS^-(aq) + OH^-(aq)$. Anions that hydrolyze form basic solutions. **16.2** False. $K_a = [H^+][CN^-]/[HCN]$. The term $[H_2O]$ is not included in the K_a expression. **16.3** True. **16.4** True. **16.5** False. $K_w = [H^+][OH^-]$. **16.6** False. pOH = 2; pH = 14 − 2 = 12. **16.7** True. **16.8** False. H_2O is a Lewis base because it donates an electron pair to SO_3. **16.9** False. The opposite is true. In a series of acids with the same central atom (P in this case) but a differing number of attached groups, acid strength increases with increasing oxidation number of the central atom. The oxidation number of P in H_3PO_4 is +5, which is greater than +3, the oxidation number of P in H_3PO_3. **16.10** False. It acts as a Bronsted base, accepting a proton to form NH_4^+. **16.11** False. NH_3 does not contain an ionizable OH unit, yet it is a base. A Bronsted base will produce OH^- in its reaction with water. **16.12** True. **16.13** True. **16.14** False. Although CH_4 contains hydrogen atoms, the hydrogen atoms do not ionize in water; in fact, CH_4 is insoluble in water. Carbon is not sufficiently electronegative to cause the hydrogens to become acidic. **16.15** True. **16.16** True. **16.17** False. The substances forming a conjugate acid-base pair differ only by the presence or absence of a single proton. Thus, H_2O and OH^- form a conjugate acid-base pair. **16.18** False. An amphoteric substance can act both as an acid and a base. The hydroxide ion can gain a proton to form water, and lose a proton to form the oxide ion. These occur in different types of reactions. In water, the more common reaction is as a base. **16.19** True. **16.20** False. The hydrogen atom is attached to an oxygen atom of a OH group in —COOH. This is the acidic hydrogen in carboxylic acids. **16.21** (a) strong acid; (b) weak acid; (c) weak base; (d) strong base (e) strong acid; (f) weak acid; (g) somewhat weak acid. **16.22** Add a H^+ to form a conjugate acid: (a) NH_3; (b) HCO_3^-; (c) H_2S; (d) H_2O. **16.23** Remove H^+ to form a conjugate base: (a) Cl^-; (b) HPO_4^{2-}; (c) NH_3; (d) HCO_3^-. **16.24** (a) NH_4^+—acid, H_2O—base; (b) $HC_2H_3O_2$—acid, H_2O—base; (c) HF—acid, NaOH—base. **16.25** (a) (1) Both. (2) Lewis. The fluoride ion donates electron density to the boron atom in BP_3. (3) Lewis. (b) Water acts as an electron-pair donor, and the sulfur atom in SO_3 acts as an electron-pair acceptor. After the initial addition compound is formed, a proton from an oxygen atom of water migrates to an oxygen atom in SO_3.

16.26 Each substance listed is either a strong acid or base. **(a)** pH = $-\log(0.15) = 0.82$; pOH = $14 - $ pH $= 13.18$. **(b)** pH = $-\log(0.0025) = 2.60$; pOH = $14 - 2.60 = 11.40$. **(c)** pH = $14 - $ pOH $= 14 - [-\log(0.075)] = 14 - 1.12 = 12.88$; pOH = $-\log(0.075) = 1.12$. **(d)** Note that there are two OH^- units per $Ba(OH)_2$ formula. pOH = $-\log(2 \times 0.0036) = 2.14$; pH = $14 - 2.14 = 11.86$. **16.27** $HNO_2(aq) \rightleftarrows H^+(aq) + NO_2^-(aq)$. The equilibrium concentration of $H^+(aq)$ is calculated from the given pH: $\log[H^+] = -2.17$; $[H^+] = 6.76 \times 10^{-3}$. $[NO_2^-] = 6.76 \times 10^{-3}$, because at equilibrium equal quantities of NO_2^- and H^+ are formed. The equilibrium concentration of HNO_2 is its initial concentration minus the amount ionized: $0.10\ M - 6.76 \times 10^{-3}\ M = 9.3 \times 10^{-2}\ M$. Substituting these values into the K_a expression yields $K_a = [H^+][NO_2^-]/[HNO_2] = (6.76 \times 10^{-3})(6.76 \times 10^{-3})/9.3 \times 10^{-2} = 4.9 \times 10^{-4}$. **16.28** This problem is solved in a fashion similar to problem 16.27, except that we now want to calculate $[HC_2H_3O_2]$ given K_a and the pH. From pH = 3.50, we can calculate $[H^+]$ as $3.16 \times 10^{-4}\ M$. As explained in the solution to excercise 12, the concentration of $C_2H_3O_2^-$ is also $3.16 \times 10^{-4}\ M$, the same as $[H^+]$. Thus, $K_a = [H^+][C_2H_3O_2^-]/[HC_2H_3O_2] = (3.16 \times 10^{-4})(3.16 \times 10^{-4})/[HC_2H_3O_2] = 1.8 \times 10^{-5}$; $[HC_2H_3O_2] = (3.16 \times 10^{-4})^2/1.8 \times 10^{-5} = 5.6 \times 10^{-3}$. Solving for $[HC_2H_3O_2]$ gives $[HC_2H_3O_2] = 5.6 \times 10^{-3}\ M$. The initial molarity of acetic acid is the sum of the ionized and unionized portions: $5.6 \times 10^{-3}\ M + 3.16 \times 10^{-4}\ M = 5.9 \times 10^{-3}\ M$. **16.29** Using a procedure analogous to the one in the solution to Exercise 12, $HClO(aq) \rightleftarrows H^+(aq) + ClO^-(aq)$; $K_a = [H^+][ClO^-]/[HClO] = 3.2 \times 10^{-8}$.

$HClO$	$\rightleftarrows$	$H^+(aq) +$	$ClO^-(aq)$
Initial (M):	0.20	0	0
Equilibrium (M):	$(0.20 - x)$	x	x

Assume x is much smaller than 0.20. $K_a = (x)(x)/0.20 = 3.2 \times 10^{-8}$ or $x = [H^+] = 8.0 \times 10^{-5}\ M$. The small size of x as compared to 0.02 justifies the assumption. The pH is $-\log(8.0 \times 10^{-5}) = 4.10$. **16.30** $NH_4Cl(aq) \xrightarrow{100\%} NH_4^+(aq) + Cl^-(aq)$; $NH_4^+(aq) + H_2O(l) \rightleftarrows NH_3(aq) + H^+(aq)$.

NH_4^+	$\rightleftarrows$	NH_3	$+$	H^+
Initial (M):	0.35	0		0
Equilibrium (M):	$(0.35 - x)$	x		x

Assume that x is significantly less than 0.35; therefore $0.35 - x \simeq 0.35$. $K_a = [NH_3][H^+]/[NH_4^+] = K_w/K_b(NH_3) = 1.0 \times 10^{-14}/1.8 \times 10^{-5} = 5.6 \times 10^{-10}$; $K_a = (x)(x)/0.35 = 5.6 \times 10^{-10}$; $x = [H^+] = 1.4 \times 10^{-5}$; pH = 4.85. **16.31** $NaCN(aq) \xrightarrow{100\%} Na^+(q) + CN^-(aq)$; $CN^-(aq) + H_2O(l) \rightleftarrows HCN(aq) + OH^-(aq)$.

CN^-	$\rightleftarrows$	$HCN +$	OH^-
Initial (M):	0.055	0	0
Equilibrium (M):	$(0.055 - x)$	x	x

Assume that x is significantly less than 0.055; therefore $0.055 - x \simeq 0.055$. $K_b = [HCN][OH^-]/[CN^-] = K_w/K_a(HCN) = 1.0 \times 10^{-14}/4.9 \times 10^{-10} = 2.0 \times 10^{-5}$; $K_b = (x)(x)/0.055 = 2.0 \times 10^{-5}$; $x = [OH^-] = 1.0 \times 10^{-3}$; pH = $14 - $ pOH $= 14 - 3.0 = 11.0$. **16.32 (a)** Neutral. $K^+(aq)$ and $Cl^-(aq)$ are not sufficiently acidic or basic to react with water. **(b)** Acidic: $Fe(H_2O)_6^{3+} \rightleftarrows Fe(H_2O)_5(OH)^{2+} + H^+(aq)$. **(c)** Basic: $F^-(aq) + H_2O \rightleftarrows HF(aq) + OH^-(aq)$. **16.33** The two possible reactions of HCO_3^- in

water are $HCO_3^-(aq) \rightleftarrows H^+(aq) + CO_3^{2-}(aq)$, with $K_{a2} = 5.6 \times 10^{-11}$, and $HCO_3^-(aq) + H_2O \rightleftarrows H_2CO_3(aq) + OH^-(aq)$, with an unknown K_b. The solution will be acidic or basic depending on which equilibrium constant has the larger value. The value of K_b can be calculated from the relation $K_a(H_2CO_3) \times K_b(HCO_3^-) = K_w$. Thus, $K_b = 1 \times 10^{-14}/(4.3 \times 10^{-7}) = 2.3 \times 10^{-8}$. Since K_b is larger than K_a for HCO_3^-, the reaction to form the OH^- ion predominates, producing a basic solution. **16.34** The ethyl group, CH_3CH_2, bonded to the OH group in ethanol (CH_3CH_2OH), only weakly attracts the electron pair it shares with oxygen. Thus, the O—H bond is only slightly polarized, and there is only a weak tendency for the hydrogen of OH to be transferred to the solvent water to form $H^+(aq)$. **16.35 (a)** A solution of NaCl will be more acidic because the other salt, Na_2S contains a basic anion, S^{2-}. **(b)** The solution of $FeCl_3$ is more acidic because it contains iron with a higher charge. **(c)** The solution of $Cr(NO_3)_3$ is more acidic because chromium has a 3+ charge where as calcium has a 2+ charge. In general, the higher the charge of the metal ion, the more acidic the solution. **(d)** HBr is more acidic; it is a strong acid. NaH contains the basic ion H^-. **16.36 (a)** Ascorbic acid has two ionizable hydrogen atoms. You can calculate the $[H^+]$ of the solution by using the first ionization only: If K_a values differ by a factor of 10^3 or more, you can obtain a reasonable estimate of the pH by using only K_{a1}.

$$H_2C_6H_6O_6(aq) \rightleftarrows H^+(aq) + C_6H_6O_6^-(aq)$$

Initial (M):	2.0×10^{-3}	0	0
Change (M):	$-x$	$+x$	$+x$
Equil (M):	$2.0 \times 10^{-3} - x$	x	x

You can assume that $x \ll 2.0 \times 10^{-3}$ because of the small value of K_{a1}. Substituting the equilibrium values into the equilibrium-constant expression gives

$$K_{a1} = \frac{(x)(x)}{2.0 \times 10^{-3}} = 8.0 \times 10^{-5}$$

Solving for the hydrogen ion concentration, x, gives

$$x = 4.0 \times 10^{-4}\ M \text{ and pH} = 3.40$$

16.37 (e). 16.38 (a). 16.39 (b).

$$M_{NaOH} = \frac{(2.5\ \text{g})\left(\dfrac{1\ \text{mol}}{40.01\ \text{g}}\right)}{(100\ \text{mL})\left(\dfrac{1\text{L}}{1000\ \text{mL}}\right)} = 0.62$$

pOH = $-\log(0.62) = 0.21$; pH = $14 - 0.21 = 13.79$. **16.40 (e).** K_b for S^{2-} is larger than K_b for either F^- or CO_3^{2-}. **16.41 (a). 16.42 (e).** H_2S has the most polar covalent H—X bond along with the weakest H—X bond. **16.43 (a). 16.44 (b).** The boron atom accepts a pair of electrons; yet BBr_3 has no ionizable protons. **16.45 (e).** Al^{3+} has the largest charge/ionic radius ratio; therefore it should react to the greatest extent with polar water to form $H^+(aq)$. **16.46 (e).** Alkali and alkaline-earth binary hydrides contain the H^- ion, which is a basic ion.

16.47 (d). $\dfrac{\left[H^+\right]}{C_0(HCN)} \times 100 = \dfrac{8.57 \times 10^{-6}}{0.15} \times 100 = 5.7 \times 10^{-3}\%$

HCN is an extremely weak acid as shown by the percent dissociation.

16.48 (a). $\dfrac{\left[H^+\right]\left[C_2H_3O_2^-\right]}{\left[HC_2H_3O_2\right]} = 1.8 \times 10^{-5}$

$$[HC_2H_3O_2] = \left(1.26 \times 10^{-3}\right)^2 \big/ 1.8 \times 10^{-5}$$
$$= 8.82 \times 10^{-2}$$

The initial concentration of acetic acid, C_0, is calculated by adding to the equilibrium concentration of acetic acid the amount that ionized, which is equal to the hydrogen ion concentration:

$$C_0 = [HC_2H_3O] + [H^+] = 8.82 \times 10^{-2} + 1.26 \times 10^{-3}$$
$$= 8.95 \times 10^{-2}\,M$$

Moles of acetic acid = 8.95×10^{-2} *mol/liter* $\times$ 0.200 *liter* = 1.79×10^{-2} *mol*. **16.49 (d).** The lower the pH, the more acidic the solution. Only NH_4^+ is acidic of those ions given. **16.50 (d)** $Ba(OH)_2$ produces two OH^- ions per $Ba(OH)_2$ formula. Thus, $[OH^-] = 2 \times 0.030\,M = 0.060\,M$. pOH = 1.22; pH = 12.78. **16.51 (e)** $HX \rightleftharpoons H^+ + X^-$. If 2.2 percent of HX is ionized, then $(0.022)(0.25\,M) = 5.5 \times 10^{-3}\,M$ of hydrogen ion is formed: pH = 2.26. **16.52 (b)** $B + H_2O \rightleftharpoons BH^+ + OH^-$. From pH, pOH is obtained: pOH = 4.75. Therefore $[OH^-] = 1.78 \times 10^{-5}$. The equilibrium reaction tells you that $[OH^-] = [BH^+] = 1.78 \times 10^{-5}$, and $[B] = 0.15 - 1.78 \times 10^{-5} \simeq 0.15$.

$$K_b = \frac{\left[\,BH\,\right]\left[OH^-\right]}{[B]} = \frac{\left(1.78 \times 10^{-5}\right)^2}{0.15} = 2.04 \times 10^{-9}$$

17 | Additional Aspects of Aqueous Equilibria

OVERVIEW OF THE CHAPTER

17.1 COMMON-ION EFFECT

Review: Net ionic equations (4.4); LeChatelier's principle (15.6)

Objective: You should be able to predict qualitatively and calculate quantitatively the effect of an added common ion on the pH of an aqueous solution of a weak acid or base.

17.2 TITRATIONS OF ACIDS AND BASES: TITRATION CURVES

Review: Neutralization (4.7); titrations and indicators (4.7); pH calculations involving strong and weak acids and bases (16.4, 16.5, 16.6).

Objectives: You should be able to:

1. Describe the form of the titration curves for titration of a strong acid by a strong base, a weak acid by a strong base, or a strong acid by a weak base.
2. Calculate the pH at any point, including the equivalence point, in acid-base titrations.

Review: pH (16.3); weak acids and bases (16.5, 16.6)

17.3 BUFFERS

Objectives: You should be able to:

1. Calculate the concentrations of each species present in a solution formed by mixing an acid and a base.
2. Describe how a buffer solution of a particular pH is made and how it operates to control pH.
3. Calculate the change in pH of a simple buffer solution of known composition caused by adding a small amount of strong acid or base.

17.4 SLIGHTLY SOLUBLE SALTS: SOLUBILITY-PRODUCT CONSTANTS

Review: Heterogeneous equilibria (15.3); hydrolysis of ions (16.8); LeChatelier's principle (15.6); solubility principles (13.1, 13.3).

Objectives: You should be able to:

1. Set up the expression for the solubility-product constant for a salt.

2. Calculate K_{sp} from solubility data and solubility from the value for K_{sp}.

3. Calculate the effect of an added common ion on the solubility of a slightly soluble salt.

Review: Hydration (13.1); Lewis acid-base concepts (16.10); concept of reaction quotient (15.5).

Objectives: You should be able to:

1. Predict whether a precipitate will form when two solutions are mixed, given appropriate K_{sp} values.

2. Explain the effect of pH on a solubility equilibrium involving a basic or acidic ion.

3. Formulate the equilibrium between a metal ion and a Lewis base to form a complex ion of a metal.

4. Describe how complex formation can affect the solubility of a slightly soluble salt.

5. Calculate the concentration of a metal ion in equilibrium with a ligand with which it forms a soluble complex ion, from a knowledge of initial concentrations and K_f.

6. Explain the origin of amphoteric behavior and write equations describing the dissolution of an amphoteric metal hydroxide in either an acidic or basic medium.

7. Explain the general principles that apply to the groupings of metal ions in the qualitative analysis of an aqueous mixture.

17.5, 17.6 CRITERIA FOR PRECIPITATING OR DISSOLVING SLIGHTLY SOLUBLE SALTS

TOPIC SUMMARIES AND EXERCISES

COMMON-ION EFFECT

A 0.10 M HC$_2$H$_3$O$_2$ solution has a pH equal to 2.9, whereas a solution containing 0.10 M HC$_2$H$_3$O$_2$ and 0.10 M NaC$_2$H$_3$O$_2$ has a pH equal to 4.7. Why has the addition of NaC$_2$H$_3$O$_2$ to a 0.10 M HC$_2$H$_3$O$_2$ solution caused a decrease in hydrogen ion concentration (increase in pH)? Section 17.1 in the text provides answers to this question.

■ When C$_2$H$_3$O$_2^-$ (from the strong electrolyte NaC$_2$H$_3$O$_2$) is added to an acetic acid solution that is at equilibrium, the equilibrium condition is displaced. According to LeChatelier's principle, the system restores itself to an equilibrium state by removing the added stress, that is, some of the added acetate ion.

$$HC_2H_3O_2(aq) \rightleftharpoons H^+(aq) + C_2H_3O_2^-(aq)$$

Equilibrium   Stress
←————————————   C$_2$H$_3$O$_2^-$ added
shifts to left
to remove some of
the added C$_2$H$_3$O$_2^-$

This simultaneously removes H^+ ions from solution with the formation of $HC_2H_3O_2$ and thereby increases the pH.

- The above change in equilibrium is an example of the **common-ion effect:** *addition of one of the ions involved in the ionic equilibria represses the ionization.*

The common ion effect is often produced by two methods:

1. Addition of a salt containing an ion common to the weak acid or base. Examples are the addition of NaF to a HF solution, KNO_2 to a HNO_2 solution and NH_4Cl to an NH_3 solution.

2. Addition of a strong base to partially neutralize a weak acid (HX) and thereby form its conjugate base (X^-):

$$HX(aq) + OH^-(aq) \xrightarrow{(100\%)} H_2O(l) + X^-(aq)$$

or addition of a strong acid (B) to partially neutralize a weak base to form its conjugate acid (BH^+):

$$B(aq) + H^+(aq) \xrightarrow{(100\%)} BH^+(aq)$$

EXERCISE 1   What would be the effect of adding the following to a solution of NH_3: **(a)** NH_4Cl; **(b)** KOH; **(c)** HCl?
  Solution: **(a)** Addition of NH_4^+ from the strong electrolyte NH_4Cl causes the following equilibrium to shift to the left:

$$NH_3(aq) + H_2O(l) \rightleftharpoons NH_4^+(aq) + OH^-(aq)$$

added

The concentration of OH^- ion decreases; pH decreases. **(b)** KOH is a strong base forming $K^+(aq)$ and $OH^-(aq)$. The addition of OH^- ion also drives the above equilibrium to the left, reducing the concentration of NH_4^+. Note that the pH is increased because not all of the added OH^- ion from KOH is removed. **(c)** Hydrogen ion from the strong acid HCl reacts with the weak base NH_3 to form NH_4^+.

$$NH_3(aq) + H^+(aq) \longrightarrow NH_4^+(aq)$$

NH_4^+ is weakly acidic. Therefore the total $[H^+]$ is increased and the pH decreases.

EXERCISE 2   Calculate the pH of a 0.15 M formic acid ($HCHO_2$) solution that also contains 0.050 M sodium formate ($NaCHO_2$). $K_a = 1.8 \times 10^{-4}$ for formic acid.
  Solution: Formic acid is a weak acid and forms the following equilibrium:

$$HCHO_2(aq) \rightleftharpoons H^+(aq) + CHO_2^-(aq)$$

Sodium formate is a strong electrolyte and completely dissociates:

$$NaCHO_2(aq) \xrightarrow{100\%} Na^+(aq) + CHO_2^-(aq)$$

Note that the ionization of $NaCHO_2$ forms CHO_2^-, an ion common to the equilibria involving $HCHO_2$. Therefore, both $HCHO_2$ and CHO_2^- are initially present in the solution. As you did in Chapter 16, develop a concentration table of species involved in the formic acid equilibria.

	HCH_2O	$\rightleftharpoons$	H^+	$+$	CHO_2^-	
Initial (M):	0.15		0		0.050	$\Big\{$ from $NaCHO_2$
Reaction (M):	$-x$		$+x$		$+x$	
Equilibrium (M):	$0.15 - x$		x		$0.050 + x$	

You can assume $x \ll 0.15$ and 0.050 because K_a is reasonably small. Solve for x by using the equilibrium constant expression.

$$K_a = \frac{\left[H^+\right]\left[CHO_2^-\right]}{\left[HCHO_2\right]} = \frac{(x)(0.05)}{0.15}$$

$$x = \frac{K_a(0.15)}{(0.050)} = \frac{(1.8 \times 10^{-4})(0.15)}{(0.050)}$$

$$x = \left[H^+\right] = 5.4 \times 10^{-4} \, M$$

$$pH = -\log\left[H^+\right] = 3.27$$

Note: A 0.15 M $HCHO_2$ solution *without* 0.050 M sodium formate also present has a pH = 2.28. The effect of the added sodium formate is to reduce $[H^+]$ and increase pH.

In Section 4.7 of the text, titrations were briefly described. The use of acid-base indicators to signal the presence of the equivalence point of an acid-base reaction was also discussed. In Section 17.2 of the text, you learn how to determine the appropriate acid-base indicator for a titration by examining titration curves. A **titration curve** is a plot of pH against the volume of added titrant.

Figure 17.5 in the text shows the general shape of a titration curve for a strong acid-strong base titration.

- Note that the equivalence-point pH occurs at 7.
- Also, the titration curve shows a large pH change for a small change in the volume of added titrant near the equivalence point.
- To detect the equivalence point in a strong acid-base titration, you need to choose an indicator whose color changes in the pH range of this rapidly rising section of the titration curve. Since this pH range is large, many indicators are available, and they do not need to change color precisely at pH = 7.
- You should be able to calculate pH at various points along the titration curve. There are four types of calculations:

1. Initial pH of the strong acid or strong base solution before the titrant is added.

2. pH of the strong acid or strong base solution that exists after *partial* neutralization of initial acid or base by the added titrant.

3. Equivalence-point pH, which in this case is 7, because water and a neutral salt are produced.

4. pH of the solution formed by addition of excess titrant beyond the equivalence point.

- See Sample Exercise 17.3 in the text for examples of these calculations.

Titration curves for weak acid-strong base and weak base-strong acid titrations are shown in Figures 17.7 and 17.9 in the text.

- Note the characteristics of these curves compared to those in Figure 17.2. Observe in these weak acid-base titration curves that a smaller pH changes occurs near the equivalence points. A suitable indicator to detect the equivalence point in these titrations must show a color change very near the equivalence-point pH.

- Also note that the equivalence-point pH for the titration of a weak acid with a strong base occurs at a pH that is greater than 7, and that for the titration of a weak base with a strong acid it occurs at a pH that is less than 7.

The calculations of pH changes during the titration of a weak acid with a strong base or the titration of a weak base with a strong acid are more complicated than the calculations of pH changes during the titration of a strong acid or base. The dissociation of the weak acid or base must be considered. There are four distinct types of calculations required:

1. At the start of the titration, the pH of the solution is determined by the initial concentration of the weak acid or base being titrated and its dissociation constant.

2. The pH values up to, but not including the equivalence point, are calculated from the equilibrium concentrations of the salt produced and the residual weak acid or base after addition of titrant. (See Sample Exercises 17.4 and 17.5 in the text for an example.)

3. At the equivalence point, the solution contains the conjugate base or acid of the titrated weak acid or base; the pH is calculated from the concentration of the conjugate base or acid and the value of the respective K_b or K_a.

4. Beyond the equivalence point, the pH is primarily controlled by the equilibrium concentration of excess titrating agent.

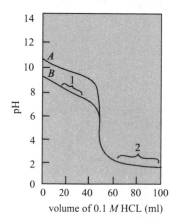

Figure 17.1 The pH of solutions of two weak bases, A and B, as a function of added HCl.

EXERCISE 3   In Figure 17.1 a graph of the pH of solutions of two weak bases, A and B, as a function of added 0.1 M HCl is shown. Both bases have the same initial concentrations. **(a)** Which weak base, A or B, has associated with it the larger K_b value? **(b)** Is segment 1 on the graph a buffer region? **(c)** Why is segment 2 on the graph the same for both bases?

   Solution: **(a)** For a series of weak bases with the same initial concentrations, the larger the K_b value of a weak base, the greater is the extent of its reaction with H_2O to produce OH^-, and the larger is the initial pH of the solution. Weak base A has associated with it the larger K_b value because its initial pH is larger. **(b)** Yes, because changes in pH are small as a function of added HCl. **(c)** After the equivalence point has been reached, the pH is determined by the concentration of excess titrant, HCl, which is the same for the titrations of both weak acids.

EXERCISE 4   Referring to the graph in Exercise 3, explain why phenolphthalein is a poor choice as an indicator to detect the equivalence point of the titration of a weak base with a strong acid. The color transition range as a function of pH for phenolphthalein is 9.0–9.6.

   Solution: The equivalence point in the titration of a weak base by a strong acid occurs at a pH less than 7. The use of phenolphthalein as an indicator would result in an apparent end point in the basic region (pH = 8.0–9.6) before the equivalence point is reached. Thus the titration would be ended too early, and the error in the volume of 0.1 M HCl required to neutralize the weak acid would be large.

EXERCISE 5   To help you understand how to do titration problems, we will examine the titration of a weak base using a strong acid. Sample Exercises 17.4 and 17.5 in the text show you how to solve problems involving the titration of a weak acid using a strong base. The problem is: Calculate the pH in the titration of 50.00 mL of 0.200 M ammonia by 0.200 M hydrochloric acid: **(a)** at the start of the titration; **(b)** after 20.00 mL of HCl has been added; and **(c)** at the equivalence point. K_b for ammonia is 1.8×10^{-5}.

   Solution: **(a)** At the start of the titration, the solution contains only aqueous ammonia; thus, the pH is determined from the following equilibrium and concentrations:

$$NH_3(aq) \rightleftharpoons NH_4^+(aq) + OH^-(aq)$$

Initial (M):	0.200	0	0
Reaction (M):	$-x$	$+x$	$+x$
Equilibrium (M):	$0.200 - x$	x	x

You can solve for x by using the K_b expression for NH_3 and by making the assumption that $x \ll 0.200\ M$:

$$K_b = \frac{\left[NH_4^+\right]\left[OH^-\right]}{\left[NH_3\right]} = \frac{(x)(x)}{0.200} = 1.8 \times 10^{-5}$$

Solving for the value of x gives

$$x^2 = 3.60 \times 10^{-6} \quad \text{or} \quad x = [OH^-] = 1.90 \times 10^{-3}$$

$$pOH = 2.72 \quad \text{or} \quad pH = 11.28$$

(b) The addition of 20.00 mL of HCl partially neutralizes the ammonia. Also, because the addition of HCl changes the volume of the final solution, you need to account for the dilution when doing calculations. In this problem, you will first calculate the molarities of HCl and ammonia in the mixture before they react. The total volume of the solution of HCl and ammonia is 20.00 mL + 50.00 mL = 70.00 mL. To calculate the molarities in the mixture, you can use the following relationship:

$$M_i V_i = M_f V_f$$

where i refers to the components before mixing and f refers to the final mixture.

$$M_{HCl}(\text{mixture}) = \frac{M_{HCl} V_{HCl}}{V_{\text{mixture}}} = \frac{(0.200\ M)(20.00\ \text{mL})}{70.00\ \text{mL}} = 0.0571\ M$$

$$M_{NH_3}(\text{mixture}) = \frac{M_{NH_3} V_{NH_3}}{V_{\text{mixture}}} = \frac{(0.200\ M)(50.00\ \text{mL})}{70.00\ \text{mL}} = 0.143\ M$$

After you have calculated the molarities of the components after mixing, you need to consider what happens to their concentrations after the HCl reacts with the base in the solution, NH_3. This is a neutralization reaction. You should note that the concentration of HCl is less than that of NH_3. This means that HCl will be a limiting reacting and will be entirely consumed. Write the neutralization reaction and underneath write the initial concentrations (the concentrations after dilution). Assume the reaction goes to completion and calculate the molarities after the reaction occurs:

	$H^+(aq)$	$+$	$NH_3(aq)$	$\xrightarrow{100\%}$	$NH_4^+(aq)$
Initial (M):	0.0571		0.143		0
Reaction (M):	-0.0571		-0.0571		$+0.0571$
Final (M):	0		0.086		0.0571

Note that HCl (H⁺) is indeed the limiting reactant in the above reaction. The resulting solution contains 0.086 M ammonia and 0.0571 M ammonium ion, which is a common ion solution. You can now solve for pH using the equilibrium used in (a):

	$NH_3(aq)$	$\rightleftharpoons$	$NH_4^+(aq)$	+	$OH^-(aq)$
Initial (M):	0.086		0.0571		0
Reaction (M):	$-x$		$+x$		$+x$
Equilibrium (M):	$0.086 - x$		$0.0571 + x$		x

You can solve for x by using the K_b expression and assuming that $x \ll 0.086$ M or 0.0571 M:

$$K_b = \frac{(x)(0.0571)}{0.086} = 1.8 \times 10^{-5}$$

Solving for x gives: $x = [OH^-] = 2.7 \times 10^{-5}$ $\qquad$ Thus, pOH = 4.57 and pH = 9.43. Note that the pH decreased upon addition of HCl. This is to be expected because acid is added to the original solution. If the pH had increased, you would have known that an error had been made. Also, the pH should be above 7 since the ammonia solution was not completely neutralized. **(c)** At the equivalence point the ammonia is completely neutralized. You need to calculate the volume of HCl required for complete neutralization of 50.00 mL of 0.200 M NH_3 ammonia. You can use the following relationship to calculate the volume of HCl required because HCl and ammonia reaction in a 1:1 mole ratio:

$$M_{HCl}V_{HCl} = M_{NH_3}V_{NH_3} \qquad [\text{Remember: moles } = M \times V]$$

$$V_{HCl} = \frac{M_{NH_3}V_{NH_3}}{V_{HCl}} = \frac{(0.200\ M)(50.00\ mL)}{0.200\ M} = 50.00\ mL$$

The solution formed at the equivalence point has a volume of 50.00 mL + 50.00 mL = 100.00 mL. You can now calculate the new molarities of HCl and ammonia upon mixing as you did in **(b)**; however, you should observe that the final solution is simply twice the volume of either starting solution. This means that the molarities after dilution are simply one-half of their original molarities.

$$M_{HCl} = \frac{1}{2}(0.200\ M) = 0.100\ M$$

$$M_{NH_3} = \frac{1}{2}(0.200\ M) = 0.100\ M$$

Next, calculate the concentration of ammonium ion that forms at the equivalence point after all of the ammonia is neutralized.

	$H^+(aq)$	+	$NH_3(aq)$	$\xrightarrow{100\%}$	$NH_4^+(aq)$
Initial (M):	0.100		0.100		0
Reaction (M):	-0.100		-0.100		$+0.100$
Final (M):	0		0		0.100

The solution now only contains ammonium ion, a weak acid. The new equilibrium that is responsible for the pH of the solution is:

	$NH_4^+(aq)$	$\rightleftharpoons$	$NH_3(aq)$	+	$H^+(aq)$
Initial (M):	0.100		0		0
Reaction (M):	$-x$		$+x$		$+x$
Equilibrium (M):	$0.100 - x$		x		x

Solve for x by using K_a for ammonium ion and assuming that $x \ll 0.100$. K_a is calculated from the relationship $K_a \times K_b = K_w$.

$$K_a = \frac{[H^+][NH_3]}{[NH_4^+]} = \frac{K_w}{K_b(NH_3)} = \frac{1.0 \times 10^{-10}}{1.8 \times 10^{-5}} = 5.6 \times 10^{-10}$$

$$5.6 \times 10^{-10} = \frac{(x)(x)}{0.100}$$

Solving for x gives:

$$x^2 = 5.6 \times 10^{-11} \quad \text{or} \quad x = [H^+] = 7.48 \times 10^{-6} \, M \quad \text{or} \quad pH = 5.13$$

Note that the pH at the equivalence point of the titration of a weak base is in the acid region.

A solution containing 0.1 M $HC_2H_3O_2$ and 0.1 M $NaC_2H_3O_2$ resists *large* changes in pH when *small* amounts of acid or base are added. Similarly, a solution of 0.1 M NH_3 and 0.1 M NH_4Cl resists large changes in pH under similar conditions. Solutions that resist large changes in pH when small amounts of acid or base are added are called **buffer solutions.** You will encounter two types of buffers:

| BUFFER SOLUTIONS

1. *Acid buffer:* A weak acid and a salt containing the conjugate base of the weak acid (for example, $HC_2H_3O_2$ and $NaC_2H_3O_2$);

2. *Base buffer:* A weak base and a salt containing the conjugate acid of the weak base (for example, NH_3 and NH_4Cl).

▪ A buffer solution must have an acidic component that can neutralize added base and a basic component that can neutralize added acid.

▪ *Note that buffer equilibria are the same as those found in solutions that exhibit the common-ion effect.*

Buffer capacity refers to the effectiveness of a buffer system in resisting changes in pH.

▪ Buffer solutions are the most effective when the concentrations of the conjugate acid-base pair are about equal, that is, when the pH of the buffer solution is close to pK_a.

▪ Buffer capacity also depends on the concentration of the conjugate acid-base pair. For example, a 0.1 M $NaNO_2$ and 0.1 M HNO_2 solution has a larger buffer capacity than a solution containing 0.001 M $NaNO_2$ and 0.0001 M HNO_2 because the former solution has more acid and base particles that can neutralize added acid or base.

EXERCISE 6   Write the net ionic equations describing the reactions that occur in the $HC_2H_3O_2 = NaC_2H_3O_2$ buffer system upon addition of **(a)** HCl; **(b)** NaOH; **(c)** NaCl.
    Solution: First identify the acid and base components of the buffer. Then identify the acid and/or base component of the substance added to the buffer. The net ionic equation is written by recognizing that the acid (base) component of the buffer reacts with the base (acid) component of the substance added. **(a)** The base component of the buffer, $C_2H_3O_2^-$, reacts with the added acid (HCl):

$$C_2H_3O_2^-(aq) + H^+(aq) \longrightarrow HC_2H_3O_2(aq)$$

(b) The acid component of the buffer, $HC_2H_3O_2$, reacts with the added base (NaOH):

$$HC_2H_3O_2(aq) + OH^-(aq) \longrightarrow H_2O + C_2H_3O_2^-(aq)$$

(c) There is no reaction because $Na^+(aq)$ is not acidic and $Cl^-(aq)$ is not basic.

EXERCISE 7   A 1.00-L buffer solution contains 0.100 M NH_4Cl and 0.020 M NH_3. K_b for NH_3 is 1.8×10^{-5}. **(a)** What is the pH of the solution? **(b)** What is the pH of the solution if 0.0010 mol of NaOH is added?

    Solution: **(a)** The base dissociation reaction and the corresponding equilibrium-constant expression are written first:

$$NH_3(aq) + H_2O(l) \rightleftharpoons NH_4^+(aq) + OH^-(aq)$$

$$K_b = \frac{[NH_4^+][OH^-]}{[NH_3]} = 1.8 \times 10^{-5}$$

If you let x equal the molarity of NH_3 that ionizes, you can construct the following concentration table:

	H_2O	+	NH_3	$\rightleftharpoons$	NH_4^+	+	OH^-
Initial:			0.020 M		0.100 M		0 M
Ionization effect:			$-x$ M		$+x$ M		$+x$ M
Equilibrium:			$(0.020 - x)$ M		$(0.100 + x)$ M		x M
Equilibrium*:			0.020 M		0.100 M		x M

*The value of x is small because the presence of NH_4^+ from NH_4Cl represses the ionization of NH_3; so you can assume that the equilibrium concentrations are 0.020 M for NH_3, and 0.100 M for NH_4^+.

    Solving for x by appropriate substitutions into the K_b expression yields

$$\frac{[NH_4^+][OH^-]}{[NH_3]} = 1.8 \times 10^{-5}$$

$$\frac{(0.100 \ M)(x)}{0.020 \ M} = 1.8 \times 10^{-5}$$

$$x = \left(\frac{0.020}{0.100}\right)(1.8 \times 10^{-5}) = 3.6 \times 10^{-6} M = [OH^-]$$

$$pOH = -\log[OH^-] = 5.44$$

The pH is calculated using the relation

$$pH + pOH = 14.00$$
$$pH = 14 - pOH = 14.00 - 5.44 = 8.56$$

(b) The added NaOH reacts with the acid component of the buffer as shown below:

$$OH^-(aq) + NH_4^+(aq) \longrightarrow H_2O(l) + NH_3(aq)$$

This reaction proceeds in the forward direction to a very large extent. You can effectively assume that all of the added $OH^-(aq)$ reacts with an equivalent amount of $NH_4^+(aq)$ to form an equivalent amount of $NH_3(aq)$. Thus, you must change the initial concentrations of NH_4^+ and NH_3 to account for this reaction. You must increase the concentration of NH_3 by the amount of NaOH that has reacted and decrease correspondingly the initial concentration of NH_4^+. The result is shown in the following concentration table.

	H$_2$O	+	NH$_3$	$\rightleftharpoons$	NH$_4^+$	+	OH$^-$
Initial:			0.020 M		0.100 M		0.0010 M*
Change in conc. because of added OH$^-$:			+0.0010 M		−0.0010 M		−0.0010 M
New initial conc.:			0.021 M		0.099 M		0 M
Ionization effect:			−x M		+x M		+x M
Equilibrium:			(0.021 − x) M		(0.099 + x) M		x M

Assume that $x \ll 0.021$ and 0.099:

Equilibrium:	0.021 M	0.099 M	x M

*Concentration of added NaOH

Note that the added base (OH$^-$) has increased the equilibrium concentration of the base component (NH$_3$) of the buffer. Solving for x using the K_b expression yields

$$\frac{[NH_4^+][OH^-]}{[NH_3]} = 1.8 \times 10^{-5}$$

$$\frac{(0.099)(x)}{0.021} = 1.8 \times 10^{-5}$$

$$x = \left(\frac{0.021}{0.099}\right)(1.8 \times 10^{-5})\,M = 3.8 \times 10^{-6}\,M = [OH^-]$$

$$pOH = -\log[OH^-] = 5.42$$

The pH is 8.58. The effect of the added OH$^-$ is to change the pH by only 0.02 pH units, which is a very small change.

EXERCISE 8   Which buffer solution has the larger buffer capacity toward added acid, 0.010 M HC$_2$H$_3$O$_2$ and 0.010 M NaC$_2$H$_3$O$_2$ or 0.010 M HC$_2$H$_3$O$_2$ and 0.0020 M NaC$_2$H$_3$O$_2$?

   Solution: The buffer capacity of a solution is defined as its ability to resist changes in pH when an acid or a base is added. The smaller the change in pH upon addition of an acid or a base, the greater the buffer capacity. The buffer solution containing 0.010 M HC$_2$H$_3$O$_2$ and 0.010 M NaC$_2$H$_3$O$_2$ has the larger buffer capacity toward added acid because it contains more base particles to neutralize added acid.

EXERCISE 9   What concentrations of acetic acid and sodium acetate are required to prepare a buffer solution with a pH = 4.60? $K_a = 1.8 \times 10^{-5}$ for acetic acid.

   Solution: Sodium acetate is a strong electrolyte and completely ionizes to form sodium ion and acetate ion.

$$NaC_2H_3O_2(aq) \xrightarrow{100\%} Na^+(aq) + C_2H_3O_2^-(aq)$$

Acetic acid is a weak acid,

$$HC_2H_3O_2(aq) \rightleftharpoons H^+(aq) + C_2H_3O_2^-(aq)$$

It is this equilibrium that determines pH. In this problem, you are given the pH of the buffer solution. From the pH, the equilibrium concentration of H$^+$ is determined:

$$[H^+] = 10^{-4.60} = 2.51 \times 10^{-5}$$

The required concentrations of acetic acid and acetate ion to produce a pH = 4.60 are calculated by first developing a concentration table and then using K_a. Since you do not know the initial concentrations of acetic acid and acetate ion, you can let x = *initial* molarity of HC$_2$H$_3$O$_2$ and y = *initial* concentration of acetate ion, C$_2$H$_3$O$_2^-$.

$$\begin{array}{lccc} & HC_2H_3O_2(aq) \rightleftharpoons & H^+(aq) & + & C_2H_3O_2^-(aq) \\ \text{Initial } (M) & x & 0 & y \\ \text{Reaction } (M) & -2.51 \times 10^{-5} & +2.51 \times 10^{-5} & +2.51 \times 10^{-5} \\ \text{Equilibrium } (M) & x - 2.51 \times 10^{-5} & 2.51 \times 10^{-5} & y + 2.51 \times 10^{-5} \end{array}$$

Choose x and y so that $x, y > 2.51 \times 10^{-5}$. Therefore,

$$\begin{array}{lccc} \text{Equilibrium } (M) & x & 2.51 \times 10^{-5} & y \end{array}$$

$$K_a = \frac{[H^+][C_2H_3O_2^-]}{[HC_2H_3O_2]} = \frac{(2.51 \times 10^{-5})(y)}{x} = 1.8 \times 10^{-5}$$

$$\frac{y}{x} = \frac{K_a}{[H^+]} = \frac{1.8 \times 10^{-5}}{2.51 \times 10^{-5}} = 0.72$$

Note that you have one equation with two unknowns. You cannot solve for x and y unless you assign a value to one of them. You can design a buffer by choosing arbitrarily $[HC_2H_3O_2] = x = 0.10\ M$ and then solving for $[C_2H_3O_2^-](=y)$.

$$\frac{y}{0.10\ M} = 0.72 \quad \text{or} \quad y = (0.10\ M)(0.72) \quad \text{or} \quad y = 0.072\ M$$

The acetate ion concentration (and therefore the sodium acetate concentration) must be 0.072 M in a 0.10 M $HC_2H_3O_2$ buffer solution to produce a pH = 4.60.

SLIGHTLY SOLUBLE SALTS: SOLUBILITY-PRODUCT CONSTANTS

When a saturated solution of the slightly soluble salt $PbCl_2$ is in contact with solid $PbCl_2$, there is an equilibrium between the solid $PbCl_2$ and the dissolved Pb^{2+} and Cl^- ions. This equilibrium is formulated as

$$PbCl_2(s) \rightleftharpoons Pb^{2+}(aq) + 2Cl^-(aq)$$

- The concentration of ions in a saturated solution are determined by the **solubility-product constant, K_{sp},** which for $PbCl_2$ is given by

$$K_{sp} = [Pb^{2+}][Cl^-]^2 = 1.7 \times 10^{-5}$$

- The solubility-product expression for the generalized equilibrium

$$M_aX_b(s) \rightleftharpoons aM^{m+}(aq) + bX^{n-}(aq)$$

is

$$K_{sp} = [M^{m+}]^a[X^{n-}]^b$$

- The concentration units for dissolved particles are moles per liter and the concentration of the solid is included in the value of K_{sp}.

Solubility and solubility product are not the same.

- Solubility refers to the amount of substance that dissolves in a given quantity of solvent. The solubility of a slightly soluble salt is affected by the presence of an ion common to its equilibrium.

• The common-ion effect, which represses the dissociation of $M_aX_b(s)$, causes the solubility of $M_aX_b(s)$ to decrease.

• For example, the solubility of $CaSO_4$ is reduced in the presence of a Na_2SO_4 solution. The presence of SO_4^{2-} ion from Na_2SO_4 shifts the equilibrium

$$CaSO_4(s) \rightleftharpoons Ca^{2+}(aq) + SO_4^{2-}(aq)$$

to the left, thereby decreasing the amount of calcium sulfate that dissolves.

EXERCISE 10 (a) Calculate the solubility of $La(IO_3)_3$ in water in moles per liter using its K_{sp} value of 6×10^{-10}. (b) What concentration of La^{3+} is necessary to have a concentration of $0.02\ M\ IO_3^-$?

Solution: (a) The solubility equilibrium is formulated as

$$La(IO_3)_3(s) \rightleftharpoons La^{3+}(aq) + 3IO_3^-(aq)$$

and the corresponding solubility-product constant is

$$K_{sp} = [La^{3+}][IO_3^-]^3 = 6 \times 10^{-10}$$

The dissolving of $La(IO_3)_3$ in water yields three IO_3^- ions for each La^{3+} ion. Therefore, if the solubility of $La(IO_3)_3$ equals x, then x mol/L of La^{3+} forms and $3x$ mol/L of IO_3^- forms:

$$[La^{3+}] = x \qquad [IO_3^-] = 3x$$

Substituting these terms into the K_{sp} expression for $La(IO_3)_3$ yields

$$K_{sp} = [La^{3+}][IO_3^-]^3 = (x)(3x)^3 = 6 \times 10^{-10}$$
$$27x^4 = 6 \times 10^{-10}$$
$$x^4 = 2 \times 10^{-11}$$
$$x = 2 \times 10^{-3}\ M$$

Thus, the solubility of $La(IO_3)_3$ is calculated to be 2×10^{-3} mol/L. (b) The K_{sp} expression for $La(IO_3)_3$ is used to solve for the concentration of La^{3+}. The given concentration of $IO-_3$, $0.02\ M$, is substituted for the term $[IO_3^-]$

$$K_{sp} = [La^{3+}] (0.02)^3 = 6 \times 10^{-10} = [La^{3+}] (8 \times 10^{-6})$$

Solving for $[La^{3+}]$ gives

$$\left[La^{3+}\right] = \frac{6 \times 10^{-10}}{8 \times 10^{-6}} = 8 \times 10^{-5}\ M$$

EXERCISE 11 (a) What is the solubility of $BaSO_4$ in moles per liter in a solution containing $0.020\ M\ Na_2SO_4$? K_{sp} is 1.1×10^{-10} for $BaSO_4$. (b) Does this solution illustrate the common-ion effect?

Solution: (a) The solubility of $BaSO_4$ is controlled by its K_{sp} value, even if common ions are present in the solution. In this problem, there are two sources of the SO_4^{2-} ion, $BaSO_4$ and Na_2SO_4. If we let x be the solubility of $BaSO_4$, then x mol/L of Ba^{2+} and x mol/L of SO_4^{2-} form from the dissolution of $BaSO_4$. However, the equilibrium concentration of SO_4^{2-} is x plus the amount of SO_4^{2-} from Na_2SO_4—that is, $x + 0.020\ M$. The equilibrium is formulated as

$$BaSO_4(s) \rightleftharpoons Ba^{2+}(aq) + SO_4^{2-}(aq)$$
$$K_{sp} = [Ba^{2+}][SO_4^{2-}] = 1.1 \times 10^{-10}$$

The equilibrium concentrations are

$$[Ba^{2+}] = x\ M \qquad \text{and} \qquad [SO_4^{2-}] = (x + 0.020)\ M$$

The value of x is very small and can be assumed negligible compared to 0.020 M. Therefore, we can say

$$(x + 0.020)\, M \approx 0.020\, M$$

Solving for the value of x by appropriate substitutions into the K_{sp} expression for BaSO$_4$ yields

$$K_{sp} = [\text{Ba}^{2+}][\text{SO}_4^{2-}] = (x)(0.020) = 1.1 \times 10^{-10}$$
$$x = 5.5 \times 10^{-9}\, M$$

The solubility of BaSO$_4$ in a 0.020 M Na$_2$SO$_4$ solution is 5.5×10^{-9} mol/L. Note that 5.5×10^{-9} is indeed small compared to 0.020. **(b)** Yes. The common-ion effect refers to a solution in which there is present an ion common to a slightly soluble salt equilibrium. In this case the common ion is SO$_4^{2-}$, from Na$_2$SO$_4$. The common ion represses the ionization of BaSO$_4$ and thereby reduces its solubility.

CRITERIA FOR PRECIPITATING OR DISSOLVING SLIGHTLY SOLUBLE SALTS

A precipitate of a slightly soluble salt can be prepared by mixing appropriate salt solutions. One solution should contain a soluble salt containing the cation of the slightly soluble salt, and the other solution should contain a soluble salt containing the anion.

- A precipitate forms after the solutions are mixed if the **ion product, Q,** for the slightly soluble salt is greater than the K_{sp} value for the slightly soluble salt.

- You should remember that the ion product has the same form as the K_{sp} expression but it is evaluated using the initial salt-solution concentrations. Summarized in Table 17.1 are the possible relationships between the values of Q and K_{sp}.

Table 17.1   Relationships Between Q and K_{sp}

Relationship	Solution condition
$Q > K_{sp}$	Precipitation occurs until $Q = K_{sp}$
$Q = K_{sp}$	Equilibrium and a saturated solution
$Q < K_{sp}$	Solution is not saturated, and no precipitate forms

Sometimes chemists need to dissolve slightly soluble salts other than by the procedure of sufficiently diluting a saturated solution. In Section 17.5 of the text, you explore different methods for dissolving such salts. These are summarized below.

1. *If a slightly soluble salt contains a basic anion, such as BaCO$_3$, the addition of a strong acid causes its solubility to increase.* Acid (H$^+$) reacts with a basic anion and removes it from the salt equilibrium, thereby causing more of the slightly soluble salt to dissolve. For example, the addition of HNO$_3$ to a saturated solution of BaCO$_3$ that is in equilibrium with solid BaCO$_3$ causes solid BaCO$_3$ to dissolve along with evolution of CO$_2$(g). The net ionic equation describing the process is

$$\text{BaCO}_3(s) + 2\text{H}^+(aq) \longrightarrow \text{H}_2\text{O}(l) + \text{CO}_2(g) + \text{Ba}^{2+}(aq)$$

2. *Insoluble metal hydroxides and oxides that are amphoteric dissolve in both acids and bases.* The concept of amphoterism was introduced in Section 16.2 of the text, and you should review this section. Examples are amphoteric hydroxides and oxides containing the metal ions Al^{3+}, Cr^{3+}, Zn^{2+}, and Sn^{2+}. You should know these examples.

3. *A slightly soluble salt containing a metal ion that reacts with Lewis bases to form complex ions can have its solubility increased by addition of the appropriate Lewis base.* Metal ions that are good Lewis acids will often react with Lewis bases to form complex ions. The term **complex ion** refers in general to an ion that contains several different atoms. In this section, it refers to an ion that contains a metal ion bonded to Lewis bases. Some examples of complex ions that you should know are $Ag(NH_3)_2^+$, $Zn(OH)_4^{2-}$, HgS_2^{2-}, and $Cu(CN)_2^-$. For example, the net ionic describing the dissolving of $AgCl$ using $Na_2S_2O_3$ is as follows:

$$AgCl(s) + 2S_2O_3^{2-}(aq) \longrightarrow Ag(S_2O_3)_2^{3-}(aq) + Cl^-(aq)$$

The stability of a complex metal ion is measured by its **formation constant, K_f.** A formation constant for a complex metal ion is associated with the reaction of the metal ion with Lewis bases to form the complex metal ion. For the formation of $Ag(S_2O_3)_2^{3-}$, the reaction is written as

$$Ag^+(aq) + 2S_2O_3^{2-}(aq) \rightleftharpoons Ag(S_2O_3)_2^{3-}(aq)$$

$$K_f = \frac{\left[Ag(S_2O_3)_2^{3-}\right]}{\left[Ag^+\right]\left[S_2O_3^{2-}\right]^2}$$

EXERCISE 12   Predict the solubility of $Mg(OH)_2$ when the following substances are added to a saturated solution of $Mg(OH)_2$: **(a)** $NaOH$; **(b)** HCl; **(c)** $MgCl_2$.

Solution: The equilibrium of the slightly soluble salt $Mg(OH)_2$ is formulated as

$$Mg(OH)_2(s) \rightleftharpoons Mg^{2+}(aq) + 2OH^-(aq)$$

(a) Adding $NaOH$, which contains an ion common to $Mg(OH)_2$, drives the equilibrium to the left, thus decreasing the solubility of $Mg(OH)_2$. **(b)** The proton from HCl reacts with the OH^- ion to form water. This removes OH^-, thus increasing the solubility of $Mg(OH)_2$. **(c)** $MgCl_2$ is a soluble salt and a strong electrolyte. The additional magnesium ion from $MgCl_2$ drives the equilibrium to the left and thus decreases the solubility of $Mg(OH)_2$.

EXERCISE 13   Which of the following is a more informative description of the formation of $Al(OH)_4^-$ from an aluminum ion and hydroxide ions?

(a) $Al(H_2O)_6^{3+}(aq) + 4OH^-(aq) \rightleftharpoons Al(OH)_4(H_2O)_2^-(aq) + 4H_2O(l)]$
(b) $Al^{3+}(aq) + 4OH^-(aq) \rightleftharpoons Al(OH)_4^-(aq)$

Solution: Reaction **(a)** is a better description because Al^{3+} exists in water as the hydrated ion, not as a free ion. The reaction clearly shows that four hydroxide ions are displacing water molecules; or it can be viewed as hydroxide ions removing protons from bound water molecules and thus forming OH^- ions bonded to Al^{3+}.

EXERCISE 14   **(a)** Write a reaction equation that describes the addition of two hydroxide ions to the slightly soluble metal hydroxide $Zn(OH)_2(H_2O)_2$ and the addition of one proton to this same solution. **(b)** What is the term used to describe the behavior of $Zn(OH)_2(H_2O)_2$?

Solution: **(a)** The reaction of $Zn(OH)_2(H_2O)_2$ with two hydroxide ions is formulated as

$$Zn(OH)_2(H_2O)_2(s) + 2OH^-(aq) \rightleftharpoons Zn(OH)_4^-(aq) + 2H_2O(l)$$

and its reaction with one proton is formulated as

$$Zn(OH)_2(H_2O)_2(s) + H^+(aq) \rightleftharpoons Zn(OH)(H_2O)_3^+(aq)$$

Note that the total number of groups attached to zinc does not change in both cases. **(b)** Amphoterism. $Zn(OH)_2(H_2O)_2$ is an amphoteric hydroxide.

EXERCISE 15   Some slightly soluble salts can be dissolved through formation of a complex ion. For example, AgCl dissolves in water when CN^- is added because the very stable complex ion $Ag(CN)_2^-$ is formed. **(a)** Write the overall reaction equation for formation of $Ag(CN)_2^-$ from CN^- and AgCl. **(b)** Write two stepwise reaction equations that, when summed, yield the reaction equation you wrote in **(a)**. **(c)** Write the equilibrium-constant expression associated with each stepwise reaction in **(b)**.
Solution: **(a)** The overall reaction is

$$AgCl(s) + 2CN^-(aq) \rightleftharpoons Ag(CN)_2^-(ag) + Cl^-(aq)$$

Note that Cl^- is displaced from AgCl.
(b) The two equilibria involved when AgCl dissolves in aqueous cyanide solution are

$$AgCl(s) \rightleftharpoons Ag^+(aq) + Cl^-(aq) \qquad \text{[solubility equilibrium]}$$
$$Ag^+(aq) + 2CN^-(aq) \rightleftharpoons Ag(CN)_2^-(aq) \qquad \text{[complexation]}$$

When these two reaction equations are added together, the result is the overall reaction equation

$$AgCl(s) + 2CN^-(aq) \rightleftharpoons Ag(CN_2)^-(aq) + Cl^-(aq)$$

(c) The equilibrium-constant expressions are

$$K_{sp} = [Ag^+][Cl^-]$$

for the first reaction and

$$K_f = \frac{\left[Ag(CN)_2^-\right]}{\left[Ag^+\right]\left[CN^-\right]^2}$$

for the second reaction.

EXERCISE 16   Will $AgIO_3$ precipitate when 100.00 mL of 0.0100 M $AgNO_3$ is mixed with 50.00 mL of 0.0100 M KIO_3? K_{sp} of $AgIO_3$ is 3.0×10^{-8}.
Solution: KIO_3 and $AgNO_3$ undergo a reaction to form $AgIO_3$, which precipitates, and the soluble salt KNO_3. To determine whether $AgIO_3$ precipitates you must calculate the ion product, Q, for $AgIO_3$ and compare it with K_{sp}. The ion product is evaluated using the initial concentrations of silver ion and iodate ion: $Q = C_{Ag^+}C_{IO_3^-}$, where C means initial concentration. Note that Q has the same form as K_{sp} but it is evaluated using the starting concentrations of the ions. You cannot use the given concentrations because the solutions are mixed; thus, you must first calculate the molarities of the ions after mixing. The new volume is 100.00 mL + 50.00 mL = 150.00 mL. You can calculate the molarities after mixing using the following relationship:

$$M_iV_i = M_fV_f \qquad \text{or} \qquad M_f = \frac{M_iV_i}{V_f}$$

where i means initial and f means final, after mixing.

$$M_f(\text{AgNO}_3) = \frac{(0.0100\ M)(100.00\ \text{mL})}{150.00\ \text{mL}} = 0.00667\ M$$

$$M_f(\text{KIO}_3) = \frac{(0.0100\ M)(50.00\ \text{mL})}{150.00\ \text{mL}} = 0.00333\ M$$

Therefore, the starting concentrations of the ions in the mixture are $C_{\text{Ag}^+} = 0.00667\ M$ and $C_{\text{IO}_3^-} = 0.00333\ M$. You can now solve for Q,

$$Q = C_{\text{Ag}^+} C_{\text{IO}_3^-} = (0.00667)(0.00333) = 2.22 \times 10^{-5}$$

Because $Q > K_{sp}$, AgIO_3 precipitates.

EXERCISE 17   A sodium hydroxide solution with a pH of 6.00 has sufficient cerium nitrate added to make the total cerium ion concentration 0.00300 M. Will Ce(OH)_3 precipitate? K_{sp} of Ce(OH)_3 is 1.5×10^{-20}. Assume the addition of cerium nitrate does not change the pH.

Solution: The equilibrium of cerium ion with hydroxide ion is

$$\text{Ce(OH)}_3(s) \rightleftharpoons \text{Ce}^{3+}(aq) + 3\text{OH}^-(aq)$$

To determine whether cerium hydroxide precipitates, you must solve for the value of the ion product, Q. You are given the concentration of cerium ion and the pH of the solution. From the pH of the solution you can determine the hydroxide ion concentration:

$$\text{pOH} = 14 - \text{pH} = 14 - 6.00 = 8.00 \qquad \text{or} \qquad [\text{OH}^-] = 1.00 \times 10^{-8}\ M$$

Q can now be calculated:

$$Q = C_{\text{Ce}^{3+}} C_{\text{OH}^-}^3 = (0.00300)(1.00 \times 10^{-8})^3 = 3.00 \times 10^{-27}$$

Because $Q < K_{sp}$, cerium hydroxide does not precipitate.

SELF-TEST QUESTIONS

Key Terms

Having reviewed the key terms listed at the end of Chapter 17 and their applications, identify the following statements as true or false. If a statement is false, indicate why it is incorrect. Key terms are italicized in the statements.

17.1   The *common-ion effect* of NaCl in a saturated solution of AgCl increases the solubility of AgCl compared to that of AgCl in a saturated solution with no NaCl present.

17.2   Addition of NH_3 to a precipitate of AgCl in a saturated AgCl solution causes AgCl to dissolve via the formation of the *complex ion* $\text{Ag(NH}_3)_4^{2+}(aq)$.

17.3   Only the most insoluble metal sulfides form in an acidified H_2S solution during *qualitative-analysis* procedures.

17.4   A solution of 0.1 M $\text{NaC}_2\text{H}_3\text{O}_2$ and 0.001 M $\text{HC}_2\text{H}_3\text{O}_2$ should act as a very effective *buffer solution* toward added base.

17.5   The *solubility-product constant* expression for the equilibrium

$$\text{BaCO}_3(s) \rightleftharpoons \text{Ba}^{2+}(aq) + \text{CO}_3^{2-}(aq)$$

is

$$K_{sp} = \frac{[\text{Ba}^{2+}][\text{CO}_3^{2-}]}{[\text{BaCO}_3]}$$

17.6   The *formation-constant* value for $\text{Ni(NH}_3)_6^{2+}$ is 2×10^8. From this information we know that the following reaction occurs extensively:

$$\text{Ni(NH}_3)_6^{2+}(aq) \rightleftharpoons \text{Ni}^{2+}(aq) + 6\text{NH}_3(aq)$$

17.7   The *titration curve* for a weak acid, with a K_a of 3×10^{-9}, shows a steeply rising curve near the equivalence point.

17.8 A buffer solution containing 0.00100 M acetic acid and 0.00100 M sodium acetate will have a high *buffer capacity*.

17.9 The *Henderson-Hasselbach* equation shows for a buffer solution that changing the total volume of the solution does not change the pH.

17.10 *Quantitative analysis* requires the use of an accurate balance, or an accurate volume dispensing device.

Problems and Short-Answer Questions

17.11 Blood is buffered to a pH of 7.4 by several substances. One conjugate acid-base pair involved in the buffering action in blood is $H_2PO_4^-$ and HPO_4^{2-}.
  (a) Which component of the conjugate acid-base pair reacts with added acid? Write the net-ionic equation describing this buffering effect.
  (b) Calculate the ratio of $[HPO_4^{2-}]/[H_2PO_4^-]$ required to form a solution with a pH of 7.4. $K_{a2} = 6.2 \times 10^{-8}$ for $H_2PO_4^-$.

17.12 In Exercise 7 you calculated the pH of the buffer solution after addition of NaOH.
  (a) Calculate the pH after addition of 0.0015 mol of HCl.
  (b) Would the solution still be a buffer if 0.020 mol of HCl were added? In both problems, assume that no volume dilution occurs because of addition of HCl, which would be in solution form.

17.13 A solution initially contains 2.0×10^{-5} M $AgNO_3$ and 3.0×10^{-2} M NaCl. Given that $K_{sp} = 1.56 \times 10^{-10}$ for AgCl, is there a precipitate of AgCl in this solution?

17.14 When a small amount of 3 M NH_3 is added to a 0.1 M $Cu(NO_3)_2$ solution, a precipitate of $Cu(OH)_2$ forms. Upon addition of excess ammonia, $Cu(OH)_2$ dissolves. Write reaction equations to explain these observations.

17.15 Which conjugate acid-base pair should be used to prepare a buffer with a pH of 3.5?

Acid	Conjugate base	K_a
$HC_2H_3O_2$	$C_2H_3O_2^-$	1.8×10^{-5}
$HCHO_2$	CHO_2^-	1.7×10^{-4}
H_2CO_3	HCO_3^-	4.2×10^{-7}
H_3PO_4	$H_2PO_4^-$	7.5×10^{-3}

17.16 What is the solubility of AgCl in moles per liter in a 0.020 M $MgCl_2$ solution?

$$K_{sp} = 1.56 \times 10^{-10} \text{ for AgCl.}$$

17.17 When solid silver acetate, $AgC_2H_3O_2$, is placed in water, an equilibrium involving a slightly soluble salt is established. Experimentally, the equilibrium concentration of Ag^+ is greater than that of $C_2H_3O_2^-$. Explain this observation. (*Hint:* Consider the possibility of an acid-base reaction in addition to the process of silver acetate dissolving and forming ions.)

17.18 Calculate the pH in the titration of 50.00 mL of 0.1000 M NH_3 by 0.2000 M HCl:
  (a) after addition of 10.00 mL HCl;
  (b) at the equivalence point;
  (c) and 10.00 mL of HCl beyond the equivalence point. K_b for NH_3 is 1.8×10^{-5}.

17.19 How would you prepare one liter of a buffer solution of pH = 4.60 from one liter of 0.500 M acetic acid and one liter of 0.1000 M sodium acetate. That is, calculate the volumes of acetic acid and sodium acetate that must be diluted to one liter to form a buffer solution of pH = 4.60. $K_a = 1.8 \times 10^{-5}$ for acetic acid. See Exercise 9 for additional information.

17.20 Calculate the concentration of cobalt ion present in a solution containing 0.10 M $Co(NO_3)_2$ and 6.0 M NH_3. $K_f = 1.1 \times 10^5$ for $Co(NH_3)_6^{2+}$.

17.21 Calculate the pH of a solution that is prepared by dissolving 3.25 g of $NaC_2H_3O_2$ in 100 mL of 0.150 M HCl. $K_a = 1.8 \times 10^{-5}$ for acetic acid. Assume that no volume change occurs when $NaC_2H_3O_2$ is added.

Multiple-Choice Questions

17.22 Addition of HCl to a buffer solution containing NH_3-NH_4Cl causes which of the following to occur?
  (1) pH of solution increases
  (2) pH of solution decreases
  (3) NH_4^+ concentration increases
  (4) NH_3 concentration increases
  (a) (1)   **(d)** (2) and (3)
  (b) (2)   **(e)** (1) and (4)
  (c) (1) and (3)

17.23 A common ion solution is formed when which of the following substances is added to a solution containing HF?
  (a) HF   **(d)** NaF
  (b) HCl   **(e)** H_2O
  (c) NH_4Cl

17.24 Which of the following solutions should have the greatest buffering capacity toward added base?
  (a) 0.1 M $HC_2H_3O_2$ and 0.01 M $NaC_2H_3O_2$
  (b) 0.1 M $HC_2H_3O_2$ and 0.1 M $NaC_2H_3O_2$
  (c) 1 M $HC_2H_3O_2$ and 0.1 M $NaC_2H_3O_2$
  (d) 1 M HCl and 1 M NaOH
  (e) None of the above solutions can function as a buffer.

17.25 The net ionic equation describing the reaction of added NaOH to a buffer solution containing HCO_3^- and CO_3^{2-} is
  (a) $OH^-(aq) + CO_3^{2-}(aq) \rightarrow HCO_4^{3-}(aq)$
  (b) $OH^-(aq) + HCO_3^-(aq) \rightarrow H_2O(l) + CO_3^{2-}(aq)$
  (c) $2Na^+(aq) + CO_3^{2-}(aq) \rightarrow Na_2CO_3(aq)$
  (d) $Na^+(aq) + HCO_3^-(aq) \rightarrow NaHCO_3(aq)$
  (e) $Na^+(aq) + H_2O(l) \rightarrow H^+(aq) + NaOH(aq)$

17.26 What is the pH of a 0.150 M HCl solution at the start of its titration with 0.150 M NaOH?

(a) 0.150
(b) −0.150
(c) 0.82
(d) −0.82
(e) 13.2

(a) $Al(H_2O)_6^{3+}$
(b) $Al(H_2O)_5(OH)^{2+}$
(c) $Al(H_2O)_4(OH)^{2+}$
(d) $Al(H_2O)_3(OH)_3(s)$
(e) $Al(H_2O)_2(OH)_4^{-}$

17.27 What is the equivalence-point pH of the solution formed by the titration of 50.00 mL of 0.150 *M* HCl using 50.00 mL of 150 *M* NaOH?

(a) 1
(b) 4
(c) 7
(d) 9
(e) 10

17.28 What is the equivalence-point pH of the solution formed by the titration of 50.00 mL of 0.150 *M* $HC_2H_3O_2$ using 25.00 mL of 0.300 *M* NaOH?

(a) 3.22
(b) 4.53
(c) 7.00
(d) 8.26
(e) 8.89

17.29 What is the solubility-product expression for the equilibrium reaction

$$PbI_2(s) \rightleftharpoons Pb^{2+}(aq) + 2I^{-}(aq)$$

(a) $K_{sp} = [Pb^{2+}][I^{-}]$
(b) $K_{sp} = [Pb^{2+}]^2[I^{-}]$
(c) $K_{sp} = [Pb^{2+}][I^{-}]/[PbI_2]$
(d) $K_{sp} = [Pb^{2+}][I^{-}]^2$
(e) $K_{sp} = [Pb^{2+}][I^{-}]^2/[PbI_2]$

17.30 If *x* equals the solubility of Ag_2CO_3 in moles per liter, then how is the value of K_{sp} for Ag_2CO_3 related to the value of *x*?

(a) $K_{sp} = 4x^3$
(b) $K_{sp} = x^3$
(c) $K_{sp} = 2x^3$
(d) $K_{sp} = x^2$
(e) $K_{sp} = 2x^2$

17.31 The solubility of $BaCO_3$ in water is increased by which of the following: (1) addition of NaOH; (2) addition of HNO_3; (3) increasing pH of solution; (4) decreasing pH of solution?

(a) (1)
(b) (2)
(c) (1) and (3)
(d) (2) and (4)
(e) (3)

17.32 What is the formation-constant expression for $Fe(CN)_6^{4-}$?

(a) $K_f = [Fe^{2+}][CN^{-}]$
(b) $K_f = [Fe^{2+}][CN^{-}]^6$
(c) $K_f = [Fe^{2+}][CN^{-}]^6/[Fe(CN)_6^{4-}]$
(d) $K_f = [Fe(CN)_6^{4-}]/[Fe^{2+}][CN^{-}]^6$
(e) $K_f = [Fe(CN)_6^{4-}]/[Fe^{2+}][CN^{-}]$

17.33 Which of the following metal ions form amphoteric hydroxides: (1) Fe^{3+}; (2) Cr^{3+}; (3) Zn^{2+}; (4) Ca^{2+}; (5) Na^+?

(a) (1) and (2)
(b) (2) and (3)
(c) (3) and (4)
(d) (4) and (5)
(e) (2) and (4)

17.34 The pH of a standard solution of H_2S is too low to precipitate ZnS. Which of the following could be added to the solution to cause ZnS to precipitate?

(a) HCl
(b) HNO_3
(c) H_2S
(d) NaCl
(e) NaOH

17.35 Which of the following is the primary form of aluminum in a strongly basic solution?

SELF-TEST SOLUTIONS

17.1 False. According to LeChatelier's principle, addition of Cl^{-} from NaCl to the equilibrium $AgCl(s) \rightleftharpoons Ag^{+}(aq) + Cl^{-}(aq)$ drives the equilibrium to the left, thereby reducing the amount of AgCl that dissolves. **17.2** False. Although AgCl does dissolve in the presence of NH_3, the appropriate formula of the ammonia-silver metal-ion complex is $Ag(NH_3)_2^{+}$. **17.3** True. **17.4** False. Very little acid component exists to neutralize any base added to the solution. **17.5** False. No solid concentration appears in the denominator of a K_{sp} expression: $K_{sp} = [Ba^{2+}][CO_3^{2-}]$. **17.6** False. The reaction shown is for the dissociation of $Ni(NH_3)_6^{2+}$, not its formation. Thus for the given reaction, $K = 1/K_f = 1/(2 \times 10^8) = 5 \times 10^{-9}$, $Ni(NH_3)_6^{2+}$ is a very stable metal-complex ion, and the reaction $Ni(NH_3)_6^{2+}(aq) \rightleftharpoons Ni^{2+}(aq) + 6NH_3(aq)$ lies far to the left. **17.7** False. Weak acids with very small K_a's show only a small rise near the equivalence point. **17.8** False. The concentrations are small, and thus the ability of the two components to neutralize added acid or base is limited. **17.9** True. **17.10** True. **17.11** (a) HPO_4^{2-} is the base component of the conjugate acid-base pair because it has a more negative charge and thus more strongly attracts a proton (H^+). The net ionic equation is $HPO_4^{2-}(aq) + H^+(aq) \rightarrow H_2PO_4^{-}(aq)$. (b) The equilibrium constant for the equilibrium between the conjugate acid-base pair is formu-lated as $K_{a2} = [H^+][HPO_4^{2-}]/[H_2PO_4^{-}] = 6.2 \times 10^{-8}$. The required ratio is $[HPO_4^{2-}]/[H_2PO_4^{-}] = K_{a2}/[H^+]$. $[H^+]$ is calculated from the required pH of 7.4. Thus $[H^+] = $ antilog $(-7.4) = 4.0 \times 10^{-8}$. Substituting into the above equation the value of $[H^+]$ and K_{a2} yields the required ratio: $[HPO_4^{2-}]/[H_2PO_4^{-}] = 6.2 \times 10^{-8}/4.0 \times 10^{-8} = 1.6$. **17.12** (a) The added HCl reacts with the base component of the buffer: $H^+(aq) + NH_3(aq) \rightarrow NH_4^+(aq)$. This reaction essentially goes to completion. Therefore, the additional amount of NH_4^+ formed equals the amount of acid reacting with NH_3. Using similar logic, the amount of NH_3 being removed from solution also equals the amount of acid reacting with NH_3. The result of this effect is shown in the following concentration table for the equilibrium

$$NH_3(aq) + H_2O(l) \rightleftharpoons NH_4^+(aq) + OH^-(aq):$$

	NH_3	NH_4^+	$+ OH^-$
Initial (*M*):	0.020	0.100	0
Effect of added			
H^+ (*M*):	(0.020 − 0.0015)	(0.100 + 0.0015)	0
Equilibrium (*M*):	(0.0185 − *x*)	(0.1015 + *x*)	*x*

Assuming that the value of *x* is negligible compared to 0.0185 and 0.1015 and substituting appropriate values into the K_b

expression yields

$$K_b = \frac{\left[NH_4^+\right]\left[OH^-\right]}{\left[NH_3\right]} = \frac{(0.1015)(x)}{0.0185} = 1.8 \times 10^{-5}$$

$$x = \left[OH^-\right] = 3.3 \times 10^{-6}$$

$$pOH = 5.48$$

$$pH = 14 - 5.48 = 8.52$$

(b) Addition of 0.020 mol of HCl, which is equivalent in quantity to the amount of NH_3 in the buffer solution, would completely neutralize NH_3 to form NH_4^+. Thus the solution is no longer a buffer because it contains primarily NH_4^+ and only a very small quantity of NH_3 from the reaction of NH_4^+ with water. **17.13** There is a precipitate if the ion product for AgCl, Q, is greater in value than the value of K_{sp}: $Q = (2.0 \times 10^{-5})(3.0 \times 10^{-2}) = 6.0 \times 10^{-7}$. A precipitate of AgCl forms because $Q > K_{sp}$. **17.14** An ammonia solution contains the hydroxide ion and ammonia as shown by the reaction $NH_3(aq) + H_2O(l) \rightleftharpoons NH_4^+(aq) + OH^-(aq)$. The equilibrium lies to the left because ammonia is a weak base. In the presence of a small amount of ammonia and the hydroxide ion, Cu^{2+} preferentially reacts with the hydroxide ion to form $Cu(OH)_2$; $Cu^{2+}(aq) + 2OH^-(aq) \rightarrow Cu(OH)_2(s)$. When excess ammonia is added to the solution containing $Cu(OH)_2$, formation of a complex ion containing Cu^{2+} and NH_3 dominates. The excess ammonia reacts with $Cu(OH)_2$ to form the complex ion $Cu(NH_3)_4^{2+}$: $Cu(OH)_2(s) + 4NH_3(aq) \rightarrow Cu(NH_3)_4^{2+}(aq) + 2OH^-(aq)$. The formation of the complex ion causes the insoluble $Cu(OH)_2$ to dissolve. **17.15** The optimal buffer system is the one whose pK_a is the closest to the required pH. The $HCHO_2$—CHO_2^- pair has a pK_a equal to 3.8, which is the closest one to the required pH of 3.5. **17.16** Let x equal the molar solubility of AgCl. Then $[Ag^+] = x$, and $[Cl^-] = x + 2(0.02\ M)$. The factor of $2(0.02\ M)$ in the Cl^- expression occurs because 1 mol of $MgCl_2$ dissolves and releases 2 mol of chloride ion. The value of x is solved for as follows: $K_{sp} = [Ag^+][Cl^-] = (x)(x + 0.04) \simeq (x)(0.04) = 1.56 \times 10^{-10}$; $x = 3.9 \times 10^{-9}$ mol/L. **17.17** When solid silver acetate is placed in water we have the following equilibrium involving a slightly soluble salt: $AgC_2H_3O_2(s) \rightleftharpoons Ag^+(aq) + C_2H_3O_2^-(aq)$. If no other reactions occur, the equilibrium concentrations of Ag^+ and $C_2H_3O_2^-$ should be equal. Since experimentally $[Ag^+] > [C_2H_3O_2^-]$, some of the acetate ions must further react so that the concentration of silver ions increases relative to that of acetate ions. (Remember: $K_{sp} = [Ag^+][C_2H_3O_2^-] = $ constant; if $[C_2H_3O_2^-]$ decreases, $[Ag^+]$ must correspondingly increase.) The acetate ion is basic; therefore, it is the conjugate base of the weak and acetic acid. It reacts with water as follows: $C_2H_3O_2^-(aq) + H_2O \rightleftharpoons HC_2H_3O_2(aq) + OH^-(aq)$. This hydrolysis reaction removes $C_2H_3O_2^-$ from solution, thereby increasing Ag^+ concentration. **17.18 (a)** HCl reacts with 50.00 mL of ammonia as follows:

$$H^+(aq) + NH_3(aq) \xrightarrow{100\%} NH_4^+(aq)$$

$$\text{Initial molarity of } H^+ = \frac{(0.01000\ L)(0.2000\ M)}{(0.06000\ L)}$$

$$= 0.03333\ M$$

$$\text{Initial molarity of } NH_3 = \frac{(0.05000\ L)(0.1000\ M)}{(0.0600\ L)}$$

$$= 0.08333\ M$$

Note: The new molarities of H^+, NH_3, and NH_4^+ must be calculated after HCl and NH_3 are mixed together. Remember that molarity × volume (liters) = moles. In the above, the moles of H^+ and NH_3 are divided by 0.06000 L because after mixing, the new volume is 60.00 mL. H^+ is the limiting reagent and is used up:

	$H^+(aq)$	+	$NH_3(aq)$	$\xrightarrow{100\%}$	$NH_4^+(aq)$
Initial (M):	0.03333		0.08333		0
Reaction (M):	−0.03333		−0.03333		+0.03333
Equilibrium (M):	0		0.05000		0.03333

Next, calculate the pH of a solution composed of 0.05000 M acetic acid and 0.03333 M ammonium ion.

	$NH_3(aq)$	$+ H_2O(l) \rightleftharpoons$	$NH_4^+(aq)$	$+ OH^-(aq)$
Initial (M):	0.05000		0.0333	0
Reaction (M):	$-x$		$+x$	$+x$
Equilibrium (M):	$0.05000 - x \approx 0.05000$		$0.03333 + x \approx 0.0333$	$+x$

$$K_b = \frac{\left[NH_4^+\right]\left[OH^-\right]}{\left[NH_3\right]} = \frac{(0.0333)(x)}{(0.05000)} = 1.8 \times 10^{-5}$$

$$x = \left[OH^-\right] = \left(\frac{0.05000}{0.0333}\right)\left(1.8 \times 10^{-5}\right) = 2.7 \times 10^{-5}$$

$$pOH = 4.57 \qquad pH = 9.43$$

(b) To calculate the equivalence point pH, you first must calculate the volume of HCl required to neutralize the ammonia completely. At the equivalence point,

$$\text{Moles HCl = moles } NH_3$$
$$M_{HCl}V_{HCl} = M_{NH_3}V_{NH_3}$$
$$(0.2000\ M)(V_{HCl}) = (0.1000\ M)(0.05000\ L)$$
$$V_{HCl} = 0.02500\ L \text{ or } 25.00\ mL$$

At the equivalence point, both HCl and NH_3 are in equal molar quantities and completely react to form NH_4^+. The concentration of NH_4^+ is

$$M_{NH_4^+} \frac{(0.02500\ L)(0.2000\ M)}{0.07500\ L} = 0.06666\ M$$

The top term represents the moles of HCl or NH_3 that reacted to form NH_4^+. The bottom term is the new volume of the solution, 50.00 mL + 25.00 mL. Because NH_4^+, a weak acid forms, the new equilibrium is

	$NH_4^+(aq) + H_2O(l)$	$\rightleftharpoons$	$NH_3(aq)$	$+ H^+(aq)$
Initial (M)	0.06666 M		0	0
Reaction (M)	$-x$		$+x$	$+x$
Equilibrium (M)	$0.0666 - x \approx 0.0666$		x	x

$$K_a = \frac{K_w}{K_b} = \frac{[NH_3][H^+]}{[NH_4^+]} \qquad \frac{1 \times 10^{-14}}{1.8 \times 10^{-5}} = \frac{x^2}{0.06666}$$

$$x^2 = 0.06666\left(5.6 \times 10^{-10}\right) = 3.7 \times 10^{-11}$$

$$x = [H^+] = 6.1 \times 10^{-6}$$

$$pH = 5.21$$

(c) The pH when 10.00 mL of HCl is added beyond the equivalence point is controlled by the excess HCl. The molarity of excess HCl is

$$M_{HCl} = \frac{(0.01000 \text{ L})(0.2000 \text{ } M)}{(0.07500 \text{ L} + 0.01000 \text{ L})} = 0.02353$$

$$pH = -\log(0.02353) = 1.63$$

17.19 In Exercise 9, you found that the buffer solution of pH = 4.60 had to have components consisting of 0.10 M acetic acid and 0.072 M sodium acetate. To prepare one liter of the buffer you must first calculate the aliquots of sodium acetate and acetic acid that you will need to dilute with water to form one liter of buffer solution. Use the relationship

$$M_i V_i = M_f V_f$$

Acetic acid:

$$(0.5000 \text{ } M)(V_L) = (0.10 \text{ } M)(1.0 \text{ } L)$$

$$V_L = \frac{(0.010 \text{ } M)(1.0 \text{ } L)}{(0.5000 \text{ } M)} = 0.20 \text{ L}$$

Sodium acetate:

$$(0.1000 \text{ } M)(V_L) = (0.072 \text{ } M)(1.0 \text{ } L)$$

$$V_L = \frac{(0.072 \text{ } M)(1.0 \text{ } L)}{(0.1000 \text{ } M)} = 0.72 \text{ L}$$

Mix together 200 mL of 0.5000 M acetic acid and 720 mL of 0.1000 M sodium acetate to form 920 mL of solution. Add 80 mL of H_2O to bring the total volume to one liter. The concentrations at one liter will be those needed. **17.20** Because the concentration of NH_3 is much larger than that of Co^{2+}, you can initially assume that all the Co^{2+} is in the complex ion form. Note that the initial concentration of NH_3 then must also be reduced. Then calculate how much Co^{2+} forms from the dissociation of $Co(NH_3)_6^{2+}$.

	Co^{2+} +	$6NH_3$	$\rightleftharpoons$	$Co(NH_3)_6^{2+}$
*In (M)	0	$60 - 6(0.10)$		0.10
**Rxn (M)	$+x$	$+6x$		$-x$
Eq (M)	x	$5.4 + 6x \simeq 5.4$		$0.10 - x \simeq 0.10$

$$K_f = \frac{[Co(NH_3)_6^{2+}]}{[Co^{2+}][NH_3]^6} = \frac{(0.10)}{(x)(5.4)^6} = 1.1 \times 10^5$$

$$x = [Co^{2+}] = 3.7 \times 10^{-11} \text{ } M$$

*The data for the first line is based on the assumption that all the cobalt ion is completely complexed. The concentration of ammonia is reduced by 6.0(0.10) M because six moles of ammonia reacts for every mole of cobalt(II) ion reacting.
**The data for the second line is developed by recognizing that if an equilibrium is to exist, some cobalt(II) ion must now reform. If cobalt(II) ion forms, then ammonia must also form.

17.21 The acetate ion from sodium acetate chemically combines with the hydrogen ion from HCl to form unionized acetic acid. If excess sodium acetate remains after this reaction, then we have a common ion solution. First, determine the number of moles of $NaC_2H_3O_2$ and HCl in the solution. From this information, we can determine which species, $C_2H_3O_2^-$ or H^+, is in excess after the following reaction occurs: $H^+(aq) + C_2H_3O_2^-(aq) \rightarrow HC_2H_3O_2(aq)$. The number of moles of acetate ion initially present in solution is calculated as follows: 82.0 g $NaC_2H_3O_2 = 1$ mol NaC_2H_3O; 1 mol $C_2H_3O_2 = 1$ mol $NaC_2H_3O_2$; moles $C_2H_3O_2^- = (3.25 \text{ g } NaC_2H_3O_2)(1 \text{ mol } NaC_2H_3O_2/82.0 \text{ g } NaC_2H_3O_2)$ (1 mol $C_2H_3O_2^-/1$ mol $NaC_2H_3O_2$) = 0.0396 mol. The number of moles of hydrogen ion present is determined as follows:

0.150 mol HCl $\simeq$ 1 L of HCl solution

(from molarity statement)

1 mol H^+ = 1 mol HCl (HCl is a

strong acid in solution)

$$\text{Moles } H^+(aq) = (100 \text{ mL HCl})\left(\frac{1 \text{ L}}{1000 \text{ mL}}\right)$$

$$\times \left(\frac{0.150 \text{ mol HCl}}{1 \text{ L}}\right) \times \left(\frac{1 \text{ mol } H^+}{1 \text{ mol HCl}}\right)$$

$$= 0.0150 \text{ mol } H^+$$

Since mol $C_2H_3O_2^-$ is greater in magnitude than mol H^+, excess $C_2H_3O_2^-$ exists after the reaction between H^+ and $C_2H_3O_2^-$. This excess number of acetate ions is calculated as follows: excess moles of $C_2H_3O_2^- = $ moles $C_2H_3O_2^- - $ moles H^+ = 0.0396 mol $-$ 0.0150 mol = 0.0246 mol. The molarity of the remaining $C_2H_3O_2^-$ in solution is

$$M = \frac{0.0246 \text{ mol } C_2H_3O_2^-}{(100 \text{ mL})\left(\dfrac{1 \text{ L}}{1000 \text{ mL}}\right)} = 0.246 \text{ } M$$

Also note that the reaction between H^+ and $C_2H_3O_2^-$ forms $HC_2H_3O_2$. The amount of $HC_2H_3O_2$ formed equals the amount of HCl consumed during the reaction because 1 mol HCl $\simeq$ 1 mol $HC_2H_3O_2$ from the stoichiometry of the reaction. Therefore the molarity of $HC_2H_3O_2$ in the solution is

$$M = \frac{(0.0150 \text{ mol HCl})\left(\dfrac{1 \text{ mol } HC_2H_3O_2}{1 \text{ mol HCl}}\right)}{(100 \text{ mL})\left(\dfrac{1 \text{ L}}{1000 \text{ mL}}\right)} = 0.150 \text{ } M$$

You now can solve for pH using procedures previously developed. First write the equilibrium and concentration table:

	$HC_2H_3O_2$	$\rightleftharpoons$	H^+	$+$	$C_2H_3O_2^-$
Initial (M):	0.150		0 M		0.246
Reaction (M):	$-x$		$+x\ M$		$+x$
Equilibrium (M):	$(0.150-x)$		$x\ M$		$(0.246+x)$

Substituting equilibrium concentrations (with the assumption that the value of x is negligible compared to 0.150 and 0.246) into the K_a expression yields

$$K_a = \frac{\left[H^+\right]\left[C_2H_3O_2^-\right]}{\left[HC_2H_3O_2\right]} = \frac{(x)(0.246)}{0.150} = 1.8 \times 10^{-5}$$

$$x = \left[H^+\right] = 1.1 \times 10^{-5}$$

$$pH = 4.96$$

17.22 (d). Adding an acid to a buffer increases $[H^+]$ and thereby decreases the pH of the solution. Also, NH_3 reacts to form NH_4^+. **17.23 (d).** NaF contains the conjugate base, F^-, of HF. **17.24 (c).** This solution contains the acid component in the greatest amount, thereby buffering most effectively the added base. The solution described by response **(d)** contains only H_2O and NaCl because NaOH and HCl react together. **17.25 (b). 17.26 (c).** $pH = -\log [H^+] = -\log (0.150) = 0.82$. **17.27 (c).** The equivalence-point pH of the titration of a strong acid and strong base is always 7 because water and a neutral salt are formed. **17.28 (e).** $HC_2H_3O_2(aq) + OH^-(aq) \rightarrow H_2O(l) + C_2H_3O_2^-(aq)$ (neutralization). After neutralization, you need to consider reaction of $C_2H_3O_2^-$ with water:

$$C_2H_3O_2^-(aq) + H_2O(l) \longrightarrow OH^-(aq) + HC_2H_3O_2(aq)$$

$$\begin{aligned}
M_{C_2H_3O_2^-} &= \frac{\text{moles } C_2H_3O_2^-}{V} \\
&= \frac{\text{moles } HC_2H_3O_2 \text{ that reacted}}{V} \\
&= \frac{\left(M_{HC_2H_3O_2}\right)\left(V_{HC_2H_3O_2}\right)}{V} \\
&= \frac{(0.150\ M)(0.05000\ L)}{(0.05000\ L + 0.02500\ L)} \\
&= 0.100\ M
\end{aligned}$$

$$K_b = \frac{\left[OH^-\right]\left[C_2H_3O_2^-\right]}{\left[HC_2H_3O_2\right]} = \frac{x^2}{0.100} = \frac{K_w}{K_a}$$

$$= \frac{1.0 \times 10^{-14}}{1.8 \times 10^{-5}} = 5.6 \times 10^{-10}$$

$$x = \left[OH^-\right] = 7.5 \times 10^{-6}$$

$$pOH = 5.12;\ pH = 14 - 5.12 = 8.88$$

17.29 (d). 17.30 (a). $Ag_2CO_3(s) \rightleftharpoons 2Ag^+(aq) + CO_3^{2-}(aq)$; $K_{sp} = [Ag^+]^2[CO_3^{2-}] = (2x)^2(x)^3 = 4x^3$. **17.31 (d).** Acid addition (decreasing pH) causes the reaction $H^+(aq) + CO_3^{2-}(aq) \rightarrow HCO_3^-(aq)$ to occur, thereby shifting the following equilibrium to the right: $BaCO_3(s) \rightleftharpoons Ba^{2+}(aq) + CO_3^{2-}(aq)$. **17.32 (d).** **17.33 (b). 17.34 (e). 17.35 (e).**

Sectional Test 4

CHAPTERS 14–17

You should make this test as realistic as possible. **Tear out this test so that you do not have access to the information in the** *Student's Guide*. Use only data provided in the questions and a periodic table. Do not check your answers until you are finished. If you answer questions incorrectly, review the section in the text indicated after each question.

Choose the best response for each question.

1. What are the units of rate for gaseous reactions when concentration is measured in atmospheres of pressure?

(a) atm-time (b) atm/time (c) atm · time (d) 1/atm-time

[14–1]

2. Which of the following rate relationships is valid for the reaction

$$2NO_2(g) \longrightarrow 2NO(g) + O_2(g)?$$

(a) $\dfrac{\Delta[NO_2]}{\Delta t} = \dfrac{\Delta[NO]}{\Delta t}$
(c) $-\dfrac{1}{2}\dfrac{\Delta[NO_2]}{\Delta t} = \dfrac{\Delta[O_2]}{\Delta t}$

(b) $\dfrac{-\Delta[NO_2]}{\Delta t} = \dfrac{\Delta[O_2]}{\Delta t}$
(d) $\dfrac{\Delta[O_2]}{\Delta t} = \dfrac{\Delta[NO]}{\Delta t}$

[14–1]

3. For the reaction in problem 2, the concentration of $NO_2(g)$ changes from $4.2 \times 10^{-5}\ M$ to $2.8 \times 10^{-5}\ M$ in 100 seconds. Which statement is true?

(a) rate = $1.4 \times 10^{-7}\ M/s$
(c) Instantaneous rate = $1.4 \times 10^{-7}\ M/s$

(b) rate = $-1.4 \times 10^{-7}\ M/s$
(d) Instantaneous rate = $-1.4 \times 10^{-7}\ M/s$

[14–1]

4. The rate law for the reaction in problem 2 is rate = $k[NO_2]^2$. If the concentration of NO_2 is tripled, how does the rate change?

(a) Decreases by a factor of $\frac{1}{2}$ (b) Decreases by a factor of $\frac{1}{3}$
(c) Increases by a factor of three (d) Increases by a factor of nine

[14–2]

5. What is the rate law for the reaction

$$4H_2(g) + 2NO(g) \longrightarrow 2H_2O(g) + N_2(g)$$

if the rate doubles when the concentration of H_2 doubles and the rate increases by a factor of four if the concentration of NO doubles?

(a) Rate = $k[H_2][NO]$ (c) Rate = $k[H_2]^2[NO]$
(b) Rate = $k[H_2][NO]^2$ (d) Rate = $k[H_2]^2[NO]^2$
[14–2]

6. For the reverse reaction in problem 2,

$$2NO(g) + O_2(g) \longrightarrow 2NO_2(g)$$

the following initial rate data was found:

$[NO]_0$ (M)	0.10	0.10	0.40
$[O_2]_0$ (M)	0.10	0.20	0.20
rate (M/s)	2.5×10^{-4}	5.0×10^{-4}	8.0×10^{-3}

What is the rate law for the reaction?

(a) Rate = $k[NO][O_2]$ (c) Rate = $k[NO]^2[O_2]$
(b) Rate = $k[NO][O_2]^2$ (d) Rate = $k[NO]^2[O_2]^2$
[14–2]

7. The decomposition of N_2O_5 is a first-order reaction. What is k for this reaction given that 1.94×10^{-3} M decomposes to 1.03×10^{-3} M after 2.47 minutes at 64°C?

(a) 0.256 min^{-1} (b) 0.256 min (c) 3.91 min^{-1} (d) 3.91 min
[15–3]

8. What is the half-life for the decomposition of N_2O_5 in problem 7?

(a) 2.71 min (b) 0.369 min (c) 0.177 min (d) 0.0368 min
[14–3]

9. What is the concentration of azomethane remaining after 10 minutes if the initial concentration was 1.21×10^{-2} M and k is 1.60×10^{-2} min^{-1}. The reaction is first-order in azomethane.

(a) 1.20×10^{-2} M   (c) 1.04×10^{-2} M
(b) 1.12×10^{-2} M   (d) 9.82×10^{-2} M
[14–3]

10. What is the highest energy point in a reaction pathway termed?

(a) Product (b) Metastable (c) Activated complex (d) Activation energy
[14–4]

11. Which statement is true?

(a) Activation energy is a negative quantity.
(b) All collisions lead to reactions.
(c) Rates increase with temperature in a nonlinear manner.
(d) Reaction rates decrease as activation energy decreases.
[14–4]

12. The activation energy for the reaction $2N_2O_5(g) \rightarrow 4NO_2(g) + O_2(g)$ is 104.6 kJ, and k is 4.0×10^{-5} s^{-1} at 27°C. What is the new value of the rate constant if the temperature is raised 10°C? ($R = 8.314$ J/mol-K)?

(a) 6500 s^{-1} (c) 1.5×10^{-4} s^{-1}
(b) 1.2×10^{-2} s^{-1};   (d) 2.3×10^{-5} s^{-1}
[14–4]

13. Which rate law is consistent with the following mechanism?
$H_2O_2 + HI \rightarrow H_2O + HIO$ (fast and reversible)
$HIO + HI \rightarrow H_2O + I_2$ (slow)

(a) Rate = $k[H_2O_2][HI]$ (c) Rate = $k[H_2O][HI]/[H_2O_2]$
(b) Rate = $k[HIO][HI]$ (d) Rate = $k[H_2O_2][HI]^2/[H_2O]$
[14–5]

14. Which species of the following mechanism *does not* appear in the rate law?

$$O_3(g) \longrightarrow O_2(g) + O(g)$$
$$O_3(g) + O(g) \longrightarrow 2O_2(g)$$

(a) O_3 (b) O_2 (c) O   (d) (b) and (c)
[14–5]

15. Which of the following is true about a catalyst?
(a) It alters elementary steps in a mechanism.
(b) It slows down a chemical reaction.
(c) It acts as an inhibitor.
(d) It affects the position of equilibrium.
[14–6]

16. What is the K_c expression for the reaction $N_2(g) + 3H_2(g) \rightleftharpoons 2NH_3(g)$?

(a) $K_c = \dfrac{[NH_3]}{[N_2][H_2]}$ (c) $K_c = \dfrac{[N_2][H_2]}{[NH_3]}$

(b) $K_c = \dfrac{[NH_3]^2}{[N_2][H_2]^3}$ (d) $K_c = \dfrac{[N_2][H_2]^3}{[NH_3]^2}$

[15–1]

17. What is the K_p expression for the equilibrium $CaCO_3(s) \rightleftharpoons CaO(s) + CO_2(g)$?

(a) $K_p = P_{CO_2}$ (c) $K_p = \dfrac{[CaO]P_{CO_2}}{[CaCO_3]}$

(b) $K_p = [CaO]P_{CO_2}$ (d) $K_p = \dfrac{P_{CO_2}}{[CaCO_3]}$

[15–2, 15–3]

18. What is K_c for the reaction $N_2O_4(g) \rightleftharpoons 2NO_2(g)$ if at 9°C a one liter solution of chloroform contains 0.250 mol of N_2O_4 and 1.64×10^{-3} mol of NO_2?
(a) 9.34×10^4 (c) 1.07×10^{-5}
(b) 6.54×10^{-3} (d) 3.26×10^{-6}
[15–4]

19. What is K_c for the reaction $2HI(g) \rightleftharpoons H_2(g) + I_2(g)$ if 22.3% of the HI decomposes at 731 K?
(a) 9.61×10^{-2} (c) 2.06×10^{-2}
(b) 8.24×10^{-2} (d) 6.31×10^{-3}
[15–4]

20. $K_c = 0.0112$ at 25°C for the reaction $N_2(g) + O_2(g) \rightleftharpoons 2NO(g)$.

What change in the state of the reaction occurs (if any) if $[N_2]_0 = 2.00$ M, $[O_2]_0 = 1.00\ M$, $[NO]_0 = 0.500\ M$?

(a) Reaction produces more NO, $Q < K_c$.
(b) Reaction produces more reactants, $Q > K_c$.
(c) No change occurs in concentrations, $Q = K_c$.
(d) Temperature increases.
[15–5]

21. The value of K_c is 4.00 at 25°C for the following equilibrium $CH_3CO_2H(l) + C_2H_5OH(l) \rightleftharpoons CH_3CO_2C_2H_5(l) + H_2O(l)$. What is the concentration of $CH_3CO_2C_2H_5$ at equilibrium if the reactants are initially 0.200 M each? (*Note:* Include H_2O in the equilibrium expression and calculation because it is not the solvent.)

(a) 0.080 M (b) 0.133 M (c) 0.200 M (d) 0.800 M
[15–5]

22. What is the pH of a 0.25 M HBr solution?

(a) –0.25 (b) 0.25 (c) –0.60 (d) 0.60
[16–2, 16–3]

23. What is the pH of a 0.030 M $Ba(OH)_2$ solution?

(a) 1.22 (b) 12.78 (c) 1.52 (d) 12.48
[16–3, 16–4]

24. The pH of a HCl solution is 2.87. What is $[H^+]$?

(a) 0.46 (b) 5.67×10^{-2} (c) 1.35×10^{-3} (d) 17.63
[16–3, 16–4]

25. Which of the following does *not* describe an acidic solution?

(a) pH = 5 (c) $[H^+] = 1.3 \times 10^{-3}\ M$
(b) pH = 1 (d) $[H^+] = 3 \times 10^{-9}\ M$
[16–3]

26. What is the pH of a 0.200 M phenol ($K_a = 1.3 \times 10^{-10}$) solution?

(a) 3.87 (b) 4.15 (c) 4.36 (d) 5.29
[16–5]

27. K_b for trimethylamine (($CH_3)_3N$), a weak base, is 7.4×10^{-5}. What is $[OH^-]$ in a 0.0450 M trimethylamine solution?

(a) $3.3 \times 10^{-6}\ M$ (c) $1.2 \times 10^{-3}\ M$
(b) $1.8 \times 10^{-3}\ M$ (d) $2.6 \times 10^{-2}\ M$
[16–6]

28. What is the pH of a 0.010 M NaC_3H_7COO (sodium propionate) solution? $K_a = 1.4 \times 10^{-5}$ for the weak acid, propionic acid (C_3H_7COOH).

(a) 3.28 (b) 5.57 (c) 8.43 (d) 10.27
[16–6, 16–8]

29. Which of the following characterizes a solution of NaF?

(a) Acidic (b) Basic (c) Neutral (d) Weak electrolyte
[16–8]

30. Which of the following characterizes a solution of NH_4NO_3?

(a) Acidic (b) Basic (c) Neutral (d) Weak electrolyte
[16–8]

31. Which of the following characterizes a solution of $Al(NO_3)_3$?

(a) Acidic (b) Basic (c) Neutral (d) Weak electrolyte
[16–8]

32. Which substance is most acidic?

(a) H_2S (b) HClO (c) HBrO (d) HIO
[16–9]

33. Which species is a Lewis acid in the reaction

$$FeCl_3 + Cl^- \longrightarrow FeCl_4^-?$$

   (a) $FeCl_3$ (b) Cl^- (c) $FeCl_4^-$ (d) None are Lewis acids.
   [16–10]
34. In which situation is Br acting as a Lewis base?
   (a) Reaction with sodium metal to form $NaBr(s)$
   (b) Formation of $Br^-(g)$ from $Br(g)$
   (c) Dissociation of $Br_2(l)$ to bromine atoms
   (d) Reaction of Br^- with Co^{3+} to form $CoBr_4^-$
   [16–10]
35. Which of the following would reduce the pH of a 0.1 M $BaCl_2$ solution?
   (a) NaOH   (b) $MgCl_2$ (c) HCl   (d) $Ba(OH)_2$
   [17–1]
36. What would be the effect of adding NH_4Cl to a 0.10 M NH_3 solution?
   (a) Increase pH   (c) No change in pH
   (b) Decrease pH   (d) Shift position of equilibrium to form
                                        more OH^-
   [17–1]
37. What could you add to a Na_3PO_4 solution to form a buffer?
   (a) NaCl   (b) K_3PO_4 (c) NaOH   (d) HCl
   [17–3]
38. What is the pH of a solution composed of 0.20 M NH_3 and 0.15 M NH_4Cl? K_b for NH_3 is 1.8×10^{-5}.
   (a) 2.15   (b) 4.62   (c) 8.26   (d) 9.38
   [17–3]
39. What is the pH of the solution in problem 38 if 0.020 moles of HCl is added to 1 liter of solution? Assume no volume change.
   (a) 2.05   (b) 4.72   (c) 9.28   (d) 10.48
   [17–3]
40. What type of solution remains after 0.010 moles of NaOH is added to 1 liter of a buffer consisting of 0.20 M $NaC_2H_3O_2$ and 0.010 M $HC_2H_3O_2$?
   (a) A buffer solution   (c) A solution containing only
   (b) A solution containing only $NaC_2H_3O_2$
        $HC_2H_3O_2$ (d) A NaOH solution
   [17–3]
41. What quantity of sodium acetate must be added to 1.00 liter of a 0.200 M acetic acid ($K_a = 1.8 \times 10^{-5}$) solution to form a buffer of pH = 4.30?
   (a) 7.18×10^{-2} moles   (c) 3.62×10^{-5} moles
   (b) 1.90×10^{-3} moles   (d) 4.51×10^{-9} moles
   [17–3]
42. In the titration of 50.00 mL of 0.250 M NH_3 ($K_b = 1.8 \times 10^{-5}$) with 0.500 M HCl, what is the equivalence point pH? (*Note:* remember to account for dilution.)
   (a) 2.52   (b) 5.02   (c) 6.49   (d) 10.03
   [17–2]

43. What is the equivalence point pH in the titration of 50.00 mL of 0.100 M HCl with 0.100 M NaOH?

(a) pH = 3 (b) pH = 6 (c) pH = 7 (d) pH = 8

[17–2]

44. What is the K_{sp} expression for $Ca_3(PO_4)_2(s)$?

(a) $K_{sp} = [Ca^{2+}][SO_4^{2-}]$

(b) $K_{sp} =$
  $3[Ca^{2+}] + 2[PO_4^{3-}]$

(c) $K_{sp} =$
  $[Ca^{2+}]^3[PO_4^{3-}]/[CaSO_4]$

(d) $K_{sp} = [Ca^{2+}]^3[PO_4^{3-}]^2$

[17–4]

45. $K_{sp} = 2.6 \times 10^{-5}$ for $MgCO_3$ at 25°C. Neglecting the reaction of CO_3^{2-} ion with water, calculate the solubility of $MgCO_3$ in a saturated solution?

(a) 2.6×10^{-5}

(b) 2.6×10^{-4}

(c) 5.1×10^{-4}

(d) 5.1×10^{-3}

[17–4]

46. $K_{sp} = 1.39 \times 10^{-8}$ at 25°C for PbI_2. What is the concentration of lead ion (Pb^{2+}) in a solution containing 0.15 M NaI?

(a) $3.25 \times 10^{-2} M$

(b) $1.18 \times 10^{-4} M$

(c) $6.18 \times 10^{-7} M$

(d) $9.27 \times 10^{-8} M$

[17–4]

47. Which of the following will increase significantly the solubility of AgCl?

(a) $HC_2H_3O_2$ (b) NaOH (c) NH_3 (d) HF

[17–5]

48. Which of the following will increase significantly the solubility of $MgCO_3$?

(a) NaOH (b) K_2CO_3 (c) HCN (d) HNO_3

[17–5]

ANSWERS

1. (b). **2.** (c). **3.** (a). **4.** (d). **5.** (b). **6.** (c). **7.** (a). **8.** (a). **9.** (c).
10. (d). **11.** (d). **12.** (c). **13.** (d). **14.** (c). **15.** (a). **16.** (b). **17.** (a). **18.** (c).
19. (c). **20.** (b). **21.** (b). **22.** (d). **23.** (b). **24.** (c). **25.** (d). **26.** (d). **27.** (b).
28. (c) **29.** (b). **30.** (a). **31.** (a). **32.** (b). **33.** (a). **34.** (d). **35.** (c). **36.** (b).
37. (d). **38.** (d). **39.** (c). **40.** (c). **41.** (a). **42.** (b). **43.** (c). **44.** (d). **45.** (d).
46. (c). **47.** (c). **48.** (d).

Chemistry of the Environment | 18

OVERVIEW OF THE CHAPTER

Review: Bond-dissociation energy (8.9); ionization (2.5, 6.3); radiant energy (6.1); nature of photons (6.2); quantum of light and $E = h\nu$ (6.1, 6.2).

Objectives: You should be able to:

1. Sketch the manner in which the atmospheric temperature varies with altitude and list the names of the various regions of the atmosphere and the boundaries between them.

2. Sketch the manner in which atmospheric pressure decreases with elevation and explain in general terms the reason for the decrease.

3. Describe the composition of the atmosphere with respect to its four most abundant components.

4. Explain what is meant by the term *photodissociation,* and calculate the maximum wavelength of a photon that is energetically capable of producing photodissociation, given the dissociation energy of the bond to be broken.

5. Explain what is meant by *photoionization,* and relate the energy requirement for photoionization to the ionization potential of the species undergoing ionization.

18.1, 18.2 THE EARTH'S ATMOSPHERE: REGIONS, COMPOSITION, AND REACTIONS

Review: Structure of ozone (8.7); bond energies (8.9).

Objectives: You should be able to explain the presence of ozone in the mesophere and stratosphere in terms of appropriate chemical reactions.

18.3 OZONE IN THE ATMOSPHERE

Review: Lewis structures (8.6, 8.7); nomenclature (2.6, 8.10).

Objectives: You should be able to:

1. List the names and chemical formulas of the more important pollutant substances present in the troposphere and in urban atmospheres.

2. List the major sources of sulfur dioxide as an atmospheric pollutant.

18.4 CHEMISTRY OF ELEMENTS AND COMPOUNDS IN THE TROPOSPHERE

3. List the more important reactions of nitrogen oxides and ozone that occur in smog formation.

4. Explain why carbon monoxide constitutes a health hazard.

5. Explain why the concentration of carbon dioxide in the troposphere has an effect on the average temperature at the earth's surface.

18.5 THE WORLD OCEAN: COMPOSITION AND DESALINATION

Review: Concentration units (13.2).

Objectives: You should be able to:

1. List the more abundant ionic species present in seawater.

2. Explain the principles involved in desalination of water by distillation and by reverse osmosis.

18.6 FRESH WATER: IMPROVEMENT OF WATER QUALITY

Objectives: You should be able to:

1. List and explain the various stages of treatment that are used to treat fresh-water sources and waste water.

2. Describe the chemical principles involved in the lime-soda process for reducing water hardness.

TOPIC SUMMARIES AND EXERCISES

THE EARTH'S ATMOSPHERE: REGIONS, COMPOSITION, AND REACTIONS IN

The temperature of the atmosphere above the earth's surface changes in a complex manner as a function of altitude. Table 18.1 lists four regions in which temperature tends to rise or fall with increasing altitude. The direction of the temperature change, the approximate altitude ranges for these regions, and the average pressure exerted by the earth's atmosphere in each region are indicated in the table.

Table 18.1   Regions of the Atmosphere

Region	Direction of temperature change with increasing altitude	Approximate altitude range (km)	Approximate average pressure (mm Hg)
Thermosphere	Increasing	85–200	10^{-5}
Mesosphere	Decreasing	50–85	10^{-1}
Stratosphere	Increasing	12–50	10
Troposphere	Decreasing	0–12	100

At the junctions of the temperature regions, temperature attains a minimum or maximum value. These junctions are, in order of increasing altitude:

- The **tropopause,** at 12 km (temperature minimum)
- The **stratopause,** at 50 km (temperature maximum)
- The **mesopause,** at 85 km (temperature minimum)

The molecular and atomic composition of the atmosphere differs from region to region.

- Near the earth's surface, the four most abundant components, in order of decreasing concentration, are N_2, O_2, Ar, and CO_2. N_2 and O_2; they make up about 99 percent of the entire atmosphere.

- With increasing altitude we observe a decreasing average molecular weight. Only the lightest atoms, molecules, or ions exist in the higher altitudes (for example, He, O, N_2, and N_2^+).

- In the upper atmosphere, electromagnetic radiation and high-energy particles from the sun bombard species such as O_2, N_2, and H_2O. Two types of reactions in the upper atmosphere are discussed in detail in Section 18.2 in the text:

Name of reaction	Description of reaction	Example
Photodissociation	Solar photons cause molecules to split into smaller molecules or atoms.	$O_2 + hv \longrightarrow 2O$
Photoionization	Solar photons cause atoms or molecules to form ions.	$N_2 + hv \longrightarrow N_2^+ + e^-$

EXERCISE 1 Refer to the following data to answer the questions in this exercise:

$$NO + hv \longrightarrow NO^+ + e^- \quad E = 890 \text{ kJ/mol}$$
$$N_2 + hv \longrightarrow N_2^+ + e^- \quad E = 1495 \text{ kJ/mol}$$

(a) Of the two reactions

$$N_2^+ + NO \longrightarrow N_2 + NO^+$$
$$N_2 + NO^+ \longrightarrow N_2^+ + NO$$

which reaction is the one that occurs? **(b)** What is the energy change in the reaction that occurs?

Solution: **(a)** The exothermic reaction,

$$N_2^+ + NO \longrightarrow N_2 + NO^+$$

occurs. This will involve the transfer of an electron from NO to N_2^+ because more energy is released when N_2^+ adds an electron than when NO^+ does. **(b)** Using Hess's law, we can write the reaction as the sum of two reactions:

$$N_2^+ + e^- \longrightarrow N_2 + hv \quad E = -1495 \text{ kJ/mol}$$
$$NO + hv \longrightarrow NO^+ + e^- \quad E = 890 \text{ kJ/mol}$$

$$\overline{NO + N_2^+ \longrightarrow NO^+ + N_2 \quad E = -605 \text{ kJ/mol}}$$

The reverse reaction is endothermic:

$$E = 605 \text{ kJ/mol.}$$

EXERCISE 2 After nitrogen and oxygen, argon is the most abundant element in the air near sea level. Calculate the concentration of argon in air in ppm (parts per million) given that its concentration in percent by volume is 0.93.

Solution: To help you convert from percent by volume to ppm, first note that percentages are parts per hundred. Thus, the concentration of argon percent-by-volume can be restated as

$$0.93\% \text{ means } \frac{0.93}{100}$$

Also, x parts per million can be stated as

$$\frac{x}{1,000,000}$$

You can use a ratio to calculate parts per million of argon

$$\frac{0.93}{100} = \frac{x}{1,000,000}$$

Solving for x gives

$$x = \frac{(0.93)(1,000,000)}{100} = (0.93)(1 \times 10^4) = 9300 \text{ ppm}$$

This example demonstrates that you can always convert from a percent to ppm by taking the original percent composition and multiplying by 10,000 (1×10^4).

Note: For gases, one ppm means one part by volume in one million volume units of the whole. Also, mole fraction and volume fraction are the same; therefore, ppm = mole fraction $\times 10^6$.

OZONE IN THE ATMOSPHERE

The concentration of oxygen molecules in the upper atmosphere is low because of their photodissociation into oxygen atoms.

- In the mesosphere and stratosphere, atomic oxygen produced by the photodissociation of O_2 reacts with available O_2 to form ozone, O_3.

- Ozone absorbs harmful ultraviolet radiation with wavelengths in the range of 200 nm to 310 nm. This harmful radiation is not absorbed at lower altitudes. The layer of ozone in the mesosphere and stratosphere, which is called the "ozone shield," is thus very important for us.

- Depletion of ozone in the stratosphere may result from chlorfluorocarbons (CFC's) rising from the troposphere to the stratosphere and undergoing reaction with ozone. In the presence of light of wavelength $190 - 225$ *nm*, CFC's lose chlorine atoms which then react with ozone to form ClO and O_2. The ClO can regenerate a Cl atom to once again react with ozone. Be sure you review reactions $18.6 - 18.10$ in the text.

EXERCISE 3 In what regions of the atmosphere are molecular oxygen and ozone most abundant?
Solution: Molecular oxygen is found primarily in the troposphere and stratosphere, whereas ozone is found primarily in the stratosphere.

EXERCISE 4 What are the primary reactions of O_2 above 90 km and below 90 km?
Solution: Above 90 km, photodissociation of O_2 to form oxygen atoms dominates:

$$O_2 + h\nu \longrightarrow O + O \text{ (photodissociation)}$$

Below 90 km two reactions are important:

$$O + O_2 \longrightarrow O_3^* \text{ (ozone formation)}$$
$$O_3^* + M \longrightarrow M^* + O_3$$

In the latter reaction M represents a third atom or molecule that removes excess energy from O_3. (The asterisk tells us that a species has excess energy.) The presence of M is necessary in the reaction because if excess energy is not removed from O_3 it will dissociate back to O_2 and O.

EXERCISE 5   Why does temperature rise with altitude in the stratosphere?
   Solution: The exothermic reaction

$$O + O_2 + M \longrightarrow M^* + O_3$$

increases in rate of occurrence from 10 km to 50 km and reaches its maximum rate at 50 km, the stratopause. As the rate of this reaction increases, so does the amount of heat liberated. This results in an increasing temperature with increasing altitude in the stratosphere. Above 50 km in the mesosphere, the rate of the reaction decreases with increasing altitude, which results in a decreasing temperature. The reaction of ozone with NO also contributes to the temperature profile.

EXERCISE 6   The Anglo-French Concorde airplane can cruise economically at an altitude of 16–17 km. It emits NO during combustion of its aviation fuel. Why is there concern about the emission of NO at this altitude?
   Solution: At a cruising altitude of 16–17 km, the Concorde flies in the ozone shield in the stratosphere. The NO emitted is trapped in the stratosphere and reacts with ozone:

$$O_3 + NO \longrightarrow NO_2 + O_2$$

A significant reduction of the ozone concentration would allow harmful ultraviolet radiation to penetrate to the surface of the earth because the reaction rate of

$$O_3 + h\nu \longrightarrow O_2 + O$$

would be reduced.

Several minor constituents of the atmosphere can create health hazards if they are localized in the environment. This localization effect is called **air pollution.** The five substances that account for more than 90 percent of the worldwide air pollution are carbon monoxide (CO), nitrogen oxides (NO_x), hydrocarbons (compounds containing carbon and hydrogen), sulfur oxides (SO_x), and particulates:

- **Sulfur oxides:** SO_2 is considered one of the most harmful of polluting gasses. Coal mined in the eastern United States often has a high sulfur content, and when it is burned one of the products is SO_2. SO_2 is oxidized in the atmosphere to SO_3. Acid rain forms during a rainstorm when SO_3 is present in the air. Raindrops containing H_2SO_4 are a menace to human health and the environment.
- **Nitric oxide,** NO, is formed during the combustion of gasoline in an internal-combustion engine. In the presence of air, NO is slowly oxidized to nitro-

CHEMISTRY OF
ELEMENTS AND
COMPOUNDS IN THE
TROPOSPHERE

gen dioxide, NO_2. Nitric oxides, along with carbon monoxide and other products formed during the combustion of gasoline, yield a mixture of polluting gases that can produce photochemical smog in the presence of sunlight.

• **Carbon monoxide,** CO, is formed during the combustion of coal, gasoline, tobacco, and other carbon-containing compounds. It is a relatively unreactive species. However, it binds readily to hemoglobin in the human blood system. Hemoglobin is a very large iron-containing molecule that carries oxygen to muscle tissue. When CO is bound to hemoglobin, the ability of the hemoglobin to carry oxygen is diminished.

Water and CO_2 play a critical role in absorbing outgoing infrared radiation from the earth's surface.

• This absorption of infrared radiation holds heat at the earth's surface and thus keeps our nights from becoming extremely cold. No other molecules in the atmosphere have this capacity to absorb and radiate back infrared heat.

• Unfortunately, with excess CO_2 production now occurring as a product of industrial processes, including coal burning, there is a danger of too much heat being held at the earth's surface. This will raise the average temperature of the troposphere and may cause significant changes in our climate.

EXERCISE 7   Pollution from sulfur oxides is severe near industries that have as one of their processes the reaction of a metal sulfide with oxygen at high temperatures. **(a)** Write the balanced chemical reaction describing the conversion of PbS to PbO by oxidation with $O_2(g)$. **(b)** H_2SO_4 can be formed near industries emitting sulfur dioxide if water vapor is present. Explain how H_2SO_4 is formed under these conditions.
Solution: **(a)** SO_2 is a product in the oxidation of metal sulfides:

$$2PbS + 3O_2 \longrightarrow 2PbO + 2SO_2$$

(b) SO_2 is converted to SO_3 by oxidative processes. The SO_3 reacts with H_2O to form H_2SO_4:

$$SO_3 + H_2O \longrightarrow H_2SO_4$$

EXERCISE 8   Write the chemical reactions that describe the cycling of NO and NO_2 to maintain their relative concentrations, assuming no hydrocarbons are present.
Solution: The following reactions describe the cyclic process:

$$NO_2 + h\nu \longrightarrow NO + O \text{ (photodissociation of } NO_2)$$
$$O + O_2 + M \longrightarrow O_3 + M^* \text{ (ozone formation)}$$
$$NO + O_3 \longrightarrow NO_2 + O_2 \text{ (ozone reacts with NO to re-form } NO_2).$$

EXERCISE 9   CO is a relatively unreactive molecule, yet when present in the air in concentrations higher than 50 ppm it is dangerous to humans. Why?
Solution: The CO in the air we breathe binds readily to hemoglobin, reducing the ability of hemoglobin to transfer O_2 from the lungs to other areas of the body.

THE WORLD OCEAN: COMPOSITION AND DESALINATION

Seawater consists of salts, gases, and organic molecules dissolved in water.

• There are 11 ionic species that make up 99.9 percent of the dissolved salts in seawater, with sodium and chloride ions as the dominant constituents.

▪ A measure of the amount of salts in seawater is **salinity.** It is defined as the mass in grams of dry salts present in one kilogram of seawater. The average value of salinity for the world ocean is 35. Because of mixing, the relative concentrations of ionic species are constant in the world ocean.

Desalination is the process of the removal of salts to form useable water.

▪ One approach to desalination is flash-distillation.

▪ **Reverse osmosis** is also used to desalinate seawater. Refer to Section 13.5 of the text for a discussion of osmosis which involves a solvent passing through a semipermeable membrane from a dilute solution into a more concentrated one. In reverse osmosis, the opposite process occurs. If a sufficient external pressure is applied, a solvent will move from a concentrated solution to a more dilute solution.

EXERCISE 10   Oyster shells in the oceans are a source of $CaCO_3$. By roasting oyster shells we can obtain lime, CaO and CO_2. If lime is placed in seawater, a suspension of $Mg(OH)_2(s)$ is obtained. (Milk of magnesia is a suspension of $Mg(OH)_2$ in water.) Write reactions showing the roasting of oyster shells to produce lime and the subsequent reaction of lime with Mg^{2+} in seawater to form $Mg(OH)_2$.
    Solution: The roasting of oyster shells causes the decomposition of $CaCO_3(s)$:

$$CaCO_3(s) \longrightarrow CaO(s) + CO_2(g)$$

Seawater contains magnesium ions which react with lime as follows:

$$Mg^{2+}(aq) + CaO(s) + H_2O(l) \longrightarrow Mg(OH)_2(s) + Ca^{2+}(aq)$$

This reaction is possible because CaO is a basic oxide; it reacts with water to form OH^- ions.

EXERCISE 11   What is the concentration of magnesium, in grams per kilogram, in seawater with a salinity of 34 parts per million (ppm)? The concentration of magnesium is 19.35 g/kg in seawater with a salinity of 35 ppm.
    Solution: The salinity of seawater is equal to the mass in grams of dry salts of seawater dissolved in 1 kg of water. The concentration of an ion in seawater is usually reported with respect to seawater of standard salinity of 35 ppm. Since the relative concentrations of ionic species are constant throughout the world, the relative concentrations of magnesium in seawater of 34 and 35 ppm is the same. Thus the ratio of magnesium concentrations in seawater of different salinities is the same as the ratio of the seawater salinities:

$$\frac{19.35 \text{ g kg}}{x \text{ g kg}} = \frac{35 \text{ ppm}}{34 \text{ ppm}}$$

Solving for x gives 18.79 g of magnesium in 1000 kg of seawater with a salinity of 34 ppm.

Water is normally thought of as an inexhaustible natural resource because it is recycled in the natural environment. However, because some water has a high mineral content or has been polluted, it is not always suitable for human use and must be purified. Sewage water must be treated to remove harmful organisms and a variety of other substances such as organic chemicals and metals before it is returned to the environment.

FRESH WATER: IMPROVEMENT OF WATER QUALITY

• Sewage water is usually treated by a primary treatment, in which solids are allowed to settle and some organic materials are removed, followed by a secondary treatment that involves aeration and decomposition of biodegradable organic matter by aerobic bacteria.

• Some water departments also provide a tertiary treatment to remove other material in the water, such as excess phosphorus or other substances potentially harmful to the environment.

• You should look at the reactions in A CLOSER LOOK: Water Softening in your text. Water may also be treated to reduce the concentrations of Ca^{2+} and Mg^{2+}, which are responsible for water hardness. The lime-soda process is used to treat water-hardness in large-scale municipal operations. Study reactions [18.19 through 18.20] in the text.

EXERCISE 12 What types of chemical compounds are usually formed by anaerobic organisms acting on sewage water? Why are these compounds objectionable?

Solution: Anaerobic organisms produce compounds that do not contain oxygen. The two most common compounds formed during anaerobic decay are CH_4 and H_2S. Such decay produces sludges and noxious gases such as H_2S or flammable ones such as CH_4.

EXERCISE 13 What is hard water and why is it objectionable?

Solution: Hard water contains calcium, magnesium, and iron salts; the negative ions are usually chloride, sulfate, and hydrogen carbonate. The presence of the cations causes the formation of insoluble soaps that have no cleansing ability and leave a "ring" in tubs and sinks. Also, under appropriate conditions, such as very hot water, insoluble salts such as calcium sulfate and calcium carbonate form.

EXERCISE 14 An aqueous solution of ammonia may be used to remove hardness resulting from the presence of calcium and hydrogen carbonate ions. Write a net ionic equation that shows how ammonia removed this hardness.

Solution: Ammonia makes the water basic, which increases the carbonate ion concentration so that calcium carbonate precipitates. The reaction is similar to the lime-soda process:

$$Ca^{2+}(aq) + 2HCO_3^-(aq) + 2NH_3(aq) \longrightarrow CaCO_3(s) + 2NH_4^+(aq) + CO_3^{2-}(aq)$$

SELF-TEST QUESTIONS

Key Terms

Having reviewed the key terms listed at the end of Chapter 18 and their applications, identify the following statements as true or false. If a statement is false, indicate why it is incorrect. Key terms are italicized in the statements.

18.1 In the *troposphere*, atmospheric pressure decreases rapidly with increasing altitude.

18.2 In the *stratosphere*, temperature decreases with increasing altitude.

18.3 *Hemoglobin* is an iron-containing molecule in the blood that readily binds to CO_2.

18.4 Upon photolysis using light with wavelengths in the range of 190–225 nm, *chlorofluoromethanes* primarily form fluorine atoms and $CF_{3-x}Cl_x$ ($x = 1, 2,$ or 3) species.

18.5 *Acid rains* may occur when rainwater absorbs $SO_3(g)$.

18.6 The breaking apart of H_2O by light with the appropriate energy to form H and OH is an example of a *photodissociation* reaction.

18.7 Formation of nitrogen atoms from the reaction of N_2^+ ions with electrons in the atmosphere is an example of *photoionization* reactions.

18.8 Water with a dissolved salt concentration of 10 g/kg has a lower *salinity* than that of seawater.

18.9 In the *reverse osmosis* process of desalination, solute molecules pass through a semipermeable membrane, thereby removing salts from seawater.

18.10 *Desalination* refers to the general process of removing salts from seawater and brine. In general, desalination processes involves small energy use.

18.11 In the *lime-soda process* for removing water hardness resulting from the presence of Ca^{2+} and Mg^{2+} ions, soda ash is added to remove Mg^{2+} ions by precipitating $MgCO_3$.

18.12 *Biodegradable* organic material in water provides a source of oxygen.

18.13 The photodissociation of $NO_2(g)$ initiates the reactions associated with *photochemical smog.*

Problems and Short-Answer Questions

18.14 Methane (CH_4) comprises 1.5 ppm of our atmosphere. What is this concentration in percent by volume?

18.15 The compound CCl_2F_2 has been used as a propellant in aerosol cans. Solar photolysis of CCl_2F_2 occurs at 213.9 nm at low stratospheric temperatures.
   (a) What is the energy in kJ/mol required for photolysis?
   (b) Write a reaction equation that describes the photodissociation reaction.
   (c) Why are scientists concerned about CCl_2F_2 in the atmosphere?

18.16 Why does N_2 undergo very little solar photolysis in the upper atmosphere?

18.17 At twilight, the concentration of oxygen atoms decreases rapidly below 80 km in the atmosphere. Why?

18.18 The troposphere region in the atmosphere is transparent to visible light but not to infrared radiation. What gases in the troposphere cause the lack of transparency in the infrared region of radiation?

18.19 Atmospheric reactions are too slow to account for the removal of CO from the atmosphere; in the lower atmosphere, these reactions only account for approximately 0.1 percent of CO removal. What other mechanisms are available to remove CO from the atmosphere?

18.20 In an urban atmosphere the level of NO increases rapidly from 6 to 8 A.M. but decreases rapidly after 8 A.M. Why?

18.21 Give an example of an oxidant formed by a photochemical process.

18.22 Acute injury to plants by SO_2 results from sulfate salts forming at the tips or edges of leaves. These salts eventually cause leaf-dropping in plants. Suggest a mechanism for the formation of sulfate salts in plants.

18.23 How can limestone remove SO_2 from the atmosphere?

18.24 What is meant by the term *mean free path* (see Chapter 9)? Why do the mean free paths of molecules in the atmosphere increase with increasing elevation?

18.25 Write equations to describe the following steps in the lime-soda process for softening water:
   (a) the addition of $Ca(OH)_2$ to remove carbonate (HCO_3^-) hardness;

(b) the addition of excess $Ca(OH)_2$ to remove Mg^{2+};
(c) the addition of Na_2CO_3 to remove the Ca^{2+} ion after much of the HCO_3^- ion is removed.

18.26 One of the gaseous impurities found in water is hydrogen sulfide. The process of aeration is used to remove most of the dissolved hydrogen sulfide. When water is aerated, it becomes saturated with oxygen. The dissolved oxygen reacts with hydrogen sulfide to form sulfur, thus removing hydrogen sulfide impurities. Write a chemical equation that describes the removal of hydrogen sulfide by the process of aeration.

18.27 The total ionic concentration of seawater is approximately 1.100×10^{-3} mol/L, and the total ionic concentration in the body fluids of bony saltwater fish is approximately 3.70×10^{-4} mol/L. Using the concept of osmotic pressure, explain why bony fish in the ocean swallow seawater and excrete salts through their gills.

18.28 Chlorine is used to sterilize water in water treatment plants. Explain why household bleach, which is a dilute solution of hypochlorous acid, can be used to sterilize dishes during their final rinse after they have been washed with soap.

18.29 Many years ago lead pipes were used to carry water in houses. Why is it advisable to replace lead pipes with copper or plastic pipes in such houses?

18.30 If 10.0 g of $MgCl_2$ is completely electrolyzed in a 0.500-L container at 700°C, what pressure of $Cl_2(g)$ is then exerted?

Multiple-Choice Questions

18.31 In which region of the atmosphere is the rate of production of O_3 at its maximum value?
   (a) troposphere   **(d)** thermosphere
   (b) stratosphere   **(e)** thermopause.
   (c) mesosphere

18.32 When environmentalists talk about acid rain in certain industrial regions, which of the following acids is the major species involved?
   (a) H_2SO_4 **(d)** H_3PO_4
   (b) HI   **(e)** $HClO_4$
   (c) HNO_2

18.33 Which trace pollutant is normally present in the greatest concentration in a polluted urban atmosphere?
   (a) SO_2 **(d)** NO_2
   (b) C_2H_4 **(e)** CO
   (c) NH_3

18.34 The concentration of one of the molecules in the atmosphere that prevent the loss of heat radiated from the earth's surface is appreciably affected by human activities. What is this molecule?
   (a) PAN   **(d)** CO_2
   (b) H_2 **(e)** CO
   (c) H_2O

18.35 In the stratosphere, O_3 and ultraviolet radiation from the sun are converted into which of the following?
   (a) $O_2 + O$ **(d)** $N_2 +$ heat
   (b) $O_3 +$ heat   **(e)** NO + heat
   (c) $3O -$ heat

18.36 Which of the following is a region in the atmosphere in which temperature reaches a maximum value?
(a) Stratosphere;
(b) stratopause;
(c) mesosphere;
(d) mesopause;
(e) none of the above.

18.37 A catalyst not only speeds up the rate of a reaction, but it also is usually regenerated during the process. Which of the following species acts as a catalyst in the destruction of ozone?
(a) O_2
(b) O_3
(c) CO
(d) Cl
(e) H_2O

18.38 The mole-fraction content of O_2 in dry air near sea level is 0.209. What is the partial pressure of O_2 in dry air when the total dry air pressure is 750 mm Hg?
(a) 750 mm Hg
(b) 700 mm Hg
(c) 0.00637 mm Hg
(d) 3589 mm Hg
(e) 157 mm Hg

18.39 At an altitude of 400 km, which form of oxygen is the most abundant?
(a) O_3
(b) O_2
(c) O_2^+
(d) O
(e) O^{2-}

18.40 What is the correct relative order of concentrations of ionic constituents in seawater?
(a) $Mg^{2+} > SO_4^{2-} > Na^+ > Cl^-$
(b) $Cl^- > Na^+ > SO_4^{2-} > Mg^{2+}$
(c) $Na^+ > Cl^- > Mg^{2+} > SO_4^{2-}$
(d) $Cl^- > SO_4^{2-} > Mg^{2+} > Na^+$
(e) $Mg^{2+} > Na^+ > SO_4^{2+} > Cl^-$

18.41 What is the primary form of $CO_2(g)$ dissolved in seawater?
(a) $CO_2(aq)$
(b) $H_2CO_3(aq)$
(c) $HCO_3^-(aq)$
(d) $CO_3^{2-}(aq)$
(e) $CaCO_3(s)$

18.42 Anaerobic bacteria in polluted water consume organic material and thereby form gases. What is one of the gases formed?
(a) CH_3Cl
(b) H_2S
(c) H_2O
(d) SO_2
(e) HNO_3

18.43 Which of the following elements is removed in the tertiary treatment of municipal sewage water but not usually removed in the primary and secondary treatment stages?
(a) C
(b) H
(c) O
(d) S
(e) P

18.44 Which of the following statements about ozone in the atmosphere is true?
(a) The highest rate of formation of ozone is near the tropopause.
(b) The collisions between O atoms and O_2 molecules are greater at higher altitudes than lower ones.
(c) The ozone molecule, once formed in the stratosphere, does not last long.

(d) The photodecomposition of ozone releases solar photons.
(e) Ozone reacts slowly with chlorine atoms.

18.45 Which statement is true?
(a) Water vapor absorbs and then radiates infrared radiation.
(b) Carbon dioxide plays the most critical role in maintaining surface temperature of the earth.
(c) Stagnant waste ponds provide oxygen over a long time period.
(d) Eutrophication is the addition of oxygen to water by plant matter.
(e) CaO is added to municipal water to make it acidic.

SELF-TEST SOLUTIONS

18.1 True. **18.2** False. Temperature increases with increasing altitude in the stratosphere. **18.3** False. Hemoglobin transports O_2; hemoglobin does not bind to CO_2, but it does bind to CO. **18.4** False. Photolysis yields chlorine atoms and CF_xCl_{3-x} species. **18.5** True. **18.6** True. **18.7** False. It is a dissociative-recombination reaction. **18.8** True. **18.9** False. Solvent molecules pass through the semipermeable membrane. **18.10** False. Both multistage flash distillation and reverse osmosis require energy expenditure; however, the latter process requires a lesser amount of energy. **18.11** False. Soda ash, Na_2CO_3, causes the precipitation of $CaCO_3$, not $MgCO_3$; Mg^{2+} precipitates as $Mg(OH)_2$. **8.12** False. Aerobic bacteria consume dissolved oxygen in water when they oxidize organic materials; therefore the amount of oxygen decreases. **8.13** True. **18.14** The concentration unit parts per million (ppm) for a gas is based on volume measurements and represents the volume of a compound contained in 1 million volumes of air. The concentration of methane is 1.5 ppm or 1.5 volumes of methane per 1 million volumes of air: fractional concentration of $CH_4 =$ 1.5 volumes $CH_4/(1 \times 10^6$ volumes air$) = 1.5 \times 10^{-6}$; percentage composition of $CH_4 = (1.5/1 \times 10^6)(100) = 1.5 \times 10^{-4}\%$.

18.15

(a)
$$\lambda = 213.9 \text{ nm} = (213.9 \text{ nm})\left(10^{-7}\,\frac{\text{cm}}{\text{nm}}\right)$$

$$= 213.9 \times 10^{-7} \text{ cm}$$

$$E = h\left(\frac{c}{\lambda}\right) = \left(6.625 \times 10^{-34}\text{ J-sec}\right)$$

$$\times \left(\frac{3 \times 10^{10}\text{ cm/sec}}{213.9 \times 10^{-7}\text{ cm}}\right)$$

$$= 9.292 \times 10^{-19} \text{ J}$$

The energy per mole is E times Avogadro's number:

$$E_{molar} = \left(9.292 \times 10^{-19}\,\frac{\text{J}}{\text{photon}}\right)$$

$$\times \left(6.022 \times 10^{23}\,\frac{\text{photon}}{\text{mol}}\right)$$

$$= 5.60 \times 10^5 \text{ J/mol}$$

(b) $CCl_2F_2 + hv \rightarrow CClF_2 + Cl$. (c) The chlorine atom produced when CCl_2F_2 undergoes photodissociation reacts with ozone: $Cl + O_3 \rightarrow ClO + O_2$. Chlorine atoms are regenerated by the reaction of ClO with atomic oxygen: $ClO + O \rightarrow Cl + O_2$. The removal of ozone from the stratosphere would allow high-energy radiation to reach earth. **18.16** The N_2 bond is a triple bond: $:N\equiv N:$. High-energy photons are required to break it. Also, N_2 does not readily absorb photons. **18.17** At twilight the reaction $O_2 + hv \rightarrow 2O$ is minimal; however, the reaction $O + O + M \rightarrow O_2 + M^*$ occurs rapidly, resulting in a net decrease of O atoms. **18.18** Primarily H_2O causes the lack of transparency, but CO_2 also contributes. **18.19** The two primary mechanisms are (1) microorganisms in the soil that remove CO from air and (2) the ocean which dissolves some of the CO. **18.20** The level of NO increases rapidly from 6 to 8 A.M. because of NO production from cars driven by commuters. After 8 A.M. the rate of production decreases; the NO reacts rapidly with ozone to form NO_2 and O_2. **18.21** Ozone and PAN (peroxyacylnitrates) are formed via photochemical reactions; they oxidize species not readily oxidized by O_2. **18.22** The plant absorbs SO_2, which then reacts with H_2O to form H_2SO_3. The plant metabolism then converts the H_2SO_3 to sulfate salts. **18.23** The limestone ($CaCO_3$) reacts with SO_2 according to the following equation: $CaCO_3 + SO_2 \rightarrow CO_2 + CaSO_3$. **18.24** The mean free path of a gas molecule is the average distance it travels between collisions. The value of the mean free path increases as the number of gas molecules in a unit volume decreases—that is, as the gas density decreases. It was shown in Chapter 9 that gas density is directly proportional to pressure: the greater the pressure, the greater the gas density. Therefore, the mean free paths of gas molecules increase with increasing elevation since the pressure of the atmosphere decreases with increasing altitude. **18.25 (a)** The addition of $Ca(OH)_2$ to water produces hydroxide ions, which react with HCO_3^-: $HCO_3^-(aq) + OH^-(aq) \rightarrow H_2O + CO_3^{2-}(aq)$. The carbonate ion precipitates Ca^{2+} in the form of $CaCO_3$: $Ca^{2+}(aq) + CO_3^{2-}(aq) \rightarrow CaCO_3(s)$. **(b)** The excess $Ca(OH)_2$ increases the pH of the solution sufficiently to precipitate Mg^{2+}: $Mg^{2+}(aq) + 2CO_3^{2-}(aq) + 2H_2O \rightarrow 2HCO_3^-(aq) + Mg(OH)_2(s)$. **(c)** If any Ca^{2+} ion remains, the addition of CO_3^{2-} from Na_2CO_3 precipitates Ca^{2+}: $Ca^{2+}(aq) + CO_3^{2-}(aq) \rightarrow CaCO_3(s)$. **18.26** Hydrogen sulfide and oxygen undergo a redox reaction in water as follows: $H_2S(aq) + \frac{1}{2}O_2(g) \rightarrow S(s) + H_2O(l)$. **18.27** Since the total ionic concentration in body fluids of bony fish is less than that of seawater, the osmotic pressure of body fluids in bony fish is less than that of seawater ($\pi \propto$ total ionic concentration). As a result of the difference in osmotic pressures, water leaves bony fish through their cellular tissues in order to equalize the osmotic pressures. To prevent dehydration and to maintain the low osmotic pressure of their body fluids, bony fish swallow seawater to replace water lost through osmosis. Since salts are contained in the swallowed seawater, bony fish excrete salts through their gills in

order to maintain a constant total ionic concentration. **18.28** The sterilizing action of chlorine is believed to be due to hypochlorous acid, which is the same component as occurs in bleach. Hypochlorous acid is formed when chlorine reacts with water: $Cl_2(aq) + H_2O \rightarrow HOCl(aq) + HCl(aq)$. **18.29** In houses that have lead pipes that carry drinking water, the lead slowly dissolves in the water and eventually accumulates in people who drink the water. Copper and plastic are not significantly toxic, but lead is very toxic.

18.30 $MgCl_2(l) \xrightarrow{\text{Electrolysis}} Mg(l) + Cl_2(g)$

$$95.2 \text{ g MgCl}_2 = 1 \text{ mol MgCl}_2$$

$$1 \text{ mol MgCl}_2 = 1 \text{ mol Cl}_2$$

$$\text{Moles Cl}_2 \text{ produced} = (10.0 \text{ g MgCl}_2)$$

$$\times \left(\frac{1 \text{ mol MgCl}_2}{95.2 \text{ g MgCl}_2} \right) \times \left(\frac{1 \text{ mol Cl}_2}{1 \text{ mol MgCl}_2} \right)$$

$$= 0.105 \text{ mol Cl}_2$$

From the ideal-gas equation, we can calculate the pressure exerted by 0.105 mol of Cl_2:

$$P = \frac{nRT}{V}$$

$$= \frac{(0.105 \text{ mol}) \left(0.0821 \dfrac{\text{L·atm}}{\text{K·mol}} \right) (973 \text{ K})}{0.500 \text{ L}}$$

$$= 16.8 \text{ atm}$$

18.31 (b). **18.32 (a)**. **18.33 (e)**. **18.34 (c)**. A low concentration of water vapor in the atmosphere would allow infrared heat to escape, causing the earth's surface to be cold at night. **18.35 (a)**. **18.36 (b)**.

18.37 (d). $\quad Cl + O_3 \longrightarrow ClO + O_2$ (occurs in

$\underline{\quad ClO + O \longrightarrow Cl + O_2 \text{ ozone layer)}}$

$O_3 + O \longrightarrow 2O_2$

18.38 (e). See solution to problem 10.33 in the *Student's Guide* for the derivation a similar expression: $P_{O_2} = (n_{O_2}/n_t)P_t$, where n_{O_2}/n_t is the mole fraction of O_2. Thus, $P_{O_2} = (0.209)(750 \text{ mm Hg}) = 157 \text{ mm Hg}$. **18.39 (d)**. At 400 km, 99% of elemental oxygen is in the form of O, not O_2, due to solar photolysis. **18.40 (b)**. **18.41 (c)**. **18.42 (b)**. **18.43 (e)**. **18.44 (c)**. **18.45 (a)**.

19 Chemical Thermodynamics

OVERVIEW OF THE CHAPTER

**19.1
SPONTANEOUS
CHANGES**

Review: Activation energy (14.4); chemical equilibrium (15.2); first law of thermodynamics (5.2); concept of rate of reaction (15.1); spontaneous solution processes (16.1, 16.4).

Objective: You should be able to define the term *spontaneity* and apply it in identifying spontaneous processes.

**19.2, 19.3, 19.4
ENTROPY: A
MEASURE OF THE
RANDOMNESS OF A
SYSTEM**

Review: Properties of gases, liquids, and solids (10.1, 11.1); solution processes (13.1); thermodynamic state functions (5.2, 5.3).

Objectives: You should be able to:

1. Describe how entropy is related to randomness or disorder.

2. State the second law of thermodynamics.

3. Predict whether the entropy change in a given process is positive, negative, or near zero.

4. State the third law of thermodynamics.

5. Describe how and why the entropy of a substance changes with increasing temperature or when a phase change occurs, starting with the substance as a pure solid at 0 K.

6. Calculate $\Delta S°$ for any reaction from tabulated absolute entropy values, $S°$.

**19.5, 19.6, 19.7
FREE ENERGY AND
THE EQUILIBRIUM
STATE**

Review: Enthalpy (5.3); standard state and heat of formation (5.7); equilibrium constant (15.2); reaction quotient (15.5).

Objectives: You should be able to:

1. Define free energy in terms of enthalpy and entropy.

2. Explain the relationship between the sign of the free-energy change, ΔG, and whether a process is spontaneous in the forward direction.

3. Calculate the standard free-energy change at constant temperature and pressure, $\Delta G°$, for any process from tabulated values for the standard free energies of reactants and products.

4. List the usual conventions regarding standard states in setting the *values* for standard free energies.

5. Calculate $\Delta G°$ from K and perform the reverse operation.

6. Describe the relationship between ΔG and the maximum work that can be derived from a spontaneous process, or the minimum work required to accomplish a nonspontaneous process.

7. Calculate the free-energy change under nonstandard conditions, ΔG, given $\Delta G°$, temperature, and the data needed to calculate the reaction quotient.

Review: Melting and freezing processes (11.4).

Objectives: You should be able to:

1. Predict how ΔG will change with temperature, given the signs for ΔH and ΔS.

2. Estimate $\Delta G°$ at any temperature, given $\Delta S°_{298}$ and $\Delta H°_{298}$.

19.5 THE DEPENDENCE OF FREE ENERGY ON TEMPERATURE

TOPIC SUMMARIES AND EXERCISES

SPONTANEOUS CHANGES

Thermodynamic principles can be used to answer the question "Does a particular chemical reaction spontaneously proceed in the forward direction?" This chapter focuses on how two important thermodynamic quantities, entropy and free energy, are used to answer the previous question. First, what do we already know from our prior studies about spontaneous processes?

- Processes that occur spontaneously in a specific direction are not spontaneous in the opposite direction.
- Systems not at equilibrium tend to proceed toward a state of equilibrium.
- Most exothermic reactions at room temperature are spontaneous; however, there are endothermic processes that are also spontaneous, such as KNO_3 dissolving in water. What can we conclude? An important driving force for spontaneous processes is their tendency to proceed toward a state of minimum energy, but it can not be the only factor favoring spontaneity. In the next section, you will learn about another thermodynamic term that is involved in predicting the direction of a spontaneous process, entropy.

EXERCISE 1   At 25°C, the ΔH value for the reaction

$$2H_2(g) + O_2(g) \longrightarrow 2H_2O(l)$$

is −572 kJ. Can you conclude from the sign of ΔH that the reaction is spontaneous?

Solution: Most exothermic reactions ($\Delta H < 0$) are spontaneous. However, because there are exceptions to this statement, the prediction that the reaction is spontaneous because it is exothermic may be in error. In this case the reaction *is* spontaneous.

EXERCISE 2 Which of the following processes are spontaneous: **(a)** the dissolving of solid NaCl in water to form a solution; **(b)** the decomposition of carbon dioxide into its elements at 25°C and 1 atm pressure; **(c)** a ball rolling up a hill; **(d)** the freezing of water at −3°C and 1 atm pressure?

Solution: **(a)** Spontaneous. Table salt (NaCl) readily dissolves in H_2O. **(b)** Not spontaneous. The decomposition of most compounds into their constituent elements is not spontaneous. **(c)** Not spontaneous. The rolling of a ball downhill under the influence of gravity is spontaneous. **(d)** Spontaneous. We know from our experiences that the freezing of water at 0°C or lower at 1 atm pressure forms ice.

ENTROPY: A MEASURE OF THE RANDOMNESS OF A SYSTEM

To predict whether a process is spontaneous, you will need to know the change in both the enthalpy and the entropy of the system. **Entropy** is a thermodynamic state function represented by the symbol S.

- Entropy is a measure of the number of ways energy is distributed in a system. The more ways energy is distributed in a system, the more random or the more disordered the system is said to be. *The greater the entropy of a system, the greater is its randomness or disorder.*
- Entropies are positive for both elements and compounds.
- The units of entropy are J/K-mol for one mole of a substance.
- $\Delta S = S_{final} - S_{initial}$
- $\Delta S > 0$ if the system changes to a state of greater disorder.
- $\Delta S < 0$ if the system changes to a state of greater order.

What factors can be considered in determining the relative disorder or randomness of a system or process?

- The state of a system: For a given substance, the following trend in entropy exists:

$$S_{solid} < S_{liquid} \ll S_{gas}.$$

- Changes in state that lead to an increase in entropy of a system: (1) Melting ($S_{liquid} > S_{solid}$); (2) Dissolving ($S_{solution} > [S_{solvent} + S_{solute}]$); (3) Vaporization ($S_{vapor} > S_{liquid}$); (4) Heating of a substance ($S_{T_2} > S_{T_1}$).
- You should be able to use your knowledge of the structural properties of solids, liquids and gases to explain the trends given above.

The **second law of thermodynamics** involves entropy:

- *In any spontaneous process the entropy of the universe always increases.* For all spontaneous processes:

$$\Delta S_{universe} = \Delta S_{system} + \Delta S_{surroundings} > 0.$$

• For a system that is at equilibrium, $\Delta S_{universe} = 0$.

Note: A process that involves an increase in order ($\Delta S_{system} < 0$) can be spontaneous provided it causes an even greater positive entropy changes in the surroundings.

Note: An isolated system (no exchange of energy between the surroundings and system) always increases its entropy when it undergoes a spontaneous change. For such a change, $\Delta S_{surroundings} = 0$.

Entropy changes for chemical reactions can be calculated from standard entropy values.

• Standard entropies (symbol, $S°$) have been determined for many substances, and these values are based on the **third law of thermodynamics:** *The entropy of a perfect, pure substance at 0 K is zero.* As the temperature of a substance increases from 0 K, its entropy also increases. Increasing temperature causes an increase in thermal motions, which results in a greater disorder.

• Entropy, like enthalpy, is a state function; therefore you can calculate the entropy change for a reaction as follows:

$$aA + bB + cC \rightleftharpoons pP + qQ + rR$$

$S°$ = (sum of standard entropies for products) − (sum of standard entropies for reactants)

$$S° = [pS°(P) + qS°(Q) + rS°(R)] - [aS°(A) + bS°(B) + cS°(C)]$$

or

$$S° = \Sigma n S°_{products} - \Sigma m S°_{reactants}$$

where n and m are the coefficients in front of the species in the balanced chemical reaction.

EXERCISE 3   For each of the following pairs, which has the higher molecular entropy if all substance are at 1 atm pressure: **(a)** $CO_2(s)$ or $CO_2(g)$; **(b)** $NH_3(l)$ or $NH_3(g)$; **(c)** a crystal of pure magnesium at 0 K or a crystal at 200 K?
    Solution: **(a)** $CO_2(s)$ molecules are held rigidly in a crystal and undergo only limited thermal motion. $CO_2(g)$ molecules are distributed throughout a much larger volume and are more free to move about in this volume than an equivalent amount of solid CO_2. This random movement of $CO_2(g)$ molecules compared to the rigid arrangement of $CO_2(s)$ molecules results in $CO_2(g)$ molecules being more randomly distributed and thus having a higher entropy than the $CO_2(s)$ molecules. **(b)** $NH_3(g)$ has a higher entropy than $NH_3(l)$ for reasons similar to those discussed in (a). $NH_3(l)$ molecules occupy a smaller volume than $NH_3(g)$ molecules. Therefore $NH_3(l)$ molecules move less randomly about

the container in which they are enclosed; their motions are more restricted. Thus $NH_3(g)$ molecules are more randomly distributed and have a higher entropy. **(c)** $Mg(s)$ at 0 K contains magnesium atoms in a rigid lattice with no vibrational motion. At 200 K, the atoms have vibrational motion, which results in a more random arrangement of atoms and thus a higher entropy.

EXERCISE 4   The standard entropy for a perfect, pure, crystalline substance is zero at 0 K. Why is the entropy for an impure substance greater than zero at 0 K?

Solution: All molecular motion has stopped for an impure substance at 0 K, but the impurities can be distributed in more than one way in the crystal. Consequently, the impure substance has an entropy greater than zero.

EXERCISE 5 ΔS represents the entropy change for a chemical or physical process. For a chemical reaction, ΔS is the sum of the absolute entropies of products minus the sum of the absolute entropies of reactants. The sign of ΔS is positive when the change involves an increase in randomness and is negative when it involves a decrease in randomness. Predict for each of the following changes whether ΔS is greater than zero or less than zero:

(a)   sugar + water $\rightarrow$ sugar dissolved in water

(b) $Na_2SO_4(s) \rightarrow 2Na^+(aq) + SO_4^{2-}(aq)$

(c) $2H_2(g) + O_2(g) \rightarrow 2H_2O(g)$

Solution: **(a)** The entropy change in this process is greater than zero because the sugar solution contains a more random arrangement of sugar and water molecules. In the solid state, sugar molecules are more rigidly confined than in the solution state. **(b)** The entropy change for the dissolving of a salt to form solvated ions is greater than zero. In the liquid phase, the solvated ions Na^+ and SO_4^{2-} are free to move about the entire volume of the liquid, whereas in the solid ionic crystal lattice the ions are confined. Therefore dissolved ions have more random movements than ions in a crystal. **(c)** Note that 2 mol of products are formed for every 3 mol of reactants. Therefore, the entropy change is less than zero because fewer moles of gas exist after the reaction than before.

EXERCISE 6   Calculate the entropy change for the following reaction at 25°C;

$$2SO_2(g) + O_2(g) \longrightarrow 2SO_3(g)$$

given the following standard entropy values at 25°C: $SO_2(g)$, 248.5 J/K-mol; $O_2(g)$, 205.0 J/K-mol; $SO_3(g)$, 256.2 J/K-mol.

Solution: To calculate the entropy change for the reaction, use the relation

$$\Delta S^\circ_{rxn} = \Sigma nS^\circ \text{ (products)} - \Sigma mS^\circ \text{ (reactants)} = 2S^\circ(SO_3) - [2S^\circ(SO_2) + S^\circ(O_2)]$$

Using the provided values of absolute entropies, calculate ΔS°_{rxn}:

$$\Delta S^\circ_{rxn} = \left[(2 \text{ mol}) \left(256.2 \frac{J}{K\text{-mol}} \right) \right] - \left[(2 \text{ mol}) \left(248.5 \frac{J}{K\text{-mol}} \right) \right.$$
$$\left. + (1 \text{ mol}) \left(205.0 \frac{J}{K\text{-mol}} \right) \right]$$

$$\Delta S^\circ_{rxn} = 512.4 \text{ J/K} - 702.0 \text{ J/K} = -189.6 \text{ J/K}$$

The negative sign of ΔS° tells you that this reaction involves a decrease in randomness. You can infer this result from the reaction equation because three mol of reactants forms fewer moles of products, two.

You have learned that two factors must be considered in determining whether a process or reaction is spontaneous:

1. Exothermic processes ($\Delta H < 0$) are more likely to be spontaneous than endothermic processes ($\Delta H > 0$).

2. Processes that involve an increase in entropy ($\Delta S > 0$) are more probable than those showing a decrease in entropy ($\Delta S < 0$).

Enthalpy and entropy are connected together by the thermodynamic state function **free energy** (symbol, G). The change in free energy for any process or reaction at constant temperature and pressure is given by the relation

$$\Delta G = \Delta H - T\Delta S$$

- If $\Delta G < 0$, a reaction spontaneously proceeds in the forward direction until equilibrium is reached.
- If $\Delta G = 0$, a system is at equilibrium.
- If $\Delta G > 0$, a reaction is not spontaneous as written, but the reverse direction is spontaneous.

ΔG is a state function, therefore we can tabulate standard free energies of formation for substances.

- The concepts of standard states and of heat of formation have been discussed in Chapter 5 of the text. Just as ΔH_f° for elements are set to zero, ΔG_f° for elements are also set to zero.
- Standard free-energy changes for reactions are calculated using the relation:

ΔG_{rxn}° = (sum of standard free energies of formation of products)
      − (sum of standard free energies of formation of reactants)
$\Delta G_{rxn}^\circ = \Sigma n \Delta G_f^\circ$ (products) − $\Sigma m \Delta G_f^\circ$ (reactants)

- The free-energy change for a chemical process under standard state conditions (ΔG°) and the free-energy change (ΔG) when concentrations are not all 1 M and 1 atm pressure is given by the following equation:

$$\Delta G = \Delta G^\circ + RT \ln Q$$

where $R = 8.314$ J/K-mol, T is in degrees kelvin, and Q is the reaction quotient (see Section 15.5 in the text).

ΔG° is related to the equilibrium constant at constant temperature and pressure as follows:

$$\Delta G^\circ = -RT \ln K$$

• You need to know the relationships between the signs of $\Delta G°$ and K:

$\Delta G° < 0$ $K > 1$ Products are favored over reactants at equilibrium
$\Delta G° = 0$ $K = 1$
$\Delta G° > 1$ $K < 1$ Reactants are favored over products at equilibrium

A reaction that is spontaneous has the capacity to do work. If a process is not spontaneous, the increase in free energy represents the minimum amount of work that must be done on the system to cause the process to occur. One way to provide the minimum amount of work required is to couple a nonspontaneous process with one that is spontaneous. The spontaneous process must have a negative ΔG that is larger than the positive ΔG for the nonspontaneous process.

EXERCISE 7   Differentiate between the functions ΔG and $\Delta G°$ for the generalized reaction

$$a\text{A} + b\text{B} \longrightarrow c\text{C} + d\text{D}$$

Solution: $\Delta G°$ is the standard free-energy change for the reaction when the reactants A and B in their standard states at 298 K react to form products C and D in their standard states at 298 K. Standard states for all reactant and product substances are defined as 1 atm pressure for gases, 1 M for solutions, and the pure substances for solids or liquids.

$a\text{A}$ (1 atm or 1 M, or pure solid or liquid)
   $+ b\text{B}$ (1 atm, or 1 M, or pure solid or liquid)
   $\xrightarrow{298\text{ K}}$ $c\text{C}$ (1 atm, or 1 M, or pure solid or liquid)
   $+ d\text{D}$ (1 atm, or 1 M, or pure solid or liquid)

ΔG is the free-energy change for any process in which one or more reactants or products are not in their standard states.

EXERCISE 8   At 298 K, are the following substances in their standard state: **(a)** Al(s); **(b)** Fe(s) with some oxide impurity; **(c)** H$_2$(g) at 2 atm pressure; **(d)** 1 M solution of sugar in water?
   Solution: **(a)** Yes. The standard state for a solid or liquid is the pure substance in its stable form at 298 K. **(b)** No. The iron is not pure. **(c)** No. The standard-state concentration for a gas is 1 atm pressure. **(d)** Yes. The standard state for a solute is a 1 M concentration solution.

EXERCISE 9   Given the ΔG value for the following phase changes at 1 atm, predict whether each change is spontaneous: **(a)** At 283 K, $\Delta G = -250$ J/mol for H$_2$O(s) → H$_2$O(l); **(b)** At 273 K, $\Delta G = 0$ J/mol for H$_2$O(s) ⇌ H$_2$O(l); **(c)** At 263 K, $\Delta G = 210$ J/mol for H$_2$O(s) → H$_2$O(l).
   Solution: **(a)** The reaction is spontaneous since $\Delta G < 0$. We expect this because ice spontaneously melts at 283 K (10°C). **(b)** The system is at equilibrium because $\Delta G = 0$; there is no net reaction. Ice and water coexist at the melting point of ice, 273 K (0°C). **(c)** The reaction is not spontaneous because $\Delta G > 0$. Ice does not spontaneously melt below its melting point, 273 K.

EXERCISE 10   **(a)** Given that $\Delta G°_f$ for C$_6$H$_{12}$O$_6$(s) equals -907.9 kJ/mol, $\Delta G°_f$ for CO$_2$(g) equals -394.6 kJ/mol, and $\Delta G°_f$ for H$_2$O(l) equals -237.2 kJ/mol at 298 K, calculate $\Delta G°$ for the oxidation of glucose:

$$\text{C}_6\text{H}_{12}\text{O}_6(s) + 6\text{O}_2(g) \rightleftharpoons 6\text{CO}_2(g) + 6\text{H}_2\text{O}(l)$$

(b) Is the oxidation of glucose spontaneous at standard-state conditions? **(c)** Calculate the equilibrium constant for the reaction.

Solution: **(a)** The standard free-energy change for the reaction is calculated using the expression

$$\Delta G° = [6\Delta G_f°(CO_2) + 6\Delta G_f°(H_2O)] - [\Delta G_f°(C_6H_{12}O_6) + 6\Delta G_f°(O_2)]$$

The standard free energy of formation for an element in its stable state at 298 K is zero. Substituting into this equation the appropriate $\Delta G_f°$ values yields

$$\Delta G° = [6 \text{ mol}(-394.6 \text{ kJ/mol}) + 6 \text{ mol}(-237.2 \text{ kJ/mol})]$$
$$- [(1 \text{ mol})(-907.9 \text{ kJ/mol}) + (6 \text{ mol})(0 \text{ kJ/mol})] = -2882.9 \text{ kJ}$$

(b) Yes, because $\Delta G° < 0$. **(c)** The equilibrium constant is calculated using the expression:

$$\Delta G° = -RT \ln K$$

$$\ln K = -\frac{\Delta G°}{RT} = \frac{-(-2882.9 \times 10^3 \text{ J/mol})}{(8.314 \text{ J/K-mol})(298 \text{ K})} = 1.16 \times 10^3$$

Taking the antilog of $\ln K$ yields $K = 1 \times 10^{505}$. The value of K is very large; the reaction essentially goes to completion. Note that the units of $\Delta G°$ are changed to joules from kilojoules because the units of R contain the unit joule.

EXERCISE 11 Is the following reaction spontaneous at 25°C under the given conditions?

$$N_2(g) \quad + O_2(g) \longrightarrow 2NO(g)$$
$$P(\text{atm}) = 1.0 \qquad 1.0 \qquad\quad 4.0$$
$$\Delta G_{rxn}° = +173.1 \text{ kJ}$$

Solution: To determine if the reaction is spontaneous, the sign of ΔG_{rxn} must be determined. First, calculate the reaction quotient, Q:

$$Q = \frac{P_{NO}^2}{P_{N_2} P_{O_2}} = \frac{(4.0)^2}{(1.0)(1.0)} = 16$$

Then use the relation and substitute for the value of Q:

$$\Delta G_{rxn} = \Delta G_{rxn} + RT \ln Q$$

$$= 173.1 \text{ kJ} + \left(8.314 \ \frac{J}{K}\right)(298 \text{ K}) \ln 16$$

$$= 173.1 \text{ kJ} + 6.9 \text{ kJ} = 180.0 \text{ kJ}$$

The reaction is not spontaneous because ΔG_{rxn} is greater than zero.

THE DEPENDENCE OF FREE ENERGY ON TEMPERATURE

The value of ΔG for a physical or chemical change depends on the values ΔH, ΔS, and the temperature at which the change occurs: $\Delta G = \Delta H - T\Delta S$.

▪ As the temperature increases, the importance of the term $T\Delta S$ in the expression $\Delta G = \Delta H - T\Delta S$ increases (at very high temperatures, ΔG approaches the value of $-T\Delta S$).

- As the temperature decreases, the importance of ΔH increases (at very low temperature ΔG approaches the value of ΔH).
- Look at Table 19.3 in the text and note how the sign of ΔG depends on temperature and on the signs of ΔH and ΔS.

EXERCISE 12   What is the temperature at which sodium chloride reversibly melts, that is, when the melting occurs so slowly that the solid phase is always in equilibrium with the liquid phase? The enthalpy of melting is 30.3 kJ/mol, and the entropy change upon melting is 28.2 J/mol-K.
   Solution: When sodium chloride reversibly melts, the solid phase is in equilibrium with the liquid phase. Therefore $\Delta G = 0$ for the melting process:

$$\Delta G = 0 = \Delta H - T\Delta S$$

We can solve for temperature by rearranging the equation to:

$$\Delta H = T\Delta S$$

$$T = \frac{\Delta H}{\Delta S} = \frac{30,300 \text{ J/mol}}{28.2 \text{ J/mol-K}} = 1070 \text{ K}$$

Sodium chloride reversibly melts at 1070 K.

EXERCISE 13   Is a reaction spontaneous if $T\Delta S > 0$ and $\Delta H < 0$?
   Solution: The sign of ΔG for a reaction is determined by the expression $\Delta G = \Delta H - T\Delta S$. If $T\Delta S > 0$, then $-T\Delta S < 0$, and the sign of ΔG for this set of conditions is always negative because ΔH is also less than zero. The reaction is spontaneous at all temperatures.

EXERCISE 14   At 298 K, $\Delta G° = -190.5$ kJ and $\Delta H° = -184.6$ kJ for the reaction

$$H_2(g) + Cl_2(g) \longrightarrow 2HCl(g)$$

(a) What is the standard entropy change for the reaction? **(b)** What is the primary driving force for the reaction at 298 K?
   Solution: **(a)** $\Delta S°$ is calculated using the expression $\Delta G° = \Delta H° - T\Delta S°$. Rearranging the equation and solving for $\Delta S°$ yields

$$\Delta S° = \frac{\Delta H° - \Delta G°}{T}$$

Substituting the values of $\Delta H°$ and $\Delta G°$ in the equation yields

$$\Delta S° = \frac{-184.6 \text{ kJ} - (-190.5 \text{ kJ})}{298 \text{ K}} = \frac{5.9 \text{ kJ}}{298 \text{ K}}$$

$$\Delta S° = 2.0 \times 10^{-2} \text{ kJ/K} = 20 \text{ J/K}$$

(b) The primary driving force for the reaction at 298 K is the enthalpy change because the term $-T\Delta S$, which equals -5.9 kJ, is much smaller in magnitude than ΔH, -184.6 kJ.

SELF-TEST QUESTIONS

Key Terms

Having reviewed the key terms listed at the end of Chapter 19 and their applications, identify the following statements as true or false. If a statement is false indicate why it is incorrect. Key terms are italicized in the statements.

19.1   An increase in the *entropy* of a system occurs if $\Delta S_{\text{system}} > 0$.

19.2   According to the *second law of thermodynamics,* the spontaneous freezing of a liquid in contact with surroundings below the melting point of the liquid is accompanied by a decrease in the entropy of the liquid only.

19.3   The *Gibb's free-energy* difference between products and reactants for a reaction at equilibrium is zero.

19.4 A molecule such as HCl has *vibrational* and *rotational* energies associated with the movement of hydrogen and chloride atoms relative to each other and with motion of the molecule as a whole.

19.5 Reactions for which energy is required to initiate the process cannot be classified as *spontaneous processes*.

19.6 According to the *third law of thermodynamics*, the entropy of any crystalline solid at absolute zero temperature is zero.

19.7 When heat is released in a chemical reaction in an *isolated system*, the entropy of the surroundings increases.

19.8 When a molecule moves in a linear direction it undergoes *translational motion*.

19.9 The *standard free energy* of any pure substance at standard state conditions is zero.

Problems and Short-Answer Questions

19.10 Does the word "spontaneous" as used in thermodynamics imply anything about how long it takes for a spontaneous process to reach an equilibrium state?

19.11 **(a)** Calculate ΔS_{system} for the reversible melting of ice at 273 K (assuming $\Delta G = 0$), given that $\Delta H = +6.0$ kJ/mol for the process. **(b)** From the information given in **(a)**, show quantitatively that $\Delta S_{universe}$ equals zero at the melting point of ice.

19.12 Which substance, H_2S or H_2O, has the higher molar entropy of vaporization at 1 atm and their respective boiling points? (Hint: Consider hydrogen-bonding effects in the liquid state.)

19.13 **(a)** Calculate ΔG at 298 K for the following reaction if the reaction mixture consists of 1.0 atm of SO_2, 2.0 atm of O_2, and 2.0 atm of SO_3:

$$2SO_2(g) + O_2(g) \rightleftharpoons 2SO_3(g) \quad \Delta G^\circ_{298} = -140.0 \text{ kJ}$$

(b) Does a larger "driving force" to produce SO_3 exist at these conditions compared to standard-state conditions?

19.14 Indicate for each of the following pairs of signs of ΔH and ΔS whether ΔG is negative at high or low temperatures.
    (a) ΔH and ΔS are both positive;
    (b) ΔH and ΔS are both negative;
    (c) ΔH is positive; and ΔS is negative;
    (d) ΔH is negative, and ΔS is positive.

19.15 Why are most spontaneous reactions at 298 K exothermic?

19.16 **(a)** Calculate ΔS° for the reaction:

$$4CuO(g) \longrightarrow 2Cu_2O(s) + O_2(g)$$

at 298 K, given the following:

$$\Delta H^\circ_f(CuO) = -157.3 \text{ kJ/mol}$$
$$\Delta H^\circ_f(Cu_2O) = -168.6 \text{ kJ/mol}$$
$$\Delta G^\circ_f(CuO) = -129.7 \text{ kJ/mol}$$
$$\Delta G^\circ_f(Cu_2O) = -146.0 \text{ kJ/mol}$$

(b) Is this reaction feasible for the preparation of $Cu_2O(s)$ at standard state conditions?
(c) If the reaction is not feasible, how could it be made to occur?

19.17 **(a)** ΔG° equals + 20.00 kJ/mol for the reaction

$$CuS(s) + H_2(g) \longrightarrow Cu(s) + H_2S(g)$$

at 298 K and 1 atm of pressure. Calculate the equilibrium constant for the reaction. **(b)** At 298 K, does the equilibrium lie largely in the direction of products or of reactants?

19.18 Given the following data at 25°C

	$S°$(J/mol-K)	$\Delta H°_f$(kJ/mol)
$NO_2(g)$	240.45	33.8
$N_2O_4(g)$	304.33	9.66

    (a) Calculate the value of ΔG° for the following reaction at 25°C:

$$2NO_2(g) \rightleftharpoons N_2O_4(g)$$

    (b) Is the formation of $N_2O_4(g)$ a spontaneous process at 25°C at standard-state conditions? **(c)** What is the value of the equilibrium constant for this reaction?

19.19 Calculate the equilibrium constant for the following reaction at 25°C

$$2H_2(g) + O_2(g) \rightleftharpoons 2H_2O(l)$$

given the following standard free-energy: $H_2O(l)$, −237.2 kJ/mol at 25°C.

Multiple-Choice Questions

19.20 ΔG°_f refers to the free-energy change associated with which of the following?
    (a) the formation of 1 mol of a compound from its constituent elements
    (b) the decomposition of 1 mole of a compound into its constituent elements
    (c) the freezing of 1 mol of a substance
    (d) the melting of 1 mol of a substance
    (e) all of the above

19.21 Under what conditions can we absolutely say a system is at equilibrium at constant pressure and temperature?
    (a) $\Delta H = 0$ **(d)** $\Delta G > 0$
    (b) $\Delta H > 0$ **(e)** $\Delta S > 0$
    (c) $\Delta G = 0$

19.22 Which of the following substances would you expect to have the highest entropy at 1 atm of pressure?
    (a) $H_2O(s)$(1 mol)
    (b) $H_2O(l)$(1 mol)
    (c) $H_2O(g)$(1 mol)
    (d) 0.5 mol of ethanol in 0.5 mol of $H_2O(l)$
    (e) a gaseous solution of 0.5 mol of $H_2O(g)$ and 0.5 mol of $CH_4(g)$

19.23 For any *isolated* spontaneous process, which of the following conditions are true: (1) $\Delta G_{system} < 0$; (2) $\Delta G_{system} > 0$; (3) $\Delta S_{system} < 0$; (4) $\Delta S_{system} > 0$; (5) $\Delta S_{universe} < 0$; (6) $\Delta S_{universe} > 0$?

    (a) (1), (3), and (4)   (d) (1), (4), and (6)
    (b) (2), (4), and (6)   (e) (2), (4), and (6)
    (c) (1), (4), and (5)

19.24 For which of the following changes would you expect ΔS to be less than zero?

    (a) $H_2O(l) \rightarrow H_2O(g)$
    (b) $CO(g) + \frac{1}{2}O_2(g) \rightarrow CO_2(g)$
    (c) $2SO_3(g) \rightarrow 2SO_2(g) + O_2(g)$
    (d) $CO_2(s) \rightarrow CO_2(g)$
    (e) $2AgCl(s) \rightarrow 2Ag(s) + Cl_2(g)$

19.25 When KNO_3 dissolves in water at room temperature, ΔH is positive for the dissolution process. Given this information, what can you conclude?

    (a) $\Delta G > 0$ for the dissolution process.
    (b) $\Delta G = 0$ for the dissolution process.
    (c) The dissolving of salts in water is always a spontaneous process.
    (d) $\Delta S > 0$ for the dissolution process.
    (e) $\Delta S < 0$ for the dissolution process.

19.26 Which of the following is true for the sublimation of solid CO_2 at room temperature and pressure?

    (a) ΔH and ΔS are positive.
    (b) ΔH and ΔS are negative.
    (c) ΔH is negative, and ΔS is positive.
    (d) ΔG becomes more positive as temperature increases.
    (e) ΔG is positive.

19.27 Which of the following can be used to calculate directly the value of an equilibrium constant for a reaction at a given temperature?

    (a) ΔH (d) ΔG
    (b) $\Delta H°$ (e) $\Delta G°$
    (c) ΔS

19.28 Calculate $\Delta G°$ for the following reaction at 25°C

$$4NO(g) + 6H_2O(g) \longrightarrow 5O_2(g) + 4NH_3(g)$$

given the following $\Delta G_f°$ values: NO, 86.71 kJ/mol; H_2O, −228.61 kJ/mol; O_2, 0.0 kJ/mol; NH_3, −16.66 kJ/mol.

    (a) 958.2 kJ   (d) −1318.18 kJ
    (b) −2145.14 kJ   (e) 1318.18 kJ
    (c) 2145.14 kJ

19.29 Estimate the normal boiling point for HCl, given the following information:

$\Delta H_{vaporization}° = 16.13$ kJ/mol and $\Delta S_{vaporization}° = 85.77$ J/mol-K.

    (a) 188°C   (d) 0.189 K
    (b) 188 K   (e) 100°C
    (c) 0.189°C

SELF-TEST SOLUTIONS

19.1 True. **19.2** False. Since the liquid is not isolated from its surroundings, the entropy of the surroundings must also

undergo a change. **19.3** True. **19.4** True. **19.5** False. Heat may be required to initiate a reaction, but as long as the reaction then generates enough of its own heat to continue driving the reaction, the reaction is considered spontaneous. **19.6** False. Only for pure crystalline solids does $S = 0$ at 0 K. The presence of impurities results in $S > 0$. **19.7** False. An isolated system can not exchange heat with its surroundings. Therefore not heat is transferred to the surroundings and its entropy is not changed. **19.8** True. **19.9** False. It is zero only for elements in their standard states, not all substances. **19.10** In thermodynamics, time is not a factor in calculating values such as ΔH, ΔG, or ΔS. A reaction can be spontaneous and approach an equilibrium state quickly or slowly. The time it takes for a reaction to reach an equilibrium state is the domain of chemical kinetics. **19.11** ΔS is calculated from reversible changes. Since the problem states that the melting of ice at 273 K occurs in a reversible manner we know that $\Delta G = 0$. From the relationship $\Delta G = \Delta H - T\Delta S$, and the fact that $\Delta G = 0$, we derive the relationship $\Delta S = \Delta H/T$. This latter relationship is used to calculate ΔS_{system} from the given data:

$$\Delta S_{system} = \frac{\Delta H}{T} = \frac{(6.0 \text{ kJ/mol})(1000 \text{ J/kJ})}{273 \text{ K}}$$

$$= 22 \text{ J/mol-K}$$

The units for ΔH are converted from kJ to J because the units for ΔS are J/mol-K. **(b)** The heat absorbed by the ice as it melts is taken from the surroundings. Therefore, $\Delta H_{surroundings} = -6.0$ kJ for the melting of 1 mol of water, $\Delta S_{universe}$ is calculated as follows:

$$\Delta S_{universe} = \Delta S_{system} + \Delta S_{surroundings}$$

$$= \frac{6.0 \text{ kJ}}{273 \text{ K}} + \frac{-6.0 \text{ kJ}}{273 \text{ K}} = 0$$

$\Delta S_{universe} = 0$ for the reversible melting of ice as $\Delta S_{system} = -\Delta S_{surroundings}$ during a reversible phase change. **19.12** $H_2O(l)$ molecules are highly associated because of hydrogen bonding among water molecules, whereas this is not the case for $H_2S(l)$. Therefore, $H_2O(l)$ molecules are more ordered in their arrangements than is the case for $H_2S(l)$ molecules. When $H_2O(l)$ molecules vaporize, a greater degree of molecular order is destroyed than is the case for $H_2S(l)$. Therefore $\Delta S_{vaporization}$ is greater in magnitude for water than for H_2S. The presence of hydrogen bonding is also reflected in the larger heat of vaporization of $H_2O(l)$. **19.13** Equation 19.20 in the text is used to solve for ΔG at nonstandard state conditions: $\Delta G = \Delta G° + RT \ln Q$. Q is the reaction quotient, and its value is determined as follows: $Q = P_{SO_3}^2/P_{SO_2}^2 P_{O_2} = (2.0)^2/(1.0)^2(2.0) = 2.0$. From the given value of $\Delta G_{298}°$, we calculate ΔG_{298}: $\Delta G = \Delta G° + RT \ln Q = (-140.0 \text{ kJ}) + (8.314 \text{ J/K})(298 \text{ K}) (1 \text{ kJ}/10^3 \text{ J}) \ln (2.0) = -138.3$ kJ. **(b)** The free-energy change becomes less negative, changing from −140.0 kJ to −138.3 kJ as the pressures of SO_2, SO_3, and O_2 are changed from 1.0 atm each (standard-state conditions, $\Delta G°$) to their given values. The smaller negative value for ΔG indicates a lesser, not a larger, driving force to produce SO_3. **19.14** $\Delta G = \Delta H - T\Delta S$. To answer these questions, evaluate the effect of the signs and magnitudes of ΔH and ΔS upon the sign and value of ΔG. **(a)** ΔG is negative at high temperatures because the term $-T\Delta S$ dominates and ΔS is positive. **(b)** ΔG is negative at low temperatures because ΔH dominates and ΔH is negative. **(c)** ΔG is positive at all temperatures. **(d)** ΔG is negative at all temperatures. **19.15** Inspection of the equation $\Delta G = \Delta H - T\Delta S$ shows that the term $T\Delta S$ determines the sign of ΔG at high temperatures and ΔH determines its sign at low temperatures. At 298 K, the value of term $T\Delta S$ is usually small compared to the value of ΔH; thus $\Delta G \simeq \Delta H$, and the sign of ΔH determines the sign of ΔG. Since ΔG is a negative number for a spontaneous reaction, then ΔH at 298 K is usually negative. However, exceptions do occur, as described in the text. **19.16 (a)** In order to calculate $\Delta S°$ using the relation $\Delta S° = (\Delta H° - \Delta G°)/T$, the values of $\Delta H°$ and $\Delta G°$ must first be calculated. $\Delta G° = [\Delta G_f°(O_2) + 2\Delta G_f°(Cu_2O)] - [4\Delta G_f°(CuO)] = [0$ kJ $+ (2$ mol$)(-146.0$ kJ/mol$)] - [(4$ mol$)(-129.7$ kJ/mol$] = 226.8$ kJ. And $\Delta H° = [\Delta H_f°(O_2) + 2\Delta H_f°(Cu_2O)] - [4\Delta H_f°(CuO)] = [0$ kJ $+ (2$ mol$)(-168.6$ kJ/mol$)] - [(4$ mol$) (-157.3$ kJ/mol$)] = 292.0$ kJ. Solving for $\Delta S°$, $\Delta S° = (\Delta H° - \Delta G°)/T = (292.0$ kJ $- 226.8$ kJ$)/298$ K $= 0.219$ kJ/K $= 219$ J/K. **(b)** At standard state conditions $\Delta G° = 226.8$ kJ; the reaction is not feasible because the sign of $\Delta G°$ is positive. **(c)** 226.8 kJ is the *minimum* amount of energy needed to cause the reaction to occur. We could couple this reaction with a spontaneous one that has a $\Delta G°$ more negative than -226.8 kJ; the second reaction should not interfere with the decomposition of $CuO(s)$.

19.17 (a) $\ln K = \dfrac{-\Delta G°}{RT}$

$$= \dfrac{-(20,000 \text{ J/mol})}{(8.314 \text{ J K-mol})(298 \text{ K})} = -8.07$$

$$K = 3.10 \times 10^{-4}$$

(b) The equilibrium lies far on the side of reactants because $K < 0$. **19.18 (a)** To calculate $\Delta G°$ for the reaction, first calculate $\Delta S°$ and $\Delta H°$: $\Delta S° = S°(N_2O_4) - 2S°(NO_2) = (1$ mol$)(304.33$ J/mol-K$) - (2$ mol$)(240.45$ J/mol-K$) = -176.57$ J/K. And $\Delta H° =$

$\Delta H_f°(N_2O_4) - 2\Delta H_f°(NO_2) = (1$ mol$)(9.66$ kJ/mol$) - 2(33.8$ kJ/mol$) = -57.94$ kJ. Thus $\Delta G° = \Delta H° - T\Delta S° = -57,940$ J $- (298$ K$)(-176.57$ J/K$) = -5320$ J $= -5.32$ kJ. **(b)** The reaction is spontaneous at 25°C because $\Delta G° < 0$ at standard-state conditions.

(c) $\log K = \dfrac{-\Delta G°}{RT}$

$$= \dfrac{-(-5320 \text{ J})}{(8.314 \text{ J/K-mol})(298 \text{ K})} = 2.15$$

$$K = 8.57$$

19.19 The equilibrium constant is calculated using the relationship: $\Delta G_{rxn}° = -RT \ln K$. To solve for K, you first need to calculate the standard free-energy change.

$\Delta G_{rxn}° = [2\Delta G_f°(H_2O(l))] - [2\Delta G_f°(H_2(g)) + \Delta G_f°(O_2(g))]$

$\qquad = [(2$ mol$)(-237.2$ kJ/mol$)] - [(2$ mole$)(0$ kJ/mol$)$

$\qquad\quad + (1$ mol$)(0$ kJ/mol$)]$

$\qquad = -474.4$ kJ

$\Delta G_{rxn}° = -RT \ln K$

$$-474.4 \text{ kJ} \times \dfrac{1000 \text{ J}}{1 \text{ kJ}} = -(8.314 \text{ J/K-mol})(298 \text{ K}) \ln K$$

$$\ln K = 191.5$$

$$K = e^{191.5} = 1.5 \times 10^{83}$$

19.20 (a). **19.21 (c)**. **19.22 (e)**. A mixture of gases shows the greatest random motion of particles. **19.23 (d)**. **19.24 (b)** A mixture of gases with $1\frac{1}{2}$ particles reacting yields a single gas with fewer particles of gas, one particle. **19.25 (d)**. **19.26 (a)**. Sublimation of CO_2 is a spontaneous process at room temperature and pressure. This process requires energy ($\Delta H > 0$), and a solid-to-gas conversion always shows $\Delta S > 0$. **19.27 (e)**. **19.28 (a)**. **19.29 (b)**.

20 Electrochemistry

OVERVIEW OF THE CHAPTER

20.1, 20.2
OXIDATION-
REDUCTION
REACTIONS:
BALANCING

Review: Balancing chemical equations (3.1); oxidation state (8.10).

Objectives: You should be able to:

1. Identify the oxidant and reductant in an oxidation-reduction reaction.
2. Balance simple oxidation-reduction reactions by the oxidation number method.
3. Balance simple oxidation-reduction reactions by the method of half-reactions.

20.3, 20.4 VOLTAIC
CELLS: STANDARD
ELECTRODE AND
CELL POTENTIALS

Review: Electrolyte solutions (4.2).

Objectives: You should be able to:

1. Diagram simple voltaic and electrolytic cells, labeling the anode, the cathode, the directions of ion and electron movement, and the signs of the electrodes.
2. Given appropriate electrode potentials, calculate the emf generated by a voltaic cell.
3. Use electrode potentials to predict whether a reaction will be spontaneous.

20.5, 20.6
RELATIONSHIP OF
EMF TO
CONCENTRATION,
FREE-ENERGY
CHANGES AND THE
EQUILIBRIUM
CONSTANT

Review: Equilibrium Constant (15.2); free-energy function (19.5); free energy and the equilibrium state (15.5, 15.6).

Objectives: You should be able to:

1. Interconvert $E°$, $\Delta G°$, and K for oxidation-reduction reactions.
2. Use the Nernst equation to calculate the concentration of an ion, given E, $E°$, and the concentrations of the remaining ions.
3. Use the Nernst equation to calculate an emf under nonstandard conditions.

20.8, 20.9
ELECTROLYSIS: FARADAY'S LAW

Objectives: You should be able to:

1. Diagram simple electrolytic cells, labeling the anode, the cathode, the directions of ion and electron movement, and the sign of the electrodes.

2. Given appropriate electrode potentials, calculate the minimum emf required to cause electrolysis.

3. Interrelate time, current, and the amount of substance produced or consumed in an electrolysis reaction; given two of the three quantities, you should be able to calculate the third.

4. Calculate the maximum electrical work performed by a voltaic cell, and the minimum electrical work required for an electrolytic process.

20.7, 20.10
PRACTICAL EXAMPLES OF ELECTROCHEMISTRY

Objectives: You should be able to:

1. Describe the lead storage battery, the dry cell, and the nickel-cadmium cell.

2. Describe corrosion in terms of the electrochemistry involved, and explain the principles that underlie cathodic protection.

TOPIC SUMMARIES AND EXERCISES

OXIDATION-REDUCTION EQUATIONS: BALANCING

The concept of oxidation state was introduced in Section 8.10. You should review the rules for assigning oxidation numbers to atoms. In Section 20.1 you should note how changes in oxidation numbers of atoms are used to understand a class of reactions known as oxidation-reduction reactions. To illustrate the concepts, consider the following reaction, with the oxidation state of each atom shown above its elemental symbol.

$$3\overset{+1}{\text{Ag}}\overset{+5}{\text{N}}\overset{-2}{\text{O}}_3(aq) + \overset{0}{\text{Al}}(s) \longrightarrow 3\overset{0}{\text{Ag}}(s) + \overset{+3}{\text{Al}}(\overset{+5}{\text{N}}\overset{-2}{\text{O}}_3)_3(aq)$$

- Silver undergoes **reduction** because its oxidation number is *decreasing* from +1 to 0 in the reaction.
- Aluminum undergoes **oxidation** because its oxidation number is *increasing* from 0 to +3.
- Silver is an **oxidizing agent** or **oxidant** because it causes aluminum to be oxidized.
- Aluminum is a **reducing agent** or **reductant** because it causes silver to be reduced.

You should become proficient in balancing oxidation-reduction reactions. The method discussed in the text requires you to write two half-reactions and balance them in acid or base solution.

• As a tool for balancing oxidation-reduction (redox) reactions, a redox reaction can be broken down into two separate reactions, called **half-reactions.** One half-reaction represents the oxidation process and the other half the reduction process. The reaction between $AgNO_3$ and Al is broken down as follows:

$$\text{Reduction: } 3AgNO_3(aq) + 3e^- \longrightarrow 3Ag(s) + 3NO_3^-(aq)$$
$$\text{Oxidation: } \qquad\qquad Al(s) \longrightarrow Al^{3+}(aq) + 3e^-$$

Notice (1) that the number of electrons gained in the oxidation half-reaction equals the number lost in the reduction half-reaction and (2) that the sums of atoms and charges must be the same on each side of the arrow in each half-reaction.

• In Section 20.2, the authors of the text demonstrate the use of half-reactions in balancing redox reactions. You must learn the procedures for balancing redox reactions in both acid and base solutions. Exercise 4 illustrates how to apply the rules.

EXERCISE 1   Identify the atoms undergoing oxidation and reduction in the following reactions:

(a) $2AuCl_4^-(aq) + 3Cu(s) \longrightarrow 2Au(s) + 8Cl^-(aq) + 3Cu^{2+}(aq)$
(b) $Cl_2(g) + OH^-(aq) \longrightarrow Cl^-(aq) + ClO^-(aq) + H^+(aq)$

Solution: First identify the oxidation number of each element. The atom undergoing oxidation is the one whose oxidation number increases; the atom undergoing reduction is the one whose oxidation number decreases:

(a) $2AuCl_4^- + 3Cu \longrightarrow 2Au + 8Cl^- + 3Cu^{2+}$

Oxidation number: +3 −1   0   0   −1 +2

Cu has been oxidized from 0 to +2, and Au has been reduced from +3 to 0.

(b) $Cl_2 + OH^- \longrightarrow Cl^- + ClO^- + H^+$

Oxidation number: 0 −2 +1   −1   +1 −2 +1

Cl has been oxidized from 0 to +1, and Cl has been reduced from 0 to −1.
    This problem demonstrates the fact that an element can undergo oxidation and reduction simultaneously in a chemical reaction.

EXERCISE 2   In each of the following equations, is hydrogen peroxide acting as a reducing or oxidizing agent?

(a) $2MnO_4^-(aq) + 5H_2O_2(aq) + 6H^+(aq) \longrightarrow 2Mn^{2+}(aq) + 5O_2(g) + 8H_2O(l)$
(b) $PbS(s) + 4H_2O_2(aq) \longrightarrow PbSO_4(s) + 4H_2O(l)$

Solution: Assign oxidation numbers to oxygen in H_2O_2 and all products containing oxygen and then determine the change in oxidation number. **(a)** H_2O_2 is acting as a reducing agent. Oxygen in H_2O_2 has a −1 oxidation number; it increases to 0 in elemental O_2. This is verified by observing that manganese in MnO_4^- is reduced; the oxidation number of manganese decreases from +7 to +2. **(b)** H_2O_2 is acting as an oxidizing agent. Oxygen in H_2O_2 decreases in oxidation number from -1 to -2 in H_2O; this is verified by observing that the oxidation number of sulfur in PbS increases from -2 to $+6$ in $PbSO_4$.

EXERCISE 3   Write two half-reactions that when added together form the following balanced redox reaction:

$$S_2O_3^{2-}(aq) + 4I_2(aq) + 10OH^-(aq) \longrightarrow 2SO_4^{2-}(aq) + 8I^-(aq) + 5H_2O(l)$$

Solution: First assign oxidation numbers to all atoms in the overall reaction:

$$\underset{S_2O_3^{2-}(aq)}{\overset{+2\ \ -2}{}} + \underset{4I_2(aq)}{\overset{0}{}} + \underset{10OH^-(aq)}{\overset{-2\ +1}{}} \longrightarrow \underset{2SO_4^{2-}(aq)}{\overset{+6\ \ -2}{}} + \underset{8I^-(aq)}{\overset{-1}{}} + \underset{5H_2O(l)}{\overset{+1\ \ -2}{}}$$

Next, determine which atoms are involved in oxidation and reduction. Sulfur in thiosulfate, $S_2O_3^{2-}$, is oxidized from $+2$ to $+6$ in sulfate, SO_4^{2-}. Iodine, I_2, is reduced from 0 to -1 in I^-. Write the oxidation half-reaction by including $S_2O_3^{2-}$ and other species given in the overall reaction that are necessary for the number of atoms to be the same on both sides of the arrow:

$$S_2O_3^{2-}(aq) + 10OH^-(aq) \longrightarrow 2SO_4^{2-}(aq) + 5H_2O(l)$$

(OH^- and H_2O are necessary for the oxygen atoms in $S_2O_3^{2-}$ and SO_4^{2-} to be balanced.) Then add enough electrons so that the sum of charges is the same on both sides of the arrow. In this example, there is a $12-$ total charge $[(2-) + 10(1-)]$ on the left and a $4-$ total charge $[2(2-) + 5(0)]$. Eight negative charges are missing on the right. The oxidation half-reaction requires $8e^-$ on the right for it to be properly balanced.

$$S_2O_3^{2-}(aq) + 10OH^-(aq) \longrightarrow 2SO_4^{2-}(aq) + 5H_2O(l) + 8e^-$$

The reduction half-reaction is formed using the same procedure. Iodine is assigned a zero oxidation state because it is in its elemental form. The ion I^- is in the -1 oxidation state, the same as its ion state. I_2, which contains two iodine atoms, must form two I^- ions. Thus the reaction describing the reduction of iodine is

$$I_2(aq) + 2e^- \longrightarrow 2I^-(aq)$$

The number of electrons lost in the oxidation half-reaction must equal the number of electrons gained in the reduction half-reaction. Eight electrons are lost in the oxidation half-reaction; therefore you must multiply the reduction half-reaction for iodine by four to make the number of electrons equal:

$$4I_2(aq) + 8e^- \longrightarrow 8I^-(aq)$$

EXERCISE 4   Using the procedures given in Section 20.2 of the text, balance the following redox reactions:

(a) $Zn(s) + NO_3^-(aq) \longrightarrow Zn^{2+}(aq) + N_2(g)$ (acidic solution)
(b) $Br_2(aq) \longrightarrow BrO_3^-(aq) + Br^-(aq)$ (basic solution)

Solution: First, write the skeletal half-reactions for oxidation and reduction. To do this, you must be able to identify the atoms undergoing oxidation and reduction:

$$\text{Oxidation:}\quad Zn(s) \longrightarrow Zn^{2+}(aq)$$
$$\text{Reduction:}\ NO_3^-(aq) \longrightarrow N_2(g)$$

Note that in the reduction half-reaction only species containing nitrogen are included and that the oxygen atoms are not balanced at this time.

Second, balance all atoms other than oxygen and hydrogen in each half-reaction:

$$Zn \longrightarrow Zn^{2+} \qquad 2NO_3^- \longrightarrow N_2$$

Two NO_3^- ions are required to balance the two nitrogen atoms contained in N_2.

Third, balance oxygen atoms in each reaction by adding an appropriate number of water molecules:

$$Zn \longrightarrow Zn^{2+} \qquad 2NO_3^- \longrightarrow N_2 + 6H_2O$$

Six H_2O molecules are added as products to balance the six oxygen atoms contained in two NO_3^- ions.

Fourth, balance hydrogen atoms by adding an appropriate number of hydrogen ions:

$$Zn \longrightarrow Zn^{2+} \qquad 12H^+ + 2NO_3^- \longrightarrow N_2 + 6H_2O$$

Twelve H^+ ions are added as reactants to balance the 12 hydrogen atoms contained in six H_2O molecules.

Fifth, make the sum of charges on each side of the arrow equal by adding the required number of electrons to the appropriate side:

$$Zn \longrightarrow Zn^{2+} + 2e^- \qquad 10e^- + 12H^+ + 2NO_3^- \longrightarrow N_2 + 6H_2O$$

Sixth, make the number of electrons lost equal the number of electrons gained for each half-reaction by multiplying each half-reaction by the smallest integer that makes the number of electrons in each half-reaction the same:

$$5(Zn \longrightarrow Zn^{2+} + 2e^-) \qquad 1(10e^- + 12H^+ + 2NO_3^- \longrightarrow N_2 + 6H_2O)$$

Seventh, add the two half-reactions together and cancel equal amounts of electrons, H^+ ions, and H_2O molecules that appear on both sides of the arrow:

$$5Zn \longrightarrow 5Zn^{2+} + 10e^-$$
$$\underline{10e^- + 12H^+ + 2NO_3^- \longrightarrow N_2 + 6H_2O}$$
$$5Zn + 12H^+ + 2NO_3^- \longrightarrow 5Zn^{2+} + N_2 + 6H_2O$$

Eighth, check the final equation to make sure that the sums of atoms and charges on each side of the arrow are the same.

(b) This reaction demonstrates the procedures for balancing redox reactions in basic solution. It also illustrates the phenomenon of a single substance acting both as an oxidant and reductant. Note that in this example Br_2 is reduced to Br^- and oxidized to BrO_3^-. Therefore Br_2 will be a reactant in both half-reactions.

First, when balancing a redox reaction in basic solution, follow all the steps in (a). This results in the following equations for this redox reaction:

$$5(Br_2 + 2e^- \longrightarrow 2Br^-)$$
$$\underline{1(Br_2 + 6H_2O \longrightarrow 2BrO_3^- + 12H^+ + 10e^-)}$$
$$6Br_2 + 6H_2O \longrightarrow 10Br^- + 2BrO_3^- + 12H^+$$

Dividing by 2 yields reactants and products with the smallest integral coefficients possible:

$$3Br_2 + 3H_2O \longrightarrow 5Br^- + BrO_3^- + 6H^+$$

Second, add sufficient OH^- ions to both sides of the arrow so that all the H^+ ions are converted to H_2O by the reaction

$$H^+ + OH^- \longrightarrow H_2O$$

Cancel equal numbers of water molecules on both sides of the arrow. In this exercise add $6OH^-$ to both sides of the arrow:

$$3Br_2 + 3H_2O \longrightarrow 5Br^- + BrO_3^- + 6H^+$$
$$+ 6OH^- \qquad\qquad\qquad\qquad + 6OH^-$$

$$3Br_2 + 3H_2O + 6OH^- \longrightarrow 5Br^- + BrO_3^- + 6H_2O$$

Cancel three H_2O molecules from each side and reduce the equation to the smallest coefficients:

$$3Br_2(aq) + 6OH^-(aq) \longrightarrow 5Br^-(aq) + BrO_3^-(aq) + 3H_2O(l)$$

A spontaneous oxidation-reduction reaction involves the transfer of electrons from oxidant to reductant. In principle, it is possible to design a device in which electrons are forced to move through an external circuit rather than transferring directly between the oxidant and reductant. Such a device is called a **voltaic** (or **galvanic**) **cell.**

The design of a voltaic cell prevents the oxidant and reductant from being in direct physical contact with one another. A typical voltaic cell is shown in Figure 20.5 in the text. *You should know the following aspects of such a cell.*

- One cell compartment contains the chemicals that involve the oxidation half-reaction; the other contains those needed for the reduction half-reaction.

- A solid electrical connector, called an **electrode,** is placed in each cell compartment to provide a means for electrons to be transferred in the external circuit. An electrode may participate in the chemical reaction.

- The electrode at which reduction occurs is termed the **cathode;** the one at which oxidation occurs is termed the **anode.**

- By convention, the anode is assigned a negative charge; the cathode is assigned a positive charge.

- Electrons flow from the anode to the cathode in the external circuit through a wire that connects them.

- The solutions are connected by a **salt bridge** or semi-permeable membrane that allows ions to flow into or out of each cell compartment. A salt bridge enables each cell compartment to maintain a net charge of zero without allowing the two solutions to mix.

- Positive ions move in solution away from the anode to the cathode; negative ions move in solution away from the cathode to the anode.

In a voltaic cell, electrons are pumped through the outer circuit with a force that varies depending on the nature of the species reacting.

- The driving force is called **electromotive force (emf)** and is measured in the unit of volts (symbol V); it is also referred to as the cell potential or voltage.

$$1 \text{ volt (V)} = \frac{1 \text{ joule (J)}}{1 \text{ coulomb (C)}}$$

VOLTAIC CELLS: STANDARD ELECTRODE AND CELL POTENTIALS

- The emf generated by a cell is represented by the symbol E.
- A voltaic cell operating under standard-state conditions (all gases at 1 atm pressure and all solutes at 1 M concentration) generates a cell emf called **standard emf, $E°$.** Unless otherwise stated, the temperature is assumed to be 25°C.

It is impossible to measure directly the electromotive force generated by an oxidation or reduction half-reaction, but we can easily measure the potential difference between an anode and a cathode using a voltaic cell.

- A cell emf is simply the potential difference between the anode and cathode.
- If we assign a zero potential to a reference, standard electrode reaction, we can then measure the potentials of a series of half-reactions with respect to this reference. The chosen reference electrode system is the **standard hydrogen electrode:**

$$2H^+(aq)(1\ M) + 2e^+ \xrightarrow{25°C} H_2(g)(1\ atm) \quad E° = 0.00\ V$$

- A cell potential can be calculated as follows: $E°_{cell} = E°_{ox} + E°_{red}$.

When using standard reduction potentials of half-reactions to determine the standard potential of a half-reaction or a voltaic cell at standard-state conditions, four important facts must be remembered:

1. *The standard potential for any reduction half-reaction is equal in magnitude but opposite in sign to that of its reverse reaction (an oxidation half-reaction).* For example, note the relationship between $E°_{red}$ and $E°_{ox}$ for nickel:

Reduction: $Ni^{2+}(aq) + 2e^- \longrightarrow Ni(s) \quad E°_{red} = -0.44\ V$
Oxidation: $Ni(s) \longrightarrow Ni^{2+}(aq) + 2e^- \quad E°_{ox} = -E°_{red} = -(-0.44\ V) = 0.44\ V$

2. *The more positive the $E°$ value for a reduction (oxidation) half-reaction, the greater the tendency for that reduction (oxidation) half-reaction to occur as written.* Thus, in the following series:

				$E°$ (V)
Ease of reduction	↑	$Ag^+ + e^- \longrightarrow Ag$	Ease of oxidation	0.80
		$2H^+ + 2e^- \longrightarrow H_2$	↓	0.00
		$Fe^{2+} + 2e^- \longrightarrow Fe$		-0.44

Ag^+ is more easily reduced than H^+ or Fe^{2+}. The negative emf of Fe^{2+} means that Fe^{2+} is more difficult to reduce than H^+. Note that Fe is more easily oxidized than Ag or H_2.

3. *A voltaic cell exhibits a positive emf; thus for any spontaneous redox process the sum of oxidation and reduction half-cell potentials must yield a positive voltage.*

4. *Do not change the standard reduction (oxidation) potential of a half-reaction when balancing a half-reaction to form a balanced oxidation-reduction* *reaction.* Standard potentials are not affected when a half-reaction is multiplied by a number.

EXERCISE 5 A voltaic cell is to be designed at standard-state conditions using the following two half-reactions:

$$Mn^{2+}(aq) + 2e^- \longrightarrow Mn(s) \qquad E^{\circ}_{red} = -1.18 \text{ V}$$
$$Cr^{3+}(aq) + 3e^- \longrightarrow Cr(s) \qquad E^{\circ}_{red} = -0.74 \text{ V}$$

(a) Write the chemical equation describing the spontaneous cell reaction and calculate E°_{cell}. **(b)** Draw a figure representing the voltaic cell, using these reactions. Label all components and indicate the direction of electron flow.

    Solution: **(a)** In order to produce a spontaneous cell reaction at standard-state conditions, one of the given reduction reactions must be reversed and its standard potential sign changed so that when its standard oxidation potential is added to the standard-reduction potential a positive cell potential results. *The reduction half-reaction with the more negative standard reduction potential is always the one reversed;* in this exercise, the reduction reaction of manganese is reversed:

$$3[Mn(s) \longrightarrow Mn^{2+}(aq) + 2e^-] \qquad E^{\circ} = 1.18 \text{ V}$$
$$\underline{2[Cr^{3+}(aq) + 3e^- \longrightarrow Cr(s)] \qquad E^{\circ} = -0.74 \text{ V}}$$
$$3Mn(s) + 2Cr^{3+}(aq) \longrightarrow 3Mn^{2+}(aq) + 2Cr(s) \qquad E^{\circ}_{cell} = 0.44 \text{ V}$$

Note that although the reduction half-reaction is multiplied by 2 and the oxidation half-reaction is multiplied by 3 so that the number of electrons lost and gained are equal, their standard potentials are *not* multiplied by these factors.

(b)

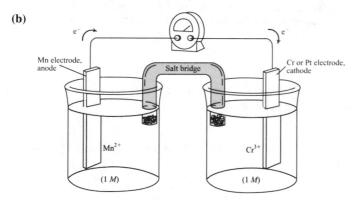

Mn is needed as an anodic electrode because Mn is a reactant in the oxidation half-reaction. It must be supplied. An inert electrode, such as Pt or a carbon rod, can be used as the cathodic electrode in this cell. In the reduction half-reaction you see that Cr is formed and deposits on an electrode. Since the electrode does not provide any reactant material, but only serves as an electron transfer surface, it does not have to be Cr. In such cases, an inert electrode is usually used

EXERCISE 6 Given the following E° values at 25°C

$$Fe^{2+} + 2e^- \longrightarrow Fe \qquad E^{\circ} = -0.44 \text{ V}$$
$$Sn^{2+} + 2e^- \longrightarrow Sn \qquad E^{\circ} = -0.14 \text{ V}$$

determine if the following reaction is spontaneous

$$Fe^{2+} + Sn \longrightarrow Fe + Sn^{2+}$$

Also identify the anode and cathode for the reaction that spontaneously occurs.

Solution: First determine if the $E°$ value for the reaction is positive. If it is, then the reaction is spontaneous as written at standard state conditions. If the $E°$ value is negative, then the reverse reaction is spontaneous.

$$Fe^{2+} + 2e^- \longrightarrow Fe \qquad E° = -0.44V$$
$$\underline{Sn \longrightarrow Sn^{2+} + 2e^+ \qquad E° = -(-0.14 \text{ V}) = 0.14 \text{ V}}$$
$$Fe^{2+} + Sn \longrightarrow Fe + Sn^{2+} \qquad E° = -0.30 \text{ V}$$

Because the $E°$ value is negative, the reverse reaction is spontaneous:

$$Fe + Sn^{2+} \longrightarrow Fe^{2+} + Sn$$

The anode is the half-reaction that is oxidation: $Fe \longrightarrow Fe^{2+} + 2e^-$. The cathode is the half-reaction that is reduction: $Sn^{2+} + 2e^- \longrightarrow Sn$.

EXERCISE 7   Determine which one of the three metals Zn, Fe, and Na is the most active metal, given the following data:

$$Na^+(aq) + e^- \longrightarrow Na(s) \qquad E°_{red} = -2.71 \text{ V}$$
$$Zn^{2+}(aq) + 2e^- \longrightarrow Zn(s) \qquad E°_{red} = -0.76 \text{ V}$$
$$Fe^{2+}(aq) + 2e^- \longrightarrow Fe(s) \qquad E°_{red} = -0.44 \text{ V}$$

Solution: An active metal is one that is easily oxidized (that is, a good reducing agent). Metals with large negative standard reduction electrode potentials (large positive standard oxidation potentials) are good reducing agents. Sodium metal is the most active metal because its reduction reaction has the largest negative standard reduction potential (largest positive oxidation potential). Na(s) is, therefore, the most active metal (best reducing agent) of those listed.

RELATIONSHIP OF EMF TO CONCENTRATION, FREE-ENERGY CHANGES AND THE EQUILIBRIUM CONSTANT

A spontaneous oxidation-reduction reaction has an accompanying negative free-energy change and a positive cell potential.

- The relationship between cell emf and the free-energy change for any redox process is given by

$$\Delta G = -n \mathscr{F} E$$

where n is the number of moles of electrons transferred in the redox reaction and $\mathscr{F}$ is faraday's constant, which equals 96,500 J/V-mol e^-. A faraday represents the charge carried by one mole of electrons.
- If all reactants and products are in their standard states, the relation is $\Delta G° = -n \mathscr{F} E°$.
- Note that a positive emf yields a negative ΔG.

Most reactions are not done at standard state conditions; thus cell emf will be expressed as E not E°. You should become proficient in using the Nernst equation to relate E to E° and concentration.

- At 298 K, the potential for a redox reaction at nonstandard conditions is related to E° as given by the Nernst equation:

$$E = E° - \frac{0.0592\ V}{n} \log Q$$

where Q is the ion-product for the redox reaction and n is the number of moles of electrons involved in the redox reaction. The concentrations of pure liquids and solids and the solvent of a dilute mixture do not appear in the reaction quotient.

If a redox reaction has achieved an equilibrium state, then $\Delta G = 0$ and $E = 0$ V. At 298 K, the equilibrium constant for a redox reaction is related to $E°$ for the reaction by the relation

$$E° = \frac{0.0592\ V}{n} \log K$$

or

$$\log K = \frac{nE°}{0.0592\ V}$$

EXERCISE 8 Calculate $\Delta G°$, K, and E at 298 K for the reaction

$$2Br^-(aq) + Cl_2(g) \longrightarrow Br_2(l) + 2Cl^-(aq)$$

given $E° = 0.300$ V and the following concentrations: $[Br^-] = 0.10$ M, $P_{Cl_2} = 0.50$ atm, and $[Cl^-] = 0.010$ M.

Solution: Inspection of the redox reaction shows that two moles of electrons are transferred. The standard free-energy change is calculated as follows:

$$\Delta G° = -n\mathscr{F}E° = -\left(2\ \text{mol e}^-\right)\left(96{,}500\ \frac{J}{V - \text{mol e}^-}\right)(0.300\ V)$$

$$= -5.79 \times 10^4\ J = -57.9\ kJ$$

The value of log K is obtained using the relation

$$\log K = \frac{nE°}{0.0592\ V} = \frac{(2)(0.300\ V)}{0.0592\ V} = 10.2$$

Taking the antilog of both sides yields $K = 2 \times 10^{10}$. The value of E is calculated using the Nernst equation:

$$E = E° - \left(\frac{0.0592\ V}{n}\right)\log \frac{\left[Cl^-\right]^2}{\left[Br^-\right]^2 P_{Cl_2}}$$

The concentration of Br_2 is not shown because it is a pure liquid.

$$E = 0.300\ V - \left(\frac{0.0591\ V}{2}\right)\log \frac{(0.010)^2}{(0.10)^2(0.50)}$$

$$E = 0.300\ V - (0.0295\ V)(-1.70) = 0.300\ V + 0.050\ V = 0.350\ V$$

EXERCISE 9   Calculate the voltage generated by a hydrogen-concentration cell utilizing the following concentrations:

$$H_2(2 \text{ atm}) + 2H^+ (0.1 \ M) \rightleftharpoons H_2(0.5 \text{ atm}) + 2H^+ (0.5 \ M)$$

A concentration cell involves a reaction having the same reactants as products, but with differing concentrations. In which direction is the reaction spontaneous?

Solution: The voltage is calculated using the Nernst equation:

$$E = E^\circ - \left(\frac{0.0592 \text{ V}}{n} \right) \log Q$$

Inspection of the redox equation shows $n = 2$. Substituting the value of E° into the Nernst equation and remembering that $Q = $ [products]/[reactants], we have

$$E = 0.00 \text{ V} - \left(\frac{0.0592 \text{ V}}{2} \right) \log \frac{P_{H_2} \left[H^+ \right]^2}{P_{H_2} \left[H^+ \right]^2}$$

Substituting the given concentration values yields

$$E = 0.00 \text{ V} - \left(\frac{0.0592 \text{ V}}{2} \right) \log \frac{(0.5)(0.5)^2}{(2)(0.1)^2} = 0.00 \text{ V} - (0.0295 \text{ V})(0.80)$$

$$= -0.02 \text{ V}$$

The concentration cell is spontaneous toward the *left* of the reaction equation because $E < 0$ for the given reaction.

ELECTROLYSIS: FARADAY'S LAW

A nonspontaneous oxidation-reduction reaction can be made to occur by the application of a direct current of electricity to the reactants. This process is called **electrolysis,** and the special apparatus in which it takes place is called an electrolytic cell.

The minimum applied emf that causes an electrolytic reaction to proceed is calculated using standard electrode potentials and the Nernst equation.

- The reduction half-reaction with the least negative E_{red} will occur first at the cathode.

- Similarly, at the anode, the oxidation half-reaction with the least negative E_{ox} will occur first.

- If a half-reaction has a large **overvoltage,** an additional voltage required to cause electrolysis, then another half-reaction with a lower overvoltage may preferentially occur. For example, the overvoltages for the oxidation of water to $H^+(aq)$ and its reduction to $H_2(g)$ are high. Therefore, electrolysis of aqueous salt solutions often yields an oxidized or reduced form of the salt rather than both $H^+(aq)$ and $H_2(g)$.

Faraday's law, a result of the work of Michael Faraday, states that the quantity of a substance undergoing a change during electrolysis is directly proportional to the quantity of electricity that passes through the cell.

- The quantity of electricity passing through a cell is expressed in number of faradays ($\mathscr{F}$). One faraday is equal to 96,500 coulombs or one mole of electrons.

Therefore,

$$\text{Number of faradays} = (\text{number of coulombs})\left(\frac{1\ \mathscr{F}}{96{,}500\ \text{C}}\right)$$

$$\text{Number of moles of electrons} = (\text{number of coulombs})\left(\frac{1\ \text{mol electrons}}{96{,}500\ C}\right)$$

- The number of coulombs (abbreviated as C) is equal to the product of current passing through the cell (in amperes) and the amount of time (in seconds) that the current is passed through the cell:

$$\text{Coulombs} = \text{amperes} \times \text{seconds} = it \qquad 1\ \text{C} = 1\ \text{amp-sec}$$

- Remember when doing Faraday's law problems that one mole of electrons equals one faraday. This equivalence enables you to convert between mass (moles) of a substance and coulombs passed through a cell.

There is a direct relationship between electrochemical processes and work:

$$W_{max} = -n\mathscr{F}E = -(\text{mol})\left(\frac{C}{\text{mol}}\right)\left(\frac{J}{C}\right) = J$$

Conversely, the minimum amount of work required for an electrolytic process is

$$W_{min} = -n\ \mathscr{F}\ E$$

where E is the cell emf for the electrolysis reaction.

- Remember that work done *by* the system *on* its surroundings is indicated by a *negative* sign. Thus a negative W_{max} corresponds to a spontaneous process.
- Electrical work is more commonly expressed as watts × time, where a watt is the rate of energy expenditure:

$$1\ \text{watt} = \frac{1\ J}{\text{sec}}$$

Note: 3.6×10^6 J = 1 kilowatt-hour.
  1 J = 1 watt-s

EXERCISE 10   What are the minimum requirements for an electrolytic cell?
   Solution: A battery (voltaic cell) or the rectified current from an electric generator is required as a direct-current source of electrons. The electrons are forced onto one electrode and removed from another. Electrodes are composed of substances that are good conductors; they are in contact with a liquid or solution containing the reactants. Reduction occurs at the surface of the electrode to which electrons are pumped, and oxidation occurs at the electrode from which electrons are removed.

EXERCISE 11   The electrodes in an electrolytic cell are given the names anode and cathode. What processes occur at the surface of these electrodes in the electrolytic solution?
   Solution: Oxidation occurs at the anode (the positive electrode in an electrolytic cell), and reduction occurs at the cathode (the negative electrode in an electrolytic cell). Notice that the signs of the electrodes are opposite to those found in a voltaic cell.

EXERCISE 12   What mass of zinc metal is produced at the cathode in an electrolysis cell when a constant current of 10.00 amp is passed for 1 hr?

Solution: First calculate how many moles of electrons are passed through the cell by calculating the number of faradays (remember that 1 hr = 3600 sec):

$$\text{Number of moles of electrons} = (\text{number of coulombs})\left(\frac{1 \text{ mol electrons}}{96,500 \ C}\right)$$

$$= (\text{amperes})(\text{seconds})\left(\frac{1 \text{ mol electrons}}{96,500 \ C}\right)$$

$$= (10.00 \ amps)(3600 \ s)\left(\frac{1 \text{ mol electrons}}{96,500 \ C}\right)$$

$$= (3.600x\ 10^4\ C)\left(\frac{1 \text{ mol electrons}}{96,500 \ C}\right) = 0.373 \text{ mol electrons}$$

The half-reaction for zinc tells you that two moles of electrons passing through a cell containing zinc ions causes one mole of zinc metal to form. You can calculate the number of moles of zinc metal deposited from this information:

$$\text{Number of moles of Zn} = (0.373 \text{ mol electrons})\left(\frac{1 \text{ mol Zn}}{2 \text{ mol electrons}}\right) = 0.187 \text{ mol Zn}$$

Thus the mass of zinc metal produced is

$$\text{Mass of Zn} = (0.187 \text{ mol Zn})\left(\frac{65.4 \ g \text{ Zn}}{1 \text{ mol Zn}}\right) = 12.2 \ g$$

EXERCISE 13   (a) A solution of $CuSO_4$ is electrolyzed between two copper electrodes. What are the expected electrolysis products given the following data?

	E° (V)
$2H_2O(l) + 2e^- \longrightarrow H_2(g) \longrightarrow 2OH^-(aq)$	−0.83
$Cu^{2+}(aq) + 2e^- \longrightarrow Cu(s)$	+0.337
$4H^+(aq) + O_2(g) + 4e^- \longrightarrow 2H_2O(l)$	+1.23
$S_2O_8^{-2}(aq) + 2e^- \longrightarrow 2SO_4^{2-}(aq)$	+2.01

(b) Could this process be used to refine impure copper?

Solution: To answer this question you need the E° data given in the problem. If it were not provided you would have to examine a table of E° values in your text or in a reference source. **(a)** Only $H_2O(l)$ and $Cu^{2+}(aq)$ can possibly be reduced because only H_2O, Cu^{2+}, SO_4^{2-} are present. The reduction emf of $Cu^{2+}(aq)$ is more positive than that of $H_2O(l)$; thus Cu^{2+} will be electrolyzed at the copper cathode:

$$Cu^{2+}(aq) + 2e^- \longrightarrow Cu(s)$$

At the anode there are three possible reactions:

	E°(V)
$Cu(s) \longrightarrow Cu^{2+}(aq) + 2e^-$	−0.337
$2H_2O(l) \longrightarrow 4H^+(aq) + O_2(g) + 4e^-$	−1.23
$2SO_4^{2-}(aq) \longrightarrow S_2O_8^{2-}(aq) + 2e^-$	−2.01

The oxidation of $Cu(s)$ requires the least amount of applied potential; therefore, at the anode the reduction of the copper electrode occurs. **(b)** Yes. Impure copper could be used as the anode, and a solution of $CuSO_4$ electrolyzed. The copper in the anode electrode would dissolve, with pure copper plating out at the cathode.

EXERCISE 14   Calculate the mass of Mg produced in an electrolytic cell if 2.67×10^2 kwh of electricity is passed through a solution of $MgCl_2$ at 4.20 volts.

Solution: The number of joules of work required is calculated from the equivalence 3.6×10^6 J = 1 kwh:

$$J = \left(2.67 \times 10^2 \text{ kwh}\right)\left(\frac{3.6 \times 10^6 \text{ J}}{1 \text{ kwh}}\right) = 9.6 \times 10^8 \text{ J}$$

From the joules of work and volts you can then calculate the number of coulombs passed through the cell from the relationship

$$joule = coul \times volt$$

or

$$coul = \frac{J}{volt} = \left(\frac{9.6 \times 10^8 \text{ J}}{4.20 \text{ V}}\right)\left(\frac{1 \text{ C-V}}{1 \text{ J}}\right) = 2.3 \times 10^8 \text{ C}$$

This corresponds to

$$Faradays = \left(2.3 \times 10^8 \text{ C}\right)\left(\frac{1 \mathscr{F}}{96,500 \text{ C}}\right) = 2.4 \times 10^2 \mathscr{F}$$

This also corresponds to the number of moles of electrons passed through the cell because $1 \mathscr{F} = 1$ mol electrons. Therefore, from the half-reaction

$$Mg^{2+}(aq) + 2e^- \longrightarrow Mg(s)$$

you can state $2 \mathscr{F} \simeq 1$ mol Mg.

The grams of $Mg(s)$ produced are thus

$$Mass \text{ Mg} = \left(2.4 \times 10^2 \mathscr{F}\right)\left(\frac{1 \text{ mol Mg}}{2 \mathscr{F}}\right)\left(\frac{24.3 \text{ g Mg}}{1 \text{ mol Mg}}\right)$$

$$= 2.9 \times 10^3 \text{ g Mg} = 2900 \text{ g Mg}$$

Several practical applications of electrochemical principles are discussed in Chapter 20. Of particular importance are the lead-acid storage battery, the dry cell, the rechargeable nickel-cadmium battery, fuel cells, corrosion of metals and electroplating. Exercises 15-20 are designed so that you can check your understanding of some of these applications.

PRACTICAL EXAMPLES OF ELECTROCHEMISTRY

EXERCISE 15   Explain why in the Downs cell metallic sodium is produced from the electrolysis of molten NaCl and not from the electrolysis of aqueous NaCl.

Solution: The electrolysis of aqueous NaCl forms $H_2(g)$ and $Cl_2(g)$ because water is more easily reduced than sodium ions.

EXERCISE 16   When an object is to be electroplated, why is it made the cathode?

Solution: In electroplating, a metal ion in solution is reduced to the metal at the surface of the object. Reduction occurs at the cathode.

EXERCISE 17   The density of the sulfuric acid solution in a lead storage battery is found to be 1.19 g/cm³. Does the battery need recharging?

Solution: Yes. During the discharge of a lead storage battery the sulfuric acid is consumed and the density of the solution in the battery decreases. If the density is below 1.20 g/cm³, it needs recharging.

EXERCISE 18   Differentiate between a fuel cell and a battery.

   Solution: Batteries store electrical energy, and their lifetime is limited by the amount of reactants initially present. Fuel cells do not store electrical energy because the reactants are continuously supplied to the electrodes; they are energy-conversion devices.

EXERCISE 19   Why is magnesium often used in the cathodic protection of iron?

   Solution: The standard reduction electrode potentials for iron and magnesium are

$$Mg^{2+} + 2e^- \longrightarrow Mg \qquad E^{\circ}_{red} = -2.37 \text{ V}$$
$$Fe^{2+} + 2e^- \longrightarrow Fe \qquad E^{\circ}_{red} = -0.44 \text{ V}$$

Because $E^{\circ}_{red}(Mg)$ is more negative than $E^{\circ}_{red}(Fe)$, Mg is more easily oxidized than Fe. Thus when iron and magnesium are connected, magnesium is preferentially oxidized, and iron becomes the site at which the following reduction reaction occurs:

$$O_2(g) + 4H^+(aq) + 4e^- \longrightarrow 2H_2O$$

Thus iron is acting as a cathodic electrode and is inert to oxidation until all the magnesium is used.

EXERCISE 20   Explain why the corrosion of iron becomes less favorable as the pH is increased.

   Solution: The initial reaction of iron in corrosion is

$$Fe(s) \longrightarrow Fe^{2+}(aq) + 2e^-$$

This reaction is coupled the cathodic reaction

$$O_2(g) + 4H^+(aq) + 4e^- \longrightarrow 2H_2O(l)$$

Note that H^+ takes part in the reaction with oxygen. As the pH is increased, the amount of H^+ present in solution decreases (more OH^-). According to LeChatelier's principle, reducing the concentration of H^+ will cause the cathodic reaction to shift to the left, thereby making the reduction of O_2 less favorable. If the cathodic reaction becomes less favorable, then the anodic oxidation of iron also becomes less favorable.

SELF-TEST QUESTIONS

Key Terms

Having reviewed the key terms listed at the end of Chapter 20 and their applications, identify the following statements as true or false. If a statement is false, indicate why it is incorrect. Key terms are italicized in the statements.

20.1   The total charge required for the reduction of 1 mol of Al^{3+} ions to 1 mol of solid Al is 1 *faraday*.

20.2   The attachment of Cu(s) ($E^{\circ}_{red} = 0.34$ V) to Zn(s) ($E^{\circ}_{red} = -0.76$ V) would provide *cathodic protection* for zinc.

20.3   The *corrosion* of iron in nature is believed to be an electrochemical process in which oxygen is consumed.

20.4   By convention, the *standard electrode potential* for

$$H_2(g) \longrightarrow 2H^+(aq) + 2e^-$$

is chosen to be 1.0 V.

20.5   In the voltaic cell reaction

$$Ni(s) + 2Ag^+(aq) \longrightarrow Ni^{2+}(aq) + 2Ag(s)$$

the *anode* is Ni(s).

20.6   In the reaction in problem 20.5, oxidation occurs at the *cathode*, Ag.

20.7   An *electrolytic cell* produces a potential that yields useful work.

20.8   For the sign of ΔG to be negative for an electrochemical process, the sign of *electromotive force* must be positive.

20.9 According to the *Nernst Equation,* the cell emf, *E,* has a direct dependence upon temperature in degrees Kelvin.

20.10 A *fuel cell* is a type of voltaic cell in which the reactants are supplied continuously.

20.11 The anode in a *voltaic cell* is assigned a positive sign because electrons are produced at the cathode.

20.12 *Electrochemistry* studies the production of electricity using oxidation-reduction reactions.

20.13 A process involving an oxidation-reduction reaction that requires an outside source of electrical energy is termed *electrolysis.*

20.14 The following *half-reaction* is an example of oxidation

$$Zn^{2+}(aq) + 2e^- \longrightarrow Zn(s)$$

20.15 The following is an example of an *oxidation-reduction reaction*

$$CuCl_2(aq) + 4NH_3(aq) \longrightarrow [Cu(NH_3)_4]Cl_2(aq)$$

20.16 Oxygen is the *oxidant,* or *oxidizing agent,* in the reaction

$$4NH_3(g) + 3O_2(g) \longrightarrow 2N_2(g) + 6H_2O(l)$$

20.17 A *reductant,* or *reducing agent,* is always reduced in a chemical reaction.

20.18 The term *standard cell potential,* or *standard emf,* means that all reagents are at standard state conditions at the given temperature.

20.19 A *cell potential* represents the difference between the potential energy per electrical charge between two electrodes in a voltaic cell.

20.20 The *standard oxidation potential* for the oxidation of metallic zinc is 0.76 *V.* From this information, you know that the *standard reduction potential* for the zinc ion is −0.76 *V.*

Problems and Short-Answer Questions

20.21 Complete and balance the following redox reactions by the half-reaction method:
  (a) $Cr_2O_7^{2-} + I^- \longrightarrow Cr^{3+} + I_2$ (acid solution)
  (b) $MnO_4^- + C_2O_4^{2-} \longrightarrow Mn^{2+} + CO_2$ (acid solution)
  (c) $HSO_3^- + NO_3^- \longrightarrow HSO_4^- + NO$ (acid solution)
  (d) $H_2O_2 + ClO_2 \longrightarrow ClO_2^- + O_2$ (base solution)
  (e) $Br_2 + AsO_2^- \longrightarrow Br^- + AsO_4^{3-}$ (base solution)
  (f) $Bi(OH)_3 + SnO_2^- \longrightarrow Bi + SnO_3^{2-}$ (base solution)

20.22 Explain why the electrolysis of pure water to form $H_2(g)$ and $O_2(g)$ requires the presence of a small amount of an electrolyte, such as Na_2SO_4.

20.23 Construct a table that summarizes the following information about the cathode and anode in both an electrolytic and a voltaic cell: the type of ion attracted to the electrode; the type of electrode reaction that occurs; the sign of the electrode; and the direction of electron movement in the outer circuit relative to the electrode.

20.24 How many grams of cadmium are consumed in a nickel-cadmium battery if it operates at a constant current of 0.20 amp for 30.00 sec? The half-reaction of interest is

$$Cd(s) + 2OH^-(aq) \longrightarrow Cd(OH)_2(s) + 2e^-$$

20.25 Given the following data, explain why metallic copper can be dissolved by nitric acid but not by hydrochloric acid at standard-state conditions:

$$\begin{aligned} Cu^{2+} + 2e^- &\longrightarrow Cu & E° &= 0.34 \text{ V} \\ NO_3^- + 4H^+ + 3e^- &\longrightarrow NO + H_2O & E° &= 0.96 \text{ V} \\ Cl_2 + 2e^- &\longrightarrow 2Cl^- & E° &= 1.36 \text{ V} \end{aligned}$$

20.26 Given the following data:

$$\begin{aligned} Fe^{2+} + 2e^- &\longrightarrow Fe & E° &= -0.44 \text{ V} \\ Ag^+ + e^- &\longrightarrow Ag & E° &= 0.80 \text{ V} \end{aligned}$$

Answer the following questions with respect to the reaction

$$Fe^{2+}(aq) + 2Ag(s) \longrightarrow Fe(s) + 2Ag^+(aq)$$

  (a) What is $E°$ for the reaction?
  (b) Is the reaction spontaneous at standard-state conditions?
  (c) What is the value of E at equilibrium?
  (d) What is the value of the equilibrium constant at 25°C?
  (e) If $[Fe^{2+}] = 0.100$ *M* and $[Ag^+] = 0.0100$ *M*, what is the magnitude of E at 25°C?
  (f) For the reaction that is spontaneous, what is the maximum amount of work that can be performed?

20.27 Explain why a steel wool pad does not rust as rapidly in a soap solution as in an acidic solution. Soap is a basic substance.

20.28 A solution is 1 *M* in $AgNO_3$ and $Fe(NO_3)_2$. In the electrolysis of this solution using inert electrodes, the voltage is gradually increased. What is the order in which substances in the solution would plate out or react at the cathode? Assume that the overvoltage of hydrogen is 0.30 V.

20.29 Calculate $E°$ for the following unbalanced cell reactions. Indicate if the reactions are spontaneous as written.
  (a) $Al^{3+}(aq) + Mg(s) \longrightarrow Al(s) + Mg^{2+}(aq)$
  (b) $ClO_4^-(aq) + Mn^{2+}(aq) + H_2O(l) \longrightarrow$
$$MnO_4^-(aq) + H^+(aq) + ClO_3^-(aq)$$
  Given data:

$$\begin{aligned} Al^{3+} + 3e^- &\longrightarrow Al & E° &= -1.66 \text{ V} \\ Mg^{2+} + 2e^- &\longrightarrow Mg & E° &= -2.37 \text{ V} \\ MnO_4^- + 8H^+ + 5e^- &\longrightarrow Mn^{2+} + 4H_2O \\ && E° &= 1.51 \text{ V} \\ ClO_4^- + 2H^+ + 2e^- &\longrightarrow ClO_3^- + H_2O \\ && E° &= 1.19 \text{ V} \end{aligned}$$

20.30 Given the following reaction,

$$2Al(s) + 3Mn^{2+}(aq) \longrightarrow 2Al^{3+}(aq) + 3Mn(s)$$

What is the concentration of Al^{3+} if $E°_{cell} = 0.48$ *V,* $E_{cell} = 0.47$ *V* and the concentration of Mn^{2+} is 0.49 *M* at 25°C?

Multiple-Choice Questions

Use the following data for questions 20.31–20.34.

$$Ag^+ + e^- \longrightarrow Ag \quad E° = 0.80V$$

$$Al^{3+} + 3e^- \longrightarrow Al \quad E° = -1.66 \text{ V}$$

20.31 What is $E°_{cell}$ for a voltaic cell using the two half-reactions at 25°C?

(a) –2.46 V (d) 0.86 V
(b) –0.86 V (e) 2.46 V
(c) 0.00 V

20.32 What is E_{cell} if $[Ag^+] = 0.50$ M and $[Al^{3+}] = 1.0$ M at 25°C?

(a) 2.40 V (d) 2.48 V
(b) 2.44 V (e) 2.52 V
(c) 2.46 V

20.33 What would happen to the cell emf if NH_3 were added to the silver cell and $Ag(NH_3)_2^+$ forms?

(a) no change (c) decreases
(b) increases

20.34 What is K for the reaction at 25°C?

(a) 10^{249} (d) 10^{-42}
(b) 10^{42} (e) 10^{-249}
(c) 1

20.35 Which of the following is the cathodic reaction in the electrolysis of aqueous NaCl?

(a) $Na^+ + e^- \longrightarrow Na$
(b) $Na \longrightarrow Na^+ + e^-$
(c) $2Cl^- \longrightarrow Cl_2 + 2e^-$
(d) $Cl_2 + 2e^- \longrightarrow 2Cl^-$
(e) $2H_2O + 2e^- \longrightarrow H_2 + 2OH^-$

20.36 Chromium metal can be electrolytically plated out from an acidic solution containing CrO_3. Assuming that all of the CrO_3 is in a soluble form, how many coulombs are required to cause 3.68 g of Cr to be deposited on the cathodic electrode?

(a) 20,500 coul (d) 61,500 coul
(b) 41,000 coul (e) 96,500 coul
(c) 10,250 coul

20.37 Which of the following is the half-cell potential for

$$H_2(2 \text{ atm}) \longrightarrow 2H^+(0.001 \text{ } M) + 2e^-$$

(a) 0.19 V (d) –0.04 V
(b) 0.25 V (e) –0.10 V
(c) 0.01 V

20.38 Given the following data:

$$Ca^{2+}(aq) + 2e^- \longrightarrow Ca(s) \quad E° = -2.87 \text{ V}$$
$$Zn^{2+}(aq) + 2e^- \longrightarrow Zn(s \quad E° = -0.76 \text{ V}$$
$$Co^{2+}(aq) + 2e^- \longrightarrow Co(s) \quad E° = -0.28 \text{ V}$$
$$Sn^{2+}(aq) + 2e^- \longrightarrow Sn(s) \quad E° = -0.14 \text{ V}$$
$$Pb^{2+}(aq) + 2e^- \longrightarrow Pb(s) \quad E° = -0.13 \text{ V}$$

which of the following correctly describes the ease of oxidation of the substances listed under standard state conditions?

(a) $Ca^{2+} > Zn^{2+} > Co^{2+} > Sn^{2+} > Pb^{2+}$

(b) $Pb^{2+} > Sn^{2+} > Co^{2+} > Zn^{2+} > Ca^{2+}$
(c) $Ca > Zn > Co > Sn > Pb$
(d) $Pb > Sn > Co > Zn > Ca$
(e) $Pb > Ca^{2+} > Zn > Sn^{2+} > Pb$

20.39 For a redox reaction at equilibrium, which of the following statements is always true: (1) $\Delta H = 0$; (2) $\Delta G = 0$; (e) $\Delta G° = 0$; (4) $E = 0$; (5) $E° = 0$?

(a) (1) only (d) (2) and (4) only
(b) (2) only (e) (3) and (5) only
(c) (1), (2), and (3) only

20.40 Given that $K = 3.76 \times 10^{14}$ for the reaction

$$2Fe^{3+}(aq) + Cu(s) \longrightarrow 2Fe^{2+}(aq) + Cu^{2+}(aq)$$

at 25°C, which of the following corresponds to $E°$ for the reaction?

(a) –0.43 V (d) –0.22 V
(b) 0.43 V (e) 0.00 V
(c) 0.22 V

20.41 According to LeChatelier's principle, which of the following effects would addition of NaOH have upon the reduction of NO_3^- to $NO_2(g)$ in acid solution: (1) increase emf; (2) decrease emf; (3) increase concentration of NO_3^-; (4) increase concentration of NO_2?

(a) (1) only (d) (2) and (3) only
(b) (2) only (e) (1) and (4) only
(c) (1) and (3) only

20.42 Which of the following metals could provide cathodic protection for Mn? (Refer to Appendix E in the text for emf data.)

(a) Zn (d) Cu
(b) Co (e) Mg
(c) Sn

20.43 Which of the following is formed during the discharging of a battery?

(a) $Pb(s)$ (d) $H^+(aq)$
(b) $PbO_2(s)$ (e) $HSO_4^- (aq)$
(c) $PbSO_4(s)$

20.44 From your knowledge of complex ions (see Section 17.5 of the text), which of the following metal-complex ions would you expect to form during the chemical reaction that occurs within a common dry cell?

(a) $Zn(NH_3)_4^{6+}$ (d) $MnCl_4^{2-}$
(b) $Zn(NH_3)_4^{2+}$ (e) $Mn(H_2O)_6^{2+}$
(c) $Zn(OH)_4^{2-}$

20.45 What species is formed at the Pt electrode in Figure 20.1?

(a) Fe (c) Ni
(b) Pt (d) Ni^{2+}

20.46 Which electrode is the cathode in Figure 20.1 on page 349?

(a) Fe (b) Pt

20.47 At which electrode in Figure 20.1 does oxidation occur?

(a) Fe (b) Pt

20.48 Which cation exists initially in the cathodic compartment in Figure 20.1?

(a) Ni^{2+} (b) Fe^{2+}

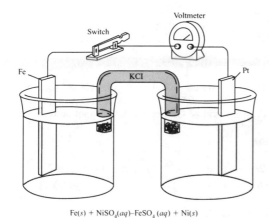

Fe(s) + NiSO₄(aq)–FeSO₄(aq) + Ni(s)
$E° = 0.16V$

Figure 20.1

20.49 In which compartment does the oxidizing agent exist in Figure 20.1?
  (a) anodic compartment
  (b) cathodic compartment

20.50 In which direction in the external circuit do electrons migrate in Figure 20.1?
  (a) anode to cathode
  (b) cathode to anode

20.51 In Figure 20.1, toward which electrode do Fe^{2+} ions migrate?
  (a) Fe   (b) Pt

20.52 Into which compartment do Cl^- ions in the salt bridge in Figure 20.1 migrate?
  (a) anodic compartment
  (b) cathodic compartment

20.53 In Figure 20.1, what is the polarity of the anode?
  (a) +   (b) –

20.54 For the unbalanced redox reaction

$$Ag_2S(s) + NO_3^-(aq) \longrightarrow Ag^+(aq) + S(s) + NO(g)$$

in acidic solution, what is the balanced reduction half-reaction?
  (a) $Ag_2S(s) \longrightarrow 2Ag^+(s) + S(s) + 2e^-$
  (b) $S^{2-} \longrightarrow S(s) + 2e^-$
  (c) $NO_3^-(aq) + 4H^+(aq) + 3e^- \longrightarrow NO(g) + 2H_2O$
  (d) $2H^+(aq) + 2e^- \longrightarrow H_2(g)$
  (e) $NO_3^-(aq) + 2H^+(aq) + e^- \longrightarrow NO(g) + 2OH^-(aq)$

20.55 For the unbalanced redox reaction

$$Cl_2(g) \longrightarrow ClO_4^-(aq) + Cl^-(aq)$$

in basic solution, what is the balanced oxidation half-reaction?
  (a) $Cl_2 + 2e^- \longrightarrow Cl^-$
  (b) $Cl_2 + 4O_2 + 2e^- \longrightarrow 2ClO_4^-$
  (c) $Cl_2 + 8H_2O \longrightarrow 2ClO_4^- + 16H^+ + 14e^-$
  (d) $Cl_2 + 16OH^- \longrightarrow 2ClO_4^- + 8H_2O + 14e^-$
  (e) $2Cl_2 + 8OH^- \longrightarrow 2ClO_4^- + Cl_2 + 6e^-$

20.56 What is the coefficient in front of the underlined substance in the balanced form of the following redox reaction?

$$Pb(NO_3)_2(s) \longrightarrow PbO(s) + \underline{NO_2(g)} + O_2(g)$$

  (a) 1   (d) 4
  (b) 2   (e) 5
  (c) 3

20.57 What is the coefficient in front of the underlined substance in the balanced form of the following redox reaction?

$$\underline{MnO_4^-(aq)} + SO_2(g) + H_2O(l) \longrightarrow Mn^{2+} + SO_4^{2-} + H^+$$

  (a) 1   (d) 4
  (b) 2   (e) 5
  (c) 3

20.58 Which of the following is capable of acting both as an oxidizing and reducing agent?
  (a) $KMnO_4$;   (d) NaI;
  (b) Cl_2;   (e) $SnCl_2$
  (c) H_2O_2;

20.59 Which of the following is the oxidizing agent in the reaction

$$SO_3^{2-}(aq) + Br_2(l) + H_2O(l) \longrightarrow$$
$$SO_4^{2-}(aq) + 2Br^-(aq) + 2H^+(aq)$$

  (a) SO_3^{2-};   (d) SO_4^{2-};
  (b) Br_2;   (e) H^+
  (c) H_2O

SELF-TEST SOLUTIONS

20.1 False. $Al^{3+} + 3e^- \longrightarrow Al$. One mole of Al^{3+} requires 3 mol of electrons, or 3 $\mathscr{F}$ of charge to form 1 mol of Al. **20.2** False. Zn ($E_{ox}° = 0.76$ V) is more readily oxidized than Cu($E_{ox}°$ = −0.34 V); thus Cu cannot protect Zn from oxidation. **20.3** True. **20.4** False. It is chosen to be 0.00 V. **20.5** True. **20.6** False. Reduction always occurs at the cathode of any type of cell. **20.7** False. Work must be supplied to an electrolytic cell to make a non-spontaneous reaction occur. **20.8** True. **20.9** True. **20.10** True. **20.11** False. The anode is assigned a negative sign because it is the source of electrons in the outer circuit. **20.12** True. **20.13** True. **20.14** False. The reaction shows reduction, the gain of electrons. **20.15** False. An oxidation-reduction reaction involves atoms changing oxidation numbers. The reaction shown is a complex ion reaction, with all atoms retaining the same oxidation state. **20.16** True. An oxidizing agent is reduced; oxygen is reduced. **20.17** False. A reducing agent causes another species to gain electrons; thus it is oxidized. **20.18** True. **20.19** True. **20.20** True.
20.21

$$3(2I^+ \longrightarrow I_2 + 2e^-)$$
$$\textbf{(a)} \frac{(Cr_2O_7^{2-} + 14H^+ + 6e^- \longrightarrow 2Cr^{3+} + 7H_2O)}{6I^- + Cr_2O_7^{2-} + 14H^+ \longrightarrow 3I_2 + 2Cr^{3+} + 7H_2O}$$

(b)

$$2(MnO_4^- + 8H^+ + 5e^- \longrightarrow Mn^{2+} + 4H_2O)$$
$$5(C_2O_4^{2-} \longrightarrow 2CO_2 + 2e^-)$$

$$\overline{2MnO_4^- + 5C_2O_4^{2-} + 16H^+ \longrightarrow 2Mn^{2+} + 10CO_2 + 8H_2O}$$

(c)

$$3(H_2O + HSO_3^- \longrightarrow HSO_4^- + 2H^+ + 2e^-)$$
$$2(3e^- + 4H^+ + NO_3^- \longrightarrow NO + 2H_2O)$$

$$\overline{3HSO_3^- + 2NO_3^- + 2H^+ \longrightarrow 3HSO_4^- + 2NO + H_2O}$$

(d)

$$1(H_2O_2 \longrightarrow O_2 + 2H^+ + 2e^-)$$
$$2(ClO_2 + e^- \longrightarrow ClO_2^-)$$

$$\overline{H_2O_2 + 2ClO_2 \longrightarrow O_2 + 2ClO_2^- + 2H^+}$$
$$+ 2OH^- \qquad\qquad + 2OH^-$$

$$H_2O_2 + 2ClO_2 + 2OH^- \longrightarrow O_2 + 2ClO_2^- + 2H_2O$$

(e)

$$(Br_2 + 2e^- \longrightarrow 2Br^-)$$
$$(2H_2O + AsO_2^- \longrightarrow AsO_4^{3-} + 4H^+ + 2e^-)$$

$$\overline{2H_2O + Br_2 + AsO_2^- \longrightarrow 2Br^- + AsO_4^{3-} + 4H^+}$$
$$+ 4OH^- \qquad\qquad + 4OH^-$$

$$Br_2 + AsO_2^- + 4OH^- \longrightarrow 2Br^- + AsO_4^{3-} + 2H_2O$$

(f)

$$3e^- + 3H^+ + Bi(OH)_3 \longrightarrow Bi + 3H_2O$$
$$3(H_2O + SnO_2^- \longrightarrow SnO_3^{2-} + 2H^+ + e^-)$$

$$\overline{Bi(OH)_3 + 3SnO_2^- \longrightarrow Bi + 3SnO_3^{2-} + 3H^+}$$
$$+ 3OH^- \qquad\qquad + 3OH^-$$

$$Bi(OH)_3 + 3SnO_2^- + 3OH^- \longrightarrow Bi + 3SnO_2^{2-} + 3H_2O$$

20.22 Pure water is not a good conductor. In order to complete the circuit in the electrolysis cell, a small amount of an electrolyte must be added to water so that the solution is able to conduct a charge.

20.23

	Electrolytic		Voltaic	
	Cathode	**Anode**	**Cathode**	**Anode**
Ions attracted	Cations	Anions	Cations	Anions
Electrode reactions	Reduction	Oxidation	Reduction	Oxidation
Electrode sign	–	+	+	–
Electron movement	Toward	Away	Toward	Away

20.24 The anodic and cathode reaction in the nickel-cadmium battery is: $Cd(s) + 2OH^-(aq) \longrightarrow Cd(OH)_2(s) + 2e^-$. One mole of cadmium is consumed for every 2 mol of electrons (2 $\mathscr{F}$ of charge) transferred during the reaction. The number of faradays of charge transferred and grams of cadmium consumed are:

$$\text{Faradays} = (0.20\ \text{amp})(30.00\ \text{sec})\left(\frac{1\ C}{1\ \text{amp-sec}}\right)\left(\frac{1\ \mathscr{F}}{96{,}500\ C}\right)$$

$$= 6.2 \times 10^{-5}\ \mathscr{F}$$

$$\text{Grams Cd consumed} = \left(6.2 \times 10^{-5}\ \mathscr{F}\right)\left(\frac{1\ \text{mol Cd}}{2\ \mathscr{F}}\right)\left(\frac{112.40\ g}{1\ \text{mol Cd}}\right)$$

$$= 3.5 \times 10^{-3}\ g$$

20.25 The NO_3^- ion can oxidize metallic Cu at standard-state conditions with $E° = 0.96\ V - 0.34\ V = 0.62\ V$. However, the Cl^- ion cannot oxidize Cu because Cl^- cannot be reduced further.

20.26 (a)

$Fe^{2+} + 2e^- \longrightarrow Fe$		$E° = -0.44\ V$
$2(Ag \longrightarrow Ag^+ + e^-)$		$E° = -0.80\ V$
$Fe^{2+} + 2Ag \longrightarrow 2Ag^+ + Fe$		$E° = -1.24\ V$

(b) The reaction is not spontaneous because $E° < 0$. **(c)** $E = 0$ for any reaction at equilibrium. **(d)** You can calculate log K by using the equation $\log K = nE°/0.0592\ V = (2)(-1.24\ V)/0.0592\ V = -42.0$. Taking the antilog of both sides yields $K = 1.0 \times 10^{-42}$. **(e)** E for the reaction is calculated using the Nernst equation:

$$E = E° - \left(\frac{0.0592\ V}{2}\right)\log\frac{[Ag^+]^2}{[Fe^{2+}]}$$

$$= -1.24\ V - (0.0295\ V)\log\frac{(0.0100)^2}{(0.100)}$$

$$= -1.24\ V - (-0.887\ V) = -1.15\ V$$

(f) The reaction that is spontaneous is $2Ag^+ + Fe \longrightarrow Fe^{2+} + 2Ag$ with $E = 1.24\ V$.

$$W_{max} = -nFE = -(2\ \text{mol}\ e^-)\left(\frac{96{,}500\ C}{1\ \text{mol}\ e^-}\right)(1.24\ V)\left(\frac{1\ J}{1\ C\text{-}V}\right)$$

$$= -2.39 \times 10^5\ J$$

or in kwh,

$$(2.39 \times 10^5\ J)\left(\frac{1\ \text{kwh}}{3.6 \times 10^6\ J}\right) = 6.6 \times 10^{-2}\ \text{kwh}$$

20.27 The cathodic reaction that occurs when iron corrodes is $O_2(g) + 2H^+(aq) + 4e^- \longrightarrow 2H_2O$. In a soap solution, which is basic, the H^+ ion concentration is less than in an acidic or neutral solution. Therefore, according to LeChatelier's principle, the reaction is driven to the left in basic solution, and the reduction of $O_2(g)$ becomes less favorable. Consequently, the oxidation of Fe is also less rapid. **20.28** At the anode, Cu is oxidized. At the cathode, the following half-reactions should be considered: $e^- + Ag^+ \longrightarrow Ag$ ($E° = 0.80\ V$); $2e^- + Fe^{2+} \longrightarrow Fe$ ($E° = -0.44\ V$); $2e^- + 2H_2O \longrightarrow H_2(g) + 2OH^-$ ($E° = -0.41\ V$). If the overvoltage of hydrogen is taken into account, the reduction potential for water is $-0.41\ V + (-0.30\ V) = -0.71\ V$. The substance with the most positive reduction potential will plate out first; therefore Ag plates out first. The next most positive reduction potential is for iron; thus iron plates out second. As the potential is increased, water is eventually reduced, forming H_2. **20.29 (a)** Note the half-reaction for Mg must be reversed:

$$2(Al^{3+} + 3e^- \longrightarrow Al) \qquad E^\circ_{red} = -1.66 \text{ V}$$
$$\underline{3(Mg \longrightarrow Mg^{2+} + 2e^-) \qquad E^\circ_{ox} = 2.37 \text{ V}}$$
$$2Al^{3+} + 3Mg \longrightarrow 2Al + 3Mg^{2+} \qquad E^\circ_{cell} = E^\circ_{ox} + E^\circ_{red}$$
$$= -1.66 \text{ V} + 2.37 \text{ V}$$
$$= 0.71 \text{ V}$$

The reaction is spontaneous as $E^\circ_{cell} > 0$. **(b)** The MnO_4^- half-reaction must be reversed.

$$5(ClO_4^- + 2H^+ + 2e^+ \longrightarrow ClO_3^- + H_2O)$$
$$E^\circ_{red} = 1.19 \text{ V}$$
$$2(Mn^{2+} + 4H_2O \longrightarrow 2MnO_4^- + 5e^- + 8H^+)$$
$$E^\circ_{ox} = -1.51 \text{ V}$$
$$\overline{2Mn^{2+} + 3H_2O + 5ClO_4^- \longrightarrow 2MnO_4^- + 6H^+ + 5ClO_3^-}$$
$$E^\circ_{cell} = 1.19 \text{ V} - 1.51 \text{ V}$$
$$= -0.32 \text{V}$$

The reaction is *not* spontaneous as E°_{cell} is < 0.
20.30 The Nernst equation is used to solve for the unknown concentration

$$E = E^\circ - \frac{0.0592}{n} \log Q$$

$n = 6$ because the half-reactions are

$$2Al \longrightarrow 2Al^{3+} + 6e^-$$
$$3Mn^{2+} + 6e^- \longrightarrow 3Mn$$

$$Q = \frac{\left[Al^{2+}\right]^2}{\left[Mn^{2+}\right]^3} = \frac{\left[Al^{3+}\right]^2}{(0.49)^3}$$

$$0.47 \text{ V} = 0.48 \text{ V} - \frac{0.0592}{6} \log Q$$

$$-0.01 = \frac{-0.0592}{6} \log Q$$

$$\log Q = \frac{6(-0.01)}{-0.0592} = 1.02$$

$$Q = \text{antilog} (1.02)$$

$$Q = 10$$

$$\frac{\left[Al^{2+}\right]^2}{(0.49)^3} = 10$$

$$\left[Al^{3+}\right]^2 = (0.118)(10) = 1.2 \ M^2$$

$$\left[Al^{2+}\right] = \sqrt{1.2 \ M^2} = 1.1 \ M$$

20.31 (e)

$$3(Ag^+ + e^- \longrightarrow Ag) \qquad E^\circ_{red} = 0.80 \text{ V}$$
$$\underline{Al \longrightarrow Al^{3+} + 3e^- \qquad E^\circ_{ox} = 1.66 \text{ V}}$$
$$3Ag^+ + Al \longrightarrow 3Ag + Al^{3+} \qquad E^\circ_{cell} = 2.46 \text{ V}$$

20.32 (b)

$$E = 2.46 \text{ V} - \frac{0.0592 \text{ V}}{3} \log \frac{\left[Al^{3+}\right]}{\left[Ag^+\right]^3}$$

$$= 2.46 \text{ V} - \frac{0.0592 \text{ V}}{3} \log \frac{(1.0)}{(0.50)^3}$$

$$= 2.46 \text{ V} - (0.0197 \text{ V})(0.90)$$

$$= 2.46 \text{ V} - (0.018 \text{ V}) = 2.44 \text{ V}$$

20.33 (c) $Ag^+ + 2NH_3 \longrightarrow Ag(NH_3)_2^+$ Reduces Ag^+ concentration; thereby driving the reaction to the left and decreasing cell emf.

20.34 (a) $\log K = \dfrac{nE^\circ}{0.0592} = \dfrac{(4)(2.46 \text{ V})}{0.0592} = 125$

$$K = \text{antilog} (125) = 10^{125}$$

20.35 (e). Water is preferentially reduced in an aqueous NaCl solution, not $Na^+(aq)$.
20.36 (b) $CrO_3(aq) + 6H^+(aq) + 6e^- \longrightarrow Cr(s) + 3H_2O(l)$

$$\text{Coulombs} = (3.68 \text{ g Cr}) \times \left(\frac{1 \text{ mol Cr}}{52.0 \text{ g Cr}}\right)\left(\frac{6 \ \mathscr{F}}{1 \text{ mol Cr}}\right)$$
$$\times \left(\frac{96,500 \text{ C}}{1 \ \mathscr{F}}\right) = 41,000 \text{ coul}$$

20.37 (a). $E = 0.00 \text{ V} - (0.0592 \text{ V}/2) \log ([H^+]^2/P_{H_2}) = 0.19 \text{ V}$.
20.38 (c). 20.39 (d). 20.40 (b). $E^\circ = (0.0592 \text{ V}) \log K/n = (0.0592 \text{ V}/2) \log (3.76 \times 10^{14}) = 0.43 \text{ V}$. **20.41 (d).** $NO_3^-(aq) + 2H^+(aq) + e^- \longrightarrow NO_2(g) + H_2O(l)$. The added OH^- reacts with H^+ to form water, driving the reaction to the left, thereby decreasing the emf and increasing the equilibrium concentration of NO_3^-. **20.42 (e).** Of those metals listed, only Mg has a more positive oxidation emf than the oxidation emf of Mn (1.81 V), and thus it is preferentially oxidized. **20.43 (c).** **20.44 (b).** NH_3 and Zn^{2+} are formed during the reactions at the electrodes. These react together to form $Zn(NH_3)_4^{2+}$. **20.45 (c).** **20.46 (b).** Reduction of Ni^{2+} occurs at the Pt electrode; it is an inert electrode because it is not chemically involved in the redox reaction. **20.47 (a).** Oxidation occurs at the anode. **20.48 (a). 20.49 (b).** Ni^{2+} is reduced; therefore it is the oxidizing agent. **20.50 (a). 20.51 (b).** Fe^{2+} ions formed at the anode migrate away from it toward the Pt electrode. **20.52 (a).** Cl^- ions migrate out of the salt bridge into the anodic compartment to balance the charge of Fe^{2+} ions formed at the anode. **20.53 (b). 20.54. (c). 20.55 (d). 20.56 (d). 20.57 (b). 20.58 (c). 20.59 (b).** An oxidizing agent causes an element to be oxidized; thus an oxidizing agent is reduced: $2e^- + Br_2 \longrightarrow 2Br^-$.

21 | Nuclear Chemistry

OVERVIEW OF THE CHAPTER

21.1 NUCLEAR REACTIONS: RADIOACTIVITY

Review: Atoms (2.3, 2.4); atomic number (2.4); mass number (2.3); sub-atomic particles (2.3).

Objectives: You should be able to:

1. Write the nuclear symbols for protons, neutrons, electrons, alpha particles, and positrons.

2. Complete and balance nuclear equations, having been given all but one of the particles involved.

21.2, 21.3 PREPARATION OF NEW NUCLEI

Review: Location of lanthanides and actinides in the periodic table (2.4).

Objective: You should be able to write the shorthand notation for a nuclear reaction or given the shorthand notation, write the nuclear reaction.

21.2, 21.6 STABILITY OF NUCLEI AND BINDING ENERGIES

Review: Energy changes (5.4); fuel values (5.8).

Objectives: You should be able to:

1. Determine the effect of different types of decay on the proton-neutron ratio and predict the type of decay that a nucleus will undergo based on its composition relative to the belt of stability.

2. Use Einstein's relation, $\Delta E = c^2 \Delta m$, to calculate the energy change or the mass change of a reaction, having been given one of these quantities.

3. Calculate the binding energies of nuclei, having been given their masses and the masses of protons and neutrons.

21.4 RADIOACTIVITY: HALF-LIFE

Review: First-order reactions and concept of half-life (14.3).

Objectives: You should be able to:

352

1. Use the half-life of a substance to predict the amount of radioisotope present after a given period of time.

2. Explain how radioisotopes can be used in dating objects and as radiotracers.

3. Calculate half-life, age of an object, or the remaining amount of radioisotope, having been given any two of these pieces of information.

Review: Early radioactivity experiments (2.2).

Objectives: You should be able to:

1. Explain how radioactivity is detected, including a simplified description of the basic design of a Geiger counter.

2. Explain the roles played by the chemical behavior of an isotope and its mode of radioactivity in determining its ability to damage biological systems.

3. Define the units used to describe the level of radioactivity (*curie*) and to measure the effects of radiation on biological systems (*rem* and *rad*).

4. Describe various sources of radiation to which the general population is exposed, and indicate the relative contributions of each.

Review: Energy use (5.9).

Objectives: You should be able to:

1. Define fission and fusion, and state which types of nuclei produce energy when undergoing these processes.

2. Describe the design of a nuclear power plant, including an explanation of the role of fuel elements, control rods, moderator, and cooling fluid.

TOPIC SUMMARIES AND EXERCISES

A change in the structure of a nucleus may occur spontaneously, or it may be brought about artificially.

- The spontaneous emission of particles and electromagnetic radiation from unstable nuclei is referred to as **radioactivity.** An unstable nucleus that spontaneously transmutes to the nucleus of another element is said to undergo a process of **radioactive decay.**
- The types of radioactive decay discussed in Section 21.1 of the text are summarized in Table 21.1.

**21.5, 21.9
RADIOACTIVITY:
DETECTION AND
EFFECTS**

**21.7, 21.8
NUCLEAR FISSION
AND FUSION**

NUCLEAR
REACTIONS:
RADIOACTIVITY

Table 21.1 Types of radioactive decay

Type	Example	Comments
Alpha-particle emission ($_2^4$He nuclei)	$_{90}^{280}\text{Th} \longrightarrow {}_{88}^{226}\text{Ra} + {}_2^4\text{He}$	New isotope has a mass number that is four less and an atomic number that is two less than decaying nucleus
Beta-particle emission (high-speed electrons, $_{-1}^0$e)	$_{90}^{231}\text{Th} \longrightarrow {}_{91}^{231}\text{Pa} + {}_{-1}^0\text{e}$	New isotope has the same mass number as the decaying nucleus, but an atomic number one greater
Gamma-ray emission (high-energy electromagnetic radiation)	Gamma emission often accompanies alpha and beta emission	Involves energy released during organization of nucleus. Symbol for gamma ray, $_0^0\gamma$, is usually not shown in a nuclear equation
Positron emission (a particle of same mass as electron, but with opposite charge, $_1^0$e)	$_6^{11}\text{C} \longrightarrow {}_5^{11}\text{B} + {}_1^0\text{e}$	New isotope has the same mass number as the decaying nucleus, but an atomic number that is one less. Emission of positron particles involves conversion of proton to neutron: $_1^1\text{p} \rightarrow {}_0^1\text{n} + {}_1^0\text{e}$
Electron capture	$_{37}^{81}\text{Rb} + {}_{-1}^0\text{e} \longrightarrow {}_{36}^{81}\text{Kr}$	New isotope has the same mass number as the decaying nucleus, but with an atomic number that is one less. Decaying nucleus captures electron from its surrounding electron cloud: $_1^1\text{p} + {}_{-1}^0\text{e} \rightarrow {}_0^1\text{n}$

EXERCISE 1   Indicate the number of protons, neutrons, electrons, and nucleons possessed by each of the following: **(a)** $_8^{14}$O; **(b)** ^{238}U; **(c)** bismuth-214.

Solution: The superscript before each isotopic symbol is the mass number of the element, the total number of nucleons (sum of protons and neutrons). The atomic number, the number of protons, is the subscript. In a neutral atom the number of electrons equals the number of protons. **(a)** $_8^{14}$O has eight protons, eight electrons, six neutrons (the mass number minus the atomic number), and fourteen nucleons. **(b)** ^{238}U has 238 nucleons. Its atomic number, 92, can be determined from the periodic table. Therefore ^{238}U has 92 protons, 92 electrons, and 146 neutrons. **(c)** In bismuth-214 the number 214 is the mass number of bismuth. Its atomic number, determined from the periodic table, is 83. Thus, bismuth has 83 protons, 83 electrons, 131 neutrons, and 214 nucleons.

EXERCISE 2   Complete and balance the following nuclear equations by supplying the missing particle or energy ray. Identify the type of radioactive decay for each reaction.

(a) $_{88}^{224}\text{Ra} \longrightarrow {}_{86}^{220}\text{Rn} + $ _____   **(c)** $_{90}^{232}\text{Th} \longrightarrow {}_{91}^{232}\text{Pa} + $ _____

(b) $_{88}^{226}\text{Ra} \longrightarrow {}_{86}^{222}\text{Rn} + {}_2^4\text{He} + $ _____   **(d)** $_7^{13}\text{N} \longrightarrow {}_6^{13}\text{C} + $ _____

Solution: **(a)** The sum of atomic numbers and the sum of mass numbers for the reactants must equal the same sums for the products. The missing particle must have an atomic number and a mass number fulfilling this requirement. From the nuclear equation provided, you can write the following:

$$\text{Mass numbers:} \quad 224 = 220 + x \qquad x = 224 - 220 = 4$$
$$\text{Atomic numbers:} \quad 88 = 86 + y \qquad y = 88 - 86 = 2$$

The missing particle has a mass number of 4 and an atomic number of 2. A helium

Chapter 21   Nuclear Chemistry

nucleus possesses these properties. The balanced equation is thus

$$^{224}_{88}\text{Ra} \longrightarrow {}^{220}_{86}\text{Rn} + {}^{4}_{2}\text{He}$$

Alpha emission involves forming an isotope that has a mass number four less and an atomic number two less than the isotope undergoing radioactive decay. **(b)** Proceeding in a similar fashion as in (a),

Mass numbers: $226 = 222 + 4 + x$ $x = 0$
Atomic numbers: $88 = 86 + 2 + y$ $y = 0$

The nuclear equation is already balanced; therefore if there is a missing species it probably is high-energy gamma radiation.

$$^{226}_{86}\text{Ra} \longrightarrow {}^{222}_{86}\text{Rn} + {}^{4}_{2}\text{He} + {}^{0}_{0}\gamma$$

Gamma emission often accompanies alpha or beta emission. **(c)** Proceeding in a similar fashion as in **(a)**,

Mass numbers: $232 = 232 + x$ $x = 0$
Atomic numbers: $90 = 91 + y$ $y = 90 - 91 = -1$

An electron, $_{-1}^{0}\text{e}$, has a mass number of 0 and atomic number of -1. Therefore the balanced nuclear reaction is

$$^{232}_{90}\text{Th} \longrightarrow {}^{232}_{91}\text{Pa} + {}^{0}_{-1}\text{e}$$

Beta emission involves forming an isotope that has the same mass number and an atomic number one greater than those of the isotope undergoing radioactive decay. **(d)** Proceeding in a similar fashion as in **(a)**,

Mass numbers: $13 = 13 + x$ $x = 0$
Atomic numbers: $7 = 6 + y$ $y = 7 - 6 = 1$

A positron, $_{1}^{0}\text{e}$, has a mass number of 0 and an atomic number of 1. Therefore the balanced nuclear equation is

$$^{13}_{7}\text{N} \longrightarrow {}^{13}_{6}\text{C} + {}^{0}_{1}\text{e}$$

Positron emission involves forming an isotope that has the same mass number and an atomic number one less than those of the isotope undergoing radioactive decay.

PREPARATION OF NEW NUCLEI

Many radioactive nuclei have been prepared by bombarding existing nuclei with subatomic particles such as neutrons, electrons, protons, alpha particles, and deuterons ($_{1}^{2}\text{H}$) or with high-energy radiation, such as gamma rays.

- These types of reactions are referred to as induced nuclear reactions or **nuclear transmutations.** The newly formed nucleus may undergo radioactive decay and initiate a nuclear-disintegration series.
- For electrically charged subatomic particles to combine with a target nucleus so as to effect a nuclear transmutation, they first must be accelerated to high speeds using a particle accelerator, such as a cyclotron or synchrotron. Acceleration of these particles provides them with sufficient kinetic energy to overcome electrostatic repulsions between them and the target nucleus.

Less common particles such as $^{12}_{6}C$, $^{14}_{7}N$, and $^{10}_{5}B$ have been used to induce the formation of some **transuranium elements,** the elements occurring after uranium in the periodic table. Radioactive transuranium elements have short half-lives and consequently are difficult to isolate and identify.

▪ An induced nuclear reaction, such as

$$^{6}_{3}Li + ^{1}_{0}n \longrightarrow ^{3}_{1}H + ^{4}_{2}He$$

is often abbreviated using a shorthand notation system. The shorthand notation for the given reaction is $^{6}_{3}Li(n, \alpha)^{3}_{1}H$.

▪ In the shorthand notation the target nucleus is the first item listed, followed by the symbols for the bombarding particle (or radiation) and the ejected particle (or radiation), both enclosed in parentheses, with the product nucleus listed last:

Target nucleus (bombarding particle, ejected particle) product nucleus.

EXERCISE 3   The synthesis of transuranium element 104 was followed by a controversy. Both a University of California group and a Russian group claimed to have been the first to synthesize the element, and each suggested a name for the new element. Supply the missing particle in each of the following two reactions, which describe the reported syntheses:

$$\text{(a) } ^{242}_{94}Pu + ? \longrightarrow ^{260}_{104}X + 4^{1}_{0}n$$
$$\text{(b) } ^{249}_{98}Cf + ? \longrightarrow ^{257}_{104}X + 4^{1}_{0}n$$

Solution: Proceed to solve for the missing particle as in Exercise 2: **(a)** The sums of mass numbers and atomic numbers on the right are 264 and 104, respectively. The missing particle must have a mass number of 22 and an atomic number of 10 so that the sums on the left are 264 and 104, respectively. By referring to a periodic table, you will find that the nuclide with an atomic number of 10 (that is, containing 10 protons) is neon. Consequently it must be the missing particle:

$$^{294}_{94}Pu + ^{22}_{10}Ne \longrightarrow ^{260}_{104}X + 4^{1}_{0}n$$

(b) Using the same approach as in **(a),** you can obtain the following nuclear equation:

$$^{249}_{98}Cf + ^{12}_{6}C \longrightarrow ^{257}_{104}X + 4^{1}_{0}n$$

EXERCISE 4   Write a balanced nuclear equation for each of the following processes, or write the shorthand notation for the given reaction:

(a) $^{10}_{5}B(\alpha,p)^{13}_{6}C$
(b) $^{107}_{47}Ag(n, 2n)^{106}_{47}Ag$
(c) $^{23}_{11}Na + ^{1}_{0}n \longrightarrow ^{24}_{11}Na + \gamma$
(d) $^{238}_{92}U + ^{12}_{6}C \longrightarrow ^{246}_{98}Cf + 4^{1}_{0}n$

Solution: **(a)** The symbol α is the abbreviation for an alpha particle, and p is the abbreviation for a proton. The nuclear reaction is

$$^{10}_{5}B + ^{4}_{2}He \longrightarrow ^{13}_{6}C + ^{1}_{1}p$$

(b) The letter n is the abbreviation for a neutron. There are two neutrons ejected as a product. The nuclear equation is

$$^{107}_{47}Ag + ^{1}_{0}n \longrightarrow ^{106}_{47}Ag + 2^{1}_{0}n$$

(c) The notation is $^{23}_{11}\text{Na}(\text{n}, \gamma)^{24}_{11}\text{Na}$, where γ is the symbol for gamma radiation.

(d) The notation is $^{238}_{92}\text{U}\ (^{12}_{6}\text{C}, 4\text{n})^{246}_{98}\text{Cf}$.

Most known nuclear species do not occur naturally. They have been prepared by induced nuclear reactions.

- About 275 of the approximately 320 known naturally occurring isotopes are stable.
- The stability of a nucleus depends on the forces holding nuclear particles together. For a heavy nucleus (with an atomic number greater than 20) to be stable it must have more neutrons than protons. The presence of excess neutrons in a stable nucleus counteracts the increased proton-proton coulombic repulsions that normally would cause the nucleus to disintegrate.

Look at Figure 21.2 in your text and note that as the number of protons increases, the number of neutrons necessary to create a stable nucleus rapidly increases.

- The figure is also useful for predicting the initial mode of nuclear decay for unstable nuclei with atomic numbers less than 80.
- If a nucleus has a neutron/proton ratio that is not in the belt of stability, then you can use Figure 21.2 to help you decide how the unstable nucleus will decay.

Binding energy is the energy required to break apart a nucleus into its component protons and neutrons. To calculate the binding energy for a nucleus, use the following procedure:

1. Calculate the mass change, Δm, when a nucleus is separated into its constituent protons and neutrons. Note: **Mass defect** is the mass loss that occurs when a nucleus is formed from its constituent protons and neutrons.

2. Use Einstein's relationship $\Delta E = c^2 \Delta m$ to calculate binding energy, which is a positive quantity.

Energy changes for nuclear reactions are calculated in a manner similar to above, expect that

- Δm = nuclear mass of products − nuclear mass of reactants.
- Unless the total number of electrons for the products and reactants differs, you can use isotopic masses instead of nuclear masses. If they are not equal, you must subtract the mass of the electrons from the isotopic masses.

EXERCISE 5   Using Figure 21.2 in the text, predict which of the following three nuclides are likely to be radioactive: **(a)** $^{64}_{30}\text{Zn}$; **(b)** $^{90}_{35}\text{Br}$; **(c)** $^{103}_{47}\text{Ag}$. Briefly justify your choice.

Solution: Calculate the neutron-to-proton ratio for each nuclide. Those nuclides whose neutron-to-proton ratios are not within the stability region as shown in Figure 21.2 of the text are likely to be radioactive. The neutron-proton ratios are: **(a)** 34/30 = 1.1 for $^{64}_{30}$Zn; **(b)** 55/35 = 1.6 for $^{90}_{35}$Br; and **(c)** 56/47 = 1.2 for $^{103}_{47}$Ag. Inspection of Figure 21.1 shows that the neutron/proton ratio for $^{64}_{30}$Zn is in the region of stability and the neutron/proton ratios for $^{90}_{35}$Br and $^{103}_{47}$Ag lie above and below the region of stability, respectively. $^{90}_{35}$Br and $^{103}_{47}$Ag are therefore likely to be radioactive because their neutron/proton ratios do not lie in the belt of stability.

EXERCISE 6   Calculate the energy of 1 amu in joules using the relation $\Delta E = c^2 \Delta m$. 1 J is equivalent to 1 kg-m^2/sec^2. 1g = 6.022×10^{23} amu.

Solution: 1 amu must be converted to kilograms in order to solve for the energy in joules.

$$\Delta E = c^2 \Delta m = \left(3.00 \times 10^8 \, \frac{m}{sec} \right)^2 (1 \text{ amu}) \times \left(\frac{1g}{6.022 \times 10^{23} \text{ amu}} \right) \left(\frac{1 \text{ kg}}{10^3 g} \right)$$

$$\Delta E = 1.49 \times 10^{-10} \text{ kg-m}^2/\text{sec}^2 = 1.49 \times 10^{-10} \text{ J}$$

The equivalence 1 amu = 1.49×10^{-10} J is a useful equivalence for solving binding-energy problems. Sometimes binding energies are expressed in units of MeV (million electron volts): 1 amu = 931 MeV.

EXERCISE 7   For the nuclear reaction

$$^{14}_{7}N + ^{4}_{2}He \longrightarrow ^{17}_{8}O + ^{1}_{1}H$$

calculate the energy in joules associated with the reaction of one atom of $^{14}_{7}N$ with one of $^{4}_{2}He$, given that the isotopic masses (amu) are: $^{14}_{7}N$, 14.00307; $^{4}_{2}He$, 4.00260; $^{17}_{8}O$, 16.9991; $^{1}_{1}H$, 1.007825.

Solution: You must first calculate Δm, the difference in mass between the products and reactants, and then convert this quantity to energy in joules. The listed masses are for one atom of each isotopic element. Thus

Δm = mass products − mass reactants

Δm = mass $^{17}_{8}O$ + mass $^{1}_{1}H$ − mass $^{14}_{7}N$ − mass $^{4}_{2}He$

   = 16.9991 amu + 1.007825 amu − 14.00307 amu − 4.00260 amu

   = 0.0013 amu

The positive value for Δm indicates that energy is converted to mass during the reaction. From Exercise 6, you know that 1 amu = 1.49×10^{-10} J. Therefore you can convert mass to energy:

$$m = (0.0013 \text{ amu}) \left(1.49 \times 10^{-10} \, \frac{J}{\text{amu}} \right) = 1.9 \times 10^{-13} \text{ J}$$

RADIOACTIVITY: HALF-LIFE

$^{235}_{92}U$ undergoes radioactive decay very slowly in nature and thus can be isolated and stored for long periods of time. By contrast, $^{132}_{53}I$ undergoes substantial decay over a period of several hours; it therefore cannot be stored for long time periods.

▪ The rate of radioactive decay of a nucleus is characterized by its half-life.

▪ Since radioactive decay is a first-order process, you can use the concept of half-life developed earlier in Section 14.3 of the text to characterize the stability of a radioisotope. *The half-life of a decaying substance which obeys a first-order rate law is independent of the amount of substance reacting; it depends only on the rate constant:*

$$t_{1/2} = \frac{0.693}{k} \qquad\qquad [21.1]$$

where k is the rate constant and $t_{1/2}$ is the half-life.

▪ An important relationship that relates the time, t, required for an initial number, N_o, of nuclei of a particular isotope to decay to N_t number of particles is:

$$\ln \frac{N_t}{N_0} = -kt \qquad\qquad [21.2]$$

EXERCISE 8   Use Equations 21.1 and 21.2 above to show that the relationship between t and $t_{1/2}$ is

$$t = 1.44\ t_{1/2}\ \ln N_o/N_t$$

where t is the time after a radioactive decay process begins. *This equation is very useful when you are solving problems that involve dating and half-lives.*

   Solution: Use the expression for k in Equation [21.1] in the Student's Guide

$$k = \frac{0.693}{t_{1/2}}$$

to replace k in Equation [21.2] in the Student's Guide:

$$\ln (N_t/N_0) = -(0.693/t_{1/2})t$$

Rearrange the above equation to solve for t:

$$t = -(t_{1/2}/0.693)\ln (N_t/N_0) = -1.44t_{1/2}\ln (N_t/N_0)$$

Use the relationship $-\ln x = \ln (1/x)$ to eliminate the minus sign in the above expression:

$$t = 1.44t_{1/2}\ln (N_0/N_t)$$

EXERCISE 9   One application of the concept of radioactive half-life is that of radiocarbon dating. If you assume that the ratio of $^{14}_{6}C$ to $^{12}_{6}C$ in the atmosphere has remained constant for at least 50,000 years, then you can assume that the ratio of $^{14}_{6}C$ to $^{12}_{6}C$ in a plant at the time it died in the past is the same as that found in the CO_2 of the atmosphere. After the plant dies, it no longer incorporates into itself $^{14}_{6}C$ and $^{12}_{6}C$ from the atmosphere. Carbon-14 is radioactive, so the relative amount of this nuclide begins to decay with the death of the plant. Since the half-life of carbon-14 is 5730 years, the ratio of $^{14}_{6}C$ and $^{12}_{6}C$ in the plant as compared with that in the atmosphere will decrease by 50 percent in 5730 years. Thus the determination of the $^{14}_{6}C : ^{12}_{6}C$ ratio in a dead plant permits us to calculate its approximate age. Crater Lake in Oregon is believed to have formed about 6.1×10^3 years ago. Calculate the approximate ratio of the amount of $^{14}_{6}C$ contained in an old charcoal wood sample from a tree burned during the formation of Crater Lake to the amount of $^{14}_{6}C$ contained in a wood sample from a freshly cut tree.

   Solution: You can use the equation derived in EXERCISE 8 to solve for the ratio N_t/N_0.

$$t = 1.44t_{1/2}\ \ln (N_0/N)_t$$

In the above equation, the ratio N_0/N_t is equivalent to the ratio of the $^{14}_6C$ content in a freshly cut sample of wood to the $^{14}_6C$ content in the ancient charcoal sample for the region near Crater Lake. Therefore the ratio you need is N_t/N_0. This equivalency is valid as long as the $^{14}_6C:^{12}_6C$ ratio in the atmosphere has not changed since the formation of Crater Lake. Substituting the age of the charcoal sample and the half-life of carbon-14 into the equation gives

$$6.1 \times 10^3 \text{ yr} = (1.44)(5730 \text{ yr})\left(\ln \frac{N_0}{N_t}\right)$$

$$\ln \frac{N_0}{N_t} = \frac{6.1 \times 10^3 \text{ yr}}{(1.44)(5730 \text{ yr})} = 0.74$$

Solving for the ratio N_0/N_t yields $N_0/N_t = 2.1$. The reciprocal of this ratio, N_t/N_0, is $1/2.1 = 0.48$. Therefore, the approximate ratio of the $^{14}_6C$ content in the Crater Lake charcoal sample to the $^{14}_6C$ content in a freshly cut tree sample is 0.48.

EXERCISE 10   What is the original mass of ^{14}C in a sample if 10.00 mg of it remains after 20,000 years? The half-life of ^{14}C is 5730 years.
   Solution: The initial mass of ^{14}C can be calculated using the equation

$$t = 1.44\, t_{1/2} \ln \frac{N_0}{N_t}$$

Substituting the provided data ($t_{1/2} = 5730$ yr; $N_t = 10.00$ mg, $t = 20,000$ yr) into the equation yields

$$20,000 \text{ yr} = (1.44)(5730 \text{ yr})\ln \frac{N_0}{10.00 \text{ mg}}$$

$$\ln \frac{N_0}{10.00 \text{ mg}} = \frac{20,000 \text{ yr}}{(1.44)(5730 \text{ yr})} = 2.42$$

Taking the antiln of both sides of the equation yields

$$\frac{N_0}{10.00 \text{ mg}} = 11.3$$

Solving for N_0 gives

$$N_0 = (11.3)(10.00 \text{ mg}) = 113 \text{ mg}$$

The original mass of the ^{14}C in the sample was 113 mg.

RADIOACTIVITY: DETECTION AND EFFECTS

In Section 21.5 of the text, three methods for detecting emissions from radioactive substances are described: photographic plates, Geiger counters, and scintillation counters. You should note how each detects radioactive emissions.
   The operation of a Geiger counter to detect radioactivity is based on the ability of radiation to ionize matter. Ionization of cellular matter is of special concern to scientists.

   • Biological damage to cells occurs because of the ability of radiation to ionize and fragment molecules. Such processes in human tissue can lead to various forms of cancer: leukemia is the most common form observed.
   • *The relative ability of radiation to penetrate human tissue is*

gamma > beta > alpha. Although alpha rays do not deeply penetrate into human tissue, they are excellent ionizing agents. Therefore it is important to prevent ingestion of alpha emitters. Once they are within the body, extensive biological damage occurs.

- Radioactive damage is classified as either **somatic damage** (damage to organisms which results in sickness or death) or **genetic damage** (damage to genetic machinery which causes disorders in offspring).

As indicated by the previous discussion of the characteristics of alpha rays, biological effects of radiation depend on several factors.

- a measure of the energy of radiation received by tissue is a **rad,** the amount of radiation that deposits 1×10^{-2} J of energy per kilogram of tissue.
- The relative biological damage caused by a form of radiation is given by its **RBE factor.** If the rad value for a particular form of radiation is multiplied by its RBE factor, we obtain a radiation dose measurement termed rem:

$$\text{Rems} = (\text{rads})(\text{RBE})$$

- The SI unit for nuclear radioactivity is the **becquerel,** defined as one nuclear disintegration per second. An older, but more commonly used measure of nuclear activity is the **curie** (Ci), the number of disintegrations that 1 g of radium undergoes in 1 sec; this number is 3.7×10^{10} disintegrations/s.

EXERCISE 11   Cobalt-60 and cesium-137 are used as sources of radiation by the food industry to preserve food. Suggest a reason for irradiating potatoes with gamma rays before they are shipped to markets.

   Solution: The eyes on the surface of a potato will form buds if not irradiated with gamma rays. Gamma rays destroy the reproductive abilities of potatoes, causing them to be sterile and not form buds (eyes).

EXERCISE 12   Linus Pauling led a group of scientists in a relatively successful endeavor to stop nuclear bomb tests. **(a)** One of their concerns was the somatic damage caused by ^{90}Sr. What is somatic damage? **(b)** Sr is one of the group 2A elements and replaces an element of this group that is found extensively in animals. What is the element Sr replaces?

   Solution: **(a)** Somatic damage results when radiation affects an organism during its lifetime. **(b)** ^{90}Sr readily replaces calcium in the bones. ^{90}Sr emits beta rays that cause biological damage. As a result, the production of blood cells, which form in the marrow of bones, is diminished.

EXERCISE 13   Slow and fast neutrons have RBE factors of 5 and 10, respectively, for producing cataracts in the eyes. What is the relative radiation damage caused by a 0.4-rad dose of slow neutrons as compared with a 0.8-rad dose of fast neutrons?

   Solution: The biological effect is measured by number of rems, which equals number of rads times RBE.

$$\text{Rems for slow neutrons} = (0.4 \text{ rad})(5 \text{ RBE}) = 2.0 \text{ rems}$$
$$\text{Rems for fast neutrons} = (0.8 \text{ rad})(10 \text{ RBE}) = 8.0 \text{ rems}$$

The 0.8-rad dose of fast neutrons would cause damage to the eyes four times greater than that caused by a 0.4-rad dose of slow neutrons.

NUCLEAR FISSION AND FUSION

A **fission reaction** is a reaction in which a nuclide breaks apart into other nuclides of smaller mass number. In Section 21.7 of the text, the nuclear fission reactions of $^{235}_{92}U$, which is initiated by bombardment with neutrons, is described.

- Note in Equations 21.24 and 21.25 in the text that more neutrons per nuclear fission reaction are formed as products than are needed to initiate the fission reaction of $^{235}_{92}U$. Therefore the reaction can be self-sustaining if other fissionable materials capture the emitted neutrons.
- If a sufficient nuclear mass of $^{235}_{92}U$ is present, a large number of other nuclear fission reactions occur, with a violent explosion resulting from the large amount of energy released. Such a series of multiplying nuclear fission reactions is called a **branching chain reaction.**

A **fusion reaction** is one in which lightweight nuclei combine to form heavier nuclei and sometimes other subatomic particles.

- Temperatures of around 1×10^8 °C are required to give the combining atoms sufficient kinetic energy to overcome the coulombic repulsion forces that exist between nuclei.
- Once nuclei combine, a large amount of energy is evolved. Therefore considerable interest is focused upon developing techniques to control fusion reactions in order to harness this energy.

EXERCISE 14 How are the rates of fission reactions controlled in nuclear reactors?
Solution: Control rods made of cadmium, tantalum, or boron carbide readily capture neutrons. The control rods are placed into the fissionable material to regulate the number of neutrons that are available to initiate fission reactions.

EXERCISE 15 How are fast neutrons, released by a fission process, slowed down so that they may be captured by fissionable material?
Solution: A moderator, such as H_2O, D_2O, or graphite is used to slow down neutrons. A neutron collides with, but is not captured by, the moderator and transfers some of its kinetic energy to the moderator. With a decreased kinetic energy, the velocity of a neutron decreases.

EXERCISE 16 The sun gives off energy because of a thermonuclear process. What nuclide is given off by the sun during its thermonuclear reaction described by the net reaction

$$4^1_1H \longrightarrow ? + 2^0_1e + \gamma$$

Solution: The missing particle is helium, ^4_2He.

SELF-TEST QUESTIONS

Key Terms

Having reviewed the key terms listed at the end of Chapter 21 and their applications, identify the following statements as true or false. If a statement is false, indicate why it is incorrect. Key terms are italicized in the statements.

21.1 The mode of radioactive decay referred to as *electron capture* may be thought of as the following reaction occurring within the nucleus of an isotope:

$$^1_1p + ^{\ 0}_{-1}e \longrightarrow ^1_0n$$

21.2 Radiation with a high *rem* value is dangerous to human health and causes severe biological damage.

21.3 Different kinds of radiation that have similar *rad* values would be expected to cause similar biological damage in humans.

21.4 The nuclear mass of $^{59}_{27}$Co is 58.91837 amu. From the mass of a proton and neutron given in section 21.6 of the text, you can calculate the *mass defect* of $^{59}_{27}$C to be -0.55538 amu.

21.5 *Alpha particles* are identical to helium-4 atoms.

21.6 The symbol for a *beta particle* is $_{-1}^{1}$e.

21.7 Ten *curies* of radioactivity is equivalent to 3.7×10^{11} nuclear disintegrations per second.

21.8 An example of a *fission* process is the reaction

$$^{1}_{0}n + {}^{235}_{92}U \longrightarrow {}^{103}_{42}Mo + {}^{131}_{50}Sn + 2{}^{1}_{0}n$$

21.9 An example of a *fusion reaction* is

$$3{}^{4}_{2}He \longrightarrow {}^{12}_{6}C + energy$$

21.10 *Gamma radiation* is a type of electromagnetic radiation similar to X rays.

21.11 The symbol for a *positron*, $_{1}^{0}$e, tells us that it is a particle with a charge equivalent in magnitude but opposite in sign to that possessed by an electron.

21.12 An example of a *nuclear transmutation* reaction is $^{98}_{42}Mo + {}^{2}_{1}H \longrightarrow {}^{99}_{43}Tc + {}^{1}_{0}n$.

21.13 *Radioisotopes* undergo first-order nuclear decay processes.

21.14 A nuclear reactor uses a controlled *chain reaction* to produce usable energy.

21.15 16.28 kg of $^{239}_{94}$Pu is the amount of $^{239}_{94}$Pu required to maintain a chain reaction initially involving $^{239}_{94}$Pu reacting with neutrons in a particular nuclear reactor. This quantity of $^{239}_{94}$Pu is its *critical mass* under these conditions.

21.16 From the information given in problem 21.15, we know that 18 kg of $^{239}_{94}$Pu is *subcritical* and 10 kg of $^{239}_{94}$Pu is *supercritical*.

21.17 An example of a *nucleon* is an electron in an atom.

21.18 A nucleus containing a *magic number* of nucleons, such as 18 protons, is likely to be radioactive.

21.19 The *radioactive series* for ^{238}U involves 14 decay steps leading to a stable ^{206}Pb.

21.20 There are many *nuclear disintegration series* that occur in nature.

21.21 *Particle accelerators* involve shooting particles into a vacuum in the presence of alternating electrical or magnetic fields.

21.22 The *transuranium elements* occur immediately following uranium in the periodic table.

21.23 In a typical *Geiger counter*, gamma rays enter a tube through a thin window and ionize a gas.

21.24 A *scintillation counter* is an instrument used to detect radiation by fluorescence.

21.25 A *radiotracer* used to study reactions of carbon-based compounds is carbon-12.

21.26 Low temperatures can be used to induce *thermonuclear reactions*.

21.27 *Somatic radiation damage* is exemplified by Leukemia.

21.28 *Genetic damage* is short term in nature.

21.29 The SI unit of radioactivity is the *becquerel*, which is equivalent to one curie.

21.30 If a nucleus emits radiation only with the input of energy, it is said to be *radioactive*.

21.31 The *half-life* of strontium-90 is 29 years. This means that in three half-lives 10.0 g of it decays to 3.3 grams (1/3 of original mass).

21.32 A *radionuclide* is a nucleus that is radioactive and has a specified number of protons and neutrons.

21.33 The neutral OH molecule is unstable and is an example of a *free radical* because it possesses one unpaired electron.

Problems and Short-Answer Questions

21.34 Write balanced nuclear equations for the following nuclear transformations:
 (a) niobium-99 undergoes beta decay
 (b) chromium-48 undergoes electron capture
 (c) calcium-39 undergoes positron emission
 (d) carbon-16 undergoes neutron emission
 (e) gadolinium-148 undergoes alpha emission

21.35 Supply the missing subatomic particle, energy form, or nuclide in the following induced nuclear reactions:
 (a) $-$_____($^{3}_{1}$H, n)$^{4}_{2}$He
 (b) $^{253}_{99}$Es (_____ , n)$^{256}_{101}$Mv
 (c) $^{81}_{35}$Br (γ, n) $-$_____

21.36 One theory of stellar evolution has $^{16}_{8}$O forming by the reaction of helium-4 and carbon-12 in the core of a sun at very high temperatures.
 (a) Write the nuclear equation describing this nuclear reaction.
 (b) Is this a fission or fusion reaction?

21.37 Calculate the energy associated with the reaction you wrote for problem 21.36. Does the result of your energy calculation support this particular stellar-evolution theory? Required atomic mass (amu) are: $^{4}_{2}$He, 4.00260; $^{12}_{6}$C, 12.00000; $^{16}_{8}$O, 15.99491.

21.38 In 1990, you and your friends barbecue hamburgers using charcoal that was made recently from freshly cut wood. Later you bury some unburned charcoal in the ground. Suppose that many years later an archeologist digs up your charcoal and by experimental methods finds that it has a ^{14}C:^{12}C ratio only 30 percent of the ^{14}C:^{12}C ratio in charcoal made from a recently cut tree. Assuming that the ^{14}C:^{12}C ratio does not change in the atmosphere, approximately how far in the future will the archeologist have found the buried charcoal? The half-life of ^{14}C is 5730 years.

21.39 What type of decay process would you expect for unstable heavy nuclei with too many protons and neutrons?

21.40 $^{75}_{33}$As reacts with $^4_2$He to form a nuclide and one neutron. The product nuclide undergoes positron emission. What is the final nuclide formed in this two-step disintegration series?

21.41 Why is the rem considered a more accurate measure of biological damage than the rad?

21.42 How much radioactivity, in curies, does 1 µg of Ra-226 produce? What is this in units of disintegrations per second?

21.43 A small amount of Pb-212, in the form of the 2+ ion, is injected into a saturated solution of $PbCl_2$ that is in contact with the solid salt. A short time later some of the Pb-212 is found in solid $PbCl_2$ in the saturated solution. What does this information tell you about the nature of the equilibrium between solid $PbCl_2$ and its dissolved ions?

Multiple-Choice Questions

21.44 How many protons, neutrons, and electrons does a $^7_3$Li atom contain?

(a) 7p, 3n, 4e
(b) 4p, 3n, 7e
(c) 3p, 3n, 4e
(d) 4p, 3n, 3e
(e) 3p, 4n, 3e

21.45 How is the mass defect for a nucleus calculated?

(a) Atomic weight of the element minus mass number;
(b) atomic weight of the element minus mass of nucleons;
(c) nuclear mass minus mass number;
(d) nuclear mass minus mass of nucleons;
(e) none of the above.

21.46 Which of the following particles cannot be accelerated in a cyclotron?

(a) Alpha;
(b) beta;
(c) proton;
(d) neutron;
(e) positron.

21.47 How is the presence of radioactivity detected in a scintillation counter?

(a) by fluorescence of ZnS upon interaction with radiation
(b) by interaction of the radiation with a photographic plate
(c) by ionization of a gas
(d) by precipitation of a radioactive substance
(e) by gas chromatography techniques

21.48 What is the missing particle or energy in the reaction $^{209}_{83}$Bi + $^2_1$H $\longrightarrow$ $^{210}_{84}$Po + _____

(a) $^0_{-1}$e
(b) $^0_1$e
(c) $^1_0$n
(d) α
(e) $^4_2$He

21.49 What is the binding energy of $^{35}_{17}$Cl, given that the atomic mass equals 34.95953 amu?

(a) 4.77232×10^{-11} J
(b) 4.77232×10^{-10} J
(c) 5.27231×10^{-11} J
(d) 5.27231×10^{-12} J
(e) 5.27231×10^{-13} J

21.50 What isotope was used as the fissionable material in the first atomic bomb?

(a) $^{231}_{91}$Pa
(b) $^{237}_{93}$Np
(c) $^{238}_{92}$U
(d) $^{235}_{92}$U
(e) $^{233}_{92}$U

21.51 Which of the following reactions illustrates a nuclear process that occurs within a control rod in a nuclear reactor?

(a) $^1_1$H + $^1_1$H $\longrightarrow$ $^2_1$H + $^0_1$e
(b) $^9_4$Be + $^2_1$H $\longrightarrow$ $^{10}_5$B + $^1_0$N
(c) $^0_{-1}$e + $^{106}_{47}$Ag $\longrightarrow$ $^{106}_{46}$Pd
(d) $^9_4$Be + $^4_2$He $\longrightarrow$ $^{12}_6$C + $^1_0$N
(e) $^{113}_{48}$Cd + $^1_0$n $\longrightarrow$ $^{114}_{48}$Cd + γ

21.52 What is the major type of biological damage in humans following sub-lethal exposure to radiation?

(a) Leukemia;
(b) burns;
(c) skin cancer;
(d) loss of sight;
(e) loss of a human appendage.

SELF-TEST SOLUTIONS

21.1 True. **21.2** True. **21.3** False. A rad is a measure of the amount of energy that deposits on a kilogram of tissue. It does not measure the penetration of energy into interior cells and the biological effects of this penetration. **21.4** False. Mass defect = nuclear mass − mass of nucleons = 58.91837 amu − [(27)(1.00728 amu) + 32(1.00867 amu)] = −0.55563 amu. **21.5** False. Alpha particles are identical to helium *nuclei* (that is, 2+ helium ions). **21.6** False. Correct symbol is $^0_{-1}$e because the superscript number before a particle symbol reflects the sum of proton and neutrons, not electrons. **21.7** True. **21.8** True. **21.9** True. **21.10** True. **21.11** True. **21.12** True. **21.13** True. **21.14** True. **21.15** True. **21.16** False. 18 kg of $^{239}_{94}$Pu is greater than the critical mass and therefore is supercritical in amount. 10 kg of $^{239}_{94}$Pu is subcritical in amount because it is less than the critical mass. **21.17** False. Nucleons are protons and neutrons. **21.18** False. Magic numbers refer to nucleons with a closed shell set of protons or neutrons; these nuclei are more stable. **21.19** True. **21.20** False. There are only three in nature. **21.21** True. **21.22** True. **21.23** False. Alpha or beta particles, not gamma rays. **21.24** True. **21.25** False. Carbon-12 is not radioactive; thus is can not be used as a radiotracer. Carbon-14 is commonly used. **21.26** False. High temperatures must be used to overcome repulsions between nuclei. **21.27** True. **21.28** False. It involves damage to chromosones and genes, thus having an effect upon offsprings of humans. **21.29** False. The first phrase is true; however, it is equivalent to significantly less than one curie. One Bq is one disintegration per second. **21.30** False. A nucleus is radioactive if it spontaneously emits radiation. **21.31** False. For each half-life, strontium-90 loses one-half of its mass. Thus, in three half-lives, it loses $1/2 \times 1/2 \times 1/2$ of its original mass, or $1/8 \times 10.0$ g = 1.25 g. **21.32** True. **21.33** True. **21.34** (a) $^{99}_{41}$Nb $\rightarrow$ $^{99}_{42}$Mo + $^0_{-1}$e. (b) $^{48}_{24}$Cr + $^0_{-1}$e $\rightarrow$ $^{48}_{23}$V. (c) $^{39}_{20}$Ca $\rightarrow$ $^{39}_{19}$K + $^0_1$e. (d) $^{16}_{12}$C $\rightarrow$ $^{15}_{12}$C + $^1_0$n. (e) $^{148}_{64}$Gd $\rightarrow$ $^{144}_{62}$Sm + $^4_2$He.

21.35 (a) $_1^2\text{H}(_1^3\text{H,n})_2^4\text{He}$: $_1^2\text{H} + _1^3\text{H} \rightarrow _2^4\text{He} + _0^1\text{n}$.
(b) $_{99}^{253}\text{Es}(\alpha,\text{n})_{101}^{256}\text{Mv}$: $_{99}^{253}\text{Es} + _2^4\text{He} \rightarrow _{101}^{256}\text{Mv} + _0^1\text{n}$.
(c) $_{35}^{81}\text{Br}(\gamma,\text{n})_{35}^{80}\text{Br}$: $_{35}^{81}\text{Br} + \gamma \rightarrow _{35}^{80}\text{Br} + _0^1\text{n}$.
21.36 (a) $_2^4\text{He} + _6^{12}\text{C} \rightarrow _8^{16}\text{O}$. **(b)** This is a fusion reaction because lightweight nuclides combine to form a heavier nuclide. The fact that the reaction occurs at a high temperature classifies it as a thermonuclear reaction. **21.37** The mass change taking place during the reaction is Δm = mass products − mass reactants = 15.99491 amu − 4.00260 amu − 12.00000 amu = −0.00769 amu. The energy change is $\Delta E = (\Delta m)$(conversion factor that converts mass to J) = $\Delta m(1.49 \times 10^{-10}$ J/amu) = (−0.00769 amu)$(1.49 \times 10^{-10}$ J/amu) = -1.15×10^{-12} J. The fact that energy is released during the reaction shows than an oxygen-16 nucleus is more stable than the separate helium-4 and carbon-12 nuclei. This gives support to the theory that oxygen-16 can be made from a fusion of carbon-12 and helium-4 in a thermonuclear re-action. **21.38** Assuming that the ^{14}C:^{12}C ratio in charcoal made in year x is the same as in charcoal made in 1990, you can calculate the value of the ratio N_0:N_t, where N_0 is the concentration of ^{14}C in recently made charcoal and N_t is the concentration in year x. In this problem, the ratio is 1:0.30 because only 30 percent of the original value of the ^{14}C:^{12}C ratio in the charcoal made in 1990 remains in year x. The concentration of ^{12}C nuclides in the charcoal sample remains constant because ^{12}C is a stable isotope. The time required for the decay of ^{14}C nuclides from 1990 to year x is calculated as follows: $t = 1.44\ t_{1/2} \ln(N_0/N_t) = 1.44\ t_{1/2} \ln(1/0.30) = (1.44)(5730\ \text{yr})\ln(1/0.30) = 9934$ yr. The year that the buried charcoal will be discovered is 9934 +

1990 = 11,924. **21.39** Nuclei with too many protons and neutrons most often decay by emitting a particle containing protons and neutrons. Such a particle is the alpha particle $_2^4\text{He}$. **21.40** $_{33}^{75}\text{As} + _2^4\text{He} \rightarrow _{35}^{78}\text{Br} + _0^1\text{n}$; $_{35}^{78}\text{Br} \rightarrow _{34}^{78}\text{Se} + _1^0\text{e}$. **21.41** A rad measures the amount of energy absorbed from radioactive particles by tissue or other substances. Some particles do more radioactive damage to tissue than others. The rem uses an RBE factor to adjust rads for the relative biological damage of a particular particle. **21.42** One curie (Ci) is equivalent of 1 g of Ra-226 producing 3.7×10^{10} distintegrations per second. Thus, 1 μg of Ra-226 in disintegrations per second is equivalent to

$$\left(10^{-6}\ \text{g Ra-226}\right)\left(3.7 \times 10^{10}\ \frac{\text{disintegrations/sec}}{1\ \text{g Ra-226}}\right)$$
$$= 3.7 \times 10^4 \text{disintegrations/sec}$$

21.43 the fact that Pb-212 becomes incorporated into $\text{PbCl}_2(s)$ in a saturated solution after the introduction of Pb-212 as a 2+ ion shows that the following equilibrium is dynamic: $^{212}\text{PbCl}_2(s) \rightleftarrows {}^{212}\text{Pb}^{2+}(aq) + 2\text{Cl}^-(aq)$. **21.44 (e)**. **21.45 (d)**. **21.46 (d)**. Only charged particles are accelerated in a cyclotron; a neutron has no charge. **21.47 (a)**. **21.48 (c)**. **21.49 (a)**. Binding energy = (mass of nucleons − mass of nucleus)$(1.49 \times 10^{-10}$ J/amu) = [(17)(1.00728 amu) + 18(1.00867 amu) − 34.95953 amu] $(1.49 \times 10^{-10}$ J/amu) = (0.32029 amu)$(1.49 \times 10^{-10}$ J/amu) = 4.77232×10^{-11} J. **21.50 (d)**. **21.51 (e)**. **21.52 (a)**.

Sectional Test 5

CHAPTERS 18–21

You should make this test as realistic as possible. **Tear out this test so that you do not have access to the information in the** *Student's Guide*. Use only data provided in the questions and a periodic table. Do not check your answers until you are finished. If you answer questions incorrectly, review the section in the text indicated after each question.

Choose the best response for each question.

1. Which of the following is a *major* constituent of air?
   (a) CO (b) N_2 (c) H_2 (d) He
   [18–1]

2. Which of the following is an active chemical in the ozone shield that reduces UV radiation?
   (a) CH_3Cl (b) HCl (c) O_2 (d) O_3
   [18–3]

3. Which of the following is becoming an important technology for conversion of seawater to pure water in the Middle East?
   (a) Osmosis (b) Sublimation (c) Crystallization (d) Filtration
   [18–5]

4. What is the partial pressure of oxygen in dry air when the total dry air pressure is 740 mm Hg? The mole fraction of oxygen in air is 0.209.
   (a) 155 mm Hg (b) 278 mm Hg (c) 670 mm Hg (d) 740 mm Hg
   [18–1]

5. The reaction $O_2(g) + h\nu \longrightarrow 2O(g)$ is termed
   (a) Photoionization   (c) Ozonolysis
   (b) Photodissociation   (d) Photocatalysis
   [18–2]

6. In the upper atmosphere, the following reaction occurs:

$$O(g) + O_2(g) \longrightarrow O_3^*(g)$$

Which of the following statements is true concerning this process?
   (a) $O_3^*(g)$ is stable
   (b) $O_3^*(g)$ is not energetic
   (c) $O_3^*(g)$ may break apart into $O(g) + O_2(g)$

(d) $O_3^*(g)$ keeps its energy during collisions

[18–3]

7. Which of the following air pollutants is commonly produced in the smelting of ores?

(a) $SO_3(g)$ (b) $SO_2(g)$ (c) $O_2(g)$ (d) $NO(g)$

[18–4]

8. Which of the following is the most abundant ion in seawater?

(a) Cl^- (b) Br^- (c) Na^+ (d) Mg^{2+}

[18–5]

9. Which of the following is a flocculating agent commonly used in water treatment plants to aid removal of very small particles under basic conditions?

(a) $NaHCO_3$ (b) $MgCl_2$ (c) CaO (d) $Al_2(SO_4)_3$

[18–6]

10. Which ions are primarily responsible for water hardness?

(a) Mg^{2+}, K^+ (b) Ca^{2+}, Na^+ (c) Mg^{2+}, Ca^{2+} (d) Na^+, CO_3^{2-}

[18–6]

11. Which of the following substances has the lowest entropy at 20°C and 1 atm pressure?

(a) 1 mole of $H_2O(l)$ (c) 1 mole of $O_2(g)$

(b) 1 mole of $NaCl(s)$ (d) 1 mole of $CH_3OH(l)$

[19–3]

12. Which statement best characterizes the change in disorder for the following reaction at 25 °C: $2Na(s) + Cl_2(g) \longrightarrow 2NaCl(s)$?

(a) Decreases   (c) No change

(b) Increases   (d) Can't determine with given data

[19–3]

13. Which of the following does *not* characterize a spontaneous process?

(a) A reverse process can be considered.

(b) Processes that are spontaneous in one direction are not spontaneous in the other direction.

(c) They do not need to be driven by an outside source of energy.

(d) They are exothermic reactions, not endothermic.

[19–2]

14. According to the second law of thermodynamics:

(a) Spontaneous processes occur in isolated systems.

(b) Spontaneous processes occur with an increase in the entropy of the universe.

(c) Spontaneous processes are exothermic events.

(d) For any spontaneous process, $\Delta S_{universe} < 0$.

[19–2]

15. Which of the following corresponds to an increase in disorder in an *isolated* system?

(a) $\Delta S_{sys} > 0$ (b) $\Delta S_{sys} < 0$ (c) $\Delta S_{sys} = 0$ (d) $\Delta S_{surr} < 0$

[19–2]

16. Which of the following has $S° = 0$ J/K-mol?

(a) $NaCl(s)$ at 0°C   (c) He at 0 K

(b) $Fe(s)$ at 0°C   (d) Hg(Sn) alloy at 0 K

[19–3]

17. $\Delta S°$ is -198.2 J/K for the reaction $N_2(g) + 3H_2(g) \longrightarrow 2NH_3(g)$ at 25°C. Given that $S°(N_2) = 191.5$ J/K-mol and $S°(H_2) = 130.58$ J/K-mol at 25°C, what is $S°(NH_3)$?

    (a) -385 J/K-mol   (c) -192.5 J/K-mol

    (b) 385 J/K-mol   (d) 192.5 J/K-mol

    [19–4]

18. Which of the following characterizes a spontaneous process at constant pressure and temperature?

    (a) $\Delta G < 0$ (b) $\Delta G = 0$ (c) $\Delta G > 0$ (d) $G \leqslant 0$

    [19–5]

19. Which of the following has a $\Delta G_f° = 0$ kJ/mol at 25°C and at 1 atm pressure?

    (a) $H_2O(l)$ (b) $NaCl(aq)$ (c) $O_3(g)$ (d) $O_2(g)$

    [19–5]

20. What is the standard free energy change for the reaction $2H_2(g) + O_2(g) \longrightarrow 2H_2O(l)$ at 25°C, given the following data: $\Delta H_f°(H_2O(l) = -285.9$ kJ/mol; $S°(H_2O(l)) = 69.96$ J/mol; $S°(H_2(g)) = 130.6$ J/mol; $S°(O_2) = 205.0$ J/mol.

    (a) $+474.6$ kJ (b) -474.6 kJ (c) $+237.2$ kJ (d) -237.2 kJ

    [19–5]

21. Calculate the standard free energy change for the reaction $CaCO_3 (s) \longrightarrow CaO(s) + O_2(g)$ at 25°C given the following data: $\Delta G_f°(CaO) = -604.0$ kJ/mol; $\Delta G_f°(CaCO_3) = -1128.8$ kJ/mol. Also determine if the reaction is spontaneous.

    (a) -524.8 kJ, not spontaneous (b) -524.8 kJ, spontaneous

    (c) 524.8 kJ, not spontaneous (d) 524.8 kJ, spontaneous

    [19–5]

22. $\Delta G°$ is 102.5 kJ/mol at 1108 K for the dissociation of HBr. What is K_p for this dissociation? $(R = 8.314$ J/K$)$

    (a) 2.85×10^{-7} (b) 3.96×10^{-7} (c) 4.42×10^{-6} (d) 1.48×10^{-5}

    [19–6]

23. What is ΔG at 298 K for the following reaction, if the mixture contains 1.00 atm of $SO_2(g)$, 2.00 atm of $O_2(g)$ and 2.00 atm of $SO_3(g)$?

$$2SO_2(g) + O_2(g) \longrightarrow 2SO_3(g)$$

Data: $\Delta G_f°(SO_2(g)) = -300.4$ kJ/mol; $\Delta G_f°(SO_3(g)) = -370.4$ kJ/mol.

    (a) -141.7 kJ (b) -138.3 kJ (c) $-15,776$ kJ (d) $-17,577$ kJ

    [19–6]

24. What is the coefficient in front of Mn^{2+} when the following redox reaction is balanced in acid solution:

$$MnO_4^- + Cl^- \longrightarrow Mn^{2+} + Cl_2$$

    (a) 1 (b) 2 (c) 3 (d) 4

    [20–2]

25. What is the coefficient in front of NHO_3 when the following redox reaction is balanced in acid solution:

$$NHO_3 + I_2 \longrightarrow NO_2 + HIO_3$$

(a) 4 **(b)** 6 **(c)** 10 **(d)** 12
[20–2]

26. What is the coefficient in front of KBr when the following reaction is balanced in a basic solution:

$$MnBr_2 + Br_2 \longrightarrow MnO_2 + KBr$$

(a) 2 **(b)** 4 **(c)** 6 **(d)** 8
[20–2]

27. Which of the following is *not* true about the anode of a voltaic cell?

    (a) Reduction occurs at it.
    (b) It is assigned as the negative pole.
    (c) Electrons flow from it.
    (d) Cations in solution move away from the surface of the electrode as the redox reaction occurs.
    [20–3]

28. What is the cathodic half-cell reaction for
$3Zn + Cr_2O_7^{2-} + 14H^+ \longrightarrow 3Zn^{2+} + 2Cr^{3+} + 7H_2O$?

    (a) $Zn \longrightarrow Zn^{2+} + 2e^-$
    (b) $Zn^{2+} + 2e^- \longrightarrow Zn$
    (c) $6e^- + Cr_2O_2^{2-} + 14H^+ \longrightarrow 2Cr^{3+} + 7H_2O$
    (d) $2Cr^{3+} + 7H_2O \longrightarrow 14H^+ + Cr_2O_7^{2-} + 6e^-$
    [20–3]

For questions 29 through 33, use the table of standard electrode potentials at 25°C in the appendix of your text.

29. Which of the following will reduce Cu^{2+} to $Cu(s)$ at standard state conditions?

    (a) Ce^{3+} **(b)** F^- **(c)** I_2 **(d)** Cd
    [20–4]

30. Which of the following reactions is spontaneous at standard state conditions?

    (a) $F_2(g) + 2Fe^{3+}(aq) \longrightarrow 2Fe^{2+}(aq) + 2F^-(aq)$
    (b) $2Al^{3+}(aq) + 6H^+(aq) \longrightarrow 2Al(s) + 3H_2(g)$
    (c) $Mn^{2+}(aq) + Pb(s) \longrightarrow Pb^{2+}(aq) + Mn(s)$
    (d) $Sn^{2+}(aq) + H_2S(g) \longrightarrow Sn(s) + S(s) + 2H^+(aq)$
    [20–4, 20–5]

31. Which of the following is the strongest *reducing* agent?

    (a) $Cu(s)$ **(b)** $Cd(s)$ **(c)** $I^-(aq)$ **(d)** $Fe(s)$
    [20–5]

32. Using standard cell potentials, calculate $\Delta G°$ for the reaction

$$5AgI(s) + Mn^{2+}(aq) + 4H_2O(l) \longrightarrow$$
$$5Ag(s) + 5I^-(aq) + MnO_4^-(aq) + 8H^+(aq)$$

    (a) 160 kJ **(b)** 801 kJ **(c)** −160 kJ **(d)** −801 kJ
    [20–35]

33. What is the nonstandard electrode potential at 25°C for the reaction $Cd(s) + 2H^+(aq) \longrightarrow Cd^{2+}(aq) + H_2(g)$ when the concentration of

$Cd^{2+}(aq)$ is 0.100 M, P_{H_2} = 2.00 atm and concentration of H^+ is 0.200 M?

(a) 0.283 V (b) 0.365 V (c) 0.382 V (d) 0.424 V

[20–4]

34. Which of the following is the anodic reaction in the discharge of a lead-acid storage battery?

(a) $Pb(s) + SO_4^{2-}(aq) \longrightarrow PbSO_4(s) + 2e^-$

(b) $PbSO_4(s) + 2e^- \longrightarrow Pb(s) + SO_4^{2-}(aq)$

(c) $PbO_2(s) + SO_4^{2-}(aq) + 4H^+(aq) + 2e^- \longrightarrow PbSO_4(s) + 2H_2O(l)$

(d) $PbSO_4(s) + 2H_2O(l) \longrightarrow PbO_2(s) + SO_4^{2-}(aq) + 4H^+(aq) + 2e^-$

[20–6]

35. What substance is reduced in an alkaline-cell battery?

(a) $Zn(s)$ (b) $Zn^{2+}(aq)$ (c) $NH_4^+(aq)$ (d) $MnO_2(s)$

[20–6]

36. Which of the following is *not* produced in the electrolysis of $NaCl(aq)$?

(a) $H_2(g)$ (b) $Cl_2(g)$ (c)$Na(l)$ (d) $OH^-(aq)$

[20–7]

37. Which of the following is *not* initially produced in the electrolysis of an aqueous solution of $CuSO_4$? Data:

$$Cu^{2+}(aq) + 2e^- \longrightarrow Cu(s) \quad E^\circ_{red} = 0.34 \text{ V}$$
$$2H_2O(l) + 2e^- \longrightarrow H_2(g) + 2OH^-(aq) \quad E^\circ_{red} = -0.83 \text{ V}$$
$$4H^+(aq) + O_2(g) + 4e^- \longrightarrow 2H_2O(l) \quad E^\circ_{red} = 1.23 \text{ V}$$
$$Cl_2(g) + 2e^- \longrightarrow 2Cl^-(aq) \quad E^\circ_{red} = 1.36 \text{ V}$$
$$SO_4^{2-}(aq) + 4H^+(aq) + 2e^- \longrightarrow SO_2(g) + 2H_2O(l) \quad E^\circ_{red} = 0.20 \text{ V}$$

(a) $Cu(s)$ (b) $O_2(g)$ (c) $H^+(aq)$ (d) $SO_2(g)$

[20–7]

38. How long does it take to plate onto a surface 1.00 g of chromium metal using a current of 3.00 amperes? (1 $\mathscr{F}$ = 96,500 coulombs) given this half-reaction

$$CrO_3(aq) + 6H^+(aq) + 6e^- \longrightarrow Cr(s) + 3H_2O(l)$$

(a) 3.71×10^3 s (b) 4.25×10^3 s (c) 6.17×10^2 s (d) 8.45×10^2 s

[20–8]

39. How many moles of I_2 are formed when a solution of KI is electrolyzed for 122 minutes using a current of 3.20 amperes?

(a) 2.03×10^{-3} mol   (c) 0.121 mol

(b) 4.05×10^{-3} mol   (d) 0.243 mol

[20–8]

40. Which of the following would *not* inhibit the corrosion of iron?

(a) Removal of $O_2(g)$. (b) Removal of $H_2O(l)$ (c) Increasing pH (d) Decreasing pH

[20–9]

41. When iron is "galvanized," what metal is used to coat iron?

(a) Cr (b) Zn (c) Mn (d) Al

[20–9]

42. The process

$$\ce{^{35}_{17}Cl + ^{1}_{0}n -> ^{35}_{16}S + ^{1}_{1}H}$$

is an example of
   (a) Nuclear transmutation (b) Beta decay (c) Gamma emission
   (d) Radioactive decay
   [21–1]

43. Which of the following is an alpha particle?
   (a) α (b) $^{4}_{2}He$ (c) $^{1}_{1}H$ (d) $^{0}_{-1}e$
   [21–2]

44. What particle completes the following process?

$$\ce{^{11}_{6}C -> ^{11}_{5}B + ?}$$

   (a) $^{4}_{2}He$ (b) $^{1}_{1}H$ (c) $^{0}_{1}e$ (d) $^{0}_{-1}e$
   [21–2]

45. Which of the following particles is missing from the process

$$\ce{^{14}_{7}N + ^{4}_{2}He -> ? + ^{1}_{1}H}$$

   (a) $^{17}_{9}F$ (b) $^{16}_{8}O$ (c) $^{17}_{8}O$ (d) $^{18}_{7}N$
   [21–2]

46. When a particular unstable nucleus decays it emits a particle whose mass number is decreased by four units and a second particle. What is the second particle?
   (a) $^{4}_{2}He$ (b) $^{1}_{0}n$ (c) $4^{0}_{1}e$ (d) $4^{0}_{-1}e$
   [21–2]

47. A nucleus with 70 neutrons has a neutron-to-proton ratio less than one. How can it gain stability?
   (a) Beta decay (b) Alpha decay (c) Gamma emission
   (d) Electron capture
   [21–2]

48. Which of the following represents the preparation of a transuranium element?
   (a) $\ce{^{58}_{26}Fe + ^{0}_{1}n -> ^{59}_{26}Fe}$ (c) $\ce{^{20}_{11}Na -> ^{0}_{1}e + ^{20}_{10}Ne}$
   (b) $\ce{^{238}_{92}U + ^{0}_{1}n -> ^{239}_{93}Np + ^{0}_{-1}e}$ (d) $\ce{^{14}_{7}N + ^{0}_{1}n -> ^{14}_{6}C + ^{1}_{1}H}$
   [21–3]

49. The half-life of the nucleus $^{222}_{86}Rn$ is 3.88 days. How many mg of a 5000 mg sample of $^{222}_{86}Rn$ remain after sixty days?
   (a) 4000 mg (b) 2168 mg (c) 1.213 mg (d) 0.110 mg
   [21–4]

50. How long does it take for a sample of $^{216}_{84}Po$ to decay by 70% from its original mass? Its half-life is 0.16 s.
   (a) 0.082 s (b) 0.12 s (c) 0.28 s (d) 0.43 s
   [21–4]

51. Carbon-14 decays as follows:

$$\ce{^{14}_{6}C -> ^{0}_{-1}e + ^{14}_{7}N}$$

Isotopic masses are 14.00307 for nitrogen-14 and 14.00324 for carbon-

14. What energy change occurs in the beta decay of $^{14}_{6}C$? ($c = 3.00 \times 10^8$ m/s, 1.00 g $= 6.02 \times 10^{23}$ amu, 1 kg-m^2/s$^2 = 1$ J).

    (a) 2.5×10^{-10} J/atm (b) -2.5×10^{-10} J/atm (c) 2.5×10^{-14} J/atm
    (d) -2.5×10^{-14} J/atm
    [21–6]

52. $^{56}_{26}Fe$ has a mass defect of 0.52872 amu. What is its binding energy per nucleon?

    (a) 1.13×10^{-12} J (b) 1.41×10^{-12} J (c) 1.98×10^{-12} J
    (d) 2.31×10^{-11} J
    [21–6]

53. Which particle is missing in the following fission of uranium-235?

$$^{235}_{92}U + {}^{1}_{0}n \longrightarrow ? + {}^{91}_{36}Kr + 3{}^{1}_{0}n$$

    (a) $^{142}_{38}Sr$ (b) $^{4}_{2}He$ (c) $^{142}_{56}Ba$ (d) $^{142}_{55}Cs$
    [21–7]

54. Why does a fission chain reaction in a subcritical mass soon stop?
    (a) Insufficient number of protons (b) Too many protons
    (c) Insufficient number of neutrons (d) Too many neutrons
    [21–7]

55. Which of the following is a fusion reaction?
    (a) $^{3}_{2}He \longrightarrow {}^{1}_{1}H + {}^{2}_{1}H$
    (b) $^{4}_{2}He + {}^{1}_{0}n \longrightarrow {}^{2}_{1}H + {}^{3}_{1}H \longrightarrow 2{}^{1}_{1}H$
    (c) $^{2}_{1}H + {}^{0}_{1}e$
    (d) $^{3}_{2}He + {}^{3}_{2}He \longrightarrow {}^{4}_{2}He + 2{}^{1}_{1}H$
    [21–8]

56. Which form of radiation penetrates least the human skin?
    (a) Alpha rays (b) Gamma rays (c) X-rays (d) Neutron beam
    [21–9]

57. How many disintegrations per second does a 7.0 millicurie sample of Co-60 undergo?
    (a) 1.3×10^8 (b) 1.8×10^8 (c) 2.59×10^8 (d) 2.59×10^{11}
    [21–9]

ANSWERS

1. (b). **2.** (d). **3.** (a). **4.** (a). **5.** (b). **6.** (c). **7.** (b). **8.** (a). **9.** (d).
10. (c). **11.** (b). **12.** (a). **13.** (d). **14.** (b). **15.** (a). **16.** (c). **17.** (d). **18.** (a).
19. (d). **20.** (b). **21.** (c). **22.** (d). **23.** (b). **24.** (b). **25.** (c). **26.** (b). **27.** (a).
28. (c). **29.** (d). **30.** (a) **31.** (d). **32.** (b). **33.** (c). **34.** (a). **35.** (d). **36.** (c).
37. (d). **38.** (a). **39.** (c). **40.** (d). **41.** (b). **42.** (a). **43.** (b). **44.** (c). **45.** (c).
46. (a). **47.** (d). **48.** (b). **49.** (d). **50.** (c). **51.** (d). **52.** (b). **53.** (c). **54.** (c).
55. (d). **56.** (a). **57.** (c).

Chemistry of Hydrogen, Oxygen, Nitrogen, and Carbon | 22

OVERVIEW OF THE CHAPTER

Review: Atomic radii (7.3); ionic radii (8.3); electronegativity (8.5); periodic properties and electron configurations (6.8); sigma and pi bonds (9.3, 9.4).

22.1, 22.2
PERIODIC TRENDS: METALS AND NONMETALS

Objectives: You should be able to:

1. Identify an element as a metal, semimetal, or nonmetal on the basis of its position in the periodic table or its properties.

2. Give examples of how the first member in each family of nonmetallic elements differs from the other elements of the same family, and account for these differences.

3. Predict the relative electronegativities and metallic character of any two members of a periodic family or a horizontal row of the periodic table.

Review: Ionization energy (7.4); nomenclature of binary compounds (8.10).

22.3 HYDROGEN

Objectives: You should be able to:

1. Cite the most common occurrences of hydrogen and how it is obtained in its elemental form.

2. Cite at least two uses for hydrogen.

3. Describe and name the three isotopes of hydrogen.

4. Distinguish among ionic, metallic, and molecular hydrides.

Review: Allotropes (7.8); VSEPR model (9.1); nomenclature of oxides (8.10).

22.4 OXYGEN

Objectives: You should be able to:

1. Cite the most common occurrences of oxygen and how it is obtained in its elemental form.

2. Cite at least two uses for oxygen.

3. Describe the allotropes of oxygen.

4. Describe the chemical and physical properties of hydrogen peroxide and its method of preparation.

22.5 NITROGEN

Review: Haber process (15.1); nomenclature of nitrogen containing compounds (2.6, 8.10).

Objectives: You should be able to:

1. Cite the most common occurrences of nitrogen and how it is obtained in its elemental form.

2. Cite at least two uses for nitrogen.

3. Write balanced chemical equations for formation of nitric acid via the Ostwald process starting from NH_3.

4. Cite at least one example of each of a nitrogen compound in which nitrogen is in each of the oxidation states from −3 to +5 and be able to name them.

22.6 CARBON

Objectives: You should be able to:

1. Cite the most common occurrences of carbon and how it is obtained in its elemental form.

2. Cite at least two uses for carbon.

3. Describe the allotropes of carbon.

4. Distinguish among ionic, interstial, and covalent carbides.

TOPIC SUMMARIES AND EXERCISES

PERIODIC
PROPERTIES:
METALS AND
NONMETALS

Elements are grouped into three classes: metals, semimetals, and nonmetals.

- **Nonmetals** are located in the right-hand portion of the periodic table.
- **Semimetals** are located along the border connecting the shaded and unshaded areas in Figure 22.1 of the text.
- **Metals** lie to the left of semimetals.

Metals have lower electronegativity values than nonmetals.

- Remember that electronegativity decreases down a given family and increases from left to right in a period (horizontal row of periodic table).
- Substances containing metals and nonmetals with large differences in electronegativities tend to be ionic.
- Nonmetals combine to form covalent substances.

There is one general periodic trend that metallic and nonmetallic elements have in common: *The first member of a metallic or nonmetallic family often has marked physical and chemical differences from subsequent members of the family.*

• For example, beryllium forms covalent compounds, whereas magnesium forms ionic ones.

• Another example is the fact that both carbon and nitrogen form multiple bonds to themselves (for example, $:N{\equiv}N:$ and $H{-}C{\equiv}C{-}H$), but other members of their families form only σ bonds to themselves.

A large number of chemical reactions of nonmetals are presented in Chapters 22 and 23. Some principles of chemical reactions you should remember are:

• Water is produced when hydrogen-containing compounds are oxidized by $O_2(g)$. CO_2 is also produced when the compound contains carbon (in a limited amount of oxygen, CO or C can form). N_2 is usually formed when the compound also contains nitrogen. See reactions [22.1] and [22.2] in the text for examples.

• H_2 and C can be used to reduce metal oxides to the free metal and $H_2O(l)$ and $CO_2(g)$, respectively. See reactions [22.3] and [22.4] in the text.

• H^-, O^{2-}, NH_2^-, N^{3-}, CH_3^-, C^{4-}, and C_2^{2-} are all very strong bases toward protons. See reactions [22.5] and [22.6] in the text for examples.

EXERCISE 1   Carefully look at a periodic table and try to remember the relative locations of the more common elements. Without referring to the periodic table, try to classify the following elements as metals, nonmetals, or semimetals: Li; C; F; Ar; As; Fe; Ba; Sn; Sb; S; I.
  Solution: Metals: Li, Fe, Ba, Sn
             Semimetals: As, Sb
             Nonmetals: C, F, Ar, S, I

If you had trouble identifying the elements as to type, review the periodic table again. This will help you later with your studies of the properties of elements.

EXERCISE 2   Write the elements in the following groups in order of increasing (smallest to largest) value of the stated property.
Electronegativity: S, Na, Al, Cs
Metallic character: Ca, Cl, Al
Tendency to form pi bonds: C, Si, N
Atomic Radius: Sr, Ca, Ba
Electrical conductivity: Ge, Na, S
Ionic character: H_2S, KBr, F_2
  Solution: Electronegativity: S < Al < Cs < Na (increases from left to right in a period and decreases down a family)
    Metallic character: Cl < Al < Ca (decreases from left to right across the periodic table)
    Tendency to form pi bonds: Si < N < C (first member of a nonmetal family may form pi bonds)
      Atomic radius: Ca < Sr < Ba (increases down a family)
      Electrical conductivity: S < Ge < Na (increases with metallic character)
      Ionic character: F_2 < H_2S < KBr (increases with increasing difference in electronegativity values between the bonded atoms)

EXERCISE 3   Explain why tin can form the ion $SnCl_6^{2-}$, but carbon cannot form CCl_6^{2-}.

   Solution: Tin has low-energy, empty $4d$ orbitals in addition to its valence $5s$ and $5p$ orbitals, which are able to form bonds with orbitals of other atoms. Thus tin can form more than four bonds; in the case of $SnCl_6^{2-}$, tin forms six bonds to chlorine atoms. Carbon does not have accessible low-energy, empty d orbitals that can bond with orbitals of other atoms. As a result, carbon can form only four bonds, utilizing its $2s$ and $2p$ valence orbitals.

HYDROGEN

Hydrogen occurs mainly in a combined state, primarily in water and in organic compounds (C,H-containing compounds).

- Hydrogen gas (H_2) is not very reactive as an element; however it will react spontaneously with the most electropositive elements (for example, Na, K) and the most electronegative elements (for example, F_2). The low reactivity of dihydrogen is attributable to the large energy required to break H—H bonds (436 kJ/mol).
- One of the strongest covalent bonds is H—O, with a bond dissociation energy of 464 kJ/mol; thus reactions of H_2 (to form O—H-containing compounds) occur readily with many oxygen-containing compounds.

Summarized below is other important information contained in Section 22.3.

Abundance: 11% of the earth's crust by mass; 70% of the universe contains hydrogen.

Uses: In the manufacture of methanol (CH_3OH); addition of hydrogen to vegetable oils to form margarine.

Preparation:

1. Reaction of active metals with mineral acid

$$Zn(s) + 2H^+(aq) \longrightarrow Zn^{2+}(aq) + H_2(g).$$

2. As a byproduct in the production of methane (CH_4).

$$CH_4(g) + H_2O(g) \xrightarrow[\text{1200 K, 30 atm}]{\text{Ni catalyst}} \underbrace{CO(g) + 3H_2(g)}_{\text{water gas}}$$

Isotopes:

1. ^1H: (protium) 99.984% abundance.
2. ^2H (deuterium, D); 0.016% abundance.
3. ^3H (tritium, T); rare; prepared by nuclear transformation, radioactive.

Deuterium and tritium undergo reactions at slower rates than ^1H. Both are used to "label" reactions by replacing ^1H in compounds and thus enabling one to follow how hydrogen changes locations in reactions.

Binary Compounds:

1. Ionic hydrides: H^- occurs as an ion with the most electropositive metals (group 1A and all of group 2A except Be and Mg). The H^- ion is a strong base: $KH(s) + H_2O \longrightarrow H_2(g) + KOH(aq)$.

2. Molecular hydrides: Hydrogen is bonded covalently to nonmetals or semimetals; it may exist in the +1 or −1 oxidation state in these compounds. The thermal stability of molecular hydrides decreases down a family.

3. Metallic hydrides: These are compounds of transition metals and hydrogen which possess metallic properties. Many are nonstoichiometric hydrides, called interstitial hydrides (M_yH_x, where the ratio of y to x is not a ratio of whole numbers). Some metallic hydrides are solutions of hydrogen atoms in a transition metal matrix.

EXERCISE 4   Write equations to describe the following: **(a)** Al reacts slowly with steam to form hydrogen gas. **(b)** Ni reacts with a strong acid to form hydrogen gas. **(c)** Ba reacts with cold water to form hydrogen gas. **(d)** Cu will not react with H^+ to form $H_2(g)$. If HNO_3 is used, oxidation of Cu by NO_3^- occurs to form $NO_2(g)$.

Solution: **(a)** Metals such as Al, Mn, Zn, Cr, Cd, and Fe react slowly with steam. The reaction for Al is $2Al(s) + 6H_2O(g) \longrightarrow 2Al(OH)_3(s) + 3H_2(g)$. **(b)** Most metals, except Cu, Bi, Hg, and Ag, react with strong acids to form H_2. The reaction of Ni is $Ni(s) + 2H^+(aq) \longrightarrow Ni^2(aq) + H_2(g)$. **(c)** Active metals, such as Cs, K, Na, Ba, Sr, and Ca react with cold H_2O to form $H_2(g)$. The reaction of barium is

$$Ba(s) + 2H_2O(l) \longrightarrow Ba(OH)_2(s) + H_2(g).$$

(d) $Cu(s) + 2NO_3^-(aq) + 4H_3O^+(aq) \longrightarrow Cu^{2+}(aq) + 2NO_2(g) + 6H_2O(l)$.

EXERCISE 5   Identify the following hydrides as ionic, metallic, or molecular: NaH; SiH_4; HCl; PdH_2; CaH_2; SbH_3.

Solution: Ionic hydrides occur with Group 1A and most 2A elements: NaH; CaH_2. Metallic hydrides occur with transition elements: PdH_2. Covalent hydrides occur with nonmetals and semimetals: SiH_4; SbH_3.

EXERCISE 6   Which reaction occurs more readily?

$$CH_3OH + D_2O \longrightarrow CH_3OD + HDO$$

or

$$CH_3OD + H_2O \longrightarrow CDH_3OH + HDO$$

Solution: The kinetic stability of a C—H bond is greater than that of an O—H bond; therefore the first reaction occurs, with the deuterium of D_2O exchanging with H of the O—H bond.

Summarized below is important information contained in Section 22.4 | OXYGEN

Abundance: Oxygen occurs as O_2 in the atmosphere (21% by volume, 23% by weight). Air, earth, and seas contain about 50% by weight.

Allotropes: Two forms of elemental oxygen exist: O_2 (oxygen) and O_3 (ozone). Ozone is less stable than O_2 at room temperature and pressure:

$$3O_2(g) \longrightarrow 2O_3(g) \qquad [\Delta H = +287 \text{ kJ}].$$

Uses: O_2 is used as an oxidizing agent, particularly in combustion reactions, such as oxyacetylene welding. Ozone is used as a powerful oxidizing agent, with O_2 often formed as one of the products.

Preparation: O_2: fractional distillation of liquid air is the industrially important method. Thermal decomposition of oxides is commonly used in the laboratory:

$$2KClO_3(s) \xrightarrow[\Delta]{MnO_2} 2KCl(s) + 3O_2(g).$$

O_3: Pass O_2 through an electrical discharge:

$$3O_2(g) \xrightarrow[\text{discharge}]{} 2O_3(g).$$

Compounds:

1. Oxides: Compounds containing oxygen in the -2 oxidation state. Oxides of metallic and nonmetallic elements exhibit differing chemical behaviors. Metal oxides are called **basic oxides/basic anhydrides.** Soluble metal oxides, such as CaO, react with water to form basic solutions:

$$CaO(s) + H_2O(l) \longrightarrow Ca(OH)_2(s)$$

Insoluble metal oxides, such as Al_2O_3, react with acids to give a salt and water:

$$Al_2O_3(s) + 6HNO_3(aq) \longrightarrow 2Al(NO_3)_3(aq) + 3H_2O(l)$$

Soluble nonmetallic oxides, such as SO_3, react with water to form acidic solutions:

$$SO_3(g) + H_2O(l) \longrightarrow H_2SO_4(aq)$$

Insoluble nonmetallic oxides react with bases. **Acid anhydride** or **acidic oxide** are other terms for nonmetallic oxides.

2. *Peroxides:* Compounds containing O—O bonds with each oxygen in an oxidation state of -1. Ionic peroxides are Na_2O_2, CaO_2, SrO_2, and BaO_2. Hydrogen peroxide, H_2O_2, contains a covalently bonded O—O group. A 3%-by-weight aqueous H_2O_2 solution is commonly used as a bleaching agent. H_2O_2 is formed by reacting water with the persulfate ion, $S_2O_8^{2-}(aq)$. Concentrated H_2O_2 is dangerous because it can decompose to $H_2O(l)$ and $O_2(g)$ with explosive violence.

3. *Superoxides:* Compounds containing the O—O$^-$ ion with each oxygen in a $-\frac{1}{2}$ oxidation state: examples are CsO_2, RbO_2, KO_2. The superoxide ion only occurs with the most active metals. KO_2 is used as a source of O_2 (g) because it readily reacts with H_2O to form $O_2(g)$, $KOH(aq)$, and $H_2O_2(aq)$.

EXERCISE 7 Give an example of an oxygen containing compound in which oxygen: **(a)** forms a π bond; **(b)** is in a positive oxidation state; **(c)** expands its octet; **(d)** forms two covalent bonds.

  Solution: **(a)** Oxygen can form π bonds using its p orbitals such as in O=C=O. **(b)** Oxygen occurs in a positive oxidation state only with fluorine: O_2F (O_2^+

or a $+\frac{1}{2}$ oxidation state). (c) Oxygen cannot expand its octet of electrons because it has no empty d orbitals of low energy; thus there are no examples. (d) Examples are H_2O and H_2O_2.

EXERCISE 8 Identify each of the following as a basic or acidic anhydride and write its reaction with water: (a) $N_2O_5(g)$; (b) $Na_2O(s)$; (c) $SO_2(g)$; (d) $MgO(s)$.

Solution: A soluble basic anhydride reacts with water to form a basic solution; it is nearly always a metal oxide. A soluble acidic anhydride reacts with water to form an acidic solution; most acidic anhydrides are nonmetal oxides. (a) An acidic anhydride because nitrogen is a nonmetal:

$$N_2O_5(g) + H_2O(l) \longrightarrow 2HNO_3(aq)$$

(b) A basic anhydride because sodium is a metal. The oxide ion reacts with water to form hydroxide ion—that is, $O^{2-}(aq)$ combines with H_2O to form $2OH^-(aq)$:

$$Na_2O(s) + H_2O(l) \longrightarrow 2NaOH(aq)$$

(c) An acidic oxide because sulfur is a nonmetal:

$$SO_2(g) + H_2O(l) \longrightarrow H_2SO_3(aq)$$

(d) A basic anhydride because magnesium is a metal:

$$MgO(s) + H_2O(l) \longrightarrow Mg(OH)_2(s)$$

EXERCISE 9 Space-science chemists are interested in the use of the superoxide ion to remove CO_2 from the atmosphere to form the carbonate ion. There is another product of this reaction. What is it? Write the reaction.

Solution: If O_2^- reacts with CO_2 to form CO_3^{2-}, oxygen in O_2^- must be both oxidized and reduced—the oxidation states of C and O in CO_2 and CO_3^{2-} are the same. In forming CO_3^{2-}, the change in oxidation state of oxygen in O_2^- is $-\frac{1}{2}$ to -1 (reduction). The other part of the reaction must then involve oxidation—the increase in the oxidation number of oxygen. Thus O_2 (zero oxidation state) is the likely other product:

$$4O_2^- + 2CO_2 \longrightarrow 2CO_3^{2-} + 3O_2.$$

Nitrogen is a gaseous diatomic element (N_2).

| NITROGEN

- The high $N\equiv N$ bond energy of $+944$ kJ/mol causes an extremely stable molecule.

- Because of nitrogen's valence electronic structure of $2s^2 2p^3$, nitrogen tends principally to form the oxidation states of $+3$ and $+5$; however, nitrogen exhibits all integral oxidation numbers from $+5$ to -3.

- The ability of nitrogen to acquire three electrons to form N^{3-} reflects its high electronegativity value. This high electronegativity partially accounts for nitrogen's ability to be involved in hydrogen bonding.

Summarized below is other important information contained in Section 22.4.

Abundance: The greatest source of N_2 is the atmosphere, which contains 78% nitrogen by volume. It is also found as a constituent of living matter, particularly in proteins.

Uses: Nitrogen is used extensively in fertilizers such as $(NH_4)_3PO_4$. It is also used in nitrogen fixation, the formation of nitrogen compounds in certain plants.

Preparation: Obtained commercially by fractional distillation of liquid air.

Compounds:

1. Important hydrogen-containing compounds of nitrogen are ammonia (NH_3), hydrazine (N_2H_4), hydroxylamine (NH_2OH), and hydrogen azide (HN_3). Ammonia is the most stable of these; the others are highly reactive.

2. Nitrogen forms a number of oxides, the three most common are N_2O, NO, and NO_2.

3. Nitric acid (HNO_3), a commercially important strong acid, is formed by the catalytic conversion of NH_3 to NO, followed by the reaction of NO with O_2 to form NO_2, and the final step being the reaction of NO_2 with water to form nitric acid. The three-step process for the formation of nitric acid is known as the Ostwald process.

4. Nitrous acid (HNO_2) is less stable than HNO_3 and tends to disproportionate (nitrogen is both oxidized and reduced) to NO and HNO_3.

EXERCISE 10   Name the following oxides of nitrogen and determine the oxidation number of nitrogen in each: N_2O, NO, N_2O_3, NO_2, and N_2O_5.
    Solution: These five oxides of nitrogen are examples of nitrogen-containing compounds that show that nitrogen can have all integral positive oxidation numbers from +1 to +5.

Oxide	Name	Oxidation number of nitrogen
N_2O	Nitrogen(I) oxide or nitrous oxide	+1
NO	Nitrogen(II) oxide or nitric oxide	+2
N_2O_3	Nitrogen(III) oxide or dinitrogen trioxide	+3
NO_2	Nitrogen(IV) oxide or nitrogen dioxide	+4
N_2O_5	Nitrogen(V) oxide or dinitrogen pentoxide	+5

EXERCISE 11   What is the Haber process? Why is it an important industrial process?
    Solution: Hydrogen and nitrogen are passed over a heated, finely divided iron catalyst with traces of Al_2O_3 and K_2O, to form NH_3:

$$N_2(g) + 3H_2(g) \longrightarrow 2NH_3(g)$$

Ammonia produced by the Haber process is the basis for the fertilizer industry.

EXERCISE 12   Dinitrogen oxide possesses a liner, unsymmetrical structure. Write all reasonable Lewis structures for dinitrogen oxide.

Solution: N_2O possesses 16 valence electrons that are used in forming the Lewis structure. Because the structure is unsymmetrical, the skeletal arrangement of atoms must be N—N—O. Two reasonable resonance forms exist using this structure:

$$: \ddot{N}=N=\ddot{O} : \longleftrightarrow : N\equiv N—\ddot{\underset{..}{O}} :$$

Although carbon is not widely abundant as an element, it plays an important role in living organisms and in the petroleum industry. C,H-containing compounds form a large area of chemistry known as organic chemistry. Section 22.5 focuses on some of the inorganic forms of carbon.

| CARBON

Abundance: Over half of carbon occurs in carbonate compounds such as $CaCO_3$. It is also found in coal and petroleum deposits.

Uses: The use of carbon is widespread: polymers, diamonds, graphite, natural gas, and carbonates, such as baking soda ($NaHCO_3$), are just a few examples.

Allotropes: Three forms exist: diamond, graphite and buckministerfullerene. At room temperature graphite is slightly more thermodynamically stable than diamond, but the conversion between them is kinetically negligibly slow. Diamond consists of a three-dimensional network of single carbon-carbon bonds whereas graphite consists of sheets of carbon atoms connected by alternating carbon-carbon double bonds in hexagonal arrays. Buckministerfullerence is a recently discovered form of carbon. It consists of C_{60} molecules resembling the surface of soccer balls.

Preparation: Carbon black is formed by heating hydrocarbons in a small amount of oxygen: $CH_4(g) + O_2(g) \longrightarrow C(s) + 2H_2O(g)$. The burning of wood forms charcoal, a form of carbon.

Compounds:

1. Carbon monoxide (CO) is a colorless gas that acts as a weak Lewis base using the electron pair on carbon. It is used in metallurgy to reduce oxides of metals to the elemental form of the metal.

2. Carbon dioxide (CO_2) is a colorless gas formed in the combustion of organic compounds or by the action of acids on carbonates (CO_3^{2-}). When CO_2 is dissolved in water, H_2CO_3 (a weak, diprotic acid) forms. The anions of carbonic acid, HCO_3^- and CO_3^{2-} are basic.

3. Carbonate-containing compounds are plentiful. The principal form is calcite, $CaCO_3$. Carbonates dissolve in slightly acidic water.

4. Binary compounds with elements other than H are known as carbides. Some important examples are ionic acetylides containing the $: C\equiv C :^{2-}$ ion (for example, CaC_2), interstitial carbides formed by many transition elements (WC is extremely hard), and covalent carbides formed by boron and silicon (SiC is known as carborundum).

EXERCISE 13 Carbon monoxide reacts with chlorine gas in the presence of heat and a platinum catalyst to form phosgene, $COCl_2$, a toxic gas. What is the oxidation number of carbon in $COCl_2$? Draw the Lewis structure of $COCl_2$.

Solution: By assigning oxidation numbers of −2 to oxygen and −1 to chlorine in $COCl_2$, we find the oxidation number of carbon is +4: (+4) + (−2) + 2(−1) = 0 = charge of $COCl_2$. To draw the Lewis structure of $COCl_2$, we need to first count the total number of valence-shell electrons.

C: 4 valence-shell electrons

O: 6 valence-shell electrons

2Cl: 2 × (7 valence shell electrons) = 14

$COCl_2$: 24 valence-shell electrons or 12 pairs

The Lewis structure

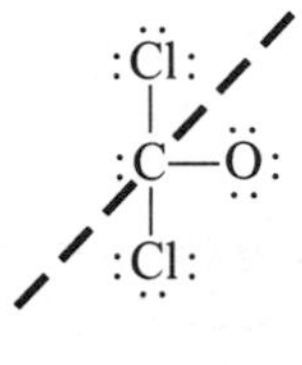

uses 13 pairs of electrons and thus is *not* an appropriate Lewis structure. To achieve an octet of electrons about each atom in $COCl_2$ and at the same time to use only 12 pairs of electrons, we must use a multiple bond between carbon and oxygen. It is logical to choose oxygen as the other atom to participate in a double bond to carbon because we have previously learned that a similar situation exists in CO_2:

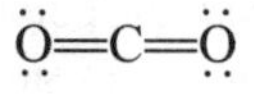

$$\ddot{O}\!=\!C\!=\!\ddot{O}$$

The Lewis structure of $COCl_2$ is

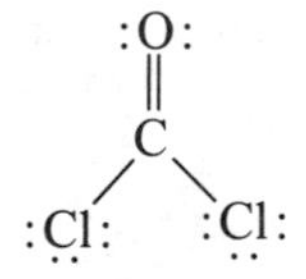

$$
\begin{array}{c}
:\!O\!: \\
\|\\
C \\
:\!\underset{\cdot\cdot}{C}\!\overset{\cdot\cdot}{l}\!:\quad:\!\underset{\cdot\cdot}{C}\!\overset{\cdot\cdot}{l}\!:
\end{array}
$$

EXERCISE 14 What types of forces hold the carbon atoms together in diamond and graphite? What is the hybridization of carbon in each?

Solution: Each carbon atom in diamond is bonded to four other carbon atoms via single covalent bonds. Each carbon uses sp^3 hybrids. The hardness of diamond results from this three-dimensional structure. In graphite, each carbon atom is bonded covalently to three other carbon atoms, all arranged in hexagonal, planar arrays to form layers of stacked sheets of hexagons. Each carbon atom is sp^2 hybridized. Van der Waals forces hold the sheets of carbon layers together; these sheets may move with respect to one another.

EXERCISE 15 Write balanced equations for the following reactions:
(a) $C(s)$ + small amount of $O_2(g)$
(b) $C(s)$ + large amount of $O_2(g)$
(c) $Al_2O_3(s)$ and $CO(g)$
(d) $MgCO_3(s)$ + HCl(aq)
(e) $MgC_2(s)$ +$H_2O(l)$
   Solution:
(a) $2C(s) + O_2(g) \longrightarrow 2CO(g)$
(b) $C(s) + O_2(g) \longrightarrow CO_2(g)$
(c) $Al_2O_3(s) + 3CO(g) \longrightarrow 2Al(s) + 3CO_2(g)$ (see 22.60 in text)
(d) $MgCO_3(s) + 2HCl(aq) \longrightarrow MgCl_2(aq) + CO_2(g) + H_2O(l)$ (see 22.71 in text)
(e) $MgC_2(s) + 2H_2O(l) \longrightarrow Mg(OH)_2(s) + C_2H_2(g)$ (see 22.75 in text)

SELF-TEST QUESTIONS

Key Terms

Having reviewed the key terms listed at the end of Chapter 22 and their applications, identify the following statements as true or false. If a statement is false, indicate why it is incorrect. Key terms are italicized in the statements.

22.1 H_2CO_2 is an example of an *acid anhydride*.

22.2 CaO is a *basic anhydride* because it reacts with bases to form salts.

22.3 One of the steps in the series of reactions involved in the *Ostwald process* is in air oxidation of NO to NO_2.

22.4 The decomposition of $H_2O_2(l)$ to form $H_2O(l)$ and $O_2(g)$ is an example of a *disproportionation reaction*.

22.5 *Water gas* is a mixture of H_2O and CO.

22.6 *Protium* is a radioactive isotope of hydrogen.

22.7 An isotope of hydrogen that contains one neutron is *deuterium*.

22.8 *Tritium* is an isotope of hydrogen which is formed continuously in the upper atmosphere and has a short half-life.

22.9 An *ionic hydride* reacts with water to form hydrogen ion.

22.10 An *interstitial hydride* may be considered to be a solution of hydrogen atoms with a transition metal as the solvent.

22.11 Ammonia is an example of a *molecular hydride*.

22.12 *Acidic oxides* typically contain a metal as the cation.

22.13 Most *basic oxides* are ionic oxides.

22.14 *Charcoal*, an amorphous form of carbon, is very dense.

22.15 When carbon is heated strongly in the absence of air *coke* is formed.

Problems and Short-Answer Questions

22.16 What nitrogen ion is analogous to the conjugate base of OH^-? Write reactions to demonstrate the analogy.

22.17 What is laughing gas? What historical role has it played in the use of chemistry in medicine?

22.18 Write the overall redox equation and the half-reactions for the oxidation of Fe^{2+} to Fe^{3+} by hydrogen peroxide in acid solution.

22.19 Use the data with equations 22.35 and 22.36 in the text to show that at standard-state conditions hydrogen peroxide is unstable because it disproportionates (that is, it undergoes self-oxidation-reduction).

22.20 Complete the following reactions:
- **(a)** $Li(s) + O_2(g) \longrightarrow$
- **(b)** $Na(s) + O_2(g) \longrightarrow$
- **(c)** $Mg(s) + O_2(g) \longrightarrow$
- **(d)** $CO_2(g) + H_2O(l) \longrightarrow$
- **(e)** $CO(g) + Cl_2(g) \longrightarrow$
- **(f)** $BaO(s) + H_2O(l) \longrightarrow$

- **(g)** $NaN_3 \xrightarrow{\Delta}$
- **(h)** $NO(g) + O_2(g) \longrightarrow$

22.21 N_2O_5 is the anhydride of what acid? Draw the Lewis structure of N_2O_5 given that it contains a N—O—N bond.

22.22 Propose some sequences of chemical reactions for the following synthesis:
- **(a)** HD from D_2O
- **(b)** CD_4 from D_2O
- **(c)** LiD from D_2O

22.23 Ozone is bubbled through 50.0 mL of an aqueous potassium iodide solution to form KOH, iodine, and dioxygen. The iodine liberated is titrated with thiosulfate ion ($S_2O_3^{2-}$) to form NaI and $Na_2S_4O_6$. Calculate the moles of O_3 passed through the solution if 60.0 mL of 0.100 M $Na_2S_2O_3$ is used in the titration. Write balanced equations describing the reactions that take place.

22.24 Using MO theory, determine the bond order of O_2^-.

22.25 What oxide of carbon is isoelectronic with N_2? How do their reactivities compare?

22.26 What species are present in an aqueous solution of CO_2? Write equations to describe their formation.

22.27 Write balanced chemical equations that exemplify the following changes in oxidation states of nitrogen.
- **(a)** 0 to −3
- **(b)** +2 to +4
- **(c)** +5 to +2

Multiple-Choice Questions

22.28 Which of the following are acidic anhydrides: (1) BaO; (2) SO_3; (3) Cl_2; (4) Na_2O_2; (5) P_4O_{10}?
- **(a)** (3) only
- **(b)** (1) and (2)
- **(c)** (2) and (5)
- **(d)** (1), (4), and (5)
- **(e)** (2), (4), and (5)

22.29 Which of the following is an example of a compound containing the superoxide ion?
- **(a)** Li_2O
- **(b)** Na_2O_2
- **(c)** H_2O_2
- **(d)** KO_2
- **(e)** CO_2

22.30 Which of the following reactions is an example of nitrogen fixation?
- **(a)** $Mg_3N_2(s) + 6H_2O(l) \longrightarrow 2NH_3(aq) + 3Mg(OH)_2(s)$
- **(b)** $2Hg^{2+}(aq) + N_2H_4(aq) \longrightarrow 2Hg(l) + N_2(g) + 4H^+(aq)$
- **(c)** $4NH_3(g) + 3O_2(g) \xrightarrow{1000°C} 2N_2(g) + 6H_2O(g)$
- **(d)** $NH_4NO_3(s) \xrightarrow{200°C} N_2O(g) + 2H_2O(g)$
- **(e)** $N_2(g) + O_2(g) \xrightarrow{\Delta} 2NO(g)$

22.31 The valence electrons of nitrogen are
- **(a)** $2s^2 2p^3$
- **(b)** $3s^2 3p^3$
- **(c)** $2s^2 2p^4$
- **(d)** $3s^2 3p^4$
- **(e)** $2s^2 2p^6$

22.32 How many protons and neutrons does deuterium possess?
- **(a)** 1 p, 1 n
- **(b)** 1 p, 2 n
- **(c)** 1 p, 3 n
- **(d)** 2 p, 1 n
- **(e)** 2 p, 2 n

22.33 Which of the following statements is *not* true about hydrogen gas?
   (a) Forces between H_2 molecules are weak.
   (b) It is a colorless gas at room temperature and pressure.
   (c) It is an effective reducing agent for many metal oxides.
   (d) The H—H bond is weak.
   (e) Igniting of H_2 in air produces H_2O.

22.34 Water gas is a mixture of
   (a) $CH_4(g)$ and $H_2O(g)$
   (b) $CO(g)$ and $H_2O(g)$
   (c) $CO(g)$ and $H_2(g)$
   (d) $C(s)$ and $H_2O(g)$
   (e) CO_2 and $H_2(g)$

22.35 CaH_2 is a(n)
   (a) Ionic hydride
   (b) Metallic hydride
   (c) Covalent hydride
   (d) An acidic substance in H_2O
   (e) Source of H^+ ion

22.36 The most thermally stable substance of those listed below is:
   (a) NH_3 (d) SbH_3
   (b) PH_3 (e) H_2Te
   (c) AsH_3

22.37 A compound containing an oxygen atom in the +2 oxidation state is
   (a) O_2F_2 (d) Li_2O
   (b) OF_2 (e) K_2O_2
   (c) H_2O

22.38 Which of the following is a basic anhydride?
   (a) SO_2 (d) $Ca(OH)_2$
   (b) H_2O (e) BaO
   (c) OF_2

22.39 Which of the following elements form superoxide compounds: (1) Li; (2) Na; (3) K; (4) Rb; (5) Cs?
   (a) (5) only   (d) (1) and (2)
   (b) (4) and (5)   (e) all
   (c) (3), (4), and (5)

22.40 Which of the following reacts with water to form ammonia?
   (a) NO_2 (d) Mg_3N_2
   (b) Mg   (e) N_2
   (c) NO_3^-

22.41 Which of the following compounds bears the same relationship to NH_3 that H_2O_2 does to water?
   (a) N_2 (d) NaN_3
   (b) N_2H_4 (e) NO
   (c) HN_3

22.42 HCN can be produced in the laboratory by reaction of
   (a) CH_4 with N_2
   (b) H_2 with NaCN
   (c) NaCN with H_2O
   (d) NaCN with HNO_3
   (e) none of the above

SELF-TEST SOLUTIONS

22.1 False. Acid anhydrides are nonmetal oxides formed by the elimination of water from nonmetal oxyacids. CO_2 is acid anhydride of $H_2CO_3(CO_2 + H_2O)$. **22.2** False. Basic anhydrides, such as CaO, react with water to form bases, and with acids to form water and a salt. **22.3** True. **22.4** True. **22.5** False. It is a mixture of $H_2(g)$ and $CO(g)$. **22.6** False. It is the most abundant isotope and very stable. **22.7** True. **22.8** True. **22.9** False. An ionic hydride reacts with water to form hydrogen gas and hydroxide ion. **22.10** True. **22.11** True. **22.12** False. They are typically composed of two nonmetals. **22.13** True. **22.14** False. It has an open structure, with a large surface area. **22.15** True. **22.16** The nitride ion (N^{3-}) is analogous to O^{2-}, the conjugate base of OH^-. The O^{2-} ion is formed by removing the proton from OH^-. An oxide ion in water acts as a base by accepting a proton from water to form the OH^- ion, as illustrated by the following chemical reaction: $CaO(s) + H_2O \rightarrow Ca(OH)_2(s)$. Similarly, the nitride ion in water acts as a base by accepting protons from water to form NH_3 and OH^- ions as illustrated by the following chemical reaction: $Mg_3N_2(s) + 6H_2O \rightarrow 2NH_3(aq) + 3Mg(OH)_2(s)$. **22.17** Nitrous oxide (N_2O) is laughing gas. It was the first substance used as a general anesthetic.

22.18

$$2(Fe^{2+} \longrightarrow Fe^{3+} + e^-)$$
$$\underline{2H^+ + H_2O_2 + 2e^- \longrightarrow 2H_2O}$$
$$2Fe^{2+} + 2H^+ + H_2O_2 \longrightarrow 2H_2O + 2Fe^{3+}$$

22.19

$2H^+ + H_2O_2 + 2e^- \longrightarrow 2H_2O$	$E° = 1.77$ V
$H_2O_2 \longrightarrow O_2 + 2H^+ + 2e^-$	$E° = 0.76$ V
$2H_2O_2 \longrightarrow 2H_2O + O_2$	$E° = 1.01$ V

The reaction is spontaneous at standard-state conditions because $E°$ for the reaction is a positive quantity. **22.20 (a)** $Li(s) + O_2(g) \rightarrow 2Li_2O(s)$ **(b)** $2Na(s) + O_2(g) \rightarrow Na_2O_2(s)$ **(c)** $2Mg(s) + O_2(g) \rightarrow 2MgO(s)$ **(d)** $CO_2(g) + H_2O(l) \rightarrow H_2CO_3(aq)$ **(e)** $CO(g) + Cl_2(g) \rightarrow COCl_2(g)$ **(f)** $BaO(s) + H_2O(l) \rightarrow Ba(OH)_2(s)$ **(g)** $2NaN_3 \rightarrow 2Na(s) + 3N_2(g)$ **(h)** $2NO(g) + O_2(g) \rightarrow 2NO(g)$. **22.21** N_2O_5 is the anhydride of HNO_3 ($2HNO_3$ less a water molecule).

2.22 (a) $LiH + D_2O \rightarrow LiO + HD$; **(b)** $Al_4C_3 + 12D_2O \rightarrow 3CD_4 + 4Al(OD)_3$; **(c)** $D_2O + Li \rightarrow \frac{1}{2}D_2 + LiOD$; $2Li + D_2 \rightarrow 2LiD$. **22.23** $2KI + O_3 + H_2O \rightarrow 2KOH + I_2 + 2O_2$; $I_2 + 2Na_2S_2O_3 \rightarrow 2NaI + Na_2S_4O_6$

$$\text{Moles O}_3 = \left(\text{moles S}_2\text{O}_3^{2-}\right) \frac{\left(1 \text{ mol I}_2\right)}{\left(2 \text{ mol S}_2\text{O}_3^{2-}\right)} \frac{\left(1 \text{ mol O}_3\right)}{\left(1 \text{ mol I}_2\right)}$$

$$= \left(0.0600\text{L S}_2\text{O}_3^{2-}\right)\left(0.100\text{M S}_2\text{O}_3^{2-}\right) \frac{\left(1 \text{ mol O}_3\right)}{\left(2 \text{ mol S}_2\text{O}_3^{2-}\right)}$$

$$= 0.00300 \text{ mol O}_3$$

22.24 $(\sigma_{1s})^2 \, (\sigma_{1s}^*)^2 \, (\sigma_{2s})^2 \, (\sigma_{2s}^*)^2 \, (\pi_{2p})^4 \, (\sigma_p)^2 \, (\pi_{2p}^*)^2$

$$\text{Bond order} = \frac{10-7}{2} = 1\tfrac{1}{2}$$

22.25 N_2 has 14 electrons as does CO. Both contain a triple bond (: N≡N :, : C≡O :), which is broken only with great difficulty. However, the carbon atom's lone pair of valence electrons is not as tightly held as those of nitrogen's because of carbon's lower electronegativity. Thus the carbon atom in CO acts as a Lewis base site toward metals. **22.26** $CO_2(g) +$ $H_2O(l) \rightleftharpoons CO_2(aq)$; $CO_2(aq) + H_2O(l) \rightleftharpoons H_2CO_3(aq)$; $H_2CO_3(aq) \rightleftharpoons H^+(aq) + HCO_3^-(aq)$; $HCO_3^-(aq) \rightleftharpoons H^+(aq)$ $+ CO_3^{2-}(aq)$. **22.27 (a)** $N_2(g) + O_2(g) \xrightarrow[\text{discharge}]{\text{elect}} 2NO(g)$;

(b) $3Mg(s) + N_2(g) \longrightarrow Mg_3N_2(s)$; **(c)** $3Cu(s) + 8HNO_3(aq)$ $\rightarrow 3Cu(NO_3)(aq) + 2NO(g) + 4H_2O(l)$. **22.28 (c)**. **22.29 (d)**. **22.30 (e)**. **22.31 (a)**. **22.32 (a)**. **22.33 (d)**. **22.34 (c)**. **22.35 (a)**. **22.36 (a)**. **22.37 (b)**. **22.38 (e)**. **22.39 (c)**. **22.40 (d)**. **22.41 (b)**. **22.42 (d)**.

23 | Chemistry of Other Nonmetallic Elements

OVERVIEW OF THE CHAPTER

23.1 GROUP 8A: NOBLE GASES

Review: Ionization energy (7.4); prediction of the geometric shape of covalent molecules from Lewis structures (9.1, 9.4).

Objectives: You should be able to:

1. Cite the most common occurrence of the noble gases.

2. Write the formulas of the known fluorides, oxyfluorides, and oxides of xenon, give the oxidation state of Xe in each, and describe the relative stabilities of the oxides as compared with the fluorides.

3. Account for the fact that xenon forms several compounds with fluorine and oxygen, krypton forms only KrF_2, and no chemical reactivity is known for the lighter noble-gas elements.

4. Describe the electron-pair and molecular geometries of the known compounds of xenon.

23.2 GROUP 7A: HALOGENS

Review: Chlorofluorocarbons in the atmosphere (18.3); emf calculations (20.4); factors affecting oxyacid strengths (16.9); nomenclature of oxyanions (2.6).

Objectives: You should be able to:

1. Predict the maximum and minimum oxidation state of each halogen discussed in this chapter. Give an example of a compound containing the element in each of these oxidation states.

2. Cite the most common occurrences of each halogen discussed in this chapter.

3. Write balanced chemical equations describing at least one means of preparation of each halogen from naturally occurring sources.

4. Describe at least one important use of each halogen element.

5. Write a balanced chemical equation describing the preparation of each of the hydrogen halides.

6. Give examples of diatomic and higher interhalogen compounds, and describe their electron-pair and molecular geometries.

7. Name and give the formulas of the oxyacids and oxyanions of the halogens.

8. Describe the variation in acid strength and oxidizing strength of the oxyacids of chlorine.

Review: Allotropes (22.4); catalysts (15.6); nomenclature of salts (2.6); polyprotic acids (16.5); use of hydrogen sulfide in qualitative analysis (17.6).

23.3 GROUP 6A: OXYGEN FAMILY

Objectives: You should be able to:

1. Predict the maximum and minimum oxidation state of any group 6A element discussed in the chapter. Give an example of a compound containing the element in each of those oxidation states.

2. Cite the most common occurrences of each group 6A element discussed in this chapter.

3. Cite the most common form of each group 6A element discussed in this chapter.

4. Indicate the formulas of the common oxides of sulfur and the properties of their aqueous solutions.

5. Write balanced chemical equations for formation of sulfuric acid from sulfur and describe the important properties of the acid.

6. Compare the chemical behaviors of selenium and tellurium with that of sulfur, with respect to common oxidation states and formulas of oxides and oxyacids.

Review: Haber process (15.1); polyprotic acids (16.5); nomenclature of inorganic compounds (2.6).

23.4 GROUP 5A: NITROGEN FAMILY

Objectives: You should be able to:

1. Predict the maximum and minimum oxidation state of each group 5A element discussed in the chapter. Give an example of a compound containing the element in each of those oxidation states.

2. Cite the most common occurrences of each group 5A element discussed in this chapter.

3. Cite the most common form of each group 5A element discussed in the chapter.

4. Describe the preparation of elemental phosphorus from its ores, using balanced chemical equations.

5. Describe the formulas and structures of the stable halides and oxides of phosphorus.

6. Write balanced chemical equations for the reactions of the halides and oxides of phosphorus with water.

7. Describe a condensation reaction and give examples involving compounds of phosphorus.

23.5 GROUP 4: CARBON FAMILY

Review: Structures of carbon and silicon (22.6); exceptions to octet rule (8.7).

Objectives: You should be able to:

1. Cite the most common occurrences of silicon and the most common form of elemental silicon.

2. Cite the most common occurrences of carbon.

3. Describe the structures possible for silicates and their empirical formulas (for example, silicate tetrahedra can combine through bridging oxygens to form a single-string silicate chain whose empirical formula is SiO_3^{2-}).

4. Correlate the physical properties of certain silicate minerals, such as asbestos, with their structures.

5. Explain the changes in composition and properties that accompany substitution of Al^{3+} for Si^{4+} in a silicate.

6. Describe what is meant by a clay mineral, and explain the role of clay minerals in soil fertility.

7. Describe the composition and manufacture of soda-lime glass.

23.6 BORON

Objectives: You should be able to:

1. Cite the most common occurrence of boron.

2. Describe the structure of diborane, and explain its unusual feature.

3. Describe a condensation reaction involving compounds of boron.

TOPIC SUMMARIES AND EXERCISES

GROUP 8A: NOBLE GASES

The noble gas family consists of helium, neon, argon, krypton, xenon, and radon. Because of their low concentration in the earth's crust and atmosphere, they are sometimes referred to as rare gases.

▪ Isotopes of radon are radioactive and are formed from the decay of other radioisotopes.

▪ Helium is found in natural gas wells under the surface of the earth. It boils at 4.2 K under one atmosphere pressure.

▪ Neon, krypton, argon, and xenon can be obtained by fractional distillation of liquid air.

Helium has a valence-shell electron configuration of $1s^2$; all other noble gases have the general valence-shell electron configuration of ns^2np^6.

▪ Note that all noble gases have "closed" outer valence shells. Such electron distributions are spherically symmetrical and particularly stable. Thus noble gases do not readily combine with other atoms.

▪ The first ionization energies for xenon and krypton are similar to those for nitrogen, bromine, and iodine. Therefore it is not unrealistic to expect xenon to exist in positive oxidation states and to form bonds with a highly electronegative element. Based on these ideas, in 1962 Neil Bartlett developed the first synthetic routes for preparing xenon fluorides. Oxygen compounds of xenon were made after Bartlett's initial work. Examples of xenon compounds are XeF_2, XeF_4, XeO_3, XeO_4, and XeO_2F_2. Only one binary compound of krypton is known, KrF_2; it decomposes at $-10°C$.

EXERCISE 1   Predict the trend in molar heats of vaporization of the noble gases.
    Solution: The molar heat of vaporization (energy required to vaporize a liquid at its normal boiling temperature) is an indicator of the magnitude of intermolecular attractive forces between atoms in the liquid state. We learned in Section 11.2 of the text that London dispersion forces between atoms increase with increasing mass of the atoms. Since noble gases exist as monatomic elements and are nonpolar, the only forces holding them together in the liquid state are London dispersion forces. Therefore molar heats of vaporization of noble gas elements should increase with increasing magnitude of London dispersion forces, which increase with increasing atomic mass. The predicted trend in molar heats of vaporization is thus $Xe > Kr > Ar > Ne > He$.

EXERCISE 2   Describe the electronic and geometrical shape of XeF_2.
    Solution: First write the Lewis structure of XeF_2, using procedures found in Section 8.5 of the text:

Xe: 8 valence-shell electrons
2F: 7 valence-shell electrons each, or a total
        of 14 electrons
———————————————————————————————
XeF_2:   22 valence-shell electrons

To construct the Lewis structure of XeF_2, we will have to expand the number of electrons about xenon beyond an octet:

$$:\ddot{F} - .\ddot{Xe}. - \ddot{F}:$$

We see xenon has five valence-shell electron pairs about it. Using procedures given in Sections 9.1 and 9.4 of the text, we predict that the valence-shell electron pairs of xenon are in a trigonal bipyramidal arrangement. To minimize lone pair-lone pair and lone pair-bonding pair electron repulsions, the fluorine atoms are placed in the axial positions in a trigonal bipyramidal structure. Therefore we expect XeF_2 to be linear in shape. Experimental evidence verifies this prediction.

EXERCISE 3   The enthalpy of formation of XeF_4 is -218 kJ/mol, and that of XeO_3 is $+402$ kJ/mol. Which compound is the stronger oxidizing agent? Why?
    Solution: A negative enthalpy of formation for XeF_4 compound compared to a positive one for XeO_3 shows that XeF_4 is stable with respect to its constituent elements. XeO_3 should be unstable and should react to form its constituent elements. Thus Xe^{+6} in XeO_3 is readily reduced to $Xe(0)$; that is, XeO_3 acts as a strong oxidizing agent.

GROUP 7A: HALOGENS

The halogen family consists of the elements fluorine, chlorine, bromine, iodine, and astatine. Since all isotopes of astatine are radioactive and have short half-lives, the discussion of the chemistry of halogens in Section 23.2 of the text concentrates of fluorine, chlorine, bromine, and iodine. Some of the properties of these elements are summarized in Table 23.1 on page 391.

The halogens are highly reactive. The relatively low bond energy of F_2 causes elemental fluorine to be extremely reactive.

- Halogens have high electronegativities and thus tend to act as oxidizing agents. A given halogen is able to oxidize the anions of halogens below it in the family. For example, Br_2 will oxidize I^- but not Cl^- and F^-.
- Elemental fluorine is prepared from the electrolysis of molten KHF_2 to form elemental fluorine and hydrogen along with potassium fluoride. It is too reactive to be prepared in an aqueous medium.
- Elemental chlorine is produced from the electrolysis of aqueous sodium chloride or molten sodium chloride.
- Both elemental chlorine and bromine can be prepared from brines containing the halide ions. Chlorine gas is used to oxidize the halide ions to the elemental halogens.

Hydrogen halides, HX, are covalent gases at room temperature and pressure.

- All hydrogen halides dissolve in water to form acidic solutions.
- All but HF are strong acids in water.
- The weak acid characteristics of HF(aq) are partly attributable to the fact that it forms hydrogen bonds to other HF molecules and water.
- Several synthetic methods are used to prepare hydrogen halides, and you should learn at least one method for preparing each as given in Section 23.2 of the text.

Oxyacids are another group of halogen compounds that act as acids in water.

- If you are not familiar with the acid characteristics of oxyacids, you will find a discussion of the relationships between their structures and acid properties in Section 16.9 and 16.10 of the text.
- Examples of oxyacids are HFO, HClO, HBrO, HIO, $HClO_2$, $HClO_3$, $HBrO_3$, HIO_3, $HClO_4$, and H_5IO_6.
- At this time you should review the nomenclature rules for oxyanions (Section 2.6 of the text).
- Both oxyacids and their oxyanions are excellent oxidizing agents, and several are thermally unstable (perchlorate salts are well known for this characteristic).
- You should also review Section 20.2 of the text, which describes how to balance half-cell reactions and calculate potentials for redox reactions. Interestingly, reduction potentials of oxyanions decrease with increasing oxidation number of the halogen. Thus ClO_4^- is a weaker oxidizing agent than ClO^-.

Table 23.1 Properties of Elements in the Halogen Family

Element	Elemental form	Highest and lowest oxidation state, with example	Common source	Example of use
Fluorine	$F_2(g)$	-1, HF 0, F_2	CaF_2	Teflon
Chlorine	$Cl_2(g)$	-1, HCl $+7$, Cl_2O_7	NaCl in seawater	Bleaching powder, $Ca(ClO)Cl$
Bromine	$Br_2(l)$	-1, HBr $+7$, $HBrO_4$	NaBr in seawater	Organic bromine compounds
Iodine	$I_2(s)$	-1, HI $+7$, IF_7	NaI in seawater	Tincture of iodine (iodine in ethyl alcohol)

Teflon structure:

$$\left[\begin{array}{c} F \quad\quad\quad F \\ \diagdown \quad\quad \diagup \\ -C-C- \\ \diagup \quad\quad \diagdown \\ F \quad\quad\quad F \end{array} \right]_n$$

Since halogens are similar in many ways, we find many binary compounds in which different halogens are bonded to one another.

- Examples are ClF, BrF_3, ClF_3, BrF_5, IF_5, and IF_7.
- For interhalogens of type XX'_n ($n = 3$, 5, or 7) where X = Cl, Br, or I. X' is nearly always fluorine.
- The structures of interhalogen compounds can be predicted using procedures found in Sections 9.1 and 9.4 of the text. These structures are consistent with bonding models utilizing d-orbital participation and hybridization of valence-shell orbitals of the central halogen in a positive oxidation state.

EXERCISE 4   Review the oxidation states of halogen compounds. Try to remember examples for each case. Now, close your text and give three examples of halogen compounds in which a halogen atom is in each of the following oxidation states: $+1$, $+3$, $+5$, $+7$.

Solution: The positive oxidation states for the halogens and examples for each are as follows:

$+1$	$+3$	$+5$	$+7$
ClF, Cl_2O, HBrO	ClF_3, BrF_3, $HClO_2$	BrF_5, IF_5, I_2O_5, $HClO_3$	IF_7, $HClO_4$, Cl_2O_7

EXERCISE 5   Give examples of halogen compounds in which: **(a)** the halogen uses d orbitals to expand its octet of electrons; **(b)** the halogen occurs as a halide ion (X^-) and forms an ionic bond; **(c)** the halogen forms polar and nonpolar bonds with other like or differing elements; **(d)** the halogen acts as electron donor by reacting as a halide ion with a Lewis acid; **(e)** a halogen forms a hydrogen bond.

Solution: **(a)** All but fluorine can use d orbitals to expand their octet of electrons. Examples are ClF_2^-, ICl_4^-, and ClF_3. You should draw the Lewis structures for these ions and compounds to check that their octets are in fact expanded. **(b)** All the halogens form ionic salts with group 1A and 1B cations. Examples are NaF, KI, $CaBr_2$, and $CaCl_2$. **(c)** Halogens form nonpolar covalent bonds in homonuclear diatomic molecules such as an F_2, Cl_2, and Br_2. Polar covalent bonds result when a halogen bonds to an element with a different electronegativity. Examples are ICl, CH_3F, and OF_2. **(d)** Halide ions act as donor atoms when reacting with compounds that are good Lewis acids. Two examples are the following:

$$BF_3 + F^- \longrightarrow BF_4^-$$
$$SnCl_4 + 2Cl^- \longrightarrow SnCl_6^{2-}$$

(e) Fluorine bonded to hydrogen in HF is the primary example of a halogen compound that forms hydrogen bonds.

EXERCISE 6   **(a)** Name the following oxychlorine acids and determine the oxidation number of chlorine in each: HClO; $HClO_2$; $HClO_3$; $HClO_4$. **(b)** List these acids in order of increasing acid strength. Briefly justify your order. **(c)** Draw the Lewis structure for ClO_3^-.

   Solution: **(a)** The names and oxidation states are as follows:

Oxyacid	Name	Halogen oxidation number
HClO	Hypochlorous	+1
$HClO_2$	Chlorous	+3
$HClO_3$	Chloric	+5
$HClO_4$	Perchloric	+7

(b) The order of acid strengths is $HClO < HClO_2 < HClO_3 < HClO_4$. As the oxidation number of the central atom in a series of oxyacids containing the same central atom increases, the acid strengths increase. **(c)** The number of valence-shell electrons possessed by ClO_3^- is

$$
\begin{array}{lc}
\text{Cl:} & 7 \\
3 \times \text{O:} & 3 \times 6 = 18 \\
\text{Charge:} & \underline{ 1 } \\
& 26 \text{ electrons or 13 pairs}
\end{array}
$$

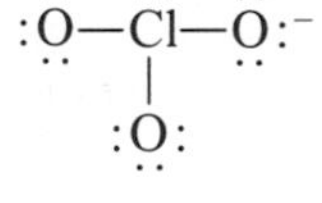

GROUP 6A: THE OXYGEN FAMILY

The oxygen family consists of the elements oxygen, sulfur, selenium, tellurium, and polonium. Since polonium is radioactive, its chemistry is not discussed. Some of the properties of these elements and their oxidation states are summarized in Table 23.2 on page 393 and below.

- Elements of the oxygen family have the general valence-shell electron configuration ns^2np^4.

- Several oxidation states are known for oxygen, but it usually occurs in the −2 oxidation state.

- Oxygen is highly electronegative and readily gains two electrons to form the oxide ion, O^{2-}.

- Sulfur commonly occurs in the −2, +4, and +6 oxidation states.

The chemistry of sulfur is the primary emphasis in this section.

- Sulfur has several allotropic forms; the more familiar form is yellow rhombic sulfur, which consists of puckered S_8 rings. Elemental sulfur occurs naturally in rocks deep below the surface of the earth and is brought to the surface and extracted by the Frasch process.

- Combustion of sulfur in air produces SO_2 and small amounts of SO_3. Both oxides dissolve in water to form acidic solutions. Gaseous SO_2 dissolves to form $SO_2(aq)$, often written as $H_2SO_3(aq)$. $SO_3(g)$ dissolves slowly to form H_2SO_4. Sulfuric acid, H_2SO_4, is a strong acid, extensively used in industrial and manufacturing processes.

- An ion related to the sulfate ion, SO_4^{2-}, is the thiosulfate ion, $S_2O_3^{2-}$. The prefix *thio-* tells you a sulfur atom has been substituted for an oxygen atom. $Na_2S_2O_3 \cdot 5H_2O$ is referred to as hypo; it is used in photography as a source of $S_2O_3^{2-}$ for complexing silver ions. An important quantitative application of $S_2O_3^{2-}$ is its use as an oxidizing agent for I_2.

Table 23.2   Properties of Elements of the Oxygen Family

Element	Common elemental form	Highest and lowest oxidation state, with example	Common source	Example of use
Oxygen	O_2 (O_3 is an allotropic form)	-2, MgO $+2$, OF_2	Air	A combustion agent as O_2
Sulfur	Rhombic sulfur, S_8 (polymeric)	-2, H_2S $+6$, SF_6	CuS	H_2SO_4, used in industrial processes
Selenium	Se (polymeric solid)	-2, H_2Se $+6$, SeO_3	Cu_2Se	Used in copying machines as conductor
Tellurium	Te (polymeric solid)	$+2$, H_2Te $+6$, TeO_3	Cu_2Te	Used in inorganic syntheses, but not commonly

---

EXERCISE 7   Give an example of an oxygen-containing compound and a sulfur-containing compound in which the oxygen and sulfur atoms: (**a**) form π bonds; (**b**) expand their octet of electrons; (**c**) are in a positive oxidation state; (**d**) form two covalent bonds.

Solution: (**a**) Oxygen can form π bonds using its p orbitals, such as in

$$\ddot{O} = C = \ddot{O}$$

Sulfur forms π bonds in compounds such as

$$\ddot{S} = C = \ddot{S}$$

(**b**) Oxygen cannot expand its octet of electrons because its has no empty d orbitals of low energy; thus there are no examples. Sulfur can expand its octet of electrons because it has empty $3d$ orbitals of sufficiently low energy; SF_6 is an example. (**c**) Oxygen occurs in the $+\frac{1}{2}$ oxidation state in O_2F. Oxygen is assigned a positive oxidation number because it has a lower electronegativity than fluorine. Sulfur occurs in the $+4$ oxidation state in SF_4. (**d**) Examples are H_2O and H_2S.

EXERCISE 8   (**a**) What type of acid or base is $SO_2(g)$? (**b**) Write its reaction with water. (**c**) Write a balanced net ionic equation for the reaction of $SO_2(g)$ with aqueous NaOH.

Solution: (a) Group 6A dioxides dissolve in water to form slightly acidic solutions. $SO_2(g)$ acts thus as a Lewis acid (it possesses no ionizable protons) when it dissolves in water. (b) $SO_2(g) + H_2O \longrightarrow H^+(aq) + HSO_3^-(aq)$. (c) Note that the weakly acidic ion HSO_3^- is formed in water. Therefore the reaction of $SO_2(g)$ with a $NaOH(aq)$ can be viewed as the reaction of $HSO_3^-(aq)$ with $OH^-(aq)$ to form $SO_3^{2-}(aq)$ and $H_2O(l)$:

$$SO_2(g) + 2OH^-(aq) \longrightarrow SO_3^{2-}(aq) + H_2O(l)$$

EXERCISE 9 (a) What chemical substance is given the name "hypo"? (b) What does the term "thio" mean? (c) Write the reaction for the reduction of iodine in the presence of aqueous thiosulfate ion.

Solution: (a) $Na_2S_2O_3 \cdot 5H_2O$, a pentahydrated salt of sodium thiosulfate, is known as "hypo" in the photography business. (b) The term "thio" means that a sulfur has been substituted for an oxygen in a compound. When an oxygen atom in the sulfate ion, SO_4^{2-}, is substituted by a sulfur atom, the thiosulfate ion is formed. (c) The thiosulfate ion is used in quantitative analysis as a reducing agent for iodine.

$$2S_2O_3^{2-}(aq) + I_2(s) \longrightarrow 2I^-(aq) + S_4O_6^{2-}(aq)$$

GROUP 5A: NITROGEN FAMILY

The nitrogen family consists of the elements nitrogen, phosphorus, arsenic, antimony, and bismuth. Some of the properties of these elements and their oxidation states are summarized in Table 23.3 and below.

- Elements of the nitrogen family have the general valence-shell electron configuration ns^2np^3.
- All elements exhibit the +5 oxidation state in compounds, but only nitrogen forms a 3− ion (nitride ion) with the more active metals.

Table 23.3   Properties of Elements of the Nitrogen Family

Element	Common elemental form	Highest and lowest oxidation state, with example	Common source	Example of use
Nitrogen	$N_2(g)$	-3, NaN_3 $+5$, HNO_3	Air	HNO_3, nitrogen-containing fertilizers
Phosphorus	White phosphorus, $P_4(s)$; red phosphorus, $P_x(s)$	-3, Be_3P_2 $+5$, P_4O_{10}	$Ca_3(PO_4)_2$	Phosphate-containing fertilizers
Arsenic	$As(s)$ $As_4(g)$	-3, Mg_3As_2 $+5$, AsF_5	As_4S_4 As_2S_3	Pesticides, poisons
Antimony	$Sb(s)$ $Sb_4(g)$	0, $Sb(s)$ $+5$, SbF_5	Sb_2S_3	Fe-Sb alloy in pewter
Bismuth	$Bi(s)$	0, $Bi(s)$ $+5$, $BiCl_5$	Bi_2S_3	Used in emetics

Phosphorus exists in several allotropic forms: white, red, and black.

- White phosphorus consists of P_4 tetrahedra. This form of the element is highly reactive.
- Red phosphorus is the most stable allotropic form; it consists of chains of phosphorus atoms.

• Phosphorus is prepared by the reduction of phosphate in $Ca_3(PO_4)_2(s)$ with coke (a special form of carbon) and SiO_2. See equation [23.30] in the text.

All of the elements form gaseous hydrides with the general formula MH_3.

• The stability of the hydrides decreases with increasing atomic mass in the family, with SbH_3 and BiH_3 being thermally unstable.
• NH_3 differs significantly in its physical properties from the other group 6A hydrides because it associates through hydrogen bonding.

Halide compounds are known for all group 6A elements.

• Both MX_3 and MX_5 halides exist for all group 6A elements except nitrogen. Nitrogen does not have low-energy, empty d orbitals available to expand its octet; therefore, it forms only MX_3 halides.
• Phosphorus halides are extensively discussed in Section 23.4 of the text. You should notice how they are prepared, what their structures are, and how they react with water.

Oxides of nitrogen and phosphorus form an important class of compounds.

• Nitrogen is represented among the nitrogen oxides in all integral oxidation numbers ranging from +1 to +5. These compounds are all covalent. Nitrogen oxides are strong oxidizing agents.
• Phosphorus forms two important oxides, P_4O_6 and P_4O_{10}. Phosphorus(III) oxide is the anhydride of phosphorous acid (H_3PO_3), and phosphorus(V) oxide is the anhydride of phosphoric acid (H_3PO_4). Phosphoric acid undergoes a condensation reaction when heated to form pyrophosphoric acid ($H_4P_2O_7$):

$$2H_3PO_4 \xrightarrow{\Delta} H_4P_2O_7 + H_2O$$

EXERCISE 10 Given the fact that a hydrogen atom attached to phosphorus is less easily replaced than a hydrogen atom attached to an oxygen atom in a P—O—H linkage in oxyacids of phosphorus, suggest a reason for the observation that phosphoric acid is triprotic (3 ionizable hydrogens) and phosphorus acid is diprotic (2 ionizable hydrogens) even though both contain three hydrogen atoms.

Solution: The formulas for phosphoric and phosphorus acids are H_3PO_4 and H_3PO_3, respectively. Hydrogen atoms attached to oxygen atoms that are bonded to phosphorus atoms in oxyacids of phosphorus are expected to be replaceable, and therefore acidic compared to hydrogen atoms attached to phosphorus. Since H_3PO_4 has three ionizable hydrogen atoms, all three must be attached to oxygen atoms. All of the hydrogen atoms cannot be attached to oxygen atoms in H_3PO_3 because only two hydrogen atoms ionize. The third hydrogen atom in H_3PO_3 must be attached to an atom other than oxygen, which only leaves the phosphorus atom. The skeletal structures for the acids are

Phosphoric acid   Phosphorous acid

EXERCISE 11 Molecular weight studies of white phosphorus indicate that it exists as tetrahedral P_4 molecules. Write equations illustrating the combination reactions of P_4 with oxygen, halogen and sulfur. Name the products formed.

Solution: All of the reactions are direct combinations:

$P_4(s) + 5O_2(g) \longrightarrow 2P_2O_5(s)$ (or P_4O_{10}, its actual form)
                                        diphosphorus pentaoxide

$P_4(s) + 10Cl_2(g) \longrightarrow 4PCl_5(g)$ phosphorus pentachloride
$P_4(s) + 10S(s) \longrightarrow 2P_2S_5$ phosphorus pentasulfide

EXERCISE 12 **(a)** What is the main source of phosphorus? **(b)** Why isn't this source used as a phosphatic fertilizer? **(c)** How can it be converted into the fertilizer "superphosphate"?

Solution: **(a)** The main source of phosphorus is rock phosphate, primarily $Ca_3(PO_4)(s)$. **(b)** The solubility of calcium phosphate in water is too low for it to be used as a source of phosphorus in fertilizer. **(c)** "Superphosphate" is formed by treating $Ca_3(PO_4)_2$ with 70 percent sulfuric acid.

$$Ca_3(PO_4)_2(s) + 2H_2SO_4(aq) \longrightarrow Ca(H_2PO_4)_2(s) + 2CaSO_4(s)$$
$$\text{"superphosphate"}$$

GROUP 4A: CARBON FAMILY

Some of the properties of the group 4A elements carbon and silicon are summarized in Table 23.4 and below.

- Note that with increasing atomic number in group 4A elements, there is a trend from nonmetallic (C), to metalloid (Si, Ge) to metallic (Sn, Pb).

- Another contrasting feature is the ability of carbon to form multiple bonds to itself and other nonmetals (N, O, and S) and to undergo catenation (to form compounds containing a chain of —C—C— bonds). Silicon and germanium undergo catenation, but to a far lesser extent.

Table 23.4   Properties of Carbon and Silicon

Element	Elemental form	Common oxidation states, with examples	Common source	Example of use
Carbon	$C(s)$, diamond structure or graphite	$0, C(s)$ $+4, CF_4$	Carbonates and CO_2 in air	Graphite is a lubricant
Silicon	$Si(s)$	$0, Si(s)$ $+4, SiCl_4,$ SiO_2	$SiO_2(s)$ silicates	Solar cells and transistors

The two important oxides of carbon are CO and CO_2.

- Oxides of silicon have different characteristics than oxides of carbon. CO and CO_2 are molecular substances, whereas silicon oxides are solids with a network of —Si—O—Si— linkages.

Silicate minerals have as their basic structural unit a silicon atom bound to four oxygen atoms that form a tetrahedron about silicon.

• The simplest structural unit is the orthosilicate ion. SiO_4^{4-}; it is found only in a few simple mineral structures.

• More complex silicate structural units are formed by a sharing of oxygens between silicate tetrahedra, yielding a network of Si—O—Si bonds. For example, the dimer of SiO_4^{4-} is $Si_2O_7^{2-}$. Varying arrangements of SiO_4^{4-} tetrahedra sharing oxygen atoms yield chains, sheets, or three-dimensional arrays. See Figures 23.27 and 23.28 in the text.

Aluminosilicates are minerals in which Al^{3+} ions have replaced some of the Si^{4+} ions in silicates. Since Al^{3+} has a lower charge than Si^{4+}, other cations (for example, K^+ or Na^+) must also be present in the mineral to maintain charge balance.

• Examples of aluminosilicates are feldspars (for example, $CaAl_2Si_2O_8$) and clays (hydrated aluminosilicates).

Also discussed in Section 23.5 of the text are glass, cements, and the nonsilicate minerals $CaCO_3$ and CaO. You should review their compositions and properties.

EXERCISE 13   The mineral orthoclase is a feldspar mineral formed by replacing one quarter of the silicon atoms in quartz, SiO_2, with aluminum ions and by maintaining charge balance with additional potassium ions. What is the empirical formula for this mineral?

Solution: If we replace a Si atom in SiO_2 with Al^{3+}, the neutral silicate group SiO_2 becomes AlO_2^-. Orthoclase is formed by replacing one AlO_2^- group for one SiO_2 group in every four SiO_2 groups. Therefore every four SiO_2 groups in quartz are converted to the $AlO_2 \cdot 3SiO_2^-$ ion in orthoclase. In order to maintain charge balance in the mineral orthoclase, one K^+ ion is required. Thus the empirical formula of orthoclase is $KAlO_2(SiO_2)_3$, or $KAlSi_3O_8$.

EXERCISE 14   The mineral plagioclase is a mixture of $NaAlSi_5O_8$ and $CaAl_2Si_2O_8$. Explain why rainwater containing CO_2 causes Ca^{2+} to be leached when the rainwater comes in contact with plagioclase.

Solution: $CO_2(g)$ dissolves in rainwater to form carbonic acid:

$$CO_2(g) + H_2O \rightleftharpoons H_2CO_3(aq)$$

H_2CO_3 is a weak acid and dissociates to form H^+ and HCO_3^- ions:

$$H_2CO_3(aq) + H_2O \rightleftharpoons H^+(aq) + HCO_3^-(aq)$$

CO_3^{2-} ions are also formed:

$$HCO_3^-(aq) + H_2O \rightleftharpoons H^+(aq) + CO_3^{2-}(aq)$$

Ca^{2+} in $CaAl_2Si_2O_8$ reacts with HCO_3^- ions in water. Thus, Ca^{2+} is slowly leached from plagioclase as described by the following overall reaction:

$$CaAl_2Si_2O_8(s) + H_2O + CO_2(aq) \rightleftharpoons$$
$$Ca^{2+}(aq) + 2HCO_3^-(aq) + Al_2Si_2O_6(OH)_2(s)$$

EXERCISE 15   Mortar is prepared by mixing one part slaked lime (produce formed from the reaction of lime and water) and three or more parts sand with sufficient water to form a mixture with a pastelike consistency. Upon exposure to air, it dries and absorbs CO_2 to form calcite. The calcite bonds together (cements) the unchanged slaked lime and the particles of sand. Write the reactions that describe the formation of slaked lime and the formation of calcite during the hardening of mortar.

Solution: When lime, CaO, reacts with water it forms slaked lime, $Ca(OH)_2$,

$$CaO(s) + H_2O(l) \longrightarrow Ca(OH)_2(s)$$

$Ca(OH)_2$ reacts with CO_2 at the surface of the mortar mix to form calcite, $CaCO_3$:

$$Ca(OH)_2(s) + CO_2(g) \longrightarrow CaCO_3(s) + H_2O(l)$$

Many years are required to convert all of the $Ca(OH)_2$ to $CaCO_3$.

BORON

Boron is the only element of group 3A that is nonmetallic.

- Boron has a valence configuration of $2s^2 2p^1$ and exhibits a valency of 3 in all of its compounds; however, it does not form the B^{3+} ion. Instead, its valence electrons are shared to form covalent bonds.

- Note that the Lewis structures of BX_3 compounds show boron to have six valence electrons, and thus boron does not always obey the octet rule.

- Boron oxide, B_2O_3, is the anhydride of boric acid, H_3BO_3. Boric acid readily undergoes condensation reactions to form metaboric acid, $(HBO_2)_x$, and other condensed forms of boric acid.

- An important anionic compound containing boron is the borohydride ion, BH_4^-. The hydrogen atoms in BH_4^- carry a partial negative charge and are said to be "hydridic." Borohydrides are good reducing agents.

- Boranes are compounds of boron and hydrogen, B_xH_y, such as B_2H_6 which is formed from the combination of two BH_3 units. They are very reactive compared to the oxides of boron.

EXERCISE 16 **(a)** What is the formula of the hydroxide of boron? **(b)** Why is it acidic rather than basic as the name implies? **(c)** Write its reaction with water.

Solution: **(a)** Boron has an oxidation number of +3; therefore, the formula of its hydroxide is $B(OH)_3$. **(b)** Because boron is nonmetallic, its hydroxide is acidic. Instead of writing $B(OH)_3$, we write H_3BO_3 (boric acid).

(c) $B(OH)_3(s) + H_2O \rightleftharpoons B(OH)_4^-(aq) + H^+(aq)$

EXERCISE 17 Write the reactions of boron with Cl_2, O_2, and S.

Solution: When boron is heated with these elements, highly exothermic reactions occur, accompanied by a green flame:

$$2B(s) + 3Cl_2(g) \longrightarrow 2BCl_3(g)$$
$$4B(s) + 3O_2(g) \longrightarrow 2B_2O_3(s) \text{ (a covalent solid)}$$
$$2B(s) + 3S(s) \longrightarrow B_2S_3(s)$$

SELF-TEST QUESTIONS

Key Terms

Having reviewed the key terms listed at the end of Chapter 23 and their applications, identify the following statements as true or false. If a statement is false, indicate why it is incorrect. Key terms are italicized in the statements.

23.1 Since iodine and fluorine form the *interhalogen* compound IF_7, it is logical to conclude that BrF_7 should also exist.

23.2 *Clay* minerals that have small particle sizes have large surface areas.

23.3 Hydrated *aluminosilicates* contain sites that enable metal ions to be stored and exchanged with other metal ions.

23.4 The simplest structural unit found in *silicate minerals* is the $Si_2O_7^{2-}$ ion.

23.5 The addition of B_2O to soda-lime *glass* results in a glass with a lower melting point.

23.6 The simplest *borane* is B_2O_3.

Problems and Short-Answer Questions

23.7 Explain why phosphorus can form two fluorides, PF_3 and PF_5, while nitrogen forms only NF_3.

23.8 What are the two common oxides of phosphorus? Each oxide has two names associated with it. Write two names associated with each oxide.

23.9 Why is elemental white phosphorus highly reactive, while elemental nitrogen is relatively inert?

23.10 Write chemical formulas for the following compounds:
(a) potassium chlorate
(b) nitric oxide
(c) thiourea
(d) orthophosphoric acid
(e) magnesium nitride.

23.11 Write names for the following compounds:
(a) $HClO_3$ (e) XeO_3
(b) $NaClO_4$ (f) ICl_3
(c) $HBrO$ (g) H_2Te
(d) PF_3

23.12 What type of bond is broken during the hydrolysis of adenosine triphosphate? Why is this bond important in certain biological systems?

23.13 Write the structure for sodium metaphosphate. What uses does this compound have? What criticisms can be put forward against widespread use of this substance?

23.14 Xenon forms compounds in which it exhibits positive oxidation states of 2, 4, 6, and 8. For each oxidation state, give an example of a xenon compound with Xe possessing that oxidation state, draw its Lewis structure, and name it.

23.15 Write balanced reaction equations describing or illustrating the following:
(a) reaction of sulfur with hot, concentrated nitric acid to form sulfuric acid and nitrogen dioxide
(b) reaction of H_2S with Fe^{3+} to form sulfur and iron (II)
(c) reaction of CaF_2 with sulfuric acid to form calcium sulfate and HF
(d) reaction of white P_4 with excess oxygen
(e) formation of a condensed phosphate

23.16 Explain why the monuments made of marble and limestone in Greece (such as the Parthenon) are eroding, particularly in regions of high SO_2 concentration.

23.17 What substance is B_2O_3 the anhydride of? What is the expected Lewis acid-base chemistry of B_2O_3? Why?

Multiple-Choice Questions

23.18 Which of the following elements has the capacity to form more than four covalent bonds with other elements?
(a) N   (d) F
(b) Na   (e) S
(c) Ca

23.19 In terms of commercial use, the most important halogen is which of the following?
(a) F   (d) I
(b) Cl   (e) At
(c) Br

23.20 Which halogens can oxidize Br^- to Br_2 in an aqueous medium at standard-state conditions?
(a) F_2 and Cl_2
(b) F_2, Cl_2, I_2
(c) I_2 only
(d) F_2 only
(e) none of the above

23.21 Which of the following is the active ingredient in many household bleaches?
(a) $NaClO_4$ (d) $NaClO$
(b) $NaClO_3$ (e) $NaCl$
(c) $NaClO_2$

23.22 Which hydrogen halide must be stored in a wax or plastic bottle because it reacts with SiO_2 in glass bottles?
(a) HF
(b) HCl
(c) HBr
(d) HI
(e) all of the above

23.23 Of the oxyacids listed below, which one possesses the greatest acid strength in H_2O?
(a) $HClO_4$ (d) $HClO$
(b) H_2CO_3 (e) $HBrO$
(c) H_3BO_3

23.24 Of the compounds listed below, which one has the following Lewis structure:

(a) CH_4 (d) NH_4^+
(b) BF_4^- (e) ClO_4^-
(c) ICl_4^-

23.25 Which of the atoms listed below has the least exothermic energy change associated with the reaction $X(g) + e^- \longrightarrow X^-(g)$?
(a) F
(b) S
(c) N
(d) O
(e) impossible to answer because electron affinities are not measurable

23.26 Which of the following substances readily act as reducing agents: (1) BaO_2; (2) $NaBH_4$; (3) Cl_2; (4) O_2; (5) HNO_3?

    (a) (1), (3), and (5) only
    (b) (2) only
    (c) (5) only
    (d) (2) and (4) only
    (e) all five substances

23.27 Whenever SiO_4^{2-} units share oxygen atoms, they do so only at the

    (a) edges of tetrahedra
    (b) faces of tetrahedra
    (c) corners of tetrahedra
    (d) faces of octahedra
    (e) corners of octahedra

23.28 Whenever SiO_4^{2-} ions condense in infinite chains, what is the empirical formula of the resulting chain?

    (a) Si_4O^{2-}
    (b) SiO_3^{2-}
    (c) $S_2O_8^{4-}$
    (d) $S_3O_8^{8-}$
    (e) $S_{10}O_{12}^{12-}$

23.29 What silicate structure is represented by the simplest formula $Si_4O_{11}^{6-}$?

    (a) single tetrahedra
    (b) double tetrahedra
    (c) single chains
    (d) double chains
    (e) sheets

23.30 What silicate ions is represented by the following structure?

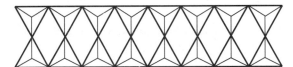

    (a) $Si_2O_6^{4-}$
    (b) $Si_4O_{11}^{6-}$
    (c) $Si_4O_{10}^{4-}$
    (d) $Si_2O_6^{4-}$
    (e) $Si_2O_7^{6-}$

23.31 Glasses are essentially mixtures of

    (a) sodium oxides
    (b) calcium carbonates
    (c) lead oxides
    (d) boron oxides
    (e) silicates

23.32 Which statement is true?

    (a) Boron is a metallic element.
    (b) Boron has a higher melting point than carbon.
    (c) The hydrogen atoms in BH_4^- are "hydridic."
    (d) Boron in diborane, B_2H_6, is surrounded by three hydrogen atoms.

23.33 Limestone is

    (a) $CaCO_3$.
    (b) Na_2O
    (c) Na_2CO_3
    (d) KCl

23.34 Silicones

    (a) consist of Si—Si—Si chains.

    (b) contain only silicon and oxygen atoms.
    (c) are rubberlike materials if crosslinking occurs.
    (d) are toxic.

SELF-TEST SOLUTIONS

23.1 False. Bromine is not sufficiently large to accommodate seven fluorine atoms, whereas iodine is. **23.2** True. **23.3** True. **23.4** False. The simplest structural unit is orthosilicate, SiO_4^{4-}. **23.5** False. It raises the melting point. **23.6** False. Boranes are compounds containing only boron and hydrogen. The simplest borane is BH_3 which readily reacts with itself to form B_2H_6. **23.7** Phosphorus has empty, low-energy $3d$ orbitals that enable it to accommodate more than eight valence electrons. Thus, not only does PF_3 (with an octet of valence electrons about phosphorus) form, but also PF_5 (with ten valence-shell electrons about phosphorus) forms. Since nitrogen does not have low-energy $3d$ orbitals available, it can accommodate only an octet of electrons. **23.8** P_4O_6: Phosphorus(III) oxide or diphosphorus trioxide (P_2O_3: empirical formula). P_4O_{10}; Phosphorus(V) oxide or diphosphorus pentoxide (P_2O_5: empirical formula). **23.9** Elemental white phosphorus, P_4, contains P—P single bonds in a tetrahedron that is sterically strained. These bonds are easily broken. The N—N bond in N_2, which is a triple bond (:N≡N:), is very difficult to rupture. **23.10 (a)** $KClO_3$; **(b)** NO; **(c)** H_2NCSNH_2; **(d)** H_3PO_4; **(e)** Mg_3N_2. **23.11 (a)** chloric acid; **(b)** sodium perchlorate; **(c)** hypobromous acid; **(d)** phosphorus trifluoride; **(e)** xenon trioxide; **(f)** iodine trichloride; **(g)** telluric acid. **23.12** The P—O—P bond of the end phosphate group in the adenosine triphosphate group is broken:

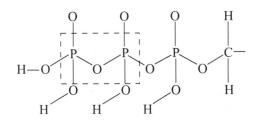

When this bond is broken, adenosine diphosphate forms, and energy is released. This energy enables endothermic biochemical reactions to occur. **23.13** The structure for sodium trimetaphosphate is

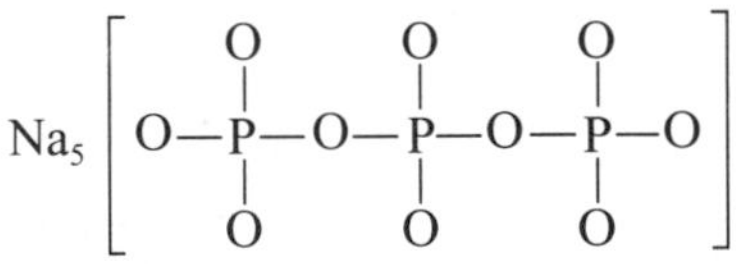

$$Na_5 \left[O-\overset{\overset{O}{\|}}{\underset{\underset{O}{|}}{P}}-O-\overset{\overset{O}{\|}}{\underset{\underset{O}{|}}{P}}-O-\overset{\overset{O}{\|}}{\underset{\underset{O}{|}}{P}}-O \right]$$

It is used in detergents because of its ability to reduce water hardness and to increase the pH of the detergent solution above 7. Its use can be criticized because the phosphorus added to natural water following its use promotes growth of algae and leads to deterioration of water quality.

Compound	Oxidation state	Lewis structure	Name
XeF_2	+2	$:\ddot{F}-\dot{X}\dot{e}-\ddot{F}:$	Xenon difluoride
XeF_4	+4	(square planar Lewis structure with Xe center and four F atoms)	Xenon tetrafluoride
XeO_3	+6	(trigonal Lewis structure with Xe center and three O atoms)	Xenon trioxide
XeO_4	+8	(tetrahedral Lewis structure with Xe center and four O atoms)	Xenon tetraoxide

23.14 23.15 (a) $S(s) + 6HNO_3(aq) \longrightarrow H_2SO_4(aq) + 6NO_2(g) + 2H_2O(l)$. (b) $2Fe^{3+}(aq) + H_2S(g) \longrightarrow 2Fe^{2+}(aq) + 2H^+(aq)$ + $S(s)$. (c) $CaF_2(s) + H_2SO_4(l) \longrightarrow CaSO_4(s) + 2HF(g)$. (d) $P_4(s) + 5O_2(g) \longrightarrow P_4O_{10}(s)$. (e) $2H_3PO_4 \longrightarrow H_4P_2O_7 + H_2O$. **23.16** $CaCO_3$ is the principal component of marble in limestone and dissolves slowly in rainwater containing CO_2: $CaCO_3(s) + H_2O + CO_2(s) \longrightarrow Ca^{2+}(aq) + 2HCO_3^-(aq)$. $CaCO_3$ also readily reacts with SO_2 as follows: $CaCO_3(s) + SO_2(g) \longrightarrow CO_2(g) + CaSO_3(aq)$. These reactions cause $CaCO_3$-containing monuments to erode. **23.17** B_2O_3 is the anhydride of H_3BO_3, (two H_3, BO_3 units minus three H_2O units). B_2O_3, like H_3BO_3, is a Lewis acid because boron is nonmetallic. **23.18 (e)**. For example, sulfur forms SF_6. It is the only element listed that has low-energy d orbitals available for bonding. **23.19 (b)**. Chlorine is extensively used in forming the strong acid HCl. **23.20 (a)**. See Table 23.3 in the text. **23.21 (d)**. **23.22 (a)**. **23.23 (a)**. **23.24 (c)**. All other compounds listed show a tetrahedral arrangement of atoms about the central element. **23.25 (d)**. The electron affinity of nitrogen is the lowest. The nitrogen atom has an especially stable half-filled valence shell. An added electron must have large repulsion energy, with the other electron already occupying the orbital to which it is added. **23.26 (b)**. **23.27 (c)**. **23.28 (b)**. **23.29 (d)**. **23.30 (c)**. **23.31 (e)**. **23.32 (c)**. **23.33 (a)**. **23.34 (c)**.

24 | Metals and Metallurgy

OVERVIEW OF THE CHAPTER

24.1 METALS: SOURCES AND PRETREATMENT

Review: Metals (2.4, 22.1); periodic table (2.4).

Objectives: You should be able to:

1. Describe what is meant by the term *mineral,* and provide a few examples of common minerals.
2. Define various terms employed in discussions of metallurgy, notably *gangue, calcination, roasting, smelting, refining,* and *leaching.*

24.2, 24.3, 24.4 METALS: ENRICHMENT PROCESSES

Review: Electrode potentials and electrochemical cells (20.3, 20.4); free energy (19.5).

Objectives: You should be able to:

1. Distinguish among pyrometallurgy, hydrometallurgy, and electrometallurgy, and provide examples of each type of metallurgical process.
2. Describe the nature of slag, and indicate the means by which it is formed in pyrometallurgical operations.
3. Describe the pyrometallurgy of iron. You should know which ores are employed; the general design of a blast furnace; the ingredients in the blast furnace; the ingredients in the blast furnace reactions; and the chemical reactions of major importance. You should know how a converter is used to refine the crude pig iron that is the product of the blast furnace operation.
4. Describe the hydrometallurgy of gold, including the chemical reactions of major importance.
5. Describe the Bayer process for purification of bauxite, including balanced chemical equations.
6. Describe the electrometallurgical purification of copper, including balanced chemical equations for electrode processes.
7. Describe the process by which sodium metal is obtained from NaCl, including balanced chemical equations for electrode processes.
8. Describe the Hall process for obtaining aluminum, including balanced chemical equations for electrode processes.

Review: Electron configuration of elements (6.9); molecular orbital theory (9.5, 9.6); delocalization of electrons (8.6).

Objectives: You should be able to:

1. Describe how the extent of metallic bonding among the first transition elements varies with the number of valence electrons.

2. Discuss the simple electron-sea model for metals, and indicate how it accounts for certain important properties of metals.

3. Describe the molecular orbital model for metals, including the idea of bands of allowed energy levels. You should also be able to distinguish between metals and insulators in terms of this model.

24.5 METALS: PHYSICAL PROPERTIES AND ELECTRONIC STRUCTURES

Review: Solutions and mixtures (1.1, 4.1).

Objectives: You should be able to:

1. Name the important types of alloys, and distinguish between them.

2. Describe the nature of intermetallic compounds, and provide examples of such compounds and their uses.

24.6 ALLOYS

Review: Aquated ions (16.10); atomic radii (7.3); *d* orbitals (6.6); electron configuration (6.8, 6.9); ionization energy (7.4); metal complexes (16.10); oxidation states (8.10); standard reduction electrode potential (20.4).

Objectives: You should be able to:

1. Identify the transition elements in a periodic table.

2. Write the electron configurations for the transition metals.

3. Describe the manner in which atomic properties such as ionization energy and atomic radius vary within each transition element series, and within each vertical group of transition elements.

4. Describe the lanthanide contraction and explain its origin.

5. Describe the general manner in which the maximum observed oxidation state varies as a function of group number among the transition elements, and account for the observed variation.

24.7 PHYSICAL PROPERTIES OF TRANSITION ELEMENTS

Review: Paramagnetism (9.7); electron spin (6.8).

Objectives: You should be able to:

1. Name and describe the types of magnetic behavior discussed in this chapter diamagnetism, paramagnetism, ferromagnetism.

2. Provide an explanation of each type of magnetic behavior in terms of the

24.7 MAGNETIC PROPERTIES

arrangements in the lattice of metal ions with unpaired spins, and their interactions with one another.

3. Describe an experimental method for determining the magnetic moment of a solid sample.

24.8 SELECTED CHEMICAL PROPERTIES: Cr, Fe, Cu

Review: Acids and bases (4.3, 16.2).

Objective: You should be able to write a balanced chemical equation corresponding to the simple aqueous solution chemistry of chromium, iron, nickel, and copper, and account for the variations in chemical properties observed among these elements in terms of the characteristics of the elements themselves.

TOPIC SUMMARIES AND EXERCISES

METALS: SOURCES AND PRETREATMENT

Metals are essential to a modern society. In this chapter of the text you explore the sources of metals, their extraction and purification, their physical properties, and the properties of mixtures of metals (alloys). A large number of new terms are introduced, and you need to pay careful attention to their definitions.

Metals are mined from that part of the solid earth that we stand upon, the **lithosphere.**

- Active metals are found in a combined state in nature. Less active metals may occur in the uncombined native state.
- Whether in a combined state or a free state, metals can be found in deposits containing several compounds that can be economically worked to extract the desired metals. These deposits are called **ores.**
- Ores contain **minerals,** which are solid substances occurring in nature with identifiable composition and well-defined crystal structures. They can be classified into one of three groups: native elements, silicates, and nonsilicates.
- A solid collection of minerals is called a **rock.** In general, silicates and aluminosilicates are the primary components of most rocks. They also make up the bulk of the earth's crust and mantle.

An ore consists of valuable pure minerals along with impurities called **gangue.**

- After an ore is mined, the desired mineral is concentrated, and the gangue is removed. Techniques for separating gangue from minerals include separation of minerals by density differences, by magnetic differences, and by flotation techniques.
- Once an ore has been enriched in the desired metal, chemical processes are used to gain further enrichment of the ore in the desired metal. These processes involve calcination, roasting, smelting, and refining.

EXERCISE 1   Give some examples of metals that are native and occur in combined states with the anions oxide and sulfide.

Solution: If you can't remember some examples, look at Table 24.1 in your text. Some examples are

Anion	Example
None (native)	Ag, Au
Oxide	Al_2O_3 (bauxite),
	Fe_3O_4 (magnetite),
	Fe_2O_3 (hematite)
Sulfide	PbS (galena)
	HgS (cinnabar)

EXERCISE 2   After silicon and oxygen, the most abundant elements in the earth's crust are aluminum, iron, and calcium if percent by weight is the basis used for comparing their relative abundances. What is the order of the abundance of these three elements if they are measured by mole fraction instead of percent by weight? The percent-by-weight abundances of aluminum, iron, and calcium are 7.5, 4.7, and 3.4, respectively.

Solution: If you assume that you have a 100-g sample of the earth's crust, the weights of these elements in the sample are

$$\text{Al: } (100 \text{ g})(0.075) = 7.5 \text{ g}$$

$$\text{Fe: } (100 \text{ g})(0.047) = 4.7 \text{ g}$$

$$\text{Ca: } (100 \text{ g})(0.034) = 3.4 \text{ g}$$

The mole fraction of a component of a mixture is defined as follows:

$$\frac{\text{Mole}}{\text{fraction}} = \frac{\text{number of moles of component}}{\begin{array}{c}\text{total number of moles of}\\\text{all components in mixture}\end{array}}$$

Since the denominator of the mole fraction is a constant for a given mixture, the order of abundance of Al, Fe, and Ca in the earth's crust based on mole fraction is simply the order of the number of moles of each in the 100-g sample:

$$\text{Al: } (7.5 \text{ g})\left(\frac{1 \text{ mol}}{26.98 \text{ g}}\right) = 0.28$$

$$\text{Fe: } (4.7 \text{ g})\left(\frac{1 \text{ mol}}{55.85 \text{ g}}\right) = 0.084$$

$$\text{Ca: } (3.4 \text{ g})\left(\frac{1 \text{ mol}}{40.08 \text{ g}}\right) = 0.085$$

The order of abundance based on mole fraction is therefore $Al > Ca \simeq Fe$.

Metallurgy involves three principal operations: concentration, reduction, and refining. The last two processes involve reducing a metal to its free form or to another compound and then final purification and isolation of the free metal. The text broadly classifies these two operations into three areas:

Pyrometallurgy. Heat is used to convert a mineral into another form or into the free metal. Three such processes are discussed:

METALS: ENRICHMENT PROCESSES

- *Calcination:* Heating an ore to an oxide, sulfate, or free metal. Carbonates readily decompose to the oxide of the metal and $CO_2(g)$.

- *Roasting:* Heating an ore below its melting point in the presence of oxygen (or CO) to produce the free metal or another compound. If oxygen is the atmosphere in the furnace, oxides or sulfates are produced. The use of CO, a reducing gas, can produce the free metal.

- *Smelting:* An ore is heated above its melting point to form a molten solution containing at least two layers. One of the layers is slag. The slag contains gangue that still remains after ore concentration—primarily silica and silicate impurities. $CaCO_3$ is commonly added to the melt to form a slag with silica—$CaSiO_3(l)$. The slag is removed from the molten mixture; thus the ore becomes more concentrated in the desired metal.

- The pyrometallurgy of iron is extensively discussed. Note the design of the blast furnace, and the role of coke (85 percent carbon) as a fuel and as the source of reducing gases, CO and H_2. The iron produced is referred to as pig iron.

Hydrometallurgy: This technique involves separating minerals from the rest of the ore by an aqueous chemical process.

- *Leaching,* the selective dissolving of the desired metal, is the most important hydrometallurgical process. The text discusses the treatment of gold ores with CN^- and oxygen to form the complex ion $Au(CN)_2^-$. Zinc dust is used to precipitate Au.

- The metallurgy of aluminum involves treating bauxite, $Al_2O_3 \cdot xH_2O$, with a concentrated NaOH solution using the *Bayer process*. The Al_2O_3 dissolves to form $Al(H_2O)_2(OH)_4^-$ a soluble complex ion. The majority of impurities, particularly silica and iron, settle out of the NaOH solution.

- Copper has been primarily obtained by pyrometallurgical techniques. A newer technique involves treating a slurry of chalcopyrite, $CuFeS_2$, with sulfuric acid and oxygen. Cu^{2+} is formed in solution, which is then electrolyzed to form free copper.

Electrometallurgy: Active metals such as sodium, aluminum, and potassium are refined by electrolysis methods. Other less active metals, such as copper, are further refined by similar electrometallurgical techniques.

- Sodium, magnesium, and aluminum must be produced from a molten salt solution. The reduction potentials of H^+ and H_2O are both more positive than those of active metals; thus they would preferentially be electrolyzed before the metal of interest. Sodium is electrolyzed in a Downs Cell. Molten NaCl is the electrolytic medium, with $CaCl_2$ added to lower the melting point of the cell medium. $Na(l)$ and $Cl_2(g)$ are produced.

- Aluminum is commercially produced by the Hall process. The cell medium contains purified Al_2O_3 (from the Bayer process) and molten cryolite, Na_3AlF_6. Graphite (C) rods are used as anodes and are consumed during electrolysis.

• Copper can be further purified in an electrolytic cell. Copper sulfate is placed in an acidic medium as an electrolyte. Impure copper is the anode, and pure copper is the cathode. Copper dissolves at the anode to form Cu^{2+}, which is reduced to pure Cu at the cathode.

EXERCISE 3   What are the major methods of reduction used in metallurgy?
   Solution: The two general methods are chemical reduction and electrolytic reduction. Electrolytic reduction of minerals is used to form the most electropositive metals, such as K, Al, Na, and Ca. Chemical reduction usually involves reacting carbon, in the form of coke or CO, with a mineral to form the desired metal or one of its compounds.

EXERCISE 4   **(a)** What are the ores used to produce the metals Hg, Cu, Fe, and Al?
(b) For these metals, write chemical equations showing a specific reduction process.
   Solution: **(a)** Mercury is produced from cinnabar, HgS; copper from $CuFeS_2$; iron from hematite, Fe_2O_3, and magnetite, Fe_3O_4; and aluminum from bauxite, $Al_2O_3 \cdot xH_2O$. **(b)** The reduction processes are *Hg*: Roasting of HgS in air:

$$HgS(s) + O_2(g) \longrightarrow Hg(g) + SO_2(g)$$

Cu: The hydrometallurgical oxidation of copper to Cu^{2+}, followed by electrolytic reduction:
(1) $2CuFeS_2(s) + 2H^+(aq) + 4O_2(g) \longrightarrow$
$$2Cu^{2+}(aq) + SO_4^{2-}(aq) + Fe_2O_3(s) + 3S(s) + H_2O(l)$$
(2) $Cu(aq)^{2+} + 2e^- \longrightarrow Cu(s)$ in an electrolytic cell
Fe: Reduced in the presence of coke and oxygen in a blast furnace:

$$2C(coke) + O_2(g) \longrightarrow 2CO(g)$$
$$Fe_2O_3(s) + 3CO(g) \longrightarrow 2Fe(s) + 3CO_2(g)$$

Al: Al_2O_3 is electrolyzed in molten cryolite, Na_3AlF_6:

$$2Al_2O_3(l) \longrightarrow 4Al(l) + 3O_2(g)$$

EXERCISE 5   **(a)** Why is the formation of a sulfate in calcination sometimes advantageous? **(b)** Why does CaO react with SiO_2 to form a slag? Write their reaction. **(c)** Why must the solution of Au and CN^- in the hydrometallurgy of gold be kept basic?
   Solution: **(a)** Sulfates are usually water soluble. The desired metal can be leached away from insoluble gangue. **(b)** CaO is a basic oxide, whereas SiO_2 is an acidic oxide. Their reaction is similar to that of an acid and base reacting to form a salt:

$$CaO(l) + SiO_2(l) \longrightarrow CaSiO_3(l)$$

(c) If the solution were to become acidic, CN^- would react to form HCN(*aq*). HCN is only moderately soluble in water, and some would escape as a toxic gas.

The physical and chemical properties of metals are varied, but most show the following characteristics.

Gray or silver luster	Electropositive
High electrical conductivity	Reducing agents
High thermal conductivity	Form cations
Malleable (hammered into thin sheets)	Low ionization energies
Ductible (drawn into wire)	Low electron affinities

METALS: PHYSICAL PROPERTIES AND ELECTRONIC STRUCTURES

The text discusses the nature of metallic bonding.

- Boiling points, melting points, heats of fusion, heats of vaporization, hardness, and densities of metals tend to increase as the strengths of the bonds between metal atoms in a metallic crystal increase.
- The strengths of metal-metal bonds are partly determined by the number and kinds of valence electrons that a metal possesses. Metals that have six outer s and d valence electrons per atom exhibit physical properties characteristic of very strong bonds between metal atoms.

A simple model for metallic bonding, the **electron-sea model,** pictures metallic bonding involving:

- a lattice of metal ions with valence electrons free to move completely about the lattice;
- electrostatic attractions between the positive metal ions and the electrons;
- electrons that are free to move under the influence of an electric field or with heating of the solid. The latter property gives rise to electrical and thermal conduction of metals.

In Sections 9.5 and 9.6 of the text, you saw that electrons can be delocalized over several nuclear centers. You also learned that atomic orbitals could overlap to form molecular orbitals. A similar situation occurs in metals.

- The interaction of metal valence atomic orbitals throughout a metallic crystal yields a large number of molecular orbitals closely spaced in energy. These molecular orbitals form a continuous band of allowed energy levels.
- A metal will be stable when more valence electrons occupy bonding molecular orbitals than antibonding molecular orbitals.
- The incomplete filling of these energy bands gives rise to characteristic metallic properties. For example, the availability of higher-energy empty orbitals in an energy band allow electrons to move throughout the metallic crystal.

EXERCISE 6   The ns and np metal orbitals of the alkali and alkaline earth metals can overlap to form a band of allowed energy states. Given this information, explain why Ca has a higher melting point (850°C) than K (64°C).

Solution: From the information provided, you know that the $4s$ and $4p$ atomic orbitals of Ca and K are involved in metallic bonding. Each metal atom contributes one $4s$ and three $4p$ atomic orbitals to the molecular orbital structure of the metal. If you assume that the molecular orbitals formed by the $4s$ and $4p$ atomic orbitals all overlap, then a continuous band of allowed energy states occurs. Because each Ca atom contributes two valence electrons ($4s^2$) to the valence band—compared to one valence electron ($4s^1$) contributed by K in its valence band—more molecular orbitals are occupied in the Ca valence band than in the K valence band. Since the melting point of a metal is directly related to the number of occupied bonding molecular orbitals in the molecular-orbital structure of the metal, Ca has a higher melting point than K.

EXERCISE 7 Beryllium has the electron configuration $1s^2 2s^2$. We might expect beryllium *not* to be a conductor of electricity because the $2s$ valence band appears to be full. Yet beryllium is a metallic conductor. Provide an explanation for this observation.

    Solution: As the problem states, if the highest filled valence band were composed *only* of the $2s$ orbitals of beryllium, the energy band would be filled. To account for the metallic conduction property of beryllium, Be must "expand" the number of allowed energy levels in the valence conduction band, so that some of the higher energy orbitals are empty. If both the $2s$ and $2p$ orbitals of beryllium form energy bands that *overlap,* then the band resulting is only one-quarter filled. Thus electrons can move from the lower filled energy levels to the higher empty orbitals, and this results in beryllium having the capacity to conduct current.

Alloys are mixtures of elements with characteristic properties of metals. Several types of alloys are discussed:

    ▪ **Solution alloys** are homogeneous mixtures with the components randomly dispersed.

    ▪ **Substitutional alloys** are homogeneous mixtures with atoms of the solute occupying positions normally occupied by the solvent atoms.

    ▪ **Interstitial alloys** are homogeneous mixture with the solute occupying interstitial positions between the solvent atoms.

    ▪ **Heterogeneous alloys** are heterogeneous mixtures.

    ▪ **Intermetallic compounds** are homogeneous alloys with specific properties and composition, such as duralumin, $CuAl_2$.

    Steels are interstitial alloys containing carbon and iron.

    ▪ As the percentage of carbon increases from less than 0.2% to 1.5%, the durability and toughness of the steel increases: mild steel (0.2%); medium steels (0.2–0.6%); high-carbon steel (0.6–1.5%).

    ▪ If an element other than carbon, such as chromium is added, an *alloy steel* is formed. Stainless steel is an alloy of 0.4% carbon, 18% chromium, 1% nickel, and the rest iron.

    Intermetallic compounds have an important role in modern technology. Examples discussed in the text include.

    ▪ Ni_3Al as a major component of jet aircraft engines.
    ▪ Cr_3Pt as a coating for the new, very hard razor blades.
    ▪ Co_5Sm as a permanent magnet.

EXERCISE 8 How could you determine whether a knife blade is made from high-carbon steel or an alloy steel?

    Solution: A knife blade made from high-carbon steel is easily sharpened but quickly loses its sharpness after being used, whereas one made from an alloy steel has the opposite properties. An alloy steel blade is more wear-resistant and harder than one of the high-carbon steel.

| ALLOYS

EXERCISE 9   Why is the interstitial element in an interstitial alloy typically a non-metal that can participate in bonding to the metal?

Solution: For the interstitial element to occupy the interstitial space between the solvent atoms, it must have a smaller radius. As we learned in Section 7.6 of the text, atomic radius tends to decrease from left to right across a period for the representative elements. Thus nonmetals have sufficiently small covalent radii to allow them to occupy interstitial positions.

PHYSICAL PROPERTIES OF TRANSITION ELEMENTS

The term **transition elements** refers to those metals whose outer d orbitals are being filled. Inserted between lanthanum and hafnium are the **inner transition elements,** the lanthanides and actinides. For these metals, the f orbitals are being filled.

Many of the physical properties of transition elements depend upon their electronic structure.

- You should review Section 6.7 to recall the order of orbital energies when electrons are added to nuclei to form atoms.

- In particular, you should note that the $(n-1)d$ and ns orbitals are very close in energy; therefore it should not be unexpected that in a few transition elements an electron may occupy a $(n-1)d$ orbital instead of a ns orbital as in Cr ($3d^5 4s^1$ instead of $3d^4 4s^2$).

- Also, you should remember that the ns electrons are lost before the $(n-1)d$ electrons when transition metals form ions.

- The first ionization energies of the first-row transition elements increase from left to right across the period; atomic radii decrease in the same direction. Both trends can be understood by recalling (Section 6.7) that the effective nuclear charge experienced by the outer electrons of an atom increases from right to left across a period and that this increase outweighs the greater repulsions among the d electrons.

The radii of the elements of the fifth period are greater than those of the fourth period, as expected.

- However, the transition elements of the sixth period have almost the same radii as the elements immediately above them in the fifth period. This effect is referred to as the **lanthanide contraction.**

- The fourteen lanthanide elements, with their poorly shielding f electrons, occur before hafnium. Therefore, the outer electrons of hafnium and the elements immediately following it experience a higher effective nuclear charge than would normally be anticipated; this results in smaller radii than expected.

Transition elements exhibit a variety of oxidation states.

- You should look at Figure 24.25 in the text and note the most common oxidation states of the first-row transition elements.

- Also you should recognize that the most common oxidation states of elements in the solid state are not always the same as those found in solution.

Manganese is used as an example in the text: Mn^{4+} is *not common in solution,* but it is found in the very stable MnO_2.

▪ Some observations and trends in the oxidation states of the first-row transition elements are:

1. The maximum oxidation state for Sc through Mn equals the total number of outer *s* and *d* electrons.

2. In general, the highest oxidation states are found when the most electronegative elements (O, F, and sometimes Cl) are combined with transition elements.

3. With rare exceptions, the +2 oxidation state for the elements is found in nature only in the combined state.

4. In water, metal ions act as Lewis acids and combine with water molecules to form aquated ions: $M(H_2O)_x^{n+}$.

5. Elements from the early part of a series exhibit higher oxidation potentials than those at the end.

EXERCISE 10   For each of the following elements state the most common oxidation state(s) and the valence-electron configuration for the element in each oxidation state: V; Mn; Co; Zr; and Hg.

  Solution: Refer to Figure 24.25 in the text to determine the most common oxidation states. Then use the principles learned in Sections 6.9 and 8.2 to determine which valence electrons are lost when transition elements form ions.

V^{5+}: All of Vanadium's $3d^3 4s^2$ valence electrons are lost to form the [Ar] electron
     configuration
V^{4+}: $3d^1$
Mn^{7+}: All of Manganese's $3d^5 4s^2$ valence electrons are lost to form the [Ar] electron
     configuration
Mn^{4+}: $3d^3$
Mn^{2+}: $3d^5$
Co^{3+}: $3d^6$
Co^{2+}: $3d^7$
Zr^{4+}: All of Zirconium's $4d^2 5s^2$ valence electrons are lost to form the [Kr] electron
     configuration
Hg^{2+}: $5d^{10}$
Hg^+: $5d^{10} 6s^1$

EXERCISE 11   Why is the density of Ta about twice as great as that of Nb, whereas the difference in densities between V and Nb is much smaller?

  Solution: Density equals mass divided by volume (proportional to atomic radius). In going down a given family of the periodic table we find atomic masses rapidly increase and atomic radii increase at a moderate rate. Thus we would expect the rate of increase in atomic masses compared to the smaller increase in atomic radii to dominate the trend in densities. We find that this is the case for Nb compared to V: The density of Nb is slightly greater. When we compare Nb to Ta we must consider the effect of the lanthanide contraction. Both Nb and Ta have the same atomic radius because of this effect. At the same time, Ta has a far greater atomic mass than Nb (178 amu compared to 91 amu). Because the atomic radii for both elements are the same, the great difference in atomic masses results in a large difference in densities: Ta is about twice as dense as Nb.

EXERCISE 12   Which should be a better oxidizing agent in H_2O, Co^{2+} or Co^{3+}?

  Solution: In general the higher the oxidation state of an element, the better it is as an oxidizing agent. Therefore Co^{3+} should be the better oxidizing agent in H_2O. Standard reduction potentials for Co^{2+} (−0.28 V) and Co^{3+} (1.8 V) in acid solution also demonstrate that Co^{3+} is more easily reduced (that is, a better oxidizing agent) than Co^{2+}.

In Section 9.7 you learned that the presence of an unpaired electron in the valence orbitals of an atom causes the atom to possess a magnetic moment.

- **Paramagnetic substances** possess unpaired electrons and are drawn into a magnetic field.
- When all electrons of an atom are paired, it exhibits diamagnetic behavior. **Diamagnetic substances** are repelled by a magnetic field.

Two types of paramagnetism are discussed in the text: simple paramagnetism and ferromagnetism.

- A substance that exhibits **simple paramagnetism** develops a temporary magnetic moment in the presence of a magnetic field. When the magnetic field is turned off, the orientations of the magnetic moments of the individual metal atoms in the substance become randomly oriented; thus, the substance no longer possesses a magnetic moment.
- Metals of the iron triad (Fe, Co, Ni) exhibit **ferromagnetism** in the uncombined state. In the presence of a magnetic field, the magnetic moments of individual metal atoms become permanently magnetized—the electron spins of the unpaired electrons permanently orient themselves. The magnitude of ferromagnetism can be as much as a million times greater than that of simple paramagnetism.

EXERCISE 13   Would you expect the atoms in a ferromagnetic substance to be extremely close to one another?
   Solution: No. If the atoms were extremely close to one another, then the unpaired electrons of adjacent atoms could pair with one another. The atoms of a ferromagnetic substance must be sufficiently far apart so unpaired electrons do not pair, but they also must not be so far apart that the unpaired electrons on adjacent atoms cannot interact cooperatively.

EXERCISE 14   Would you expect the following isolated ions to be diamagnetic or paramagnetic in the presence of a magnetic field: Ti^{3+}, Co^{2+}, and Zn^{2+}?
   Solution: Write the valence electron configuration of each ion and determine if any of the electrons are unpaired or if they are all paired. Ti^{3+}: $3d^1$. Paramagnetic—one unpaired electron. Co^{2+}: $3d^7$.   Paramagnetic—three unpaired electrons. Zn^{2+}: $3d^{10}$. Diamagnetic—all of the d electrons are paired.

Chromium
   Source: Chromite, $FeCr_2O_4$ (a mixed oxide of formula $FeO \cdot Cr_2O_3$)
   Use: Chromium-based alloys (~60 percent) and 20 percent into other uses
   Reaction in dilute H_2SO_4:

$$Cr(s) + 2H^+(aq) \longrightarrow Cr^{2+}(aq) \text{ [blue]} + H_2(g)$$
$$4Cr^{2+}(aq) + O_2(g) + 4H^+(aq) \longrightarrow Cr^{3+}(aq) + 2H_2O(l) \qquad \text{[rapid]}$$

In HCl, the green complex $Cr(H_2O)_4Cl_2^+$ forms.
   Frequent form in aqueous state: +6 oxidation state.

Basic solution: Yellow chromate ion, CrO_4^{2-}

Acidic solution: orange dichromate ion, $Cr_2O_7^{2-}$

Iron

Source: Hematite or magnetite

Uses: Steels and alloys

Reaction in warm dilute oxidizing acid (HNO_3):

$$Fe(s) + NO_3^-(aq) + 4H^+(aq) \longrightarrow Fe^{3+}(aq) + NO(g) + 2H_2O(l)$$

Frequent forms in aqueous state: Either as +2 or +3 ion. Air oxidation of +2 ion to +3 ion readily occurs:

$$4Fe^{2+}(aq) + O_2(g) + 4H^+(aq) \longrightarrow 4Fe^{3+}(aq) + 2H_2O$$

or in the presence of HCO_3^-

$$4Fe^{2+}(aq) + HCO_3^-(aq) + O_2(g) \longrightarrow$$
$$2Fe_2O_2(s)[brown] + 8CO_2(g) + 4H_2O(l)$$

A basic solution of $Fe^{3+}(aq)$ yields a gelatinous red-brown precipitate, $Fe_2O_3 \cdot n$ H_2O, although it is commonly written as $Fe(OH)_3$.

Copper

Source: Chalcopyrite, $CuFeS_2$

Use: Alloys (bronze, Cu and Sn; brass, Cu and Zn)

Reaction in dilute oxidizing acid:

$$3Cu(s) + 8HNO_3(aq) \longrightarrow 3Cu(NO_3)_2(aq) + 2NO(g) + 4H_2O(l)$$

Frequent forms in aqueous state: Primarily as the +2 ion, but Cu also exists in the +1 ion state. Salts of Cu^+ are usually insoluble in water and mostly white. Also Cu^+ readily disproportionates to Cu^{2+} and $Cu(s)$; thus Cu^{2+} is more common. Salts of Cu^{2+} are often water-soluble. $CuSO_4 \cdot 5H_2O$ (copper sulfate pentahydrate, or blue vitrol) contains four H_2O molecules strongly bound to Cu^{2+}; one is weakly attracted. $Cu(OH)_2$ [blue] forms in basic solution. It readily loses water on heating to form black $CuO(s)$. CuS [black] is one of the least soluble Cu(II) compounds. It will dissolve only in a strong oxidizing acid such as HNO_3.

$$3CuS(s) + 2NO_3^-(aq) + 8H^+(aq) \longrightarrow$$
$$3Cu^{2+}(aq) + 3S(s) + 2NO(g) + 4H_2O(l)$$

EXERCISE 15 (a) Will an aqueous solution of acidified $CrCl_2$ remain light blue with the passage of time? (b) What is the most common form of chromium in a basic solution? What ion forms when a basic solution of chromium is acidified?

Solution: (a) Chromium (II) is readily oxidized to the violet-colored chromium (III) ion in the presence of air. (b) In basic solution the chromate ion, CrO_4^{2-}, is the most stable ion form; in acid it becomes the dichromate ion, $Cr_2O_7^{2-}$.

EXERCISE 16 A solution containing Fe^{2+} is first treated with nitric acid and then evaporated. A colorless gas evolves. The residue is dissolved in water and treated with

6 M NaOH. What is the species formed upon treatment with NaOH? What is the effect of adding HNO_3? What is the colorless gas?

Solution: Addition of nitric acid causes Fe^{2+} to be oxidized to Fe^{3+}:

$$3Fe^{2+}(aq) + 4H^+(aq) + NO_3^-(aq) \longrightarrow 3Fe^{3+}(aq) + NO(g) + 2H_2O(l)$$

As the equation shows, NO should evolve upon evaporation; therefore, NO must be the colorless gas. The Fe^{3+} residue dissolves in water and forms $Fe(OH)_3(s)$ upon addition of 6 M NaOH.

EXERCISE 17   CuS dissolves in nitric acid to form sulfur and $NO(g)$. Write a chemical equation that shows this reaction.

Solution: The information in the question tells us that the NO_3^- ion is reduced to $NO(g)$ and that S^{2+} in CuS is oxidized to sulfur. A *skeleton* equation would be

$$CuS + H^+ + NO_3^- \longrightarrow Cu^{2+} + S + NO + H_2O$$

Using techniques for balancing redox equations in Sections 20.1 and 20.2, you should be able to show that the balanced equation is

$$3CuS(s) + 8H^+(aq) + 2NO_3^-(aq) \longrightarrow$$
$$3Cu^{2+}(aq) + 3S(s) + 2NO(g) + 4H_2O(l)$$

SELF-TEST QUESTIONS

Key Terms

Having reviewed the key terms listed at the end of Chapter 24 and their applications, identify the following statements as true or false. If a statement is false, indicate why it is incorrect.

24.1   The *lithosphere* contains ores that are mined.

24.2   Sterling silver is an *alloy* of gold and copper (see Table 24.6 in the text).

24.3   *Gangue* is material of value that accompanies a desired metal in a raw ore.

24.4   The *calcination* of $PbCO_3$ yields CO_2 and Pb.

24.5   The *roasting* of the ore of mercury, HgS, in oxygen produces metallic mercury and $SO_2(g)$.

24.6   *Slag* consists mostly of molten silicate minerals.

24.7   *Smelting* involves the melting of materials into a single molten layer.

24.8   *Refining* is the first stage in the recovery of a metal from an ore.

24.9   In the *pyrometallurgy* of iron, Fe_3O_4 is heated strongly in the presence of CO and H_2 to form iron.

24.10   The prefix hydro- in *hydrometallurgy* tells us that hydrometallurgy involves metallurgical reactions in aqueous media.

24.11   A process involved in pyrometallurgy is *leaching*.

24.12   Copper is recovered from ores by the *Bayer process*.

24.13   The more active metals are commonly recovered from their compounds by *electrometallurgy*.

24.14   In the *Downs Cell*, $CaCl_2$ is added to lower the melting point of the cell medium.

24.15   The *Hall process* uses cryolite to recover pure aluminum.

24.16   An *intermetallic compound* is a heterogeneous mixture of metals.

24.17   Because of the *lanthanide contraction*, Zr and Hf have similar atomic radii.

24.18   Typically *ores* contain desired metals in high concentrations.

24.19   A *mineral* is a solid compound containing a metal and is found in nature.

24.20   *Metallurgy* is the science and technology of refining metals.

24.21   Steel is a *solution alloy* because it has carbon dispersed uniformly in its matrix.

24.22   Rapid cooling a liquid alloy can result in the formation of a *heterogeneous alloy*.

Problems and Short-Answer Questions

24.23   Write chemical equations that show how gold is leached from ore by cyanide ions. The thermodynamically stable form of gold in solution is Au^{3+}.

24.24   Zinc is commonly found in nature in the form of ZnS. If ZnS is heated in the presence of oxygen, zinc oxide and sulfur dioxide form. Zinc metal is produced in a blast-furnace operation similar to the one used to reduce iron oxides to iron.

Write chemical reactions to describe the formation of zinc oxide from zinc sulfide and the reduction of zinc oxide in the presence of coke.

24.25 Why does the electrical conductivity of a metal decrease when its temperature is increased?

24.26 What are the anodic and cathodic reactions in the electrolytic cell used in the Hall process?

24.27 What is the purpose of the blast of hot air from the bottom of a blast furnace?

24.28 Magnesium is primarily mined from the sea, which contains 0.13 percent magnesium by weight. Magnesium hydroxide is first precipitated by adding slaked lime, $Ca(OH)_2$, to sea water. The slaked lime is obtained by roasting crushed oyster shells ($CaCO_3$) and then adding a small amount of water. The $Mg(OH)_2$ is neutralized with HCl; from the solution $MgCl_2$ is recovered. Write chemical equations showing how $MgCl_2$ is recovered. How could $MgCl_2$ be further refined to produce $Mg(s)$?

24.29 Why does the roasting of HgS form $Hg(g)$, whereas roasting NiS produces $NiO(s)$ instead of $Ni(g)$?

24.30 Why are scientists investigating new methods for forming metals from metal sulfides other than by roasting?

24.31 Write the valence electron configurations for
(a) Fe   (c) Ru^{3+}
(b) Ag^+ (d) Pt

24.32 Why are the physical properties of the second and third series of transition elements more similar to one another than they are to the first series of transition elements?

24.33 (a) Write a metal oxide compound for manganese in each of these oxidation states: +2, +3, +4, +6, +7. (b) List the metal oxides in order of increasing acidic character.

24.34 Which ion should exhibit a larger magnetic moment, Mn^{2+} or V^{2+}?

24.35 Will Cu^+ undergo disproportionation in an aqueous medium? Standard reduction potentials are 0.158 V for Cu^{2+} to Cu^+ and 0.522 V for Cu^+ to Cu.

24.36 Write reaction equations for (a) the reaction of $MnO_2(s)$ in a strongly basic medium to form MnO_3^{2-}; (b) the reaction of $Ag(s)$ with dilute nitric acid; (c) the decomposition of $Cu_2SO_4(s)$ upon exposure to moisture.

24.37 What are the maximum known oxidation states for
(a) Sc   (d) Co
(b) V   (e) Cu
(c) Mn   (f) Zn

Multiple-Choice Questions

24.38 What is the correct relative order of the four most abundant elements in the earth's crust?
(a) O < Si < Al < Fe
(b) Fe < Al < O < Si
(c) Al < Fe < Si < O
(d) Fe < Al < Si < O
(e) Si < Al < O < Fe

24.39 The flotation process in metallurgy involves which of the following?
(a) Mining of an ore;
(b) concentration of an ore;
(c) reduction of a metal ion;
(d) smelting of iron;
(e) leaching of a component of an ore.

24.40 Which of the following is *not* a reasonable reaction for obtaining a metal by reduction?
(a) $FeO(s) + CO(g) \longrightarrow Fe(g) + CO_2(g)$
(b) $Zn^{2+}(aq) + 2e^- \xrightarrow{electrolysis} Zn(s)$
(c) $TiCl_4(s) + 4Na(s) \longrightarrow 4NaCl(s) + Ti(s)$
(d) $WO_3(s) + 3O_2(g) \longrightarrow W(s) + 3O_3(g)$
(e) $MoO_3(s) + 3H_2(g) \longrightarrow Mo(s) + 3H_2O(g)$

24.41 During the smelting of zinc, one of the reactions in which zinc is formed is illustrated by which of the following?
(a) $ZnO(s) + O_2(g) \longrightarrow Zn(g) + O_3(g)$
(b) $2Zn(s) + O_2(g) \longrightarrow 2ZnO(s)$
(c) $2ZnS(s) + 3O_2(g) \longrightarrow 2ZnO(s) + 2SO_2(g)$
(d) $ZnO(s) + CO_2(g) \longrightarrow ZnCO_3(s)$
(e) $ZnO(s) + CO(g) \longrightarrow Zn(g) + CO_2(g)$

24.42 Which of the following processes is widely used for concentrating sulfide ores?
(a) Bayer method   (d) roasting
(b) electrolysis   (e) smelting
(c) flotation

24.43 Which of the following fourth-period elements should have the highest heat of melting?
(a) K   (d) Cr
(b) Ca   (e) Fe
(c) Ti

24.44 Which of the following elements is likely to occur in its native state?
(a) Cu   (d) Al
(b) Ca   (e) Fe
(c) Mg

24.45 Which of the following should have the highest melting point?
(a) Cs   (d) W
(b) Ba   (e) Os
(c) Ta

24.46 Which of the following types of alloys is stainless steel?
(a) Solution   (d) heterogeneous
(b) substitutional   (e) intermetallic
(c) interstitial

24.47 Which of the following has the highest electrical conductivity?
(a) Sr   (d) KCl
(b) Ni   (e) SO_2
(c) Br_2

24.48 Which of the following exhibits the greatest number of oxidation states?
(a) Zr   (d) Ni
(b) Ti   (e) Mn
(c) V

24.49 Which of the following has the largest atomic radius?

(a) Ti
(b) Cr
(c) Mn
(d) Co
(e) Cu

24.50 Which of the following pairs of atoms has the most similar atomic radii?

(a) Ti, V
(b) Ti, Zr
(c) Cr, Mo
(d) Mo, W
(e) Ni, Pd

24.51 Which of the following atoms has the largest first-ionization energy?

(a) Ti
(b) Mn
(c) Fe
(d) Ni
(e) Zn

24.52 Which of the following ions has all of its electrons paired?

(a) Ti^{2+}
(b) Cr^{3+}
(c) Mn^{2+}
(d) Ni^{2+}
(e) Cu^+

24.53 Which of the following exhibits the property of ferromagnetism?

(a) Ti
(b) Mn
(c) Co
(d) Zn
(e) Hg

24.54 Which of the following is *not* a characteristic of transition elements?

(a) Colorless ions
(b) multiple oxidation states
(c) paramagnetism occurs
(d) *d* orbitals being filled
(e) groups 5B and 6B show maximum number of metallic bonds

SELF-TEST SOLUTIONS

24.1 True. **24.2** False. It is an alloy of Ag and Cu. **24.3** False. It has little value. **24.4** False. PbO_2, not Pb, is produced. **24.5** True. **24.6** True. **24.7** False. The molten solution forms at least two layers. **24.8** False. It involves the process of converting an impure form of the metal recovered from an ore into a more pure product. **24.9** True. **24.10** True. **24.11** False. It is the most important hydrometallurgical process. **24.12** False. Al is recovered. **24.13** True. **24.14** True. **24.15** True. **24.16** False. They are homogeneous mixtures. **24.17** True. **24.18** False. Low concentrations. **24.19** False. Metallurgy involves more than just refining metals. Also involved is mining, extraction, reduction, concentrating and refining. **24.20** True. **24.21** True. **24.22** True.

24.23 $4Au(s) + 8CN^-(aq) + O_2(aq) + 2H_2O(l) \longrightarrow$
$4Au(CN)_2^-(aq) + 4OH^-(aq)$

24.24 $2ZnS(s) + 3O_2(g) \longrightarrow 2Zn(s) + 2SO_2(g)$
$2C(s) + O_2(g) \longrightarrow 2CO(g)$
$Zn(s) + CO(g) \longrightarrow Zn(g) + CO_2(g)$

24.25 With increasing temperature the metal atoms in the solid lattice vibrate more. This vibration disrupts the overlap of atomic orbitals between atoms. There are thus fewer vacant energy bands that electrons can move into under the influence of an electrical field. **24.26** Anode: $C(s) + 2O^{2-}(l) \longrightarrow CO_2(g) + 4e^-$. Cathode: $3e^- + Al^{3+}(l) \longrightarrow Al(l)$. **24.27** The blast of hot air from the bottom causes the carbon to burn to $CO(g)$. More heat is also evolved, assisting in the reduction of iron.

24.28 $CaCO_3(s) \xrightarrow{\Delta} CaO(s) + CO_2(g)$
$CaO(s) + H_2O(l) \longrightarrow Ca(OH)_2(s)$
$Ca(OH)_2(s) + Mg^{2+}(aq) \longrightarrow$
$Mg(OH)_2(aq) + Ca^{2+}(aq)$
$Mg(OH)_2(aq) + 2HCl(aq) \longrightarrow$
$MgCl_2(aq) + 2H_2O(l)$

Evaporate off water from the $MgCl_2$ solution to form $MgCl_2$. Then using an electrolytic cell and molten $MgCl_2$, electrolyze Mg^{2+} to $Mg(l)$. **24.29** Hg is low on the activity series of metals; thus in the presence of $O_2(g)$ it only very slowly forms HgO. Nickel is higher on the activity series. If native Ni were formed in roasting, it would immediately react with oxygen to form $NiO(s)$. **24.30** Roasting of metal sulfides produces $SO_2(g)$ in large quantities. In the presence of water in the atmosphere it forms sulfuric acid—known as acid rain. Federal regulations now require most of the $SO_2(g)$ in stack gases to be trapped. **24.31** (a) $3d^6 4s^2$; (b) $4d^{10}$; (c) $4d^5$; (d) $5d^9 6s^1$. **24.32** The phenomenon of the Lanthanide contraction results in the atomic radii of the second and third series of transition elements being almost the same within each group. There is a significant difference in atomic radii between the first and second series of transition elements; thus their physical properties are more different. **24.33** (a) MnO, Mn_2O_3, MnO_2, MnO_3, Mn_2O_7. (b) Acidic character increases with increasing oxidation number: $MnO < Mn_2O_3 < MnO_2 < MnO_3 < Mn_2O_7$. **24.34** The magnitude of magnetic moment will be proportional to the number of unpaired electrons. Mn^{2+} has five unpaired electrons ($3d^5$), whereas V^{2+} has three unpaired electrons ($3d^3$); thus Mn^{2+} has the larger magnetic moment.

24.35
$$Cu^+ + e^- \longrightarrow Cu \qquad E° = 0.522 \text{ V}$$
$$\underline{Cu^+ \longrightarrow Cu^{2+} + e \qquad E° = -0.158 \text{ V}}$$
$$2Cu^+ \longrightarrow Cu^{2+} + Cu \qquad E° = 0.364 \text{ V}$$

Cu^+ will disproportionate because $E°$ for the reaction is positive. **24.36** (a) $MnO_2(s) + 2OH^-(aq) \longrightarrow MnO_3^{2-}(aq) + H_2O(l)$ (b) The reaction is similar to the one for Cu: $3Ag(s) + 4HNO_3(aq) \rightarrow 3AgNO_3(aq) + NO(g) + 2H_2O(l)$ (c) Cu in Cu_2SO_4 is in the +1 oxidation state, which can disproportionate.

$$Cu_2SO_4(s) \longrightarrow Cu(s) + CuSO_4(s)$$

24.37 (a) Sc^{3+}; (b) V^{5+}; (c) Mn^{7+}; (d) Co^{4+}; (e) Cu^{3+}; (f) Zn^{2+}. **24.38** (d). **24.39** (b). **24.40** (d). **24.41** (e). **24.42** (c). **24.43** (d); Cr possesses six s and d valence electrons: $3d^5 4s^1$. **24.44** (a). **24.45** (d), a Group 6B element. **24.46** (c). **24.47** (b). **24.48** (e). **24.49** (a). **24.50** (d); effect of Lanthanide contraction. **24.51** (d); Zn has a $3d^{10}$ valence electron configuration. Removing an electron would destroy a very stable electron configuration. **24.52** (e), $3d^{10}$. **24.53** (c). **24.54** (a).

Sectional Test 6

CHAPTERS 22–24

You should make this test as realistic as possible. **Tear out this test so that you do not have access to the information in the** *Student's Guide*. Use only data provided in the questions and a periodic table. Do not check your answers until you are finished. If you answer questions incorrectly, review the section in the text indicated after each question.

Choose the best response for each question.

1. Which statement is *not* true?
   (a) Electronegativity tends to decrease down a representative family.
   (b) Metal fluorides and metal oxide compounds tend to be ionic.
   (c) Nitrogen can form a maximum of four bonds whereas phosphorus can form a maximum of six bonds.
   (d) Bonding between atoms becomes more effective with increasing atomic size.
   [22–1]

2. Which element can bond to more than four surrounding elements?
   (a) F (b) As (c) O (d) H
   [22–1]

3. When CH_3NH_2 is combusted in air, all of the following are products *except:*
   (a) $H_2(g)$ (b) $H_2O(l)$ (c) $CO_2(g)$ (d) $N_2(g)$
   [22–2]

4. Which of the following is *not* a strong base?
   (a) CH_3^- (b) H^- (c) NH_3 (d) N^{3-}
   [22–2]

5. Which of the following is tritium, an isotope of hydrogen?
   (a) $_0^1n$ (b) $_1^1H$ (c) $_1^2H$ (d) $_1^3H$
   [22–3]

6. Which statement about hydrogen and its compounds is true?
   (a) Most reactions of H_2 are slow.
   (b) The H—H bond energy is low for a single bond.
   (c) The hydride ion is acidic.
   (d) Hydrogen is an effective oxidizing agent.
   [22–3]

7. Which of the following completes the reaction

$$CaH_2(s) + 2H_2O(l) \longrightarrow Ca(OH)_2(s) + \underline{?}$$

(a) $OH^-(aq)$ (b) $H_2(g)$ (c) $H^-(aq)$ (d) $O^{2-}(aq)$
[22–3]

8. Commercial oxygen is obtained from:
   (a) Decomposition of $KClO_3$ (b) Electrolysis of water
   (c) Decomposition of ozone (d) Fractional distillation of liqui-
   fied air
   [22–4]

9. Which of the following is *not* true in describing the relationship
between O_2 and O_3?
   (a) Both are allotropes.
   (b) Passing electricity through oxygen produces ozone.
   (c) O_3 contains a π bond delocalized over the three oxygen
   atoms, whereas oxygen does not.
   (d) Oxygen is a stronger oxidizing agent.
   [22–4]

10. Which of the following nonmetal oxides is *not* an acidic anhydride?
   (a) CO_2 (b) CO (c) SO_3 (d) SO_2
   [22–4]

11. Which of the following is *not* an ionic peroxide?
   (a) H_2O_2 (b) Na_2O_2 (c) CaO_2 (d) BaO_2
   [22–4]

12. Complete the following reaction:

$$Mg_3N_2(s) + 6H_2O(l) \longrightarrow 2NH_3(aq) + \underline{?}$$

(a) $Mg(NH_2)_2(aq)$ (b) $MgH_2(s)$ (c) $Mg(OH)_2(s)$ (d) $MgO(s)$
[22–5]

13. Which of the following contains nitrogen in a negative oxidation
state?
   (a) NO_2 (b) HNF_2 (c) N_2 (d) NH_2OH
   [22–5]

14. Which nitrogen containing compound bears the same relationship
to ammonia that hydrogen peroxide does to water?
   (a) N_2O_4 (b) N_2O (c) N_2H_4 (d) NH_4^+
   [22–5]

15. Which of the following is laughing gas?
   (a) N_2O (b) NO (c) NO_2 (d) N_2O_4
   [22–5]

16. Which of the following is the first step in the Ostwald process for
producing HNO_3?

(a) $4NH_3(g) + 3O_2(g) \xrightarrow{1000°C} 2N_2(g) + 6H_2O(g)$
(b) $2NO(g) + O_2(g) \longrightarrow 2NO_2(g)$

(c) $NH_4NO_3(s) \xrightarrow{\Delta} N_2O(g) + 2H_2O(g)$

(d) $4NH_3(g) + 5O_2(g) \xrightarrow[800°C]{Pt} 4NO(g) + 6H_2O(g)$
[22–5]

17. Which substance is a soft, black, slippery solid that has a metallic luster and conducts current?

    (a) Graphite (b) Diamond (c) Silicon carbide (d) Carbon black

    [22–6]

18. Which statement is *not* true about $CO(g)$?

    (a) The carbon atom is able to act as a Lewis acid site toward some metals.

    (b) It is an important reducing agent.

    (c) It is used in blast furnaces in steel production operations.

    (d) Its Lewis structure is $:\!\ddot{C}\!\!=\!\!\ddot{O}$.

    [22–6]

19. Which of the following is carborundum, an abrasive?

    (a) CaC_2 (b) SiC (c) SiO_2 (d) C_2H_2

    [22–6]

20. Which noble-gas element has the lowest boiling point of any substance under 1 atm pressure and at 4.2 K?

    (a) He (b) Ne (c) Ar (d) Kr

    [23–1]

21. Which noble-gas element was used by Neil Bartlett in 1962 to prepare a stable fluorine-noble-gas containing compound?

    (a) He (b) Ne (c) Kr (d) Xe

    [23–1]

22. Which trend is correct for the X—X single bond dissociation energies of the diatomic halogens?

    (a) $F_2 > Cl_2 > Br_2 > I_2$ (b) $I_2 > Br_2 > Cl_2 > F_2$

    (c) $F_2 < Cl_2 > Br_2 > I_2$ (d) $F_2 > Cl_2 < Br_2 > I_2$

    [23–2]

23. Which diatomic halogen is a pale yellow-green gas at room temperature and pressure?

    (a) F_2 (b) Cl_2 (c) Br_2 (d) I_2

    [23–2]

24. Which diatomic halogen readily oxidizes water as follows:

$$X_2(aq) + H_2O(l) \longrightarrow 2HX(aq) + \tfrac{1}{2}O_2(g)$$

    (a) F_2 (b) Cl_2 (c) Br_2 (d) I_2

    [23–2]

25. Which halide ion will elemental Br_2 oxidize?

    (a) F^- (b) Cl^- (c) I^- (d) (b) and (c)

    [23–2]

26. Which of the following is the active ingredient in many liquid bleaches?

    (a) Cl_2 (b) $NaClO$ (c) Cl^- (d) $HClO_3$

    [23–2]

27. Which HX compound can *not* be prepared by the reaction of a NaX salt with H_2SO_4?

    (a) HF (b) HCl (c) HBr (d) None of the preceding

    [23–2]

28. What is the geometrical structure of ClF_3?

    (a) Trigonal planar (b) Trigonal pyramid (c) Trigonal bipyramid

    (d) Bent T

    [23–2]

29. Which is the correct trend in X—X single bond energies for group 6A elements?

(a) $O > S > Se > Te$ (b) $O < S < Se < Te$
(c) $O < S > Se > Te$ (d) $O > S < Se < Te$
[23–3]

30. The Frasch process is used to obtain which element?

(a) O (b) S (c) Se (d) Te
[23–3]

31. Which molten element occurs in the form X_8?

(a) O (b) S (c) Se (d) Te
[23–3]

32. Which of the following compounds is the anhydride of H_2SO_3?

(a) SO_2 (b) SO_3 (c) H_2SO_4 (d) Na_2SO_3
[23–3]

33. The term *thio-* means substitution of which element for oxygen?

(a) S (b) Cl (c) P (d) N
[23–3]

34. Which of the following is called "fool's gold"?

(a) AuS_2 (b) CuS (c) FeS_2 (d) NiS
[23–3]

35. Which form of phosphorus is pyrophorric?

(a) red phosphorus (b) white phosphorus (c) $Ca_3(PO_4)_2$ (d) PH_3
[23–4]

36. Which of the following is the missing compound in the reaction
$PCl_5(l) + 4H_2O(l) \longrightarrow 5HCl(aq) + \underline{?}$

(a) $HCl(aq)$ (b) $PH_3(g)$ (c) $H_3PO_4(aq)$ (d) $H_3PO_3(aq)$
[23–4]

37. Which of the following is a polymeric substance formed by a condensation reaction?

(a) H_3PO_4 (b) H_3PO_3 (c) HPO_3 (d) PH_3
[23–4]

38. Which element has a striking ability to undergo catenation?

(a) C (b) Si (c) Ge (d) Sn
[23–5]

39. Which element is the second most abundant element in the earth's crust?

(a) C (b) Si (c) Ge (d) Sn
[23–5]

40. Which of the following ions replaces up to one-half of the silicon atoms in the silicate mineral feldspar?

(a) Cr^{3+} (b) Zn^{2+} (c) Al^{3+} (d) K^+
[23–5]

41. "Photochromic" glasses contain a dispersion of which of the following in lead alkali glass?

(a) $CrCl_3$ (b) AgCl (c) $AuCl_3$ (d) $PtCl_2$
[23–5]

42. Which salt is a good reducing agent?

(a) $NaBH_4$ (b) KCl (c) NH_4Cl (d) Na_3BO_3
[23–6]

43. Which of the following is the correct order of abundance of elements in the lithosphere?

(a) Fe > Al > Si > O (b) O > Al > Fe > Si
(c) O > Si > Al > Fe (d) O > Si > Fe > Al
[24–1]

44. Which element does not occur in a natural elemental state?
(a) K (b) Au (c) Ag (d) Ir
[24–1]

45. Which of the following statements is true?
(a) Extractive metallurgy is concerned with the first step in obtaining a metal from its ore.
(b) Gangue is a type of desired mineral.
(c) Flotation is a form of pyrometallurgy.
(d) Surfaces of many mineral materials are hydrophobic.
[24–1, 24–2]

46. Which of the following represents a calcination process?
(a) $2ZnS(s) + 3O_2(g) \longrightarrow 2ZnO(s) + 2SO_2(g)$
(b) $PbCO_3(s) \longrightarrow PbO(s) + CO_2(g)$
(c) $TiC(s) + 4Cl_2(g) \longrightarrow TiCl_4(g) + CCl_4(g)$
(d) $PbO(s) + CO(g) \longrightarrow Pb(l) + CO_2(g)$
[24–2]

47. A slag containing CaO is
(a) Acidic (b) Basic (c) Neutral (d) Contains a molten metal
[24–2]

48. Major sources of iron are ores rich in
(a) Iron pyrite (b) Iron sulfates (c) Hematite (d) Iron carbonate
[24–2]

49. What is the missing compound in the following reduction reaction?
$Fe_3O_4(s) + \underline{?} \longrightarrow 3Fe(s) + 4CO_2(g)$
(a) C(s) (b) CO(g) (c) $CO_2(g)$ (d) $CO_3(g)$
[24–2]

50. The most important hydrometallurgical process is
(a) flotation (b) calcination (c) roasting (d) leaching
[24–3]

51. Which of the following ions is used to solubilize metallic gold?
(a) F^- (b) Cl^- (c) CN^- (d) SO_4^{2-}
[24–3]

52. Which of the following is bauxite?
(a) $Al(NO_3)_3$ (b) $AlCl_3$ (c) $Al_2(SO_4)_2 \cdot xH_2O$ (d) $Al_2O_3 \cdot xH_2O$
[24–3]

53. Electrometallurgy is used to produce commercially which element?
(a) Fe (b) Na (c) O_2 (d) N_2
[24–4]

54. Which of the following is *not* a property of metal?
(a) Ductile (b) Malleable (c) Lustrous appearance
(d) Low melting points
[24–5]

55. Which of the following accounts for the conductivity of metals?
(a) Movement of ions (b) Strong ionic bonds
(c) Mobility of valence electrons (d) High melting points
[24–5]

56. Which of the following is usually characteristic of a solid with a filled energy band?

(a) Insulator (b) Conductor (c) Ionic (d) Covalent
[24–5]

57. Yellow brass, an alloy, contains mostly copper and in a lesser quantity the following element:

(a) Ni (b) Zn (c) Pb (d) Sn
[24–6]

58. Which order of atomic size is correct?

(a) Sc < Y > La   (c) Ti < Zr < Hf

(b) Sc > Y > La   (d) Ti < Zr $\approx$ Hf
[24–7]

59. What is the electron configuration of Ni^{2+}?

(a) $[Ar]3d^64s^2$ (b) $[Ar]3d^8$ (c) $[Ar]3d^54s^1$ (d) $[Ar]3d^44s^2$
[24–7]

60. For which element is +3 oxidation state commonly observed?

(a) Cr (b) Mo (c) W (d) Ni
[24–7]

61. What is the maximum oxidation state observed for the elements Sc to Mn?

(a) +3 (b) +4 (c) +5 (d) +7
[24–7]

62. Which form of chromium in the +6 oxidation state is favored in strongly acidic solutions?

(a) CrO_4^{2-} (b) $HCrO_4^-$ (c) Cr^{6+} (d) $Cr_2O_7^{2-}$
[24–7]

63. Which is the most stable oxidation state of copper?

(a) +1 (b) +2 (c) +3 (d) +4
[24–7, 24–8]

64. Which of the following is characteristic of the property of paramagnetism?

(a) All electrons are paired.

(b) The substance is repelled by a magnetic field.

(c) Magnetic moments are arranged in a regular array.

(d) The interaction of a magnetic field with the substance is proportional to the number of unpaired electrons.
[24–7]

65. Which of the following elements does not commonly form ferromagnetic materials?

(a) Ba (b) Fe (c) Co (d) Ni
[24–7, 24–8]

ANSWERS

1. (d).   **2.** (b).   **3.** (a).   **4.** (c).   **5.** (d).   **6.** (a).   **7.** (b).   **8.** (d).   **9.** (d).
10. (b).   **11.** (a).   **12.** (c).   **13.** (d).   **14.** (c).   **15.** (a).   **16.** (d).   **17.** (a).   **18.** (d).
19. (b).   **20.** (a).   **21.** (d).   **22.** (c).   **23.** (b).   **24.** (a).   **25.** (c).   **26.** (b).   **27.** (c).
28. (d).   **29.** (c).   **30.** (b).   **31.** (b).   **32.** (a).   **33.** (a).   **34.** (c).   **35.** (b).   **36.** (c).
37. (c).   **38.** (a).   **39.** (b).   **40.** (c).   **41.** (b).   **42.** (a).   **43.** (c).   **44.** (a).   **45.** (d).
46. (b).   **47.** (b).   **48.** (c).   **49.** (b).   **50.** (d).   **51.** (c).   **52.** (d).   **53.** (b).   **54.** (d).
55. (c).   **56.** (a).   **57.** (b).   **58.** (d).   **59.** (b).   **60.** (a).   **61.** (d).   **62.** (b).   **63.** (b).
64. (d).   **65.** (a).

Chemistry of Coordination Compounds 25

OVERVIEW OF THE CHAPTER

Review: Complex ions (16.8); Lewis bases (16.10); Lewis structures (8.5); oxidation numbers (8.10).

Objective: You should be able to determine either the charge of a complex ion, having been given the oxidation state of the metal, or the oxidation state, having been given the charge of the complex. (You will need to recognize the common ligands and their charges.)

Review: Nomenclature (8.10).

Objective: You should be able to name coordination compounds, having been given their formulas, or write their formulas, having been given their names.

Review: Molecular shapes (9.1).

Objectives: You should be able to:

1. Describe, with the aid of drawings, the common geometries of complexes. (You will need to recognize whether the common ligands are functioning as monodentate or polydentate ligands.)

2. Describe the common types of isomerism and distinguish between structural and stereoisomerism.

3. Determine the possible number of stereoisomers for a complex, having been given its composition.

Review: Activation energy (14.4); Planck's relationship (6.2); rates (14.1).

Objectives: You should be able to:

1. Distinguish between inert and labile complexes.

2. Explain how the conductivity, precipitation reactions, and isomerism of complexes are used to infer their structures.

25.1, 25.2
COORDINATION
COMPOUNDS:
TERMINOLOGY

25.3 NAMING OF
COORDINATION
COMPOUNDS

25.4 THE
STRUCTURAL
CHEMISTRY OF
COORDINATION
COMPOUNDS:
ISOMERISM

25.5, 25.6
EXCHANGE RATES,
MAGNETISM, AND
COLORS OF
TRANSITION METAL
COMPLEXES

3. Explain how the magnetic properties of a compound can be measured and used to infer the number of unpaired electrons.

4. Explain how the colors of substances are related to their absorption and reflection of incident light.

25.7 BONDING IN TRANSITION METAL COMPLEXES: CRYSTAL-FIELD THEORY

Review: Nature of salts (8.2); shapes of $3d$ orbitals (6.6).

Objectives: You should be able to:

1. Explain how the electrostatic interaction between ligands and metal d orbitals in an octahedral complex results in a splitting of energy levels.

2. Explain the significance of the spectrochemical series.

3. Account for the tendency of electrons to pair in strong-field, low-spin complexes.

4. Sketch a representation of the d orbital energy levels in a tetrahedral complex and explain the reason for a smaller crystal-field splitting in this geometry as compared to octahedral complexes.

5. Sketch the d orbital energy levels in a square-planar complex.

TOPIC SUMMARIES AND EXERCISES

COORDINATION COMPOUNDS: TERMINOLOGY

The term **complex ion** (or simply, complex) refers to any charged species that contains a central metal atom or metal ion that is bonded to other atoms, ions, or molecules.

▪ Groups bonded to the central metal of a metal complex ion are Lewis bases: these groups are known as **ligands.**

▪ For example, the salt $K_3[Fe(CN)_6]$ contains the complex ion $Fe(CN)_6^{3-}$. Six cyanide ions (ligands) surround and bind to the Fe^{3+} ion to form $Fe(CN)_6^{3-}$. The salt $K_3[Fe(CN)_6]$ is called a **coordination compound** because it contains a metal complex ion.

The bond between each cyanide ion and the central Fe^{2+} ion in $Fe(CN)_6^{3-}$ can be pictured as a polar bond in which the bonding pair of electrons comes entirely from the ligand:

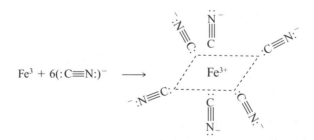

▪ The central atom and surrounding ligands in a complex form the **coordina-**

tion sphere. In $K_3[Fe(CN)_6]$, the Fe^{3+} ion and the six cyanide ions constitute the coordination sphere. In the formula $K_3[Fe(CN)_6]$, the coordination sphere is identified by placing brackets about $Fe(CN)_6^{3-}$.

▪ Atoms of a ligand that bind to the central atom of a complex are referred to as **donor atoms.** In $[Fe(CN)_6]^{3-}$, the donor atoms are the six carbon atoms from the six cyanide groups.

▪ The *total number of donor atoms* bonded to a central metal atom in a complex is known as the **coordination number** of the metal atom. In $[Fe(CN)_6]^{3-}$, iron has a coordination number of six.

Ligands are classified according to the manner in which they bind to the central atom in a complex.

▪ Ligands such as F^-, Cl^-, Br^-, and NH_3 are said to be **monodentate ligands** because only one donor atom participates in the bond to the central atom.

▪ Other ligands, such as

$$H_2NCH_2CH_2NH_2 \qquad CH_3CCHCCH_3 \qquad H_2NCH_2CH_2NHCH_2CH_2NH_2$$
$$\qquad\qquad\qquad\qquad \underset{O}{\overset{\|}{}} \;\; \underset{O^-}{\overset{|}{}}$$

   (ethylenediamine)   (anion of acetylacetone)   (diethylenetriamine)

attach themselves to the central metal atom of a complex using two or more of their donor atoms; these are referred to as **polydentate ligands.**

▪ The ligands ethylenediamine, diethylenetriamine, and the anion of acetylacetone are also referred to as **chelating agents** because their donor atoms bind simultaneously to the same central atom:

$$
\begin{array}{c}
M \\
\nearrow \;\; \nwarrow \\
D \quad\qquad D \\
\smile \\
\end{array}
$$

(D = donor atom)

▪ Note that the metal atom, the donor atoms, and all atoms bonded to one another between the donor atoms in the ligand form a ring system known as a **chelate ring.** We experimentally find that ligands with donor atoms separated by two or three other atoms make the best chelating ligands and form the most stable complexes. Therefore, chelate rings containing a total of five or six atoms tend to be most stable.

EXERCISE 1   Determine the oxidation number of the metal atom in each of the following: **(a)** $[Cr(NH_3)_6](NO_3)_3$; **(b)** $K_3[Fe(CO)(CN)_5]$; **(c)** $[Zn(NH_3)_2Cl_2]$.

   Solution: The oxidation number of the central metal atom in a complex is calculated from the relationship:

      charge of complex = oxidation number of metal atom +

                  sum of charges of ligands

or

$$\text{oxidation number of metal atom} = \text{charge of complex} -$$
$$\text{sum of charges of ligands}$$

(a) First, you need to determine the charge of the complex, which is the species containing the metal atom in brackets. The complex, $[Cr(NH_3)_6]^{n+}$ must have a positive charge (cation) because the counter ion is NO_3^-, the anion of the compound. The complex ion must have a 3+ charge because each NO_3^- group has a 1− charge: $[Cr(NH_3)_6]^{3+}$. Next, determine the charge of all ligands; in this case, the only ligand is ammonia, and it is a neutral molecule, of charge zero. Let n represent the oxidation number of chromium:

$$[Cr(NH_3)_6]^{3+}$$
$$1(n) + 6(0) = +3$$
$$n + 0 = +3$$
$$n = +3$$

The oxidation number (and ion charge) of chromium is therefore +3. **(b)** Since each potassium ion in $K_3[Fe(CN)_5(CO)]$ possesses a 1+ charge, the complex $[Fe(CN)_5(CO)]^{3-}$ must possess a 3− charge to maintain charge balance. Each cyanide ion has a 1− charge, and the CO group possesses a zero charge. You can calculate the oxidation number of iron as follows:

$$[Fe(CN)_5(CO)]^{3-}$$
$$(1)(n) + 5(-1) + 1(0) = -3$$
$$n - 5 + 0 = -3$$
$$n = +2$$

Therefore the oxidation state of iron is +2. **(c)** Proceeding in a manner similar to that used in **(a)** and **(b)**, you can calculate the oxidation number of zinc as follows:

$$[Zn(NH_3)_2Cl_2]$$
$$(1)(n) + 2(0) + 2(-1) = 0$$
$$n + 0 - 2 = 0$$
$$n = 2$$

Therefore, zinc has an oxidation number of +2.

EXERCISE 2 Draw the Lewis structures of the following ligands and indicate the donor atoms in each: **(a)** NH_3; **(b)** CO; **(c)** oxalate ion, $C_2O_4^{2-}$; **(d)** ethylenediamine; **(e)** Br^-. Also identify each ligand in **(a)**–**(d)** as a monodentate ligand or a chelating agent.

Solution: **(a)** The Lewis structure for NH_3 is

$$H-\overset{\displaystyle ..}{N}-H$$
$$|$$
$$H$$

The nitrogen atom is the donor atom because it possesses a nonbonding electron pair capable of donating electron density to another atom. Since there is only one donor atom, it is a monodentate ligand. **(b)** The Lewis structure for CO is

$$:C\equiv O:$$

Both carbon and oxygen have nonbonding electron pairs, and therefore both can act as donor atoms. However, only one atom is capable of donating to the same metal atom; it is impossible for the other atom to bend and attach itself to the same metal atom. Carbon monoxide usually acts as a monodentate ligand, with carbon behaving as the donor atom. In a few cases, carbon monoxide binds to two different metal atoms:

$$M—CO—M'$$

(c) The Lewis structure for the oxalate ion is

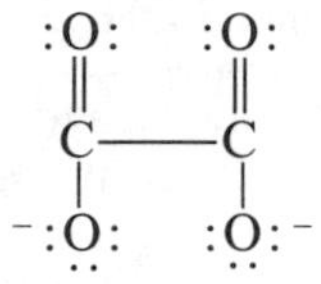

The oxalate ion acts as a chelating agent:

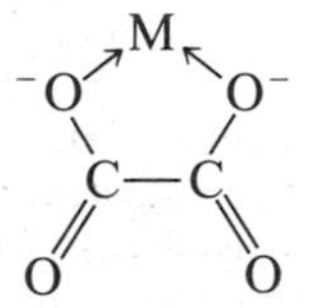

(d) The Lewis structure for ethylenediamine is

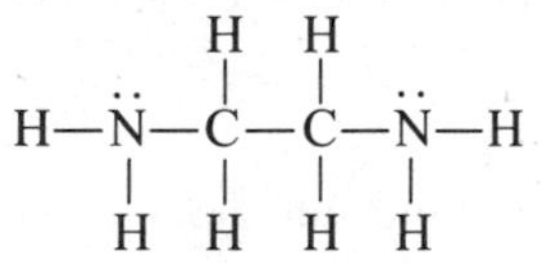

Both nitrogen atoms possess nonbonded electron pairs and therefore can behave as Lewis bases. Ethylenediamine is a chelating agent:

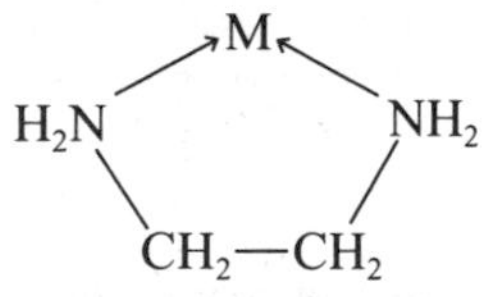

(e) The Lewis structure for Br^- is

$$:\ddot{Br}:^-$$

Although there are four lone pairs available for bonding to a metal atom, only one pair actively participates in bonding.

$$M \longleftarrow :\ddot{Br}:^-$$

EXERCISE 3 What is the coordination number of the central metal atom in each of the following complexes: **(a)** $[Zn(NH_3)_2Cl_2]$; **(b)** $[Ni(en)_3]^{2+}$, where "en" is an abbreviation for ethylenediamine; **(c)** $[Cr(H_2O)_2Cl_2]^+$?

Solution: **(a)** Four, because each ligand has only one donor atom. **(b)** Six. Each ethylenediamine has two donor atoms, and there are three ethylenediamine ligands. **(c)** Four. Each ligand has only one donor atom.

The rules for naming complex ions and compounds are given in Section 25.3. After reviewing the rules, you should do Exercises 4–7, which are designed to help you review and apply the rules to specific compounds and ions.

| NAMING OF
| COORDINATION
| COMPOUNDS

EXERCISE 4   Before you can name complexes, you must be able to name ligands. Name the following ligands: (a) F^-; (b) NO_3^-; (c) O^{2-}; (d) $C_2O_4^{2-}$; (e) CO_4^{2-}; (f) SO_4^{2-}; (g) C_5H_5N (pyridine).

Solution: Names of anionic ligands are derived from the commonly used name of the anion by dropping an *-ide* ending or the *-e* of an *-ate* or *-ite* ending and adding *-o*. Neutral ligands are named as the neutral molecule, except for H_2O, aqua, and NH_3, ammine. Two other important exceptions, not listed in Table 25.1 of the text, are CO, carbonyl, and NO, nitrosyl. (a) Fluoro (derived from *fluor- + o*); (b) nitrato (derived from *nitrat- + o*); (c) oxo (derived from *ox- + o*); (d) oxalato (derived from *oxalat- + o*); (e) carbonato (derived from *carbonat- + o*); (f) sulfato (derived from *sulfat- + o*); (g) pyridine (a neutral molecule, and not a special case).

EXERCISE 5   Name the following complex cations and neutral coordination compounds: (a) $[Ag(NH_3)_2]^+$; (b) $[Co(en)_3]^{3+}$; (c) $[Ni(CO)_4]$; (d) $[Cr(H_2O)_4Cl_2]^+$.

Solution: Complex cations and neutral coordination compounds are named by first listing the ligands, in alphabetical order and with appropriate prefixes to indicate their number, and then naming the central metal, using its name as given in the periodic table, followed by its oxidation number in parentheses. See rule 4 in Section 25.3 for use of prefixes indicating number of ligands. (a) Diamminesilver(I) ion. *Note:* NH_3 is the only ligand containing an "ammine" with two *m*'s; all others have one *m*. (b) tris(ethylenediamine)cobalt(III) ion (because ethylenediamine contains the prefix di-, its name is enclosed in parentheses and the alternate prefix tris- is used); (c) tetracarbonylnickel(0); (d) tetraaquadichlorochromium(III) ion (note that neither the terminal "a" of tetra nor the "a" of aqua is dropped).

EXERCISE 6   Name the following anion complexes: (a) $[FeCO(CN)_5]^{3-}$; (b) $[Ag(CN)_2]^-$; (c) $[PtCl_6]^{2-}$.

Solution: The rules for naming an anion complex are the same as those discussed in Exercise 5, except that the name of the central metal ends in -ate and in certain cases the Latin stem names of some metals are used. Examples of these are: Pt (platinate), Fe (ferrate), Ag (argentate), Sn (stannate), Pb (plumbate), Cu (cuprate), and Au (aurate). (a) Carbonylpentacyanoferrate(II) ion; (b) dicyanoargentate(I) ion. (c) hexachloroplatinate(IV) ion.

EXERCISE 7   Name the following coordination compounds: (a) $K[Ag(CN)_2]$; (b) $[Co(en)_3]Cl_3$; (c) $Na[Cr(en)_2(SO_4)_2]$ (en = ethylenediamine).

Solution: Remember that cations of salts are named before the anions. (a) Potassium dicyanoargentate(I); (b) tris(ethylenediamine)cobalt(III) chloride. (c) sodium bis(ethylenediamine)disufatochromate(III).

THE STRUCTURAL CHEMISTRY OF COORDINATION COMPOUNDS: ISOMERISM

Although complexes containing metal atoms with integral coordination numbers from two to six and greater are observed, the great majority of metal atoms have coordination numbers of four or six, with six the most common one.

- Two structural configurations are commonly observed for four-coordinate complexes: tetrahedral and square planar. Tetrahedral transition metal complexes occur more commonly than square-planar ones. Square-planar transition metal complexes primarily occur with a metal atom possessing eight valence-shell electrons.

- The geometry of six-coordinate complexes usually corresponds to six coordinated atoms at the center of an octahedron or a distorted octahedron. In the idealized octahedral case, all atoms are an equivalent distance from the central atom.

In Section 25.4 several classes of isomerism are discussed. These are briefly summarized in the following:

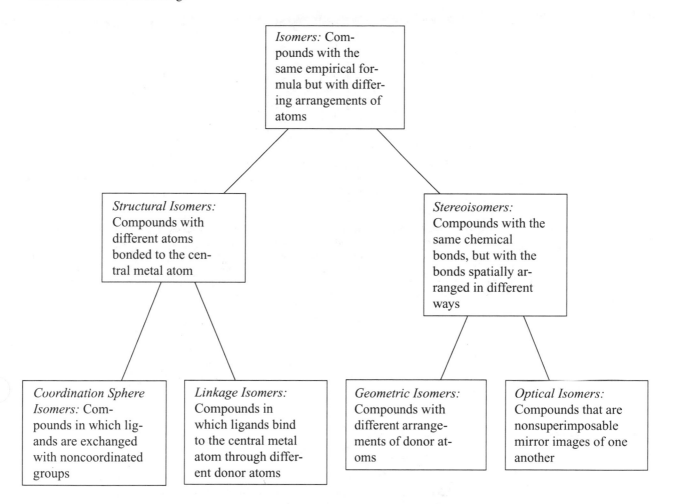

Stereoisomers form the most important class of isomers.

- An important form of geometrical isomerism is *cis-trans* isomerism. In a *cis* isomer like groups are in adjacent positions; in a *trans* isomer the like groups are across from one another in a diagonal relationship. Octahedral and square planar complexes may show geometric isomerism when two or more different ligands are present, but *cis-trans* geometrical isomerism is not observed in tetrahedral complexes.

- Optical isomerism exists when mirror images of a compound cannot be superimposed on each other. Nonsuperimposable mirror images are referred to as chiral. Exercise 10 shows you how to draw a pair of mirror images and to determine if they are superimposable.

- A solution of a chiral compound will rotate the plane of polarized light: Dextrorotatory optical isomers rotate the plane to the right, and levorotatory rotate it to the left. Equal amounts of a pair of chiral compounds is racemic; the racemic mixture does not rotate the plane of polarized light.

EXERCISE 8   Pt(II) complexes are typically square-planar. **(a)** Sketch the geometrical isomers of [Pt(NH$_3$)$_2$Cl$_2$]. **(b)** An important type of geometrical isomerism is *cis-trans* isomerism. What is *cis-trans* isomerism? Identify which exist for the structures drawn for **(a)**.

Solution: **(a)** Arrange the four ligands about platinum in a square-planar array. Only two distinct structures are possible:

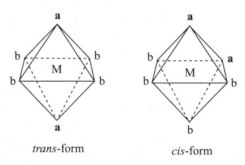

cis-form   trans-form

(b) In a *cis*-square planar complex, equivalent donor atoms are adjacent to one another. In a *trans*-square planar complex, equivalent donor atoms are opposite one another along a straight line through the central metal atom.

EXERCISE 9   Repeat questions **(a)** and **(b)** in Exercise 8 for the generalized octahedral complex [Ma$_2$b$_4$] where a and b are monodentate ligands.

Solution: **(a)** Arrange the ligands a and b at the corners of an octahedron with M at the center. Only two distinct structures are possible. **(b)** In an octahedral [Ma$_2$b$_4$] complex, the terms *cis-* and *trans-* refer to the spatial configuration of the two equivalent donor atoms; in this example the two equivalent donor atoms are a.

trans-form   *cis*-form

EXERCISE 10   Optical isomerism exists for a complex if the structure of the complex cannot be superimposed on its mirror image. Molecules that are not superimposable mirror images of one another are said to be chiral. Show that *cis*-[Co(en)$_2$(NO$_2$)$_2$]$^+$ has two optical isomers.

Solution: First sketch one of the optical forms **(a)**. Then draw the dotted line **(b)** to represent a mirror plan. Finally, sketch the mirror image **(c)** of structure **(a)** by reflecting all atoms through the mirror plan. Rotate structure **(c)** and superimpose on structure **(a)**. If you cannot superimpose structure **(c)** on **(a)**, the structures **(a)** and **(c)** are chiral. The curved lines represent ethylenediamine ligands. Note that if we rotate the molecules so that the −NO$_2$ groups of both complexes are superimposed, the curved lines of the ethylenediamine ligands are not; thus chiral forms of *cis*-[CO(en)$_2$(NO$_2$)$_2$]$^+$ exist.

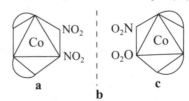

a   b   c

EXERCISE 11   **(a)** Co(NH$_3$)$_5$(NO$_2$)Cl$_2$ dissolves in water to form three moles of ions per mole of complex. One mole of the complex reacts with two moles of AgNO$_3$ to yield two moles of AgCl(*s*). Co(III) typically has a coordination number of six. Write the formula of the complex, bracketing the coordination sphere. (**b**) Does the complex exhibit *cis-trans* isomerism?

    Solution: (**a**) The coordination sphere must contain cobalt and six donor atoms because cobalt has a coordination number of six. Because two moles of AgCl(*s*) form per one mole of the complex, the two chlorine atoms cannot be tightly bound to the cobalt and consequently must be outside of the coordination sphere. The formula $[Co(NH_3)_5(NO_2)]Cl_2$ fits the experimental observation that the substance dissolves in water to form three ions: $[Co(NH_3)_5NO_2]^{2+}$ and $2Cl^-$. The predicted formula also shows Co with a coordination number of six. (**b**) The complex $[Co(NH_3)_5(NO_2)]^{2+}$ has six ligands surrounding Co(III); thus it is an octahedral complex of type Ma_5b. Cis-trans-isomerism cannot occur because there is no pair of *b* atoms.

EXERCISE 12   How many geometric isomers are possible for $[Co(NH_3)_4(NO_2)_2]$? Are there any chiral forms?

    Solution: Assume that there is an octahedral arrangement of ligands about cobalt because the coordination number of cobalt is six. In order to determine the number of isomers, you have to draw geometrical structures. Sometimes you may draw two structures that do not look identical but are identical if you can superimpose them. In this case there are two geometric isomers:

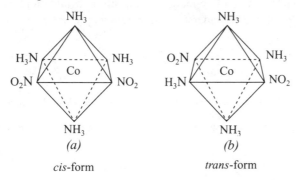

(a)	*(b)*
cis-form	*trans*-form

Ligands bonded to a transition metal ion may exchange with other ligands in the bulk solvent. This exchange of a coordinated ligand with one in the solvent is termed **ligand exchange** or **substitution reaction.** Rates may be fast or slow, depending on the ligands and the nature of the metal ion.

- If the exchange of ligands is rapid, the complex is termed **labile;** if the exchange of ligands is slow, it is termed **inert.**

- The metal ions most consistently forming inert complexes are Co(III), Cr(III), Pt(IV), and Pt(II). Complexes of these ions exist in solution for a time that is long enough to allow for identification of their structures and properties.

Another physical property of complexes studied by chemists is their magnetic behavior.

- A substance with unpaired electrons is *strongly attracted* by a magnetic field; it is said to be **paramagnetic.**

- A substance possessing no unpaired electrons is *weakly repelled* by a magnetic field and is said to be **diamagnetic.**

• The magnetic properties of a transition metal complex depend on the metal ion, on how many d electrons it possesses, and on the number and kinds of ligands bonded to the metal.

The colors of transition metal complexes also depend on the metal ion and the nature of surrounding ligands.

• For a metal in a particular oxidation state, the color of the complex depends on the number and kinds of ligands attached.

• A complex exhibits color if it absorbs energy in the region of visible light. The color observed is not the color of the visible light absorbed; rather it is the complementary color of the light absorbed. If several different energies in the visible spectrum are absorbed, the analysis is not straightforward. You should know the colors of the visible spectrum and their complementary colors.

In Section 25.7, a bonding model for transition metal complexes is presented that will further aid you in understanding magnetic properties and colors of complexes.

EXERCISE 13   The equilibrium

$$[Ni(H_2O)_6]^{2+}(aq) + 4CN^-(aq) \rightleftharpoons [Ni(CN)_4]^{2-}(aq) + 6H_2O(l)$$

lies far to the right. Tracer studies with ^{14}C-labeled cyanide ions show that during the reaction

$$[Ni(CN)_4]^{2-}(aq) + 4\,^{14}CN^-(aq) \rightleftharpoons [Ni(^{14}CN)_4]^{2-}(aq) + 4CN^-(aq)$$

$^{14}CN^-$ is almost instantaneously incorporated into the complex. Is $[Ni(CN)_4]^{2-}$ labile or inert, and stable or unstable in water? Explain.

Solution: Water does not significantly displace CN^- in $[Ni(CN)_4]^{2-}$, as is shown by the fact that the equilibrium of $[Ni(H_2O)_6]^{2+}$ with CN^- lies far to the right. This shows that $[Ni(CN)_4]^{2-}$ is stable in water. However, it is labile since $^{14}CN^-$ in the bulk solvent quickly exchanges with CN^- in the complex.

EXERCISE 14   The complex prussian blue, $[KFe(CN)_6Fe]_x$, has a deep blue color, as indicated by its name. Using Figure 25.25 of the text, predict the color region and wavelength of visible light in which the complex primarily absorbs light.

Solution: From Figure 25.25 you see that the complementary color of blue is orange, centered at a wavelength of about 600 mm. Thus, the complex absorbs light primarily in the orange region of the visible spectrum.

BONDING IN TRANSITION METAL COMPLEXES: CRYSTAL-FIELD THEORY

Crystal-field theory treats the bonding between ligands and the central metal in complexes as purely electrostatic. Key features of the crystal-field model are:

• The central metal atom with a positive charge is attracted electrostatically to the negative charge of a ligand or to the negative end of a polar ligand.

• The electrostatic field created by the surrounding ligands in a transition metal complex raises the energies of the d electrons in the central transition element compared to their energies in an isolated metal atom. This repulsive interaction between ligands and the outermost electrons of a transition metal is called the **crystal field.**

- Furthermore, the energies of the d orbitals of a transition metal are no longer equal in the presence of a crystal field. Depending on the geometrical arrangement of ligands about the central transition metal atom, different energy groupings of d orbitals occur.

- For octahedral, tetrahedral, and square-planar complexes, the groupings of d energy levels are

Energy   Octahedral   Tetrahedral   Square planar

Octahedral: d_{zy}, d_{xz}, d_{yz} (lower); $d_{z^2}, d_{x^2-y^2}$ (upper)

Tetrahedral: $d_{z^2}, d_{x^2-y^2}$ (lower); d_{zy}, d_{xz}, d_{yz} (upper)

Square planar: d_{xz}, d_{yz} (lowest); d_{z^2}; d_{xy}; $d_{x^2-y^2}$ (highest)

- The energy separation between the set of d_{z^2}, $d_{x^2-y^2}$ orbitals and the set of d_{xy}, d_{xz}, d_{yz} orbitals in the octahedral and tetrahedral diagrams is represented by Δ.

You should learn how application of Hund's rule enables you to explain the existence of high-spin and low-spin complexes and other physical properties of transition metal complexes.

- The **low-spin (strong field)** case occurs when the six ligands surrounding the metal atom create a large crystal field (large Δ); in this case, less energy is required to first pair electrons in the lower energy set of d orbitals before placing them in the higher energy set.

- For a **high-spin (weak field)** case to occur, the value of Δ must be smaller than the energy required to pair electrons (spin-pairing energy); in this situation, a lower energy state results when electrons enter the five d orbitals singularly and with the same spins before pairing of electrons occurs.

- For example, high- and low-spin complexes exist for a transition metals with four d electrons:

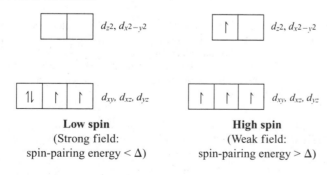

Low spin
(Strong field:
spin-pairing energy $< \Delta$)

High spin
(Weak field:
spin-pairing energy $> \Delta$)

• For tetrahedral complexes, we observe only the weak-field case because Δ for tetrahedral crystal fields is always smaller than the energy required to pair electrons.

The values of Δ for transition metal complexes are often of the same order of magnitude as the energy per photon of light in the visible spectrum.

• A transition metal complex exhibits color if an electron in a lower-energy d orbital is excited into a higher-energy d orbital and if Δ is in the range of visible light.

• The magnitude of Δ for a given transition metal atom ion depends on the number and kinds of surrounding ligands.

• The **spectrochemical series** is a list of ligands arranged in order of their ability to increase Δ in a series of complexes of type ML_x in which M is constant:

$$\xrightarrow{\quad \Delta \text{ increasing} \quad}$$

$$Cl^- < F^- < H_2O < NH_3 < en < NO_2^- < CN^-$$

Ligands to the left of this series are said to be low in the spectrochemical series, and those to the right are said to be high. Thus in many complexes CN^- causes Δ to be sufficiently large so that a low-spin complex forms (electrons pair first in lower energy orbitals) whereas Cl^- causes a high-spin complex (maximum number of unpaired electrons).

EXERCISE 15 Explain using crystal-field theory why octahedral cobalt(II) complexes exhibit different magnetic properties.
Solution: (1) First determine the electron configuration of the central metal atom. Cobalt is $[Ar]3d^7 4s^2$. Since the $4s$ electrons leave first in the formation of transition metal ions, the electron configuration of cobalt(II) is $[Ar]3d^7$. (2) Write the appropriate crystal field diagram for the given geometry. (3) Place electrons into the crystal field diagram according to the principles given. Depending on the ligand, two possible $3d$ electron populations of cobalt(II) in an octahedral environment can occur:

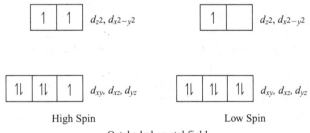

High Spin   Low Spin

Octahedral crystal field

The high-spin case occurs when the value of Δ is smaller than pairing energy. Electrons enter the $3d$ orbitals one at a time and with parallel spins; the electrons remain unpaired until all d orbitals have one electron. The low-spin case occurs when the value of Δ is significantly larger than the energy required to pair electrons in $3d$ orbitals; electrons enter the low-energy set of $3d$ orbitals until all are filled before they occupy the higher-energy set of $3d$ orbitals.

EXERCISE 16 Draw the crystal-field energy-level diagram and show the placement of electrons for: **(a)** $CoCl_4^{2-}$ (tetrahedral); **(b)** $[MnF_6]^{4-}$ (high-spin); and **(c)** $[Cr(en)_3]^{2+}$ (four unpaired electrons).

Solution: **(a)** The oxidation number of cobalt in $CoCl_4^{2-}$ is +2. Co(II) has an electron configuration of $[Ar]3d^7$. All tetrahedral complexes are high spin, even if there is the possibility of low-spin electron configuration. Place the electrons in the d orbitals singularly until all orbitals have one electron. Then begin to pair electrons in the lower energy set of orbitals first. The crystal-field energy-level diagram is

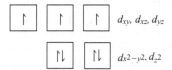

Tetrahedral crystal field

(b) Mn(II) has an electron configuration of $[Ar]3d^5$. The complex is expected to be octahedral because the coordination number is six. The F^- ion typically forms high-spin complexes; it is low in the spectrochemical series.

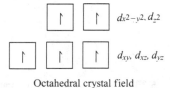

Octahedral crystal field

(c) Cr(II) has an electron configuration of $[Ar]3d^4$ $[Cr(en)_3]^{2+}$ is expected to be octahedral because the coordination number of the complex is six. Four electrons are placed in a high-spin environment to agree with the given information

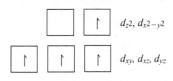

Octahedral crystal field

SELF-TEST QUESTIONS

Key Terms

Having reviewed the key terms listed at the end of Chapter 25 and their applications, identify the following statements as true or false. If a statement is false, indicate why it is incorrect. Key terms are italicized in the statements.

25.1 The *coordination sphere* of the complex ion in $[Ni(NH_3)_6]Cl_2$ contains Ni^{2+}, six NH_3 groups, and two chloride ions.

25.2 The *coordination number* of cobalt in $[Co(NH_3)_4Cl_2]^+$ is six.

25.3 An example of a ligand that is a *chelating agent,* a *polydentate ligand,* and a *bidentate ligand* is the oxalate ion.

25.4 A *ligand* acts as a Lewis acid in metal complexes.

25.5 Ethylenediamine is an example of a *monodentate ligand*.

25.6 *Isomers* exist for the compound $[Cr(H_2O)_4Cl_2]^+$. In this

case, two *geometric isomers* exist, one with a *cis* geometric arrangement of ligands and the other in a *trans* form.

25.7 *cis*-$[Cr(H_2O)_4Cl_2]^+$ is a *chiral* species.

25.8 Suppose a particular chiral molecule is *dextrorotary.* From this information we can conclude that its nonsuperimposable mirror image is also chiral and that it is *levrotatory.*

25.9 A chiral molecule and the substance that is its non-superimposable mirror image form a pair of substances are *optically active.*

25.10 The anion of acetylacetone uses one oxygen *donor atom* when binding to a metal atom.

25.11 *Stereoisomers* and *structural isomers* are terms that are equivalent in meaning.

25.12 Pentaamminenitrocobalt(III) chloride and pentaamminenitritocobalt(III) chloride are *stereoisomers.*

25.13 Two examples of *coordination sphere* isomers are $[Co(NH_3)_6][Cr(CN)_6]$ and $[Cr(NH_3)_6][Co(CN)_6]$.

25.14 Given that $[Co(NH_3)_6]^{3+}$ is an *inert* complex in the presence of the chloride ion in the solvent water, you can conclude that the following equilibrium lies to the left:

$$[Co(NH_3)_6]^{3+} + Cl^- \rightleftharpoons [Co(NH_3)_5Cl]^{2+} + NH_3$$

25.15 Complexes that exchange ligands rapidly are referred to as *labile* complexes.

25.16 A metal ion with a d^8 valence-electron configuration in an octahedral field of ligands can form *low-spin* complexes, depending on the nature of the ligands.

25.17 CN^- is high in the *spectrochemical series* because it creates a large crystal field.

25.18 Using the spectrochemical series and the principles of crystal-field theory, we predict $Co(CN)_6^{3-}$ to be a *high-spin* complex.

25.19 In the compound $[Au(CN)_2]Cl$, both ions are *complex ions*.

25.20 The complex in the compound in 25.19 is $[Au(CN)_2]^-$.

25.21 The ligand ethylenediamine has two *donor atoms*: nitrogen and oxygen.

25.22 $[Ni(en)_3]^{2+}$ has a larger formation constant than that of $[Ni(NH_3)_6]^{2+}$ because of the *chelate effect*.

25.23 Myoglobin is a globular protein containing a *porphyrin*, a large nitrogen-containing chelate.

25.24 The ligand NH_3 exhibits *linkage isomerism* because it can coordinate through either nitrogen or hydrogen.

25.25 *Coordination-sphere isomerism* occurs when a two compounds exhibit the following behavior: $[MX_4Y_2]Z$ and $[MX_4YZ]Y$.

25.26 *Optical isomerism* is a form of stereosiomerism of the following type: A pair of mirror images that are superimposeable on each other.

25.27 When equal amounts of a pair of *d*- and *l*-isomers are present, a *racemic* mixture exists.

25.28 Green is the *complementary color* of red.

25.29 When a complex absorbs the color red, it will have an *absorption spectrum* in the visible range of light.

25.30 According to *crystal-field theory*, the interaction of a *d* electron with a negatively charged ligand will lower the energy of all *d* electrons.

25.31 Energy is released when electrons pair, and it is referred to as *spin-pairing energy*.

Problems and Short-Answer Questions

25.32 Name the following compounds:
    (a) $[Co(NH_3)_6]Cl_3$
    (b) $[Fe(CO)_5]$
    (c) $K_2[ZnCl_4]$
    (d) $Ca_2[Fe(CN)_6]$
    (e) $[Fe(H_2O)_6]SO_4$
    (f) $[Ir(NH_2)_2(en)_2]Cl_3$.

25.33 Write the formula for each of the following compounds or ions:
    (a) potassium tetrachloroplatinate(II);

(b) hexaammineplatinum(II) chloride;
(c) ammonium diamminetetracyanochromate(III);
(d) hexachlorostannate(IV) ion;
(e) tetrachloroethylenediamineplatinum(IV)

25.34 Give the coordination number of the central metal ion in each of the complexes given in problem 25.32.

25.35 $[Ir(CO)(PR_3)_2Cl]$ has a square-planar arrangement of ligands about Ir. Draw and name all of its geometrical isomers. The name of the neutral ligand PR_3 is triphenylphosphine, where R is C_6H_5.

25.36 Sketch all isomeric structures for
    (a) $[Cr(H_2O)_5Cl]^+$
    (b) $[Fe(C_2O_4)_3]^{3-}$

25.37 Which of the following chelating agents would you expect to form a more stable metal ion complex?

25.25 List the ligands a, b, and c in order of increasing Δ, given that the following octahedral complexes absorb energy equivalent to Δ at the stated wavelengths: MA_6, 350 nm; MB_6, 400 nm; and MC_6, 300 nm.

25.38 The complex ion $[FeBr_4]^-$ has a tetrahedral structure. Using the crystal-field model, predict how many unpaired *d* electrons the complex possesses.

25.40 Using water and Cl^- as ligands, **(a)** give the formula of a six-coordinate Cr(III) complex that would be a nonelectrolyte in solution; **(b)** give the formula of a six-coordinate Cr(III) complex that has about the same electrical conductivity in solution as $MgCl_2$ in water; **(c)** give the formula of an octahedral complex of Cr(III) with a charge of 1−.

25.41 What magnetic properties would you predict for $[V(NH_3)_6]^{3+}$ and $[MnF_6]^{3-}$?

Multiple Choice Questions

25.42 Which type of isomerism is represented by the following pair of compounds: $[Co(en)_2BrCl]Cl$ and $[Co(en)_2Cl_2]Br_2$?
    (a) Coordination sphere;
    (b) linkage;
    (c) geometric;
    (d) optical;
    (e) none of the above.

25.43 Which of the following metal ions possesses the valence-shell electron configuration $3d^5$?
    (a) Fe^{2+} **(d)** Mn^{2+}
    (b) Cr^{3+} **(e)** Ca^{2+}
    (c) Co^{3+}

25.44 Which of the following complexes has a nonsuperimposable mirror image?
    (a) $[Zn(H_2O)_4]^{2+}$ (tetrahedral)
    (b) *trans*-$[Cr(H_2O)_4Cl_2]^+$

(c) $[Pt(en)_2]^{2+}$ (square planar)

(d) $[Cr(H_2O)_6]^{3+}$

(e) cis-$[Co(en)_2(Cl)(NO_2)]^+$

25.45 Which of the following metal ions could form both high- and low-spin complexes in an octahedral array of ligands?

(a) Cr^{3+}

(b) Mn^{3+}

(c) Cu^+

(d) Zn^{2+}

(e) Ti^{3+}

25.46 Which of the following metal ions could form diamagnetic complexes in a strong crystal-field environment of six ligands?

(a) Ni^{2+}

(b) Mn^{3+}

(c) Co^{2+}

(d) Fe^{3+}

(e) Fe^{2+}

25.47 Which of the following is the correct name for $[Co(en)_2Br_2]_2SO_4$?

(a) bromoethylenediaminecobalt(III) sulfate

(b) ethylenediaminebromocobalt(III) sulfate

(c) dibromodiethylenediaminecobalt(III) sulfate

(d) diethylenediaminedibromocobalt(III) sulfate

(e) dibromobis(ethylenediamine)cobalt(III) sulfate

25.48 Which of the following formulas is associated with the name sodium tetrafluorooxochromate(IV)?

(a) $Na_2[CrOF_4]$

(b) $Na[CrOF_4]$

(c) $[NaCrOF_4]$

(d) $[CrOF_4Na_2$

(e) $Na_2[CrF_4]$

25.49 Which of the following ligands is *incorrectly* named?

(a) H_2O, aqua

(b) F^-, fluoro

(c) NH_3, amine

(d) CN^-, cyano

(e) $CH_3COCHCOCH_3^-$, acetylacetonato

25.50 Ligands in the complex $[Fe(H_2O)_6]^{3+}$ exchange rapidly with solvent water molecules. Which of the following can be said about the complex?

(a) It is thermodynamically unstable;

(b) it is thermodynamically stable;

(c) it is inert;

(d) it is labile;

(e) it is not reactive.

25.51 Which of the following statements are true about the complexes, $Fe(H_2O)_6^{2+}$ and $Fe(CN)_6^{4-}$: (1) both contain Fe^{2+}; (2) both are low-spin complexes. (3) $Fe(H_2O)_6^{2+}$ is paramagnetic and $Fe(CN)_6^{4-}$ is diamagnetic; (4) $Fe(H_2O)_6^{2+}$ absorbs light at a lower energy than $Fe(CN)_6^{4-}$; (5) the names of the complexes are hexaaquairon(II) ion and hexacyanoiron(II) ion, respectively?

(a) (1) and (2)

(b) (1) and (3)

(c) (3), (4), and (5)

(d) (1), (3), and (4)

(e) (1), (3), (4), and (5)

25.52 Which of the following is the correct name for $[Cu(en)_2]SO_4$?

(a) copperbis(ethylenediamine)(II) sulfate

(b) copper(II)bis(ethylenediamine) sulfate

(c) bis(ethylenediamine)copper(II) sulfato

(d) bis(ethylenediamine)copper(II) sulfate

(e) bis(ethylenediamine)copper(I) sulfate

25.53 Which of the following can exist as structure(s) of tetraamminebromochlorocobalt (III) ion?

(a) *Cis*-isomer;

(b) *trans*-isomer;

(c) optical isomers;

(d) (a) and (b);

(e) (a), (b), and (c).

25.54 Which of the following ring systems is the most stable?

(a) $M-C\equiv N:$

(b)

(c)

(d)

(e) all equally stable

25.55 How many unpaired electrons does $[CoI_6]^{3-}$ possess?

(a) 0

(b) 1

(c) 2

(d) 3

(e) 4

25.56 How many unpaired electrons does $[Sc(H_2O)_3Cl_3]$ possess?

(a) 0

(b) 1

(c) 2

(d) 3

(e) 4

25.57 What is the coordination number of Ni in $[Ni(C_2O_4)_3]^{4-}$?

(a) 2

(b) 3

(c) 4

(d) 5

(e) 6

25.58 What is the coordination number of Co in $[Co(NH_3)_5Cl]SO_4$?

(a) 2

(b) 3

(c) 4

(d) 5

(e) 6

25.59 Which of the following could exhibit coordination-sphere isomerism?

(a) $Li[Al(CN)_4]$

(b) $[Pt(H_2O)_4Cl_2]$

(c) $[Co(NH_3)_4Cl_2]Br$

(d) $K_3[Fe(CN)_6]$

(e) $[Zn(H_2O)_4][CdCl_4]$

25.60 What is the observed color for a complex that primarily absorbs radiation around 700 nm?

(a) yellow

(b) green

(c) blue

(d) violet

(e) red

25.61 Which of the following complexes has two unpaired d electrons in the set of d_{z^2} and $d_{x^2-y^2}$ orbitals?

| **(a)** | $[NiF_6]^{4-}$ | **(c)** | $[IrCl_6]^{3-}$ |
| **(b)** | $[V(H_2O)_6]^{2+}$ | **(d)** | $[ZrCl_6]^{3-}$ |

25.62 The $Ni(H_2O)_6^{2+}$ ion is green whereas the $Ni(NH_3)_6^{2+}$ ion is purple. Which statement is correct?

    (a) The complementary color of green is yellow.

    (b) The complementary color of purple is red.

    (c) $Ni(H_2O)_6^{2+}$ absorbs light with a shorter wavelength than $Ni(NH_3)_6^{2+}$.

    (d) $Ni(NH_3)_6^{2+}$ absorbs light with a shorter wavelength than $Ni(H_2O)_6^{2+}$.

25.63 Which complex is diamagnetic?

| **(a)** | $Co(CN)_6^{3-}$ | **(c)** | $Ti(H_2O_6^{3+}$ |
| **(b)** | $Co(CN)_6^{2-}$ | **(d)** | $F(en)_3^{3+}$ |

SELF-TEST SOLUTIONS

25.1 False. The coordination sphere contains the metal ion, Ni^{2+}, and the surrounding ligands contain six NH_3 groups; the chloride ions are counter ions and not directly bonded to nickel. **25.2** True. **25.3** True. **25.4** False. A ligand is a Lewis base because it donates electrons to the metal in a complex. **25.5** False. Ethylenediamine is a bidentate ligand because it has two donor atoms. A monodentate ligand has only one donor atom. **25.6** True. **25.7** False. It does not have a non-superimposable mirror image. **25.8** True. **25.9** True. **25.10** False. The acetylacetonate anion uses two donor atoms. **25.11** False. Structural isomers are compounds having the same formula but possessing differing atoms bonded to the metal. In contrast, stereoisomers are substances having the same formula *and* bonding arrangements of atoms but possessing differing spatial arrangements of the bonded atoms. **25.12** False. $[Co(NH_3)_5(-NO_2)]Cl$ and $[Co(NH_3)_5(-ONO)]Cl$ differ by the fact that the first contains a NO_2 group bound to cobalt *via* a metal-nitrogen bond and the second contains the same group bound to cobalt *via* a metal-oxygen bond. This is an example of *linkage isomerism,* a form of structural isomerism. **25.13** True. **25.14** False. The terms inert and labile refer to kinetic properties of complexes, not thermodynamic properties. A complex may be inert but thermodynamically unstable. **25.15** True. **25.16** False. A d^8 transition metal ion forms only one d electron population in an octahedral field of ligands:

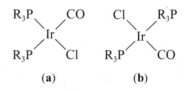

25.17 True. **25.18** False. Since CN^- is a strong-field ligand, d electrons perferentially pair in lower-energy orbitals before entering higher-energy orbitals. Therefore a low-spin complex is formed. **25.19** False. A complex ion contains two or more different elements and carries a charge. The complex ion is bracked by [ ]. Cl^- is an ion, but of one element, chlorine. **25.20** True. The term complex is another term for complex ion. **25.21** False. It has two nitrogen donor atoms. This ligand does not contain any oxygen atoms. **25.22** True. **25.23** True. **25.24** False. Ammonia can bind only through the nitrogen atom. Linkage isomerism occurs when a ligand can bind to a

metal in two different ways. The ligand, NO_2^-, exhibits linkage isomerism by using either a nitrogen or oxygen atom when binding to a metal. **25.25** True. **25.26** False. It exists when a pair of mirror images can *not* be superimposed. **25.27** True. **25.28** True. **25.29** True. **25.30** False. The interaction of two negatively charged particles will raise the energy of the d-electrons. **25.31** False. Energy is required to pair electrons in the same orbital. **25.32 (a)** hexaamminecobalt(III) chloride; **(b)** pentacarbonyliron(0); **(c)** potassium tetrachlorozincate(II); **(d)** calcium hexacyanoferrate(II); **(e)** hexaaquairon(II) sulfate; **(f)** diamminebis(ethylenediamine)iridium(III) chloride. **25.33 (a)** $K_2[PtCl_4]$; **(b)** $[Pt(NH_3)_6]Cl_2$; **(c)** $NH_4[Cr(NH_3)_2(CN)_4]$; **(d)** $[SnCl_6]^{2-}$; **(e)** $[Pt(en)Cl_4]$. **25.34 (a)** six; **(b)** five; **(c)** four; **(d)** six; **(e)** six; **(f)** six (ethylenediamine has two donor atoms). **25.35** There are two square-planar geometrical isomers:

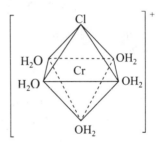

where R_3P represents triphenylphosphine. The name of isomer **(a)** is *cis*-carbonylchlorobis(triphenylphosphine)iridium (I), and isomer **(b)** is *trans*-carbonylchlorobis(triphenylphosphine) iridium(I). **25.36 (a)**

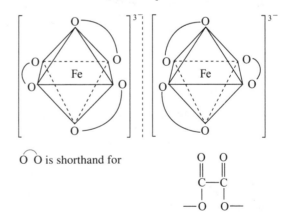

No other forms exist. **(b)** Two optical isomers exist:

O O is shorthand for

They are nonsuperimposable mirror images. **25.37** Both ligands have two donor atoms, the negatively charged oxygen atoms. In order to compare the stability of the complexes they form with metal atoms, you must determine how many atoms are in the chelate rings:

O=C——C=O   five-membered
  | |
  O   O

Five-membered chelate ring

Four-membered chelate ring

The O—M—O bond angle is smaller in the four-membered ring chelate ring system than in the five-membered one. As the O—M—O bond angle contracts, repulsions between oxygen donor atoms in the chelate ring increase, which decreases the stability of the ring. Thus

$$O=C-C=O$$
$$^-O \quad O^-$$

forms more stable metal complexes than does

(structure of carbonate: C double bonded to O, single bonded to two O^-)

25.38 The energy of absorbed radiation is inversely proportional to the wavelength: $E = hc/\lambda$. The complex that absorbs energy at the lowest wavelength has associated with it the largest Δ. The order of increasing energies absorbed by the complexes is thus $MB_6 < MA_6 < MC_6$. Since the metal is a constant factor, the differences among the ligands, a, b, and c, cause the variances in the wavelengths of absorbed radiation among the complexes. The order of increasing Δ for the ligands is thus $b < a < c$. **25.39 (a)** The Fe^{3+} ion has the electronic configuration $[Ar]3d^5$. Experimentally we observe that the crystal-field-splitting energy for tetrahedral complexes is

small because only four ligands with their negative parts interact with the d electrons, rather than six as in an octahedral case. Consequently all tetrahedral complexes are high-spin complexes. The population of Fe^{3+} $3d$ electrons in a tetrahedral environment of ligands is

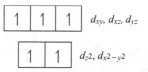

25.40 (a) $[Cr(H_2O)_3Cl_3]$; **(b)** $[Cr(H_2O)_5Cl]Cl_2$; **(c)** $[Cr(H_2O)_2Cl_4]^-$. **25.41**

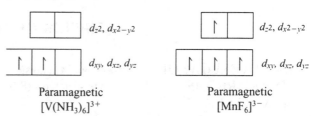

Paramagnetic $[V(NH_3)_6]^{3+}$

Paramagnetic $[MnF_6]^{3-}$

(Note: F^- is a weak-field ligand.) **25.42 (a)**. Counter ion and coordinated ligand are exchanged. **25.43 (d)**. Remember that s electrons are removed before d electrons when forming transition metal ions. **25.44 (e)**. **25.45 (b)**. Mn^{3+} is a d^4 ion; high- and low-spin complexes can exist, depending on the ligand. **25.46 (e)**. Fe^{2+} is a d^6 ion; in a strong crystal field, all six electrons are paired in the lower-energy set of orbitals (d_{xy}, d_{xz}, d_{yz}). **25.47 (e)**. **25.48 (a)**. **25.49 (c)**. NH_3, ammine. **25.50 (d)**. **25.51 (d)**. (5) is not correct because the name of $[Fe(CN)_6]^{4-}$ is hexacyanoferrate(II) ion. **25.52 (d)**. **25.53 (d)**. **25.54 (c)**. A seven-membered ring system **(d)** is far less stable than a five-membered ring system. **25.55 (e)**. **25.56 (a)**. **25.57 (e)**. **25.58 (e)**. **25.59 (c)**. **25.60 (b)**. **25.61 (a)**. **25.62 (d)**. **25.63 (a)**. Co^{3+} with a strong field ligand.

26 Chemistry of Life: Organic and Biological Chemistry

OVERVIEW OF THE CHAPTER

26.1 OVERVIEW OF HYDROCARBONS AND ALKANES

Review: Catalysis (14.6); electronegativity (8.5); geometrical isomerism (25.4); Lewis structures (8.6).

Objectives: You should be able to:

1. List four groups of hydrocarbons and draw the structural formula from each group—in this case the alkanes.

2. Write the formulas and names of the first 10 members of the alkane series.

3. Write the structural formula of an alkane given its systematic (IUPAC) name.

4. Name an alkane, given its structural formula.

5. Give an example of structural isomerism in alkanes.

26.2 ALKENES AND ALKYNES

Review: See prior review.

Objectives: You should be able to:

1. List the four groups of hydrocarbons and draw the structural formula of an example from each group—in this case as applied to alkenes and alkynes.

2. Write the structural formula of an alkene or alkyne, given its systematic (IUPAC) name.

3. Name an alkene or alkyne, given its structural formula.

4. Give an example of structural and geometrical isomerism in alkenes and alkynes.

5. Give examples of addition reactions of alkenes and alkynes, showing the structural formulas of reactants and products.

26.2 AROMATIC HYDROCARBONS

Review: Delocalization of pi electrons (8.7).

Objectives: You should be able to:

1. Explain why aromatic hydrocarbons do not readily undergo addition reactions.

2. Give two or three examples of substitution reactions of aromatic hydrocarbons.

Objectives: You should be able to:

1. Identify the groups or arrangement of atoms in a molecule that correspond to the following functional groups: alcohols; ethers; aldehydes and ketones; carboxylic acids; esters; amines and amides.

2. Give examples of the condensation reactions to alcohols to form ethers, of alcohols and carboxylic acids to form esters, and of amines and carboxylic acids to form amides.

**26.3
HYDROCARBON
DERIVATIVES**

Objectives: You should be able to:

1. List the several functions of proteins in living systems.

2. Define the terms chiral and enantiomer and draw the enantiomer of a given chiral molecule.

3. Write the reaction for formation of a peptide bond between two amino acids.

4. Explain the structures of proteins in terms of primary, secondary, and tertiary structure.

**26.4, 26.5
PROTEINS**

Objectives: You should be able to:

1. Describe two distinct reasons why energy is required by living organisms.

2. Describe the formation of cyclic structures for sugars from their open-chain forms, and distinguish between the α and β forms of the cyclic structures of glucose.

3. Describe the manner in which monosaccharides are joined together to form polysaccharides.

4. Enumerate the major groups of polysaccharides and indicate their sources and general functions.

5. Describe the structures of fats and oils and list the sources of these substances.

**26.6 ENERGY
SOURCES:
CARBOHYDRATES**

Objectives: You should be able to:

1. Draw the structures of any of the nucleotides that make up the polynucleotide DNA.

2. Describe the nature of the polymeric unit of polynucleotides.

3. Describe the double-stranded structure of DNA and explain the principle that determines the relationship between bases in the two strands.

**26.7 NUCLEIC
ACIDS: RNA AND DNA**

TOPIC SUMMARIES AND EXERCISES

OVERVIEW OF HYDROCARBONS AND ALKANES

Hydrocarbons are compounds consisting only of the elements carbon and hydrogen. Except for methane, CH_4, hydrocarbons contain stable carbon-carbon bonds; extended chains containing single, double, or triple carbon-carbon bonds are well known. These compounds and related ones are also referred to as organic compounds. In this chapter you will explore several classes of hydrocarbons; these are summarized in Table 26.1 on page 443. Note carefully the structural and carbon-carbon bonding differences among the different classes. You need to know them to understand the material in this chapter.

The most abundant and important source of hydrocarbons is **petroleum.**

- Petroleum is a complex mixture of hydrocarbons formed millions of years ago by the slow decomposition of animal and plant matter.
- Saturated hydrocarbons constitute over 90 percent of petroleum.
- Branched-chain hydrocarbons are formed by cracking crude oil that contains straight-chain hydrocarbons. The presence of branched-chain or aromatic hydrocarbons in gasoline increases its octane number. An **octane number** for a gasoline is a measure of its tendency to cause knocking in an automobile engine compared to the tendency of a standard gasoline mixture.

Alkanes contain all carbon-carbon single bonds and they are also called saturated hydrocarbons. The term saturated means that all carbon-carbon bonds are single.

- Note that hydrocarbon molecules are relatively nonpolar, dissolve in nonpolar solvents, and that their boiling points increase with increasing molar mass.
- If an alkane contains more than four carbon atoms in a continuous chain, the carbon atoms can be arranged in more than one way: a straight chain of carbon atoms or a branched chain. This is an example of **structural isomerism:** Compounds with the same molecular formula, but with different bonding arrangements.
- A chain of carbon atoms can also be cyclic, forming **cycloalkanes.** If a chain contains fewer than six carbon atoms, it will be strained, and thus more reactive. Note that the general formula for cycloalkanes, C_nH_{2n}, is different from that for straight-chain alkanes.

Alkanes are generally unreactive because C—C single bonds and C—H bonds are relatively strong, and the carbon atoms are saturated.

- Alkanes undergo combustion in air, which forms the products carbon dioxide (gas) and water (gas).
- Alkanes undergo substitution reactions with elemental fluorine, chlorine and bromine in which one or more hydrogen atoms are replaced by a halogen.

Table 26.1 Characteristics of Several Classes of Hydrocarbons

Class	Simplest general formula	Example	Characteristics
Alkane (saturated hydrocarbon)	C_nH_{2n+2}	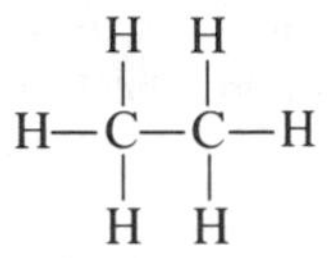 Ethane	Contains an open chain of carbon-carbon bonds; all carbon atoms in compound are attached to four other atoms *via* single σ bonds
Cycloalkanes	C_nH_{2n}	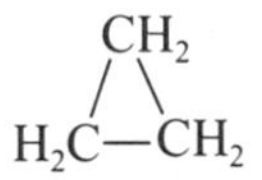	An alkane with one or more carbon rings; all bonds to carbon atoms are single bonds
Alkene (unsaturated hydrocarbon)	C_nH_{2n}	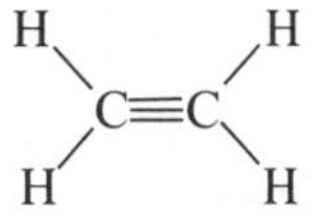 Ethylene	A hydrocarbon containing at least one carbon-carbon double bond ($>$C$=$C$<$)
Alkyne (unsaturated hydrocarbon)	C_nH_{2n-2}	Ethyne	A hydrocarbon containing one or more carbon-carbon triple bonds ($-$C$\equiv$C$-$)
Aromatic hydrocarbon		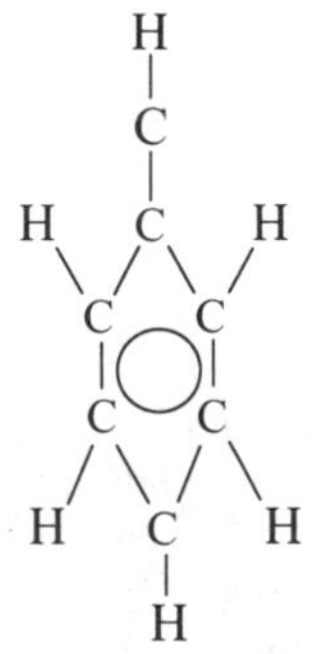 Benzene	Planar, cyclic arrangement of carbon atoms bonded to one another by both σ and π bonds (the π electrons are delocalized over the carbon atoms)

The following rules briefly summarize the systematic naming of alkanes:

1. Name the longest carbon chain in the structure (see Table 26.1 in the text). If the chain is cyclic, include the prefix cyclo- before the name of the carbon chain.

2. Number the carbon atoms consecutively from either end of the chain you named in step 1. When numbering carbon atoms in the chain, start numbering from the end that allows carbon atoms with attached groups to have the lowest numbers possible.

3. Locate and name attached groups to the longest carbon chain. See Table 26.2 of the text for names of groups. The location of any branch group is indicated by giving, before the name of the branch group, the number of the carbon atom to which it is attached. To indicate how many of each group are present, use the appropriate prefix: di- (two); tri- (three); tetra- (four); and so forth.

4. Combine the names of the branch groups, including the numbers identifying their locations in the carbon chain, with the name of the carbon chain. If two or more different branch groups are attached to the carbon chain, their names are written in alphabetical order before the name of the carbon chain; furthermore, their names are separated by a hyphen.

EXERCISE 1   Name or write structural formulas for the following:

(a) 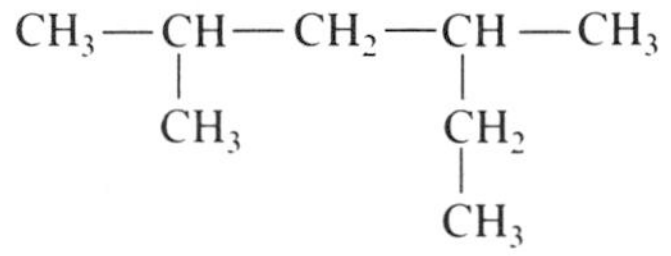 $CH_3—CH—CH_2—CH—CH_3$
$\qquad\qquad\ |\qquad\qquad\ |$
$\qquad\qquad CH_3\qquad\ CH_2$
$\qquad\qquad\qquad\qquad\quad |$
$\qquad\qquad\qquad\qquad\ CH_3$

(b) 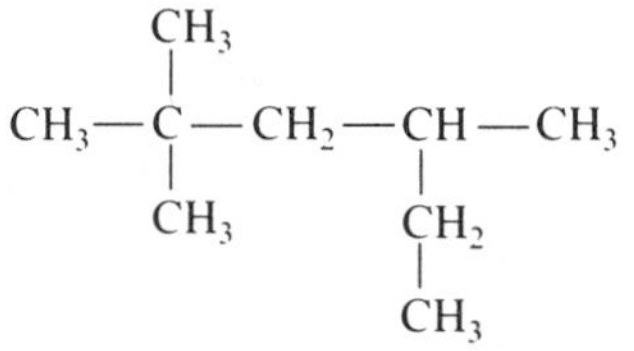
$\qquad\qquad\ CH_3$
$\qquad\qquad\ |$
$CH_3—C—CH_2—CH—CH_3$
$\qquad\quad |\qquad\qquad\ |$
$\qquad\ CH_3\qquad\ CH_2$
$\qquad\qquad\qquad\qquad |$
$\qquad\qquad\qquad\ CH_3$

(c) 1,3-Diethylcyclohexane

Solution: (a) The longest carbon chain contains six carbon atoms, as shown by this skeleton formula:

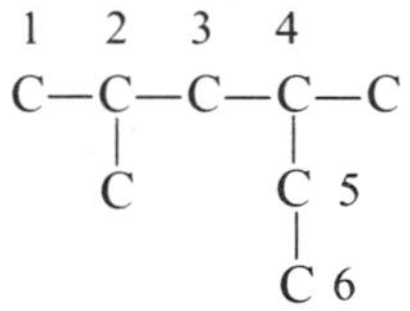

$$\begin{array}{ccccccc} 1 & & 2 & & 3 & & 4 \\ C & — & C & — & C & — & C & — & C \\ & & | & & & & | \\ & & C & & & & C & 5 \\ & & & & & & | \\ & & & & & & C & 6 \end{array}$$

The six-carbon chain is named hexane. The CH_3 (methyl) groups are at the 2 and 4 carbon positions; consequently, they are named 2,4-dimethyl. The name of the compound is 2,4-dimethylhexane. **(b)** As in **(a)**, the longest chain contains six carbon atoms and is named hexane. However, in this example, there are two methyl groups at the 2 position, and one methyl group at the 4 position. The location and number of methyl groups are indicated by 2,2,4-trimethyl, with the number 2 repeated once to indicate two methyl groups at that position. The name of the compound is 2,2,4-trimethylhexane. **(c)** The longest carbon chain is a cyclic six-carbon atom chain—the end name "cyclohexane" means six carbon atoms (hexane) arranged cyclically (cyclo-prefix). Ethyl groups occur at the carbon atoms numbered 1 and 3. The structural formula is shown at right:

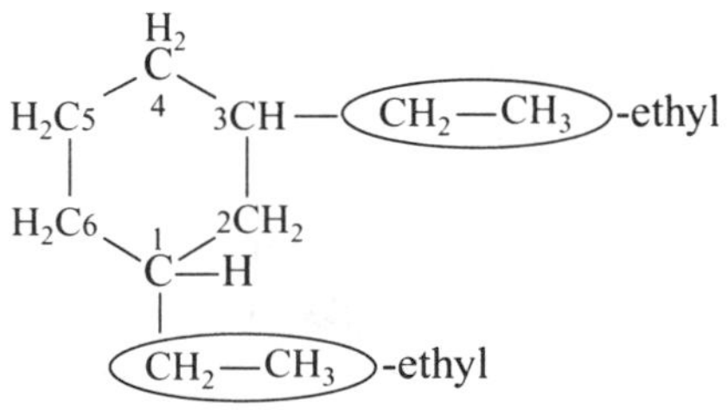

EXERCISE 2   Draw all structural isomers for a five membered alkane and name them.

Solution: When constructing a member of the alkane family you follow the rule of one bond per hydrogen atom and four bonds per carbon atom. A carbon atom may have one to four hydrogen atoms attached and one to four carbon atoms attached; however, there may not be more than a total of four bonds. For example, a carbon atom could have two hydrogen atoms and two carbon atoms attached for a total of four bonds. In the above question, you must arrange the five carbon atoms in various ways. Structures will be different if no amount of moving, twisting or rotating about the carbon-carbon bonds will cause the structures to be the same. First start with a straight chain of carbon atoms:

$$
\begin{array}{ccccc}
H & H & H & H & H \\
| & | & | & | & | \\
H-C- & C- & C- & C- & C-H \\
| & | & | & | & | \\
H & H & H & H & H
\end{array}
$$

This structure is called n-pentane. The n- prefix means it is a straight chain structure. Now, move an end carbon group to a different carbon atom in the middle of the chain:

$$
\begin{array}{cccc}
H & H & H & H \\
| & | & | & | \\
H-C- & C- & C- & C-H \\
| & | & | & | \\
H & H & C & H \\
& & H\ H\ H &
\end{array}
$$

This structure is 2-methylbutane (sometimes it is called isopentane). Finally, move the far left carbon atom to the same carbon atom that has two carbon atoms attached to it:

The name of this structure is 2,2-dimethylpropane.

ALKENES AND ALKYNES

Alkenes are hydrocarbons that have one or more $C{=}C$ bonds and simple alkenes have the general formula C_nH_{2n}. **Alkynes** have the general formula C_nH_{2n-2} and contain one or more carbon-carbon triple bonds. Both are also termed unsaturated hydrocarbons because the carbon atoms involved in double or triple bonds do not possess the maximum number of single bonds.

Alkenes often exhibit geometrical isomerism. **Geometrical isomers** consist of compounds that have the same molecular formula and the same atoms bonded to one another but they differ in the spatial arrangement of groups of atoms.

- Compounds with double bonds may be in the *cis* form (same groups are spatially on the same side of the double bond) or in the *trans* form (same groups are spatially on opposite sides of the double bond, along a diagonal).

Cis- and *trans-* geometrical isomers exist because of restricted rotation about a C-C double bond.

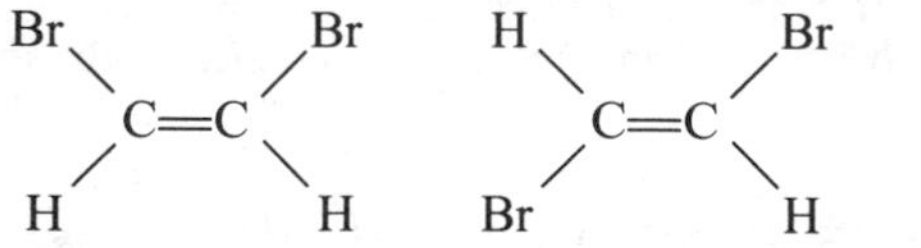

cis-dibromoethylene   *trans*-dibromoethylene

▪ Restricted rotation about a C=C bond exists because the bond consists of two types: sigma (σ) and pi (π). Rotation about a carbon-carbon double bond requires the breaking of a pi bond, a process that requires a significant quantity of energy.

▪ Alkynes do not exhibit cis-trans isomerism.

The rules for naming alkenes and alkynes are similar to the rules given in the prior section, except that rules 1 and 2 must be altered. Rules 1 and 2 for alkenes and alkynes are:

1. Name the longest continuous carbon chain which contains the carbon-carbon double or triple bonds. The name of the longest carbon chain is based on the corresponding alkane, except the name ends in -ene for an alkene; in -yne for an alkyne.

2. You indicate the position of the double or triple bond by designating the number of the carbon atom that is part of the double or triple bond using the lowest set of numbers possible. That is, start with the carbon-carbon double or triple bond that is closest to the end of the chain.

Alkenes and alkynes are more reactive than alkanes because of the presence of carbon-carbon double or triple bonds. A pi bond is weaker than a single bond and thus pi bonds are more readily broken in chemical reactions.

▪ Alkenes and alkynes undergo oxidation and substitution reactions.

▪ Alkenes and alkynes also undergo addition reactions, which involve the addition of other atoms directly to carbon atoms participating in double or triple bonds. The pi bonds are uncoupled and the electrons are available to combine with other atoms.

▪ The addition of hydrogen to form alkanes is shown in equation [26.2] in the text; this reaction is termed hydrogenation. A catalyst is necessary.

EXERCISE 3   Name or write structural formulas for the following:

(a) 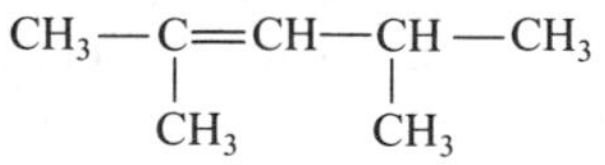 $CH_3-C=CH-CH-CH_3$
        $\quad\quad\;\; |\quad\quad\quad |$
        $\quad\quad CH_3\quad\;\; CH_3$

(b) 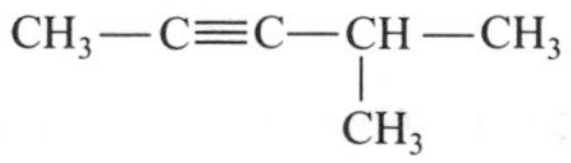 $CH_3-C≡C-CH-CH_3$
        $\quad\quad\quad\quad\quad |$
        $\quad\quad\quad\quad CH_3$

(c) 4-Methylcyclohexene

Solution: **(a)** The longest carbon chain containing the carbon-carbon double bond has five carbon atoms; it is named 2-pentene. The carbon atoms are numbered as follows:

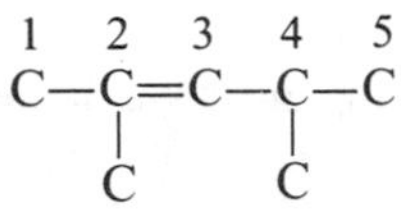

$$
\begin{array}{ccccc}
1 & 2 & 3 & 4 & 5 \\
C & C = C & C & C \\
\end{array}
$$

C—C=C—C—C with CH₃ groups at positions 2 and 4

Note that the smallest number possible is assigned to the first atom of the carbon-carbon double bond; in this situation it is assigned the number 2. Since there are two methyl groups, one each at the 2 and 4 positions of the carbon chain, the name of the compound is 2,4-dimethyl-2-pentene. **(b)** As in part **(a)**, the longest carbon chain contains five carbon atoms, but instead of containing a double bond it contains a triple bond. A triple bond is characteristic of an alkyne, whose name ends in -yne. Since one methyl group occurs at the 4 position of the carbon chain, the name of the compound is 4-methyl-2-pentyne. **(c)** The parent carbon chain is cyclohexene, which contains a six-member carbon ring with one carbon-carbon double bond. The prefix 4 before the methyl group in the name of the compound tells us that the methyl group occurs at the carbon atom numbered 4, with the first carbon atom of the carbon-carbon double bond assigned the number one position. Therefore the structural formula is

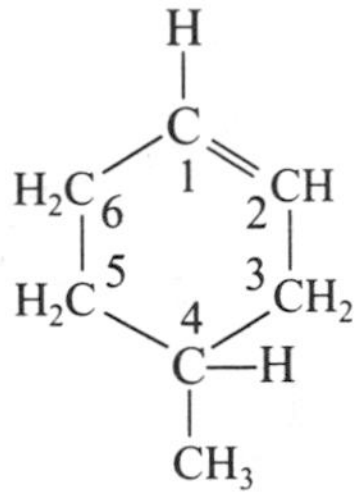

EXERCISE 4   Draw the structural formulas for *cis*-2-pentene and *trans*-2-pentene.

Solution: The *cis*-prefix indicates that the two non-hydrogen groups, each attached to a different carbon atom, are located on the same side of the double bond; in the *trans*- form they are located on opposite sides. The two geometrical isomers are:

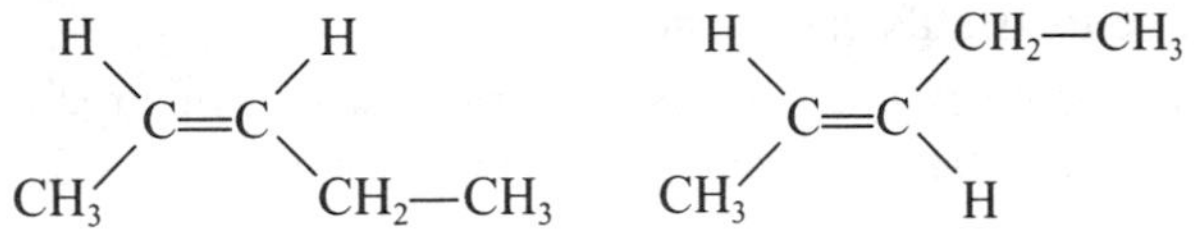

EXERCISE 5   Write the structural formulas of the organic substances formed when the following react: **(a)** 1-butene and hydrogen in the presence of catalyst; **(b)** 2-butene burning in oxygen; **(c)** 1 mol each of 1-butene and chlorine gas.

Solution: Except for **(b)**, which is a combustion reaction, all are addition reactions:

(a) $CH_2{=}CH{-}CH_2{-}CH_3 + H_2 \xrightarrow{Pt} CH_3{-}CH_2{-}CH_2{-}CH_3$

(b) $CH_3{-}CH{=}CH{-}CH_3 + 6O_2 \longrightarrow 4CO_2 + 4H_2O$

(c) 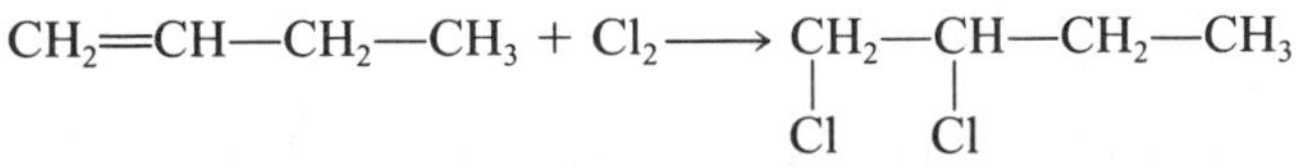 $CH_2{=}CH{-}CH_2{-}CH_3 + Cl_2 \longrightarrow \underset{\underset{Cl}{|}}{CH_2}{-}\underset{\underset{Cl}{|}}{CH}{-}CH_2{-}CH_3$

AROMATIC HYDROCARBONS

Aromatic hydrocarbons are cyclic hydrocarbons containing delocalized π electrons over several carbon atoms. The simplest example of an aromatic hydrocarbon is benzene, C_6H_6, which is discussed in section 9.5 of the text.

▪ The simplest way of showing an aromatic benzene ring is by the picture

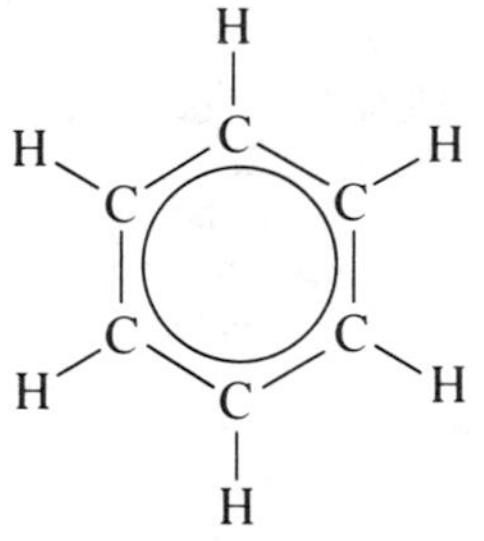

or

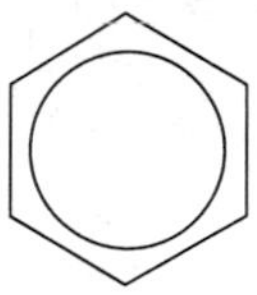

This picture is a shorthand description for the following arrangement of carbon and hydrogen atoms:

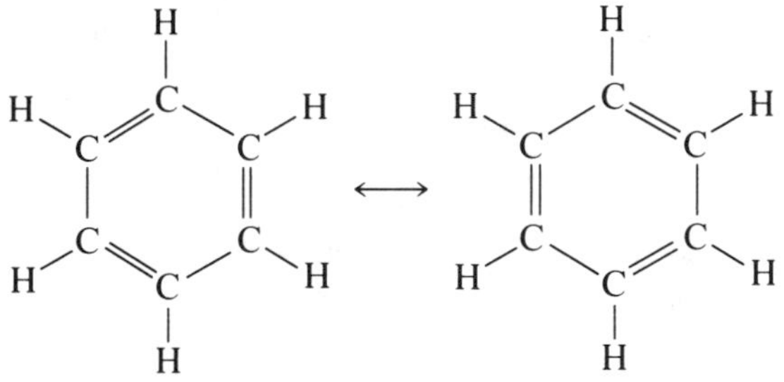

The circle in the middle of the ring indicates that the three pi bonds are delocalized over the entire ring structure.

▪ Several benzene rings can be fused together to form more complex structures. Look at Figure 26.10 in the text to see examples and the numbering system used to identify carbon atoms.

▪ There are three possible isomers of benzene when two groups are attached: ortho- (*o*-), meta- (*m*-) and para- (*p*). Examples are

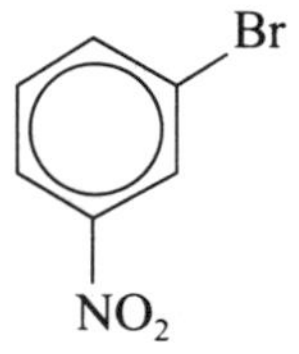

m-Bromonitrobenzene 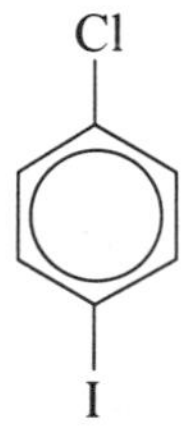   *p*-Chloroiodobenzene 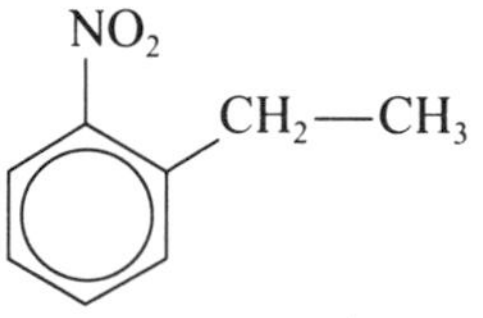   *o*-Ethylnitrobenzene

▪ Aromatic hydrocarbons do not readily undergo addition reactions; the most common type of reaction is substitution.

▪ Hydrogen atoms can be replaced by other atoms or groups such as —NO_2, —Br, —CH_3 and —OH.

▪ An important type of substitution reaction involving a catalyst are *Friedel-*

Crafts reactions. Positively charged species are created by substances such as H_2SO_4, $FeCl_3$ and $AlCl_3$. The charged reactant attacks the aromatic ring and a hydrogen ion is eventually lost from the ring.

EXERCISE 6   What are the expected products if cyclohexene and benzene are reacted with Br_2 in the solvent CCl_4?

   Solution: Cyclohexene contains one double bond that should undergo rapid addition with bromine as follows:

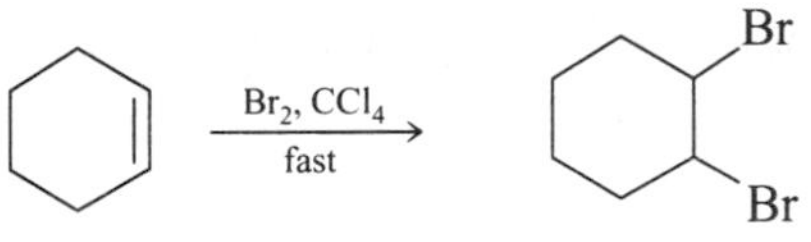

Cyclohexene   1,2-Dibromocyclohexane

A similar reaction with benzene does not occur. Remember that aromatic hydrocarbons do not readily undergo addition reactions. Br_2 reacts with benzene in the presence of $FeBr_3$, a catalyst, to form bromobenzene.

$$C_6H_6 + Br_2 \xrightarrow{\text{FeBr}_3} C_6H_5 - Br \; + H \!-\! Br$$

   Benzene   Bromobenzene

EXERCISE 7   Write formulas for the following **(a)** Ethylbenzene; **(b)** toluene (methylbenzene); **(c)** o-dichlorobenzene.

   Solution: **(a)** Most monosubstituted compounds of benzene are named by placing the name of the substituent in front of the word benzene. In this case the substituent is ethyl, C_2H_5.

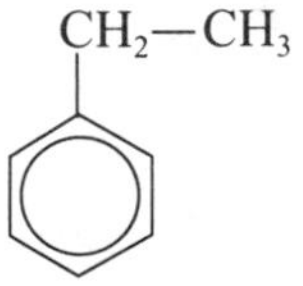

Ethylbenzene

(b) Toluene is the name given to benzene with a methyl group attached

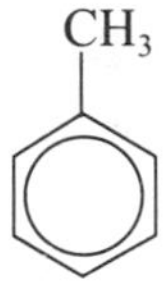

Toluene
(*not* methylbenzene)

(c) o-Dichlorobenzene is a disubstituted benzene with the chloro- groups in the ortho position

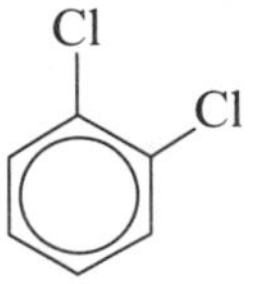

o-Dichlorobenzene
(*ortho* isomer)

HYDROCARBON DERIVATIVES

The reactivity of hydrocarbons depends on what atoms or groups are attached to carbon atoms, or on the types of carbon-carbon bonds. These reactive sites are called **functional groups.** Compounds containing functional groups other than carbon double and triple bonds are called *hydrocarbon derivatives.* The reactive sites (functional groups) discussed in the text are

$$\begin{array}{cc} \diagdown \diagup \\ \diagup C = C \diagdown \end{array} \qquad -C \equiv C-$$

—Cl (chloro)   —OH (hydroxo)

—C=O (carbonyl)—C=O (carboxyl)
                        |
                        OH

—NH$_2$ (amine)

The functional group —OH occurs in compounds called **alcohols** (R—O—H, where R is an alkyl group) and **phenols** (Ar—OH, where Ar is an aromatic group).

- The names of alcohols end in -ol, as in propanol ($CH_3CH_2CH_2OH$). Alcohols are formed by the reaction between water and an alkene in the presence of H_2SO_4.

Ethers are structurally related to alcohols; in ethers an oxygen atom holds two alkyl or aromatic groups, or one alkyl and one aromatic group (R-O-R, Ar-O-Ar, or R-O-Ar).

- Ethers are formed by alcohols in the presence of strong acids, such as sulfuric and phosphoric:

$$ROH + HOR' \xrightarrow{H_2SO_4} H_2O + ROR'$$

- The previous reaction is called a **condensation reaction.** Two reagents are combined, often with the elimination of water.

Alcohols can be oxidized to form aldehydes and ketones; both contain the **carbonyl** functional group,

$$\begin{array}{c} \diagdown \\ \diagup C = O \end{array}$$

- In a **ketone,** two organic groups are bonded to the carbon atom of the carbonyl group

$$\begin{array}{c} R-C-R' \\ \| \\ O \end{array}$$

• In an **aldehyde** at least one hydrogen atom is bonded to the carbon atom of the carbonyl group,

$$R-\underset{\underset{O}{\|}}{C}-H$$

Carboxylic acids contain the carboxyl functional group,

$$-\underset{\underset{OH}{|}}{C}=O$$

• This structure is sometimes abbreviated —COOH or —CO$_2$H.

• The carboxyl functional group has distinctive chemical properties different from those of the carbonyl functional group because the OH and C=O groups interact with each other.

• Oxidation of aldehydes produces carboxylic acids.

Esters are substances of the following sort:

$$R'-\underset{\underset{O}{\|}}{O}-O-C-R \quad \text{Ester linkage}$$

They form from the condensation reaction between an alcohol and a carboxylic acid

$$\underset{\text{acid}}{R'=\underset{\underset{O}{\|}}{C}-OH} + \underset{\text{alcohol}}{ROH} \longrightarrow \underset{\text{ester}}{R'-\underset{\underset{O}{\|}}{C}-OR} + H_2O$$

A condensation reaction between an **amine** (a basic compound containing a nitrogen with a lone pair of electrons, RNH$_2$, for example) and a carboxylic acid produces **amides:**

$$\underset{\text{acid}}{R'=\underset{\underset{O}{\|}}{C}-OH} + \underset{\text{amine}}{RNH_2} \longrightarrow \underset{\text{amide}}{R'-\underset{\underset{O}{\|}}{C}-NHR} + H_2O$$

Amides contain the characteristic **amide linkage:**

$$R'-\underset{\underset{O}{|}}{C}-\underset{\underset{H}{|}}{N}-R \quad \text{Amide linkage}$$

EXERCISE 8   For each of the following alcohols, write the structural formula and write the structural formula of the ketone or aldehyde formed (if any) when the alcohol is oxidized: **(a)** 2-butanol; **(b)** 2-methyl-2-butanol; **(c)** ethanol.

Solution: The *-ol* ending tells you that a compound is an alcohol, and the prefix number before the name containing the *-ol* ending tells you the location of the OH group. **(a)** The structural formula is

$$CH_3—CH_2—\underset{\underset{OH}{|}}{CH}—CH_3$$

Since the carbon atom with the OH group has two other carbon atoms directly bonded to it, it can be oxidized. When oxidized, it forms a ketone; in this problem, it is methyl ethyl ketone,

$$CH_3—CH_2—\underset{\underset{O}{||}}{C}—CH_3$$

(b) The structural formula is

$$CH_3—\underset{\underset{OH}{|}}{\overset{\overset{CH_3}{|}}{C}}—CH_2—CH_3$$

Since there are three carbon atoms bonded to the carbon atom with the OH group, it can *not* be oxidized to $>C{=}O$. **(c)** The structural formula is

$$CH_3—CH_2—OH$$

Since there is only one carbon atom attached to the carbon atom with the OH group, it can be oxidized. Oxidation forms an aldehyde; in this case, acetaldehyde,

$$CH_3—\underset{\underset{O}{||}}{C}—H$$

EXERCISE 9   Which is more soluble in water, 1-butanol or 1-octanol? Explain.
   Solution: 1-butanol is more soluble in water than 1-octanol. The attraction of the polar OH group in 1-octanol toward polar water molecules is not sufficient to "pull" the nonpolar, eight-carbon-atom chain of 1-octanol into solution. The nonpolar, four-carbon atom chain in 1-butanol is sufficiently small that it is surrounded by water molecules attracted to the polar OH group. This hydration of the polar and nonpolar components of 1-butanol enables it to be soluble in water.

EXERCISE 10   Write structural formulas for each of the following compounds: **(a)** phenyl acetate; **(b)** benzoic acid; **(c)** acetamide.
   Solution: **(a)** The "ate" ending of the name phenyl acetate tells you that the carboxyl group of acetic acid is in the form

$$—\underset{\underset{O}{||}}{C}—O—$$

The phenyl prefix tells us that a phenyl group has replaced the hydrogen atom of the hydroxyl unit in the carboxyl group. The formula is

$$CH_3-\overset{\overset{\textstyle O}{\|}}{C}-O-\bigcirc$$

(b) The structural formula of benzoic acid is

$$\bigcirc-\overset{\overset{\textstyle O}{\|}}{C}-OH$$

(c) The "amide" ending of acetamide tells you that the —NH_2 group has replaced the —OH group of acetic acid:

$$CH_3-\overset{\overset{\textstyle O}{\|}}{C}-NH_2$$

Analysis of organisms shows that more than 90 percent of their dry weight is comprised of macromolecules, polymers of high-molecular weight. In Chapter 26, you will study three broad classes of biopolymers: proteins, carbohydrates, and nucleic acids. Each of these groups of biopolymers is discussed and reviewed.

Amino acids, the building blocks of proteins, are carboxylic acids with an amine group at the carbon atom next to the carboxylic acid group (the α-carbon):

Generalized formula for substituents at α-carbon — α-carbon — Amine group — Carboxylic acid group

Proteins are the fundamental constituent of cells and tissues in the human body. They are large molecules with molecular weights varying from 10,000 to over 50 million amu.

- The simplest proteins are composed entirely of amino acids; conjugated proteins consist of simple proteins bound to other kinds of biochemical structures.
- The characteristic bonding linkage between amino acids in proteins is the amide linkage (see Section 26.3 of the text):

Amide linkage, referred to as **peptide linkage** in proteins

- **Polypeptides** are proteins that have molecular weights smaller than about 10,000 amu. Characteristically, polypeptides consist of about 50 or fewer amino acids bonded together via peptide linkages.

Proteins have a specific "shape" or conformation that enables them to perform specific biological functions. Section 27.5 describes three structures of proteins that together form the conformation.

- The specific sequence of amino acids in a protein is the *primary structure*. The primary structure contains the information needed by the protein to form the other two structures.

- The orientation in space of the primary structure is the *secondary structure*. Secondary structures arise from the formation of hydrogen bonds between the carbonyl group of one amino acid and an amino group of another. The alpha and triple helix structures are part of the secondary structure.

- A secondary structure undergoes twisting and folding to form a three-dimensional shape that is the *tertiary structure*. This folding leads to globular structures.

Most biological catalysts, called **enzymes,** have a protein as a principal structural component.

EXERCISE 11   Write the structural representations of the two peptides that form by condensation reactions between serine,

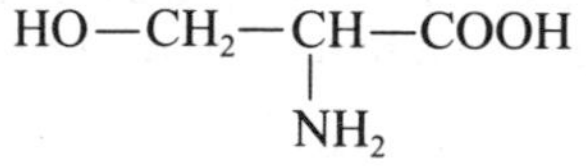

$$HO-CH_2-CH-COOH$$
$$| $$
$$NH_2$$

and alanine

$$CH_3-CH-COOH$$
$$|$$
$$NH_2$$

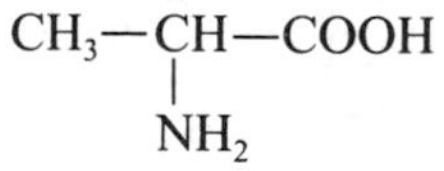

Name each dipeptide using the nomenclature rules found in Section 27.5 of the text.
   Solution: Two dipeptides are possible because a condensation reaction between the —NH$_2$ group of alanine and the —COOH group of serine (structure *a*), or between the —NH$_2$ group of serine and the —COOH group of alanine (structure *b*) can occur:

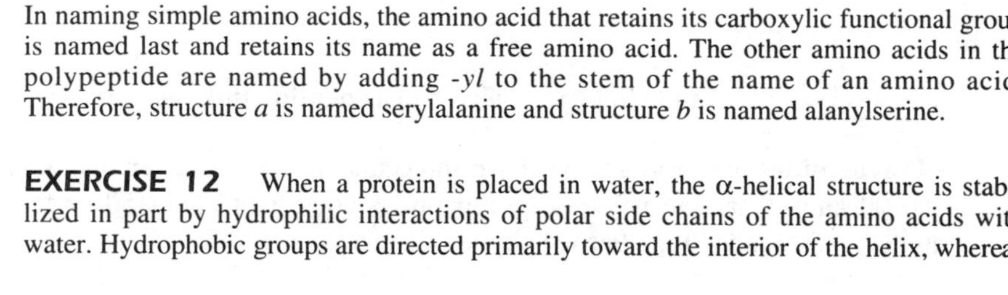

peptide linkage

In naming simple amino acids, the amino acid that retains its carboxylic functional group is named last and retains its name as a free amino acid. The other amino acids in the polypeptide are named by adding -*yl* to the stem of the name of an amino acid. Therefore, structure *a* is named serylalanine and structure *b* is named alanylserine.

EXERCISE 12   When a protein is placed in water, the α-helical structure is stabilized in part by hydrophilic interactions of polar side chains of the amino acids with water. Hydrophobic groups are directed primarily toward the interior of the helix, whereas hydrophilic groups are directed primarily toward the aqueous surroundings. Why can the folded structures of proteins be altered in weakly polar organic solvents?

Solution: When a protein is placed in weakly polar organic solvents, the hydrophobic groups in the protein polymer chain are not restricted to the interior of the α-helical structure as in the solvent water because such organic solvents are also hydrophobic substances (not attracted to polar water). The movement of hydrophobic groups in a protein structure from the interior of the α-helix to the exterior in an organic solvent causes an unfolding of the α-helix. This unfolding of the α-helix changes the structure of the protein; thus it is said to be denatured.

ENERGY SOURCES: CARBOHYDRATES

One of the steps in the process of energy production in organisms involves the breaking down of complex carbohydrates into their precursory components: sugars.

- An important simple carbohydrate is **glucose** ($C_6H_{12}O_6$), a sugar. In solution, glucose forms a hemiacetal structure: a cyclic arrangement of five carbon atoms and one oxygen atom with appropriate groups attached.

- Other simple sugars also exist. One example is **fructose,** which most commonly has a cyclic arrangement of four carbon atoms and one oxygen atom. See Equation [26.21] in the text.

In general, **carbohydrates** are polyhydroxy aldehydes or ketones with empirical formulas $C_y(H_2O)_x$ and polymers derived from the polyhydroxy aldehydes and ketones.

- **Monosaccharides** are monomeric sugars that cannot be further broken apart by hydrolysis (see Figure 26.23 in text).
- **Polysaccharides** are polymers formed from monosaccharides combining.
- Starch, glycogen, and cellulose are all polysaccharides formed from the glucose monomer.

EXERCISE 13   Identify the following structures as glucose or fructose.

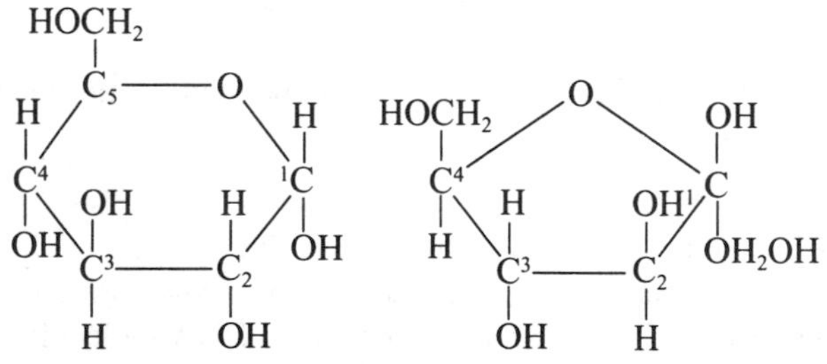

(a)   (b)

Solution: Structure **(a)** represents glucose. The glucose form of a sugar contains a cyclic arrangement of *five* carbon atoms and one oxygen atom. Structure **(b)** represents fructose. The fructose form of a sugar contains a cyclic arrangement of *four* carbon atoms and one oxygen atom.

EXERCISE 14   In this chapter you learned that a disaccharide is formed by linking two monosaccharide molecules and eliminating a water molecule. How are the disaccharides maltose and sucrose formed, and draw their structures? Indicate whether they have alpha or beta linkages.

Solution: Figure 26.23 in your text may be useful in answering this question. Sucrose is formed by the combination of glucose and fructose. The sugar units are also joined by an alpha linkage:

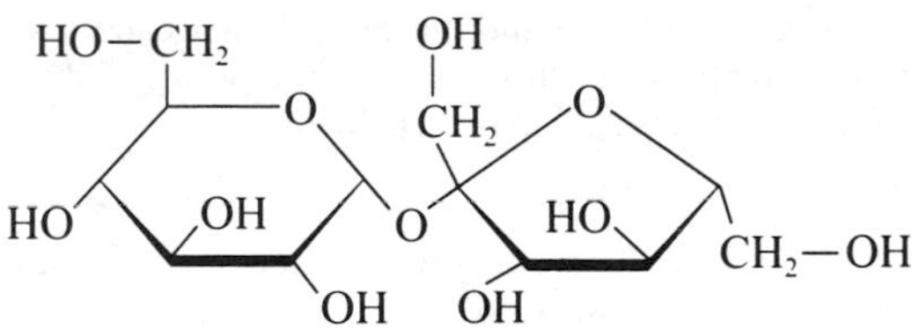

sucrose

Maltose is formed by the combination of glucose and glucose with an alpha linkage:

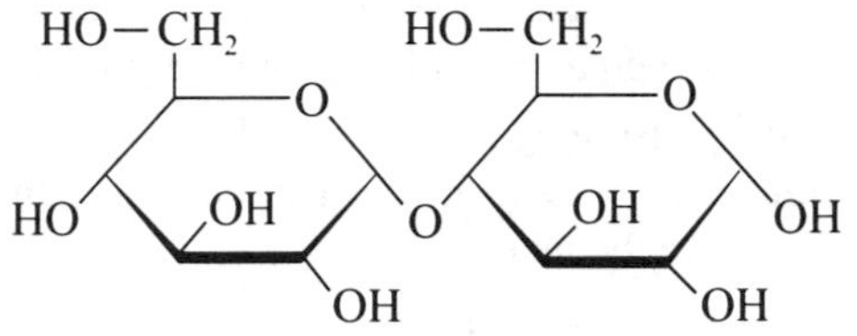

maltose

NUCLEIC ACIDS: DNA AND RNA

Chromosomes, which contain genes, consist of macromolecular substances called nucleic acids. **Nucleic acids** are bipolymers formed from repeating units called **nucleotides.**

- Look at Figure 26.28 in the text and note the three parts of nucleotide units: (a) a phosphoric acid unit, (b) a sugar in the furanose (five-membered ring) form (ribose type), and (c) a nitrogen-containing organic base adenine.
- Polynucleotides form by condensation reactions between an —OH unit of phosphoric acid with an —OH unit of a sugar unit; an ester-like linkage is formed. See Figure 26.29 in the text.

Nucleic acids are classified into two groups: **deoxyribonucleic acids** (DNA) and **ribonucleic acids** (RNA).

- DNA molecules are found almost exclusively in the nuclei of cells and are bipolymers of very high molecular weights. Nucleotide units within DNA contain a sugar unit known as deoxyribose. DNA molecules exist as double-stranded polymers wound in the form of a double helix.
- RNA molecules are found primarily in the cytoplasm (the cellular medium in which a nucleus is embedded) and are biopolymers of considerably lower molecular weights than DNA molecules. Nucleotide units within RNA molecules contain the sugar ribose. Ribose contains an —OH unit substituted for a hydrogen atom in deoxyribose.

EXERCISE 15 Identify the following rings as either ribose or deoxyribose:

(a)   (b)

Solution: Figure **(b)** represents deoxyribose, and Figure **(a)** represents ribose. Note that at carbon atom 3 in ribose there is an —OH group, whereas at the same carbon atom in deoxyribose there is a hydrogen atom.

EXERCISE 16   What effect might heating a DNA solution have upon the double-helix structure of DNA.

Solution: The strands of a DNA double helix are held together primarily by hydrogen bonds. When a solution of DNA is heated, the increased kinetic energy of the polymer causes the hydrogen bonds to break and the strands to separate. Thus, the double-helix structure is destroyed, or denatured.

SELF-TEST QUESTIONS

Key Terms

Having reviewed the key terms listed at the end of Chapter 26 and their applications, identify the following statements as true or false. If a statement is false, indicate why it is incorrect. Key terms are italicized in the statements.

26.1   *Aromatic hydrocarbons* usually undergo substitution reactions more readily than aliphatic hydrocarbons.

26.2   The compound ethane is an example of a class of aliphatic hydrocarbons known as *alkanes*.

26.3   Benzoic acid belongs to the class of hydrocarbons known as *alkynes*.

26.4   A *functional group* in an organic compound must consist of two or more atoms.

26.5   The characteristic functional group of an *alcohol* is

26.6   Under appropriate chemical conditions, ethane undergoes *addition reactions*.

26.7   Phenyl butyrate contains as a component a *cycloalkane*.

26.8   The *carbonyl* functional group is a characteristic feature of ketones.

26.9   1-Bromo-2-chloroethylene exhibits three *geometrical isomers*.

26.10   An example of an *alkyl* group is CH_3CH_2, the ethyl group.

26.11   *Ethers* have a characteristic C=O group as in

26.12   An *aldehyde* functional group is an organic compound of the following type:

26.13   A *ketone* has the following functional group representation

26.14   *Carboxylic* acids contain the —COOH functional group.

26.15   The condensation of an alcohol with a carboxylic acid gives a compound that contains an *ester* linkage.

26.16   *Saponification* is the hydrolysis of an ester in the presence of an acid.

26.17   *Amines* contain the —NR_2 group where R represents either an alkyl or aryl group only.

26.18   The presence of the functional group —$\overset{\overset{\displaystyle O}{\|}}{C}NR_2$ in a compound tells you that it is an *amide* organic compound.

26.19 *Alkenes* are less reactive than alkanes.

26.20 We view the *biosphere* as consisting of the atmosphere and solid earth.

26.21 At the two ends of the polymer chain in a *protein*, we would find a free —NH_2 unit and a free —CO_2H unit.

26.22 Simple *carbohydrates* have the general formula $C_x(H_2O)_y$.

26.23 *Nucleotides* condense together to form linear polymers.

26.24 All amino acids exist in nature in a *chiral* form.

26.25 If two molecules are nonsuperimposable mirror images of one another, they are *enantiomers*.

26.26 The sequence of amino acids along a protein chain is the *primary structure* of the protein.

26.27 Hydrogen bonds play an important role in determining the *secondary structure* of a protein.

26.28 The *tertiary structure* of a protein shows the linear arrangements of protein strands.

26.29 *DNA* molecules have lower molecular weights than *RNA* molecules.

26.30 In the *double-helix* structure for DNA, we find that adenine units on the two strands are paired for optimal hydrogen-bond interaction.

26.31 *Nucleic acids* are the primary storehouses of energy in humans.

26.32 *Cellulose* is a polysaccharide of glucose.

26.33 An important storehouse of carbohydrate energy in mammals is *glycogen*.

26.34 The *monosaccharide* fructose has the following linear form:

$$
\begin{array}{c}
CH_2OH \\
| \\
HO-C-H \\
| \\
H-C-OH \\
| \\
H-C-OH \\
| \\
CH_2OH
\end{array}
$$

26.35 Fructofuranose is an example of a *starch*.

26.36 *Biopolymers* are broadly classified into three broad groups: proteins, carbohydrates and nucleic acids.

26.37 A tripeptide consists of four *amino acids* bonded through three peptide bonds in a linear chain.

26.38 A *peptide* bond contains the following unit

$$
\begin{array}{c}
\quad\; O \\
\quad\; \| \\
-C-N- \\
\quad\;\;\; | \\
\quad\;\;\; H
\end{array}
$$

26.39 Proteins that have molar masses less than 6000 amu are called *polypeptides*.

26.40 The formula of *glucose* is $C_6O_{12}O_6$.

26.41 When *polysaccharides* are reacted in the presence of an acid catalyst, they form glucose and/or fructose.

Problems and Short-Answer Questions

26.42 Name the following compounds:

(a) $CH_3-C{=}CH_2-CH_3$
$$\qquad\;\; \begin{array}{c} | \\ CH_3 \end{array}$$

(b) $HC{\equiv}C-CH-CH-CH_3$
$$\qquad\qquad\;\; \begin{array}{cc} | & | \\ CH_3 & Br \end{array}$$

(c) HCOOH

(d) $\begin{array}{c} \quad\; O \\ \quad\; \| \\ H-C-NH_2 \end{array}$

(e) $\begin{array}{c} \quad\; O \\ \quad\; \| \\ H-C-O-CH_2-CH_2-CH_3 \end{array}$

26.43 Write and identify all geometrical isomers possible for 2-bromo-2-pentene.

26.44 Write the structural formulas for the products of the following reactions:
(a) 1 mol each of ethane and bromine in the presence of ultraviolet light;
(b) propene and hydrogen gas under pressure in the presence of platinum;
(c) 1 mol each of propyne and bromine;
(d) oxidation of CH_3CH_2CHO with potassium permanganate;
(e) 1 mol each of methanol and propionic acid;
(f) methyl-amine and propionic acid.

26.45 Explain why 1-butanol (molecular weight = 74 amu) has a higher boiling point than n-pentane (molecular weight = 72), although their molecular weights are almost the same.

26.46 Although benzene possesses delocalized π electrons, it is not considered an unsaturated hydrocarbon by most organic chemists. Why?

26.47 Draw all the structural and geometrical isomers of bromopropene.

26.48 Give the structural formula for an alcohol that is an isomer of dimethylether; for an aldehyde that is an isomer of methyl ethyl ketone.

26.49 Write a balanced chemical equation for the saponification of ethyl acetate.

26.50 Milk from cows contains about 5 percent of the sugar lactose:

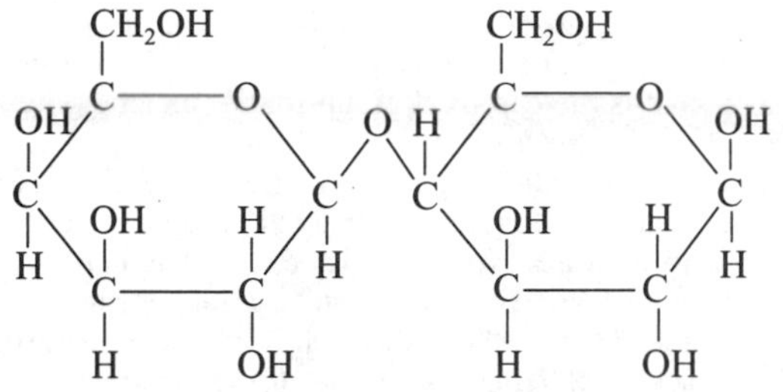

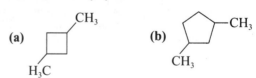

What type of saccharide is lactose?

26.51 Why is cellulose significantly less susceptible to hydrolysis in humans than starch?

26.52 Would you expect a compound indicated by the approximate formula $C_{2932}H_{4724}H_{828}S_8Fe_4O_{840}$ to consist primarily of a protein, a carbohydrate, or a small peptide?

26.53 Experimentally, the activity range for many enzymes occurs between 10°C and 50°C. As the temperature increases significantly above 50°C, the activity of enzymes decreases rapidly, and eventually all activity is destroyed. Explain why the activity of enzymes is destroyed at temperatures significantly greater than 50°C.

26.54 What is the principal difference between the structures of the sugar units in DNA and RNA molecules?

26.55 Write the structural formula of the tripeptide val-thr-gly. Refer to Figure 26.16 in the text for assistance in doing this problem.

Multiple-Choice Questions

26.56 Which of the following alcohols forms a ketone when oxidized?
- **(a)** 1-propanol
- **(b)** methanol
- **(c)** 2-methyl-2-propanol
- **(d)** 2-propanol
- **(e)** all of the above

26.57 Which of the following substances is the missing reactant in the reaction
$$\underline{\hspace{2cm}} \xrightarrow[140°C]{H_2SO_4} CH_3CH_2CH_2OCH_2CH_2CH_3 + H_2O$$
- **(a)** $CH_3CH_2CH_3$
- **(b)** $CH_3CH_2CH_2Cl$
- **(c)** $CH_3CH_2CH_2OH$
- **(d)** CH_3CH_2CHO
- **(e)** $CH_3CHOHCH_3$

26.58 Which of the following substances is an ester?
- **(a)** CH_2CH_2
- **(b)** CH_3CHO
- **(c)** $CH_3CH_2OCH_2CH_3$
- **(d)** CH_2OHCH_2OH
- **(e)** CH_3COOCH_3

26.59 Which of the following structures *best* represents a ketone unit? (X represents any single element.)

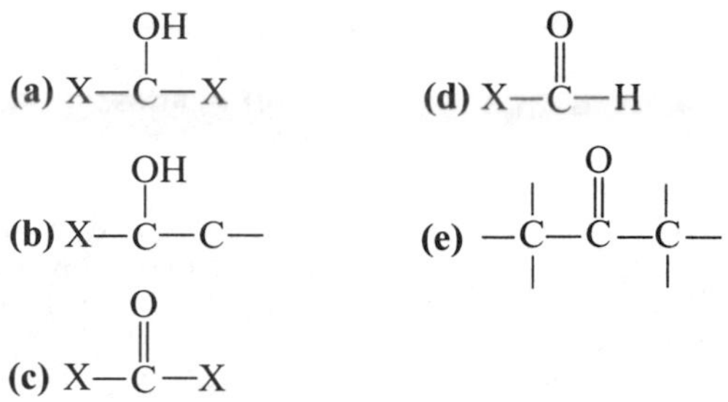

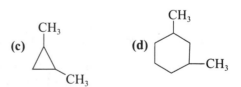

26.60 Which of the following names is associated with the compound

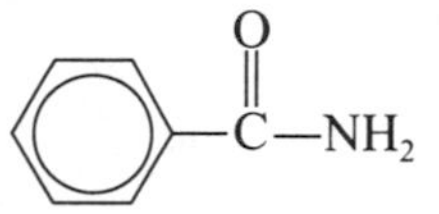

- **(a)** benzoic acid
- **(b)** ammonium benzoate
- **(c)** benzamide
- **(d)** benzoamine
- **(e)** benzene

26.61 Which of the following names best corresponds to the structural formula

$$\underset{CH_2CH_3}{\overset{H}{\,}}C=C\underset{CH_2CH_3}{\overset{H}{\,}}$$

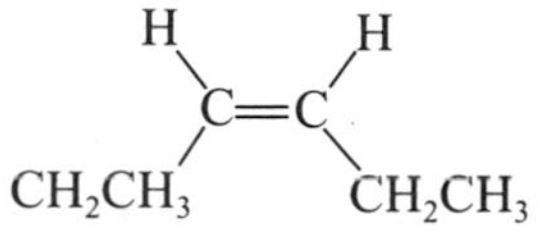

- **(a)** *trans*-3-hexene
- **(b)** *cis*-3-hexene
- **(c)** *trans*-hexane
- **(d)** *cis*-hexene
- **(e)** *cis*-diethylene

26.62 Which of the following figures best represents the structure of 1,3-dimethylcyclohexane?

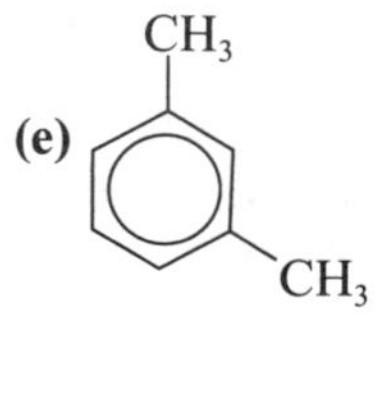

26.63 Which of the following is an expected hydrolysis product of DNA?

(a) PO_4^{3-} (d) glucose
(b) CO_2 (e) ribose
(c) H_2O

26.64 If blood-sugar concentration is too high in humans, excess glucose is converted in the liver to a substance that stores energy. Which of the following substances is formed in the liver in the described conversion?

(a) Amino acids   (d) DNA
(b) cellulose   (e) glycogen
(c) chlorophyll

26.65 Which amino acid in the polypeptide ser-phe-gly-ala-gly-lys has a terminal —CO_2H free group?

(a) ser   (d) ala
(b) phe   (e) lys
(c) gly

26.66 For glycine in water, $K_a = 1.6 \times 10^{-10}$ for the ionization of the —CO_2H group and $K_b = 2.5 \times 10^{-12}$ for the reaction of the —NH_2 group with water. In water, the expected predominant form of glycine is which of the following?

(a) 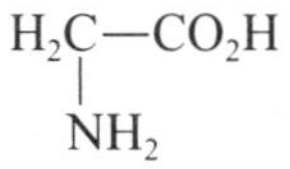 $H_2C—CO_2H$
      |
      NH_2

(e) none of the above

(b) 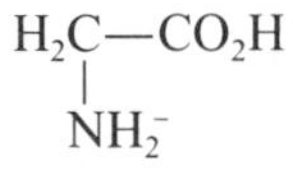 $H_2C—CO_2H$
      |
      NH_2^-

(c) 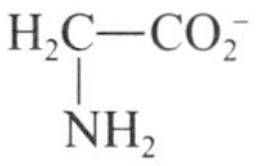 $H_2C—CO_2^-$
      |
      NH_2

(d) 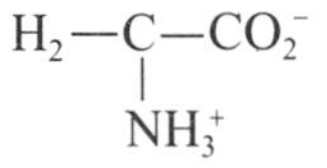 $H_2—C—CO_2^-$
        |
        NH_3^+

26.67 Natural silk from silkworms and spiders are linear biopolymers formed from simple proteins. The structure of natural silk shows a "linear" zigzag arrangement of protein molecules; the strands of protein are not coiled around one another. From this information, we can conclude that linear silk strands have which of the following structures?

(a) primary   (d) α-helix
(b) secondary   (e) double helix
(c) tertiary

26.68 Which of the following could cause breaking of internal bonds of a sulfide-containing protein: (1) reduction of internal disulfide bond; (2) acidification; (3) addition of base; (4) heating; (5) dilution with water?

(a) (1) only   (d) (1), (2), (3), (4)
(b) (4) only   (e) (1), (2), (3), (4), (5)
(c) (2), (3), (4)

SELF-TEST SOLUTIONS

26.1 True. **26.2** True. **26.3** False. Benzoic acid contains benzene and the carboxyl (—COOH) group. An alkyne contains a carbon-carbon triple bond. **26.4** False. A functional group may consist of one atom, such as Cl or Br. **26.5** False. The characteristic functional group of an alcohol is —OH; the functional group shown is the carboxyl group found in carboxylic acids. **26.6** False. Ethane does not contain a —C=C— group, a requirement for addition polymerization. **26.7** False. Phenyl butyrate is an ester containing a phenyl group (benzene minus one hydrogen atom), which is cyclic, but it is not a cycloalkane. A cycloalkane contains a cyclic arrangement of carbon atoms, but there is no delocalized π system as in an aromatic hydrocarbon. **26.8** True. **26.9** False. It exhibits two geometrical isomers, cis- and trans-. **26.10** True. **26.11** False. Ethers have the characteristic group R—O—R'. **26.12** True. **26.13** False. The carbonyl group in a ketone unit is bonded to two other carbons, not hydrogen atoms. **26.14** True. **26.15** True. **26.16** False. Saponification occurs in the presence of a base, not an acid. **26.17** False. A hydrogen group may also be present. **26.18** True. **26.19** False. A double bond in an alkene is more reactive than a single bond found in alkanes. **26.20** False. It also contains natural waters. **26.21** True. **26.22** True. **26.23** True. **26.24** False. All but glycine are chiral. **26.25** True. **26.26** True. **26.27** True. **26.28** False. The tertiary structure of a protein refers to the overall folding of protein strands upon themselves; the structure is not linear. **26.29** False. The reverse is true. **26.30** False. Adenine is paired with thymine. **26.31** False. Nucleic acids are storehouses of information that enable regulation of cell reproduction and development. **26.32** True. **26.33** True. **26.34** False. Fructose is a monosaccharide that is a polyhydroxy ketone. **26.35** False. Starch refers to a group of polysaccharides, not to a monosaccharide such as fructofuranose. **26.36** True. **26.37** False. A tripeptide consists of three amino acids linked together. **26.38** True. **26.39** False. Molar masses are in the range of 6000–50 million amu. **26.40** True. **26.41** True. **26.42** (a) 2-methyl-2-butene; (b) 4-bromo-3-methyl-1-pentyne; (c) formic acid; (d) formamide; (e) propyl formate.

26.43

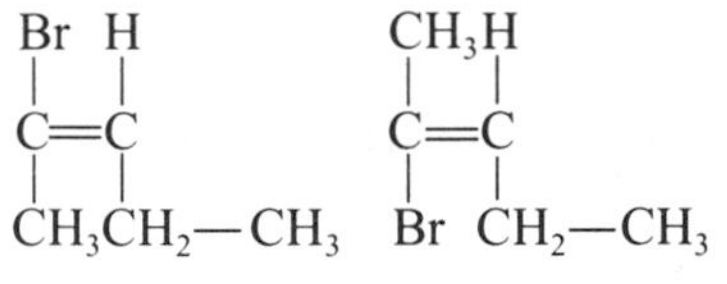

      Br H CH_3 H
       | | | |
       C = C   C = C
       | | | |
  CH_3CH_2 — CH_3 Br CH_2 — CH_3
      cis   *trans*

26.44 (a)

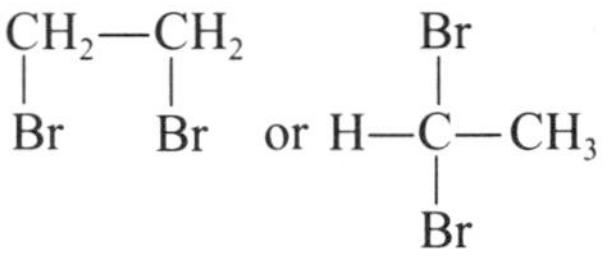

  CH_2 — CH_2 Br
   | | |
  Br   Br   or   H — C — CH_3
                           |
                          Br

(b) CH_3 — CH_2 — CH_2

(c) 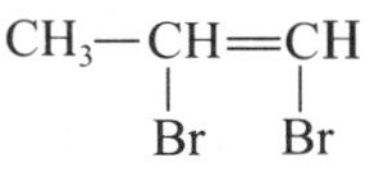 CH_3 — CH = CH
            | |
            Br   Br

(d) 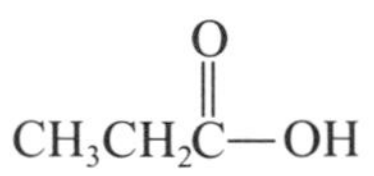   O
         ‖
   CH_3CH_2C — OH

(e) 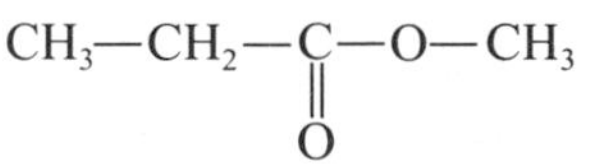 CH_3 — CH_2 — C — O — CH_3
                      ‖
                      O

(f) $CH_3-CH_2-\underset{\underset{O}{\|}}{C}-\underset{\underset{H}{|}}{N}-CH_3$

26.45 An alcohol has a higher boiling point than an alkane of about the same molecular weight because of the additional hydrogen-bond attractions between the OH functional groups in alcohol molecules. Similar attractive forces do not exist between alkanes. **26.46** The term "unsaturated" means that a carbon skeleton hold less than the maximum number of hydrogens that it might holds and implies that the skeleton could achieve a "saturated" condition by some reasonable reaction involving the π electrons localized between carbon atoms. The π electrons in benzene are delocalized around the ring and are not accessible for addition reactions characteristic of alkenes. The delocalization of the π electrons in aromatic hydrocarbons causes the carbon-carbon bonds to be quite stable toward addition reactions; they do not undergo the reactions characteristic of unsaturated compounds. **26.47** The chemical formula of bromopropene is C_3H_5Br. A double bond exists between two carbon atoms as indicated by the *-ene* ending of the chemical name. The structural and geometrical forms are

$$
\underset{\underset{H}{|}\;\;\underset{CH_2Br}{|}}{\overset{\overset{H}{|}\;\;\overset{H}{|}}{C=C}} \qquad
\underset{\underset{H}{|}\;\;\underset{CH_3}{|}}{\overset{\overset{Br}{|}\;\;\overset{H}{|}}{C=C}} \qquad
\underset{\underset{Br}{|}\;\;\underset{CH_3}{|}}{\overset{\overset{H}{|}\;\;\overset{H}{|}}{C=C}}
$$

26.48 Dimethyl ether is $H_3C-O-CH_3$. An alcohol has a $-OH$ group; thus, an isomer is H_3C-CH_2-OH, ethanol. Methyl ethyl ketone is

$$CH_3-\underset{\underset{O}{\|}}{C}-CH_2CH_3$$

An aldehyde has a carbonyl group with a hydrogen atom attached; thus an isomer that is an aldehyde is

$$H\underset{\underset{O}{\|}}{C}-CH_2CH_2CH_3$$

26.49 Saponification is the base hydrolysis of an ester to form a salt of the carboxylic acid:

$$CH_3-\underset{\underset{O}{\|}}{C}-CH_2CH_3 + NaOH \longrightarrow$$

$$CH_3-\underset{\underset{O}{\|}}{C}-O^-\,Na^+ + CH_3CH_2OH$$

26.50 Disaccharide; it contains two glucose units. **26.51** Cellulose contains glucose units connected by β-linkages, whereas starch contains glucose units connected by α-linkages. Enzymes present in the human body can hydrolyze polysaccharides containing α-linkages, but not those with β-linkages. **26.52** The compound is very large, as indicated by the large number of carbon, hydrogen, and nitrogen atoms. Thus small peptides are eliminated because they do not contain that many carbon atoms—nor iron or sulfur atoms. Carbohydrates are eliminated because they do not contain nitrogen, sulfur, or iron atoms. Proteins may contain carbon, hydrogen, nitrogen, oxygen, and sulfur atoms. Experimentally, this is the approximate formula for oxyhemoglobin, an iron-containing protein. **26.53** Proteins are denatured by the application of heat. At temperatures significantly greater than 50°C, enzymes are denatured; consequently, their activity is lost. **26.54** The sugar unit of RNA contains one $-OH$ group, whereas that of DNA contains no $-OH$ unit (see chemical diagram in Exercise 6). **26.55**

$$CH_3-\underset{\underset{CH_3}{|}}{CH}-\underset{\underset{NH_2}{|}}{CH}-\underset{\underset{O}{\|}}{C}-\underset{\underset{H}{|}}{N}-\underset{\underset{\underset{\underset{CH_3}{|}}{CH}}{|}}{CH}-\underset{\underset{O}{\|}}{C}-\underset{\underset{H}{|}}{N}-CH_2-COOH$$

$$\underbrace{\qquad\qquad}_{\text{Valine}} \quad \underbrace{\qquad\qquad}_{\text{Threonine}} \quad \underbrace{\qquad\qquad}_{\text{Glycine}}$$

26.56 (d). 26.57 (c). 26.58 (e). 26.59 (e). Structure **(c)** is not correct because if X = H, we have an aldehyde. **26.60 (c). 26.61 (b). 26.62 (d). 26.63 (a). 26.64 (e). 26.65 (e). 26.66 (d). 26.67 (a). 26.68 (e).**

Sectional Test 7

CHAPTERS 25–26

You should make this test as realistic as possible. **Tear out this test so that you do not have access to the information in the** *Student's Guide*. Use only data provided in the questions and a periodic table. Do not check your answers until you are finished. If you answer questions incorrectly, review the section in the text indicated after each question.

Choose the best response for each question.

1. What is the oxidation number of iron in $[Fe(NH_3)_4(H_2O)_2](NO_3)_3$?
   (**a**) +1 (**b**) +2 (**c**) +3 (**d**) +4
   [25–1]

2. What is the coordination number of chromium in $[Cr(en)_2Cl_2]^+$?
   (en = ethylenediamine)
   (**a**) 2 (**b**) 4 (**c**) 6 (**d**) 8
   [25–1]

3. What is the formula of a coordination compound containing potassium as the cation and a complex anion containing chromium (III) bound to two ammonia molecules and four chloride ions?
   (**a**) $K[Cr(NH_3)_2Cl_4]$ (**b**) $K_2[Cr(NH_3)_2Cl_4]$ (**c**) $K_3[Cr(NH_3)_2Cl_4]$
   (**d**) $K_4[Cr(NH_3)_2Cl_4]$
   [25–1]

4. Which of the following ligands has six donor atoms and is a chelate?
   (**a**) Ammonia (**b**) Ethylenediamine (**c**) Oxalate ion
   (**d**) Ethylenediaminetetraacetate ion
   [25–2]

5. Which statement is *not* true?
   (**a**) Bipyridine contains two nitrogen donor atoms and is a chelate.
   (**b**) In general, monodentate ligands form more stable complexes than do similar chelating agents.
   (**c**) The chelate effect results from a more favorable entropy change for complex formation involving polydentate ligands.
   (**d**) $[Cd(H_2NCH_2CH_2NH_2)_2]^{2+}(aq)$ should be more stable than $[Cd(CH_3NH_2)]^{2+}(aq)$.
   [25–2]

6. Which ligand, when it is complexed, is *incorrectly* named?

    (a) Cl^-, chloro (b) NH_3, amine (c) H_2O, aqua (d) CN^-, cyano

    [25–3]

7. Which of the following complexes is *incorrectly* named?

    (a) $[Zn(en)Cl_2]$, dichloroethylenediaminezinc(II)

    (b) $[PtCl_2Br_2]Br_2$, dibromodichloroplatinum(VI) bromide

    (c) $[V(en)_2Cl_2]^+$, dichlorodi(ethylenediamine)vanadium(I) ion

    (d) $[Ni(H_2O)_6]^{2+}$, hexaaquanickel(II) ion

    [25–3]

8. Which of the following is *correctly* named?

    (a) $[Fe(CN)_4Cl_2]^{3-}$, dichlorotetracyanoironate(III) ion

    (b) $[Co(NH_3)_6]_2(SO_4)_3$, hexaamminecobalt(III) sulfato

    (c) $Na[MnO_4]$, sodium tetraoxidemanganate(VII)

    (d) $[HgBr_4]^{2-}$, tetrabromomercurate(II) ion

    [25–3]

9. How many geometric isomers are there for $[CoBr_4Cl_2]^{3-}$?

    (a) 1 (b) 2 (c) 3 (d) 4

    [25–4]

10. Which of the following statements is true given that $[Pt(NH_3)_2ClBr]$ has two isomers?

    (a) The complex must be square planar.

    (b) The complex must be tetrahedral.

    (c) The complex must be octahedral.

    (d) The isomers are chiral.

    [25–4]

11. Which of the following complexes would exhibit optical isomerism?

    (a) $[CoCl_3Br_3]^{3-}$ (b) $[Co(ox)_3]^{3-}$ (c) *trans*-$[Co(en)_2Cl_2]^+$

    (d) $[Ag(NH_3)_2^+]$

    [25–4]

12. Which of the following ions forms inert complexes?

    (a) Cr(III) (b) Cu(II) (c) Zn(II) (d) Ni(II)

    [25–5]

13. Which of the following in the absence of a crystal field is paramagnetic?

    (a) K^+ (b) Zn^{2+} (c) Cl^- (d) Fe^{2+}

    [25–7]

14. Which pair of orbitals have equal energy in an octahedral crystal field?

    (a) $d_{x^2-y^2}$, d_{z^2} (b) $d_{x^2-y^2}$, d_{xy} (c) d_{z^2}, d_{xz} (d) d_{z^2}, d_{yz}

    [25–8]

15. Which orbital has the highest energy in a tetrahedral field?

    (a) $3p_z$ (b) $3d_{z^2}$ (c) $3d_{x^2-y^2}$ (d) $3d_{xy}$

    [25–8]

16. Which ligand would you expect to create the largest Δ in an octahedral complex of a first row transition element?

    (a) CN^- (b) en (c) Cl^- (d) H_2O

    [25–8]

17. How many unpaired electrons does copper in $K_2[Cu(CN)_4]$ possess?

    (a) 1 (b) 2 (c) 3 (d) 4

    [25–8]

18. How many unpaired electrons does cobalt in $K_3[Co(CN)_6]$ possess?

(a) 0 (b) 1 (c) 2 (d) 3

[25–8]

19. Which of the following is the correct crystal field diagram for $MnCl_6^{3-}$?

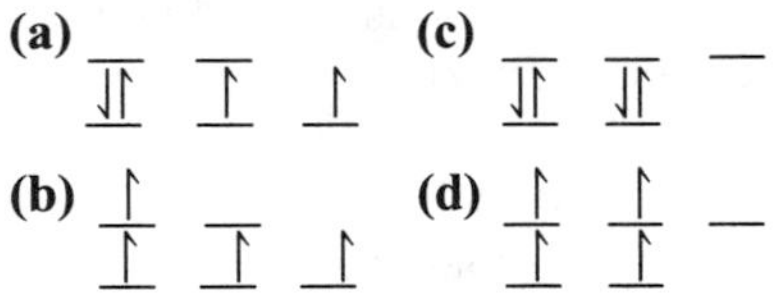

[25–8]

20. Which of the following formulas represents an alkene?

(a) $CH_3CH_2CH_3$ (b) CH_3CH_3 (c) $CH_3CH_2CHCH_2$ (d) CH_3CH_2Cl

[26–2]

21. Which of the following is 3,4-dimethylhexane?

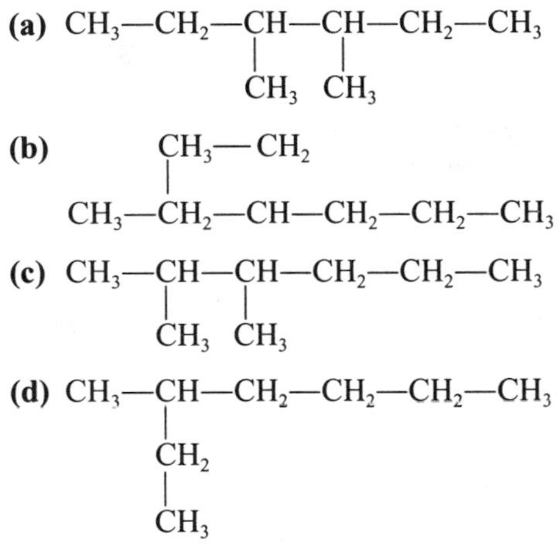

[26–1]

22. What is the structural formula for 1,3-*cis*-hexadiene?

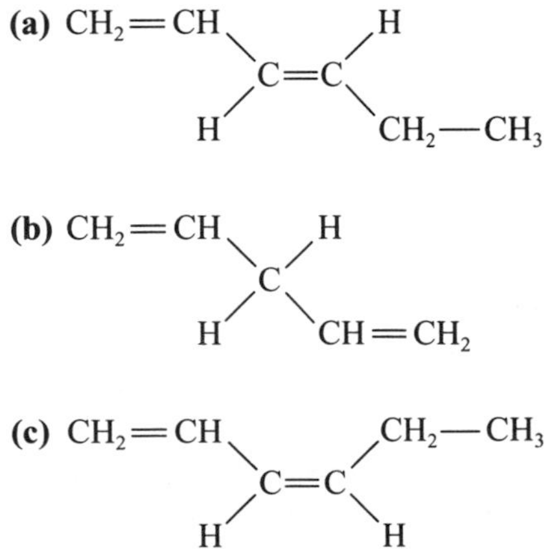

(d) all of the above

[26–2]

23. Which range of carbon atoms in petroleum fractions gives rise to the highest boiling range compounds?

24. Which hydrocarbon would you expect to have the highest octane number?

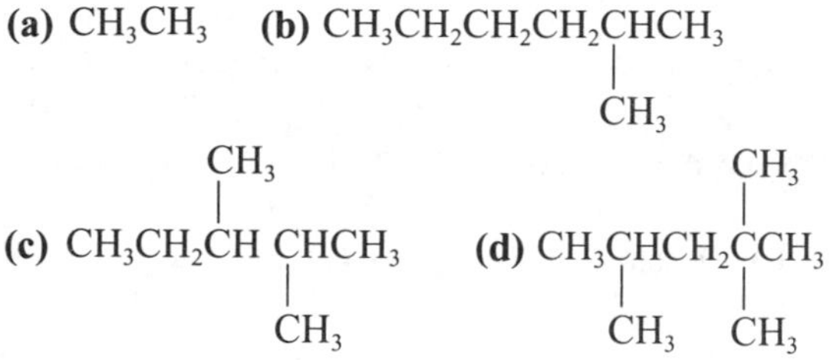

(a) CH_3CH_3 (b) $CH_3CH_2CH_2CH_2CHCH_3$
$|$
CH_3

(c) $CH_3CH_2CH\ CHCH_3$ (d) $CH_3CHCH_2CCH_3$
with CH_3 branches

[26–1]

25. What is the expected product formed from the reaction between 2-butene and Cl_2?

(a) 1-chlorobutane
(b) 2-chlorobutane
(c) 2,3-dichlorobutane
(d) 2,2-dichlorobutane

[26–2]

26. Which of the following is an alcohol?

(a) CH_3Br (b) $CH_3CH_2CH_2OH$

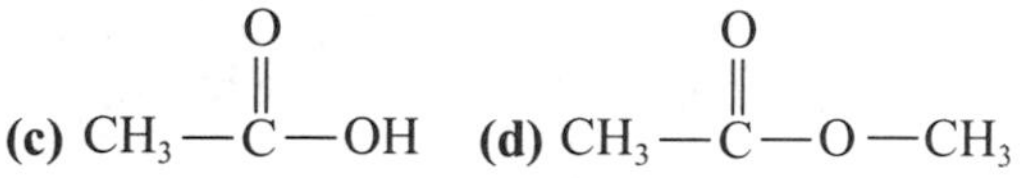

(c) $CH_3-\overset{O}{\overset{||}{C}}-OH$ (d) $CH_3-\overset{O}{\overset{||}{C}}-O-CH_3$

[26–4]

27. In question 27, which one is an ester?
[26–4]

28. What compound is formed from the reaction between ethanol and propanol in the presence of sulfuric acid?

(a) $CH_3CH_2-O-CH_2CH_3$ (b) $CH_3CH_2\overset{O}{\overset{||}{C}}-OCH_3$
(c) $CH_3CH_2-O-CH_2CH_2CH_3$ (d) $CH_3CH_2CH_2CH_2CH_2O$

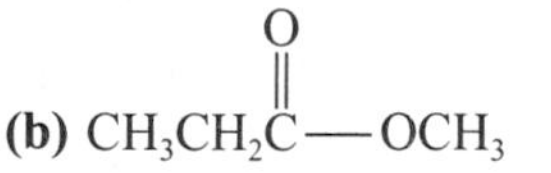

[26–4]

29. Which functional group is characteristic of a ketone?

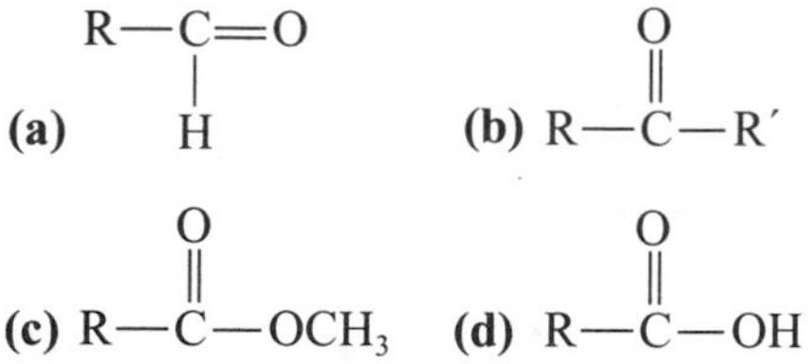

(a) $R-\underset{H}{\overset{C=O}{|}}$ (b) $R-\overset{O}{\overset{||}{C}}-R'$

(c) $R-\overset{O}{\overset{||}{C}}-OCH_3$ (d) $R-\overset{O}{\overset{||}{C}}-OH$

[26–4]

30. What is the name of ⟨benzene ring⟩$-\overset{O}{\overset{||}{C}}-OCH_3$?

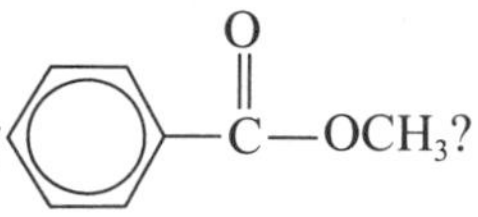

(a) Methylbenzoic acid   (b) Methylbenzocarboxylic acid
(c) Methyl benzoate   (d) Benzomethylate

[26–4]

31. Which of the following is a monosaccharide?

    (a) Maltose (b) Starch (c) Cellulose (d) Glucose

    [27–4]

32. Glucose has an aldehyde unit whereas fructose has a _____ unit instead.

    (a) Carboxylic acid (b) Ketone (c) Ester (d) Ether

    [27–4]

33. Which substance is produced in the body and is a kind of energy bank?

    (a) Starch (b) Sucrose (c) Glycogen (d) Cellulose

    [27–1, 27–4]

34. Which of the following represents a peptide linkage?

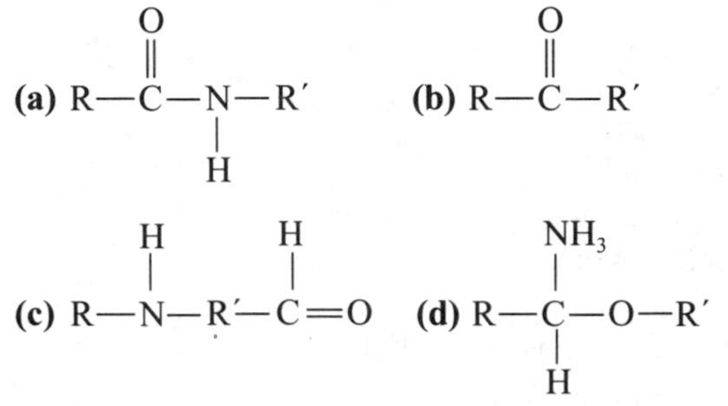

$$\text{(a) } R-\overset{\displaystyle \overset{O}{\|}}{C}-\underset{\underset{H}{|}}{N}-R' \qquad \text{(b) } R-\overset{\displaystyle \overset{O}{\|}}{C}-R'$$

$$\text{(c) } R-\underset{\overset{|}{H}}{N}-\underset{\overset{|}{H}}{R'}-C{=}O \qquad \text{(d) } R-\underset{\underset{H}{|}}{\overset{\overset{NH_3}{|}}{C}}-O-R'$$

    [27–2]

35. Which of the following is (are) an alpha amino acid?

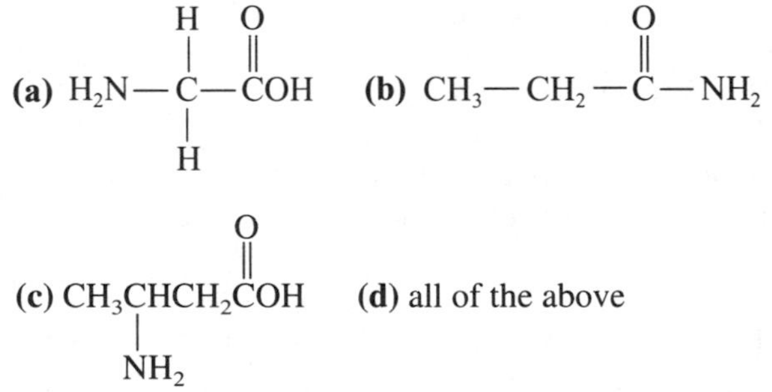

$$\text{(a) } H_2N-\underset{\underset{H}{|}}{\overset{\overset{H}{|}}{C}}-\overset{\overset{O}{\|}}{C}OH \qquad \text{(b) } CH_3-CH_2-\overset{\overset{O}{\|}}{C}-NH_2$$

$$\text{(c) } CH_3\underset{\underset{NH_2}{|}}{CH}CH_2\overset{\overset{O}{\|}}{C}OH \qquad \text{(d) all of the above}$$

    [27–2]

36. Which of the following characterizes the secondary structure of a protein?

    (a) Sequencing of amino acids

    (b) The overall shape of the molecule

    (c) Formation of chain coils

    (d) The interaction of a protein with water

    [27–2]

37. All of the following are components of a typical nucleotide except

    (a) A sugar in a furanose form

    (b) A nitrogen-containing organic base

    (c) A carboxylic acid

    (d) Phosphoric acid

    [27–6]

38. Which statement is *not* true?

    (a) DNA molecules are smaller than RNA molecules.

    (b) A nucleotide is the basic repeating unit of RNA.

(c) The sugar molecules in RNA and DNA are the same.

(d) DNA has a phosphate ester bond.

[27–6]

39. Strands of DNA:

(a) are not linked together. **(b)** form a single-helix structure.

(c) are found not to be complementary.

(d) are connected by hydrogen bonds.

[27–6]

ANSWERS

1. (c). 2. (c). 3. (a). 4. (d). 5. (b). 6. (b). 7. (c). 8. (d). 9. (b).
10. (a). 11. (b). 12. (a). 13. (d). 14. (a). 15. (d). 16. (a). 17. (b). 18. (a).
19. (b). 20. (c). 21. (a). 22. (c). 23. (d). 24. (d). 25. (c). 26. (b). 27. (d).
28. (c). 29. (b). 30. (c). 31. (d). 32. (b). 33. (c). 34. (a). 35. (a). 36. (c).
37. (d). 38. (c). 39. (d).